Lecture Notes of the Institute for Computer Sciences, Social Informatics and Telecommunications Engineering   350

Honghao Gao · Xinheng Wang ·
Muddesar Iqbal · Yuyu Yin ·
Jianwei Yin · Ning Gu (Eds.)

Collaborative Computing: Networking, Applications and Worksharing

16th EAI International Conference, CollaborateCom 2020
Shanghai, China, October 16–18, 2020
Proceedings, Part II

Springer

Editors
Honghao Gao
Shanghai University
Shanghai, China

Xinheng Wang
Xi'an Jiaotong-Liverpool University
Suzhou, China

Muddesar Iqbal
London South Bank University
London, UK

Yuyu Yin
Hangzhou Dianzi University
Hangzhou, China

Jianwei Yin
Zhejiang University
Hangzhou, China

Ning Gu
Fudan University
Shanghai, China

ISSN 1867-8211 ISSN 1867-822X (electronic)
Lecture Notes of the Institute for Computer Sciences, Social Informatics
and Telecommunications Engineering
ISBN 978-3-030-67539-4 ISBN 978-3-030-67540-0 (eBook)
https://doi.org/10.1007/978-3-030-67540-0

This Springer imprint is published by the registered company Springer Nature Switzerland AG
The registered company address is: Gewerbestrasse 11, 6330 Cham, Switzerland

Preface

We are delighted to introduce the proceedings of the 16th European Alliance for Innovation (EAI) International Conference on Collaborative Computing: Networking, Applications and Worksharing (CollaborateCom 2020). This conference brought together researchers, developers and practitioners around the world who are interested in fully realizing the promise of electronic collaboration with special attention to the aspects of networking, technology and systems, user interfaces and interaction paradigms, and interoperation with application-specific components and tools.

The technical program of CollaborateCom 2020 selected 77 papers from 211 paper submissions, comprising 61 full papers, 13 short papers and 3 workshop papers in oral presentation sessions in the main conference tracks. The conference sessions were: Collaborative Applications for Network and E-Commerce; Optimization for Collaborative Systems; Cloud and Edge Computing; Artificial Intelligence; AI Application and Optimization; Edge Computing and CollaborateNet; Classification and Recommendation; Internet of Things; Collaborative Robotics and Autonomous Systems; Resource Management; Smart Transportation; Resource Management in Artificial Intelligence; Short paper Track and Workshop Track. Apart from high-quality technical paper presentations, the technical program also featured two keynote speeches and two technical workshops. The two keynote speeches were delivered by Dr Fumiyuki Adachi from Tohoku University and Dr. Deke Guo from National University of Defense Technology. The two workshops organized were Securing IoT Networks (SITN) and Collaborative Networking Technologies towards Future Networks (CollaborateNet). The SITN workshop aims to bring together expertise from academia and industry to build secure IoT infrastructures for smart society. The CollaborateNet workshop aims to facilitate all efforts to advance current networks towards content-centric future networks using collaborative networking technologies.

Coordination with the steering chair, Imrich Chlamtac, was essential for the success of the conference. We sincerely appreciate his constant support and guidance. It was also a great pleasure to work with such an excellent organizing committee team for their hard work in organizing and supporting the conference, in particular, the Technical Program Committee, led by our General Chairs and TPC Co-Chairs, Dr. Ning Gu, Dr. Jianwei Yin, Dr. Xinheng Wang, Dr. Honghao Gao, Dr. Yuyu Yin and Dr. Muddesar Iqbal, who completed the peer-review process of the technical papers and made a high-quality technical program. We are also grateful to the Conference Manager, Karolina Marcinova, for her support and to all the authors who submitted their papers to the CollaborateCom 2020 conference and workshops.

We strongly believe that the CollaborateCom conference provides a good forum for all researchers, developers and practitioners to discuss all scientific and technical aspects that are relevant to collaborative computing. We also expect that the future CollaborateCom conferences will be as successful and stimulating, as indicated by the contributions presented in this volume.

December 2020

Honghao Gao
Xinheng Wang

Conference Organization

Steering Committee

Chair

Imrich Chlamtac   Bruno Kessler Professor, University of Trento

Members

Song Guo   The University of Aizu, Japan
Bo Li   The Hong Kong University of Science and Technology
Xiaofei Liao   Huazhong University of Science and Technology
Xinheng Wang   Xi'an Jiaotong-Liverpool University
Honghao Gao   Shanghai University

Organizing Committee

International Advisory Committee

Velimir Srića   University of Zagreb, Croatia
Mauro Pezze   Università di Milano-Bicocca, Italy
Yew-Soon Ong   Nanyang Technological University, Singapore

General Chairs

Ning Gu   Fudan University
Jianwei Yin   Zhejiang University
Xinheng Wang   Xi'an Jiaotong-Liverpool University

TPC Chair and Co-chairs

Honghao Gao   Shanghai University
Yuyu Yin   Hangzhou Dianzi University
Muddesar Iqbal   London South Bank University

Local Chairs

Zhongqin Bi   Shanghai University of Electric Power
Yihai Chen   Shanghai University

Workshops Chairs

Yusheng Xu   Xidian University
Tasos Dagiuklas   London South Bank University
Shahid Mumtaz   Instituto de Telecomunicações

Publicity and Social Media Chairs

Li Kuang	Central South University
Anwer Al-Dulaimi	EXFO Inc
Andrei Tchernykh	CICESE Research Center
Ananda Kumar	Christ College of Engineering and Technology

Publications Chairs

Youhuizi Li	Hangzhou Dianzi University
Azah Kamilah Binti Draman	Universiti Teknikal Malaysia Melaka

Web Chair

Xiaoxian Yang	Shanghai Polytechnic University

Technical Program Committee

CollaborateNet Workshop

Amando P. Singun, Jr.	Higher College of Technology
BalaAnand Muthu	V.R.S. College of Engineering & Technology
Boubakr Nour	Beijing Institute of Technology
Chaker Abdelaziz Kerrache	Huazhong University of Science and Technology
Chen Wang	Huazhong University of Science and Technology
Chi-Hua Chen	Fuzhou University
Fadi Al-Turjman	Near East University
Muhammad Atif Ur Rehman	Hongik University
Rui Cruz	Universidade de Lisboa/INESC-ID
Suresh Limkar	AISSMS Institute of Information Technology

Collaborative Robotics and Autonomous Systems

Craig West	Bristol Robotics Lab
Inmo Jang	The University of Manchester
Keir Groves	The University of Manchester
Ognjen Marjanovic	The University of Manchester
Pengzhi Li	The University of Manchester
Wei Cheah	The University of Manchester

Internet of Things

Chang Yan	Chengdu University of Information Technology
Fuhu Deng	University of Electronic Science and Technology of China
Haixia Peng	University of Waterloo
Jianfei Sun	University of Electronic Science and Technology of China

Kai Zhou	Sichuan University
Mushu Li	University of Waterloo
Ning Zhang	Texas A&M University-Corpus Christi
Qiang Gao	University of Electronic Science and Technology of China
Qixu Wang	Sichuan University
Ruijin Wang	University of Electronic Science and Technology of China
Shengke Zeng	Xihua University
Wang Dachen	Chengdu University of Information Technology
Wei Jiang	Sichuan Changhong Electric Co., Ltd
Wen Wu	University of Waterloo
Wen Xhang	Texas A&M University-Corpus Christi
Wu Xuangou	Anhui University of Technology
Xiaojie Fang	Harbin Institute of Technology
Xuangou Wu	Anhui University of Technology
Yaohua Luo	Chengdu University of Technology
Zhen Qin	University of Electronic Science and Technology of China
Zhou Jie	Xihua University

Main Track

Bin Cao	Zhejiang University of Technology
Ding Xu	Hefei University of Technology
Fan Guisheng	East China University of Science and Technology
Haiyan Wang	Nanjing University of Posts & Telecommunications
Honghao Gao	Shanghai University
Jing Qiu	Guangzhou University
Jiwei Huang	China University of Petroleum
Jun Zeng	Chongqing University
Lizhen Cui	Shandong University
Rong Jiang	Yunnan University of Finance and Economics
Shizhan Chen	Tianjin University
Tong Liu	Shanghai Univerisity
Wei He	Shandong University
Wei Du	University of Science and Technology Beijing
Xiong Luo	University of Science and Technology Beijing
Yu Weng	Minzu University of China
Yucong Duan	Hainan University
Zijian Zhang	University of Auckland, New Zealand

SITN Workshop

A. S. M. Sanwar Hosen	Jeonbuk National University
Aniello Castiglione	Parthenope University of Naples
Aruna Jamdagni	Western Sydney University

Contents – Part II

Smart Transportation

Resource Management in Artificial Intelligence

Short Paper Track

Workshop Track

Contents – Part I

AI Application and Optimization

Edge Computing and CollaborateNet

Classification and Recommendation

Internet of Things

Collaborative Robotics and Autonomous Systems

Self-organised Flocking with Simulated Homogeneous Robotic Swarm

Zhe Ban[1](✉), Craig West[2], Barry Lennox[1], and Farshad Arvin[1](✉)

[1] Swarm and Computational Intelligence Lab (SwaCIL), Department of Electrical and Electronic Engineering, The University of Manchester, Manchester, UK
{zhe.ban,farshad.arvin}@manchester.ac.uk
[2] Bristol Robotic Lab, University of West England, Bristol BS16 1QY, UK

Abstract. Flocking is a common behaviour observed in social animals such as birds and insects, which has received considerable attention in swarm robotics research studies. In this paper, a homogeneous self-organised flocking mechanism was implemented using simulated robots to verify a collective model. We identified and proposed solutions to the current gap between the theoretical model and the implementation with real-world robots. Quantitative experiments were designed with different factors which are swarm population size, desired distance between robots and the common goal force. To evaluate the group performance of the swarm, the average distance within the flock was chosen to show the coherency of the swarm, followed by statistical analysis to investigate the correlation between these factors. The results of the statistical analysis showed that compared with other factors, population size had a significant impact on the swarm flocking performance. This provides guidance on the application with real robots in terms of factors and strategic design.

Keywords: Swarm robotics · Flocking · Self-organised · Collective behaviour

1 Introduction

In nature, there are various collective motions commonly found in living organisms and social animals, such as shoals of fish [8], flocks of birds [3] and swarms of wildebeest [26]. Inspired by these collective motions, swarm robotics [20] was proposed as a research topic which provides collective strategies for a large number of simple robots to achieve collective behaviour. This collective behaviour potentially provides promising solutions to some problems in real life, such as, balancing the exploitation of renewable resources [15], fault detection [24], exploration in extreme environments [10] and coordination control of multiple autonomous cars [9]. To achieve these collective behaviours, a large and growing body of literature has investigated to model the swarm systems and to design relevant cooperation means. Considerable works have been undertaken from various angles for

© ICST Institute for Computer Sciences, Social Informatics and Telecommunications Engineering 2021
Published by Springer Nature Switzerland AG 2021. All Rights Reserved
H. Gao et al. (Eds.): CollaborateCom 2020, LNICST 350, pp. 3–17, 2021.
https://doi.org/10.1007/978-3-030-67540-0_1

different scenarios, such as distribution [6] and energy consumption [28]. Several coordination tasks have established, such as flocking [27], exploration [4], aggregation [1], foraging [23] and transportation [21]. Flocking is one the most important scenarios which has many real-world applications, e.g. in precision agriculture [5].

Cooperation strategies have a pivotal role in achieving flocking behaviour, hence, a number of strategies have been developed based on various disciplines to present collective motion and group behaviour. For example, Jia et al. [11] used a dominance matrix to compare between heterogeneous and homogeneous systems to propose a flocking framework with particles in different levels based on their contributions. In another study [7], disk graph and Delaunay graph methods were used to present connectivity with various distances. Also, mean-field game model was presented by partial differential equations to describe the system dynamics based on state and distributions [6]. The study by Thrun et al. proposed a cluster analysis that used the projection method based on the topographic map [25]. These strategies can be divided, on the basis of its framework, into two main categories: homogeneous and heterogeneous [11]. A heterogeneous group of swarm robots contains various types of robots with different roles and responsibilities, while in a homogeneous swarm, every individual follows the same strategy to achieve a common task, hence there are no behavioural or physical differences between the individuals in a swarm.

Developments of the strategies for flocking behaviour in a swarm system have led to a growing trend towards the real-world application of multi-robotic systems. Due to the limitations of real robots' hardware in practical scenarios, and based on the swarm robotics criteria defined by Şahin [19], each robot in a swarm system interacts with its direct neighbours within a specific range, *sensing radius*, to make decisions only based on its neighbours' stages. Direct communication in a swarm without having an extra observer is one of the challenges of implementing swarm scenarios using real mobile robots [13]. Some research studies [1, 17] rely on the acquisition of the location of each robot from an extra observer, whereas some [2, 11] regard each robot as an abstract particle without considering physical structures e.g. weight, size, motor speed and sensor range. Such approaches, however, can not realistically address the situation in which the group operation is influenced by the physical and hardware design constraints.

In one of the previous research studies [2], it has been theoretically demonstrated that Active Elastic Sheet (AES) is a self-propelled mechanism where swarm particles can successfully achieve collective motions. Due to the simplicity and robustness property of AES, in this paper, we chose this mechanism to demonstrate flocking behaviour of a homogeneous swarm. To combine hardware and collective control algorithms, the motion model was applied to a swarm of simulated robots by carefully considering the hardware limitations of the real robots. Hence, in this work, local communication relays on sensor values to make sure that the robots keep the desired distance from its neighbours. Each robot only detects the distance to its neighbour without acquiring the neighbour's identification, therefore, each robot is able to make decisions without an extra observer, in the whole process. Since the model of each individual robot includes all the physical properties which are carefully implemented by the control algorithm, the study here can be considered the first step from an abstract model to the progression with real robots. Simulated experiments were performed to analyse the group performance of the swarm flocking with the AES model. Followed by these qualitative experiments, the group behaviours were evaluated using a specified metric, i.e. the average distance between the robots in the swarm, and the results were statistically analysed to identify the effects of the chosen factors including time, population size and external (common) force. This information will potentially help follow-up studies to address several cautions of implementation using real swarm robots in real-world applications.

The rest of this paper was organised as follow. In Sect. 2, we introduced the collection formation and flocking mechanism. Following that, in Sect. 3, we explained the experimental setup and robotic platform. In Sect. 4, we discussed the experimental results and analysed effects of different parameters in collective swarm performance. Finally, in Sect. 5, we drew conclusions and discussed the future research direction in which the swarm robots might be involved.

2 Flocking Mechanism

The AES model [2] was originally developed and investigated using particles without hardware structures proposed. In this study, we utilised this model considering the hardware structure of physical robots. Constraints for each robot on position, $\dot{p}_i$, and rotation, $\dot{\phi}_i$, are shown in Eqs. (1) and (2). The movement of each robot is controlled by two different forces: i) the goal force, F_g, and ii) the collective force, F_i. The goal force aims to steer the entire group moving towards a desired direction, while the collective force is used to keep robots within an expected distance to avoid the collision.

$$\dot{p}_i = [F_g + \alpha(F_i + D_r\hat{\xi}_r) \cdot \hat{n}_i] \cdot \hat{n}_i, \tag{1}$$

$$\dot{\phi}_i = [F_g + \beta(F_i + D_r\hat{\xi}_r)] \cdot \hat{n}_i^{\perp}, \tag{2}$$

$$F_{gd} = \gamma_d d, \tag{3}$$

$$F_{gp} = \gamma_p \hat{v}_i, \tag{4}$$

where coefficients α and β are related to linear speed and rotation of the collective movement. $\hat{n}_i$, $\hat{n}_i^{\perp}$ are unit vectors, where $\hat{n}_i$ has the same direction as the heading of the robot, while $\hat{n}_i^{\perp}$ is perpendicular to the heading direction. D_r is the noise value in the process of detecting distances between robots. $\hat{\xi}_r$ is a unit vector with a random direction, so that noise is applied in a arbitrary direction. ϕ_i is the angle which the robot i is expected to rotate. The clockwise direction is defined as positive and counterclockwise is negative for ϕ_i. γ_d and γ_p are the corresponding weight coefficients of goal forces. $\hat{v}_i$ is related to the desired group speed along the self-propulsive direction, which is proportional to the goal force, $\boldsymbol{F}_{gp}$. $\boldsymbol{F}_{gd}$ relies on the distance between the robot location and goal $\boldsymbol{d}$. These two sub-forces are shown as Eq. (3) and Eq. (4). The goal force, $\boldsymbol{F}_g$, consists of $\boldsymbol{F}_{gp}$ and $\boldsymbol{F}_{gd}$.

Each individual robot gets information about surrounding robots using their n sensors. The summation of sensors' values is presented as a collective force, $\boldsymbol{F}_i$, shown in Eq. (5):

$$\boldsymbol{F}_i = \sum_{j=1}^{n} -\frac{k\boldsymbol{r}_{ij}}{l_{ij}\|\boldsymbol{r}_{ij}\|}(\|\boldsymbol{r}_{ij}\| - l_{ij}) , \tag{5}$$

where $\boldsymbol{r}_{ij}$ is the vector from the centre of the robot i to its neighbour j. Therefore, $\|\boldsymbol{r}_{ij}\|$ is the distance between robot i and j. l_{ij} is the desired distance between the two robots. The difference between the absolute value of $\boldsymbol{r}_{ij}$ and l_{ij} is the error of collective motion. $\frac{k}{l_{ij}}$ is a parameter which acts like a spring constant, involving the amount of force that robots generate according to the collective distances.

In terms of implementation of to the simulated robots, the process of flocking scenario is shown in Fig. 1. To begin this flocking scenario, each robot has an individual controller which is implemented by its own microcontroller. The robot uses six sensors to gain the surrounding information. These sensor readings are $\|\boldsymbol{r}_{ij}\|$ in Eq. (5). To steer the group of robots to a goal point or direction, we applied the desired distance, $\boldsymbol{d}$, for each robot. $\boldsymbol{F}_g$ and $\boldsymbol{F}_i$ have the same order of magnitude. After obtaining the force information, the controllers start to calculate total forces and correspondingly change the robot kinematics. The transformation from force to robot kinematics will be described in the Sect. 3.2. Once the motion property of each robot had been decided, the collective motion commences. Finally, the swarm collectively move to the goal position.

3   Experiments

3.1   General Foundation

Webots [14] is an open-source simulation software developed at the Swiss Federal Institute of Technology. With a 3D interface, the simulator provides numerous robotic modules and various objects, hence it is convenient to design swarm robotic systems. We used a miniature mobile robot *e-puck* [16] which is a popular swarm robotic platform and it has been utilised in many swarm robotics studies. The e-puck robot is equipped with two differential driven wheels for its actuation, and eight infra-red (IR) sensors for proximity measurements which are mainly used in our experiments for decision making. Figure 2 shows an e-puck model and its top view in Webots.

From the top view, each e-puck has eight horizontally symmetrical sensors, but not vertically symmetrical, as there are two more sensors in the front part.

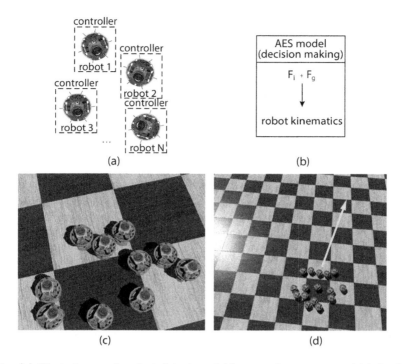

Fig. 1. (a) First step: each robot detects neighbours using sensors, which is also the data collection for calculation of collective force, F_i. Every robot has a controller to collect and calculate their own data. (b) Second step: In every individual controller, total force is calculated based on F_g and F_i using the AES model. This is also a step for each robot deciding how to change their kinematics. (c) Third step: controllers change the speed of actuators to change the robots' motion. (d) Fourth step: flocking motion achieved. The picture is a screenshot of an experiment, where the swarm collectively moved to an expected direction along the yellow arrow. (Color figure online)

Since the forces in the AES model depend on the sensor values in the experiments, the distribution of sensors has a significant impact on decision-making for each robot. Asymmetrical distribution of sensors gives rise to unbalanced forces. To balance the distribution of force, six sensors are chosen which are marked with red colour in Fig. 2 (a), correspondingly $n = 6$ in Eq. (5). Due to the fixed hardware design, the distribution of sensors is still slightly asymmetrical. According to the provided documentation[1], the infra-red sensors in each e-puck have a deviation of noise which obeys a Gaussian random distribution.

In each robot, the maximum rotation speed of the motor is 6.28 rad/s. With 20.5 mm wheel radius, the maximum speed of an e-puck is 0.25 m/s. Thanks to the fully integrated cross-compilation in Webots for the e-puck, less modification is needed for the controller from simulation to the real robots implementation.

---

[1] http://www.cyberbotics.com.

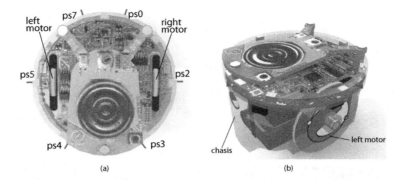

Fig. 2. (a) Top view of an e-puck robot with upward heading direction. The red lines extended from proximity sensors represent the perception ranges. Given the zero angle as the positive x direction and moving anticlockwise is taken as positive, the orientation of sensor $\{0, 2, 3, 4, 5, 7\}$ are $\{73°, 0°, 300°, 240°, 180°, 107°\}$. The intervals between them are not all the same, which will lead to the unbalanced weight from different directions. (b) 3D model of e-puck in Webots. The e-puck is differentially driven by two motors. (Color figure online)

Compared with the particle based simulations, this simulation is closer to the real-world scenarios and the controllers can be easily transferred to the real robots.

3.2   Individual Robot Test

Precise tracking performance of an individual robot is the foundation of accurate group behaviour. Since each robot basically makes decisions based on multiple forces in the AES model, we designed an individual robot test to improve the motion of a single robot before group experiments. The robot needed to make a trade-off between agility and accuracy under a force with an arbitrary direction. In this test, the relevant parameters γ_p needed to be tuned, so that the robot makes reliable decisions.

In Fig. 3(a), the pose of the robot i is described as $(x_i,\ y_i,\ \theta_i)$. In Fig. 3(b), x_i and y_i represent the expected movement of the robot which is related to the projection of $\boldsymbol{p}_i$ onto the x and y-axes. Forward velocity, $\hat{v}_i$, and angular velocity, w_i, are two variables to describe the kinematics of the robot. Equation (6) presents the transformation from the kinematics to the pose of the robot:

$$\dot{x}_i = ||\hat{v}_i|| \cos \theta_i, \quad \dot{y}_i = ||\hat{v}_i|| \sin \theta_i, \quad \dot{\theta}_i = w_i. \tag{6}$$

Since e-puck is a typical differential wheeled robot, its motion depends on the speeds of both left and right wheels. The forward velocity of the e-puck is given by the average speed of both wheels, while the rotational velocity is related to the differences between the speed of two wheels. Therefore, the position and orientation of an e-puck can be presented by:

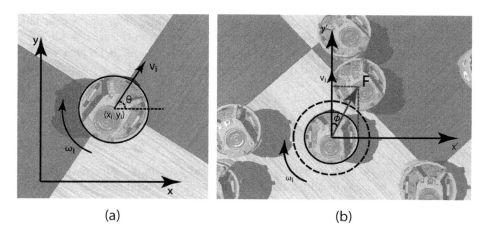

(a)   (b)

Fig. 3. (a) An e-puck and its state: the red arrow at the top of the robot indicates the heading of e-puck. v_i and w_i denote the forward velocity and angular velocity of e-puck i. The forward velocity is in the same direction as the heading of e-puck. Angular velocity is positive when e-puck rotates clockwise. (b) The coordinate of an e-puck: the x-axis is along the rotation axis of the wheels. The heading of e-puck is in the same direction as the y-axis. The black circle with the dashed line denotes the range of sensor measurement. The blue arrow denotes the force acting upon the e-puck, which is projected onto x and y-axis for calculation. All the forces are calculated according to this coordinate. In this figure, the force is from the nearby robots. By comparing this force and expected one, the robot was changing its kinematics. In the background, the swarm of mobile robots wandered to the goal as a coherent group in an open space. (Color figure online)

$$\dot{x}_i = r(\frac{w_r + w_l}{2})\cos\theta_i, \quad \dot{y}_i = r(\frac{w_r + w_l}{2})\sin\theta_i, \quad \dot{\theta}_i = r(\frac{w_r - w_l}{l}), \quad (7)$$

where w_r and w_l denote rotational speeds of right and left wheels respectively, l is the distance between the wheels, and r is the radius of the wheels.

Then, the transformation between pose goal and the angular velocity of wheels can be derived by combining Eq. (6, 7):

$$\begin{bmatrix} ||\hat{v}_i|| \\ w_i \end{bmatrix} = \begin{bmatrix} \frac{r}{2} & \frac{r}{2} \\ \frac{r}{l} & -\frac{r}{l} \end{bmatrix} \begin{bmatrix} w_r \\ w_l \end{bmatrix}. \quad (8)$$

According to the specification of the e-puck, $r = 20.5$ mm and $l = 52$ mm, the rotational speeds of the left and right wheels follow:

$$w_l = 487.8049 \, ||\hat{v}_i|| - 1.2683 w_i, \quad (9)$$

$$w_r = 487.8049 \, ||\hat{v}_i|| + 1.2683 w_i. \quad (10)$$

A coordinate of a robot designed for the change of kinematics is shown in Fig. 3(b). The origin is the centre of the e-puck and y-axis is along with the robot's heading. We projected a force onto the x-y plane. The projection has been included here for two reasons: i) the forces are vectors in the AES model and projection can transfer the information to desired pose (x_i, y_i, θ_i) which will be mentioned below and ii) it is simple to be implemented in C/C++ programming language.

According to the AES model, $\hat{v}_i$ and w_i are related to $\boldsymbol{F_i}$ and $\boldsymbol{F_g}$, which can be calculated as:

$$||\hat{v}_i|| = m\, ||\boldsymbol{F_i} + \boldsymbol{F_g}|| \quad , \quad w_i = n\angle(\boldsymbol{F_i} + \boldsymbol{F_g})\,, \tag{11}$$

where parameter m and n were tuned empirically to make sure each e-puck rapidly adjust its speed and heading, which is the fundamental of group behaviour.

3.3   Swarm Robots Test

In this section, we aimed to apply the AES model for the swarm. In order to test the group's behaviour with different kinds of initial situations, the start orientation of each agent was randomly set $\theta \in [-\pi, +\pi]$. The distances between nearby robots were less than half of the sensor range to make sure robots were able to detect their neighbours.

According to the AES model, there are two forces that mainly affect the collective motion. One is the collective force, $\boldsymbol{F_i}$, and the other is the goal force, $\boldsymbol{F_g}$. Collective force depends on the relative positions of the neighbours, which can be calculated by summing the errors between sensor values, r_{ij}, and desired distances, l_{ij}. For each robot, six forces responding to six sensor values are added up to find the resultant force, $\boldsymbol{F_i}$, and each sensor value is regarded as a vector. The direction of the vectors are from the centre of the e-puck to the corresponding sensors, which are along the red lines in Fig. 2(a). The blue arrow in Fig. 3(b) is an example which illustrates the total collective force in the group test. In terms of the goal force, the magnitude of $\boldsymbol{F_g}$ is a constant value, which pulls the swarm toward a pre-set direction.

Dot products are used to calculate the scalar projection of the forces onto a horizontal unit vector $\hat{n}_i$ and a vertical unit vector $\hat{n}_i^\perp$ in Eqs. (1) and (2). In our work, the angles between $\hat{n}_i$ and forces depend on the distribution of sensors in each e-puck coordinate system. There are seven projections calculated for each e-puck, including six forces from sensors and a goal force, $\boldsymbol{F_g}$. The component of the total force acting in the horizontal direction was calculated by summation of the component of all forces in $\hat{n}_i$ direction as presented in Eq. (5). The component of the total force in the vertical direction was calculated in the same way using $\hat{n}_i^\perp$. Prior to applying the goal force, the coordinate transformation was adopted because $\boldsymbol{F_g}$ are in the global coordinate frame, while the set of $\boldsymbol{F_i}$ controller on the e-puck is in robot coordinate frame.

Collective and goal forces have weight parameters α, β and γ_p, γ_d, w_l, respectively. These parameters influence the forces that are applied to the robot, for example, an increase in γ_p leads to a bigger force to pull the swarm toward a direction and vice versa. As a homogeneous robotic model, each robot was deployed the same controller to achieve decentralised collective motion. In this study, each experiment contained 336 positions for each e-puck. With the established individual behaviour and the calculation of forces, the flocking behaviour has been achieved. The source codes for all implementations are available on GitHub[2].

3.4   Metrics

Flocking behaviour is a simultaneous motion where a group of robots move toward a target direction. The likelihood of individuals remaining in the group depends on the coherency of the swarm. Here, we focus on the cohesiveness of the swarm. To evaluate the swarm coherency, the average distance between the swarm members is calculated as a metric in this study, which is a common method has been used in many research studies, e.g. in [17,18]. The average distance in this paper is the mean value of the distances between robots, which can be calculated as:

$$d_s = \frac{2\sum_{i=1}^{N-1}\sum_{j=1}^{N}||\boldsymbol{r}_{ij}||}{N(N-1)} \, , \tag{12}$$

where N is the number of robots in the group.

Analytical experiments were conducted to show the impact of several factors including forces, population size and desired distance on the d_s. Table 1 gives a summary of the details about factors, including different number population sizes $N \in \{4, 6, 9, 12\}$ robots, goal forces $\gamma_p \in \{2000, 3000\}$ and desired distance of $l_{ij} \in \{120, 150\}$. The desired distance l_{ij} here is dimensionless, because the value is related to sensor values. 16 sets of experiments were run and each set of experiments were repeated 10 times. To facilitate calculation of the metrics, d_s, it is important to accurately ascertain each robot's location at each sampling time.

Table 1. Sets of experiments for analysis of factors in each population size.

γ_p , l_{ij}	4	6	9	12
$\gamma_p = 2000$, $l_{ij} = 120$	Set 1	Set 2	Set 3	Set 4
$\gamma_p = 2000$, $l_{ij} = 150$	Set 5	Set 6	Set 7	Set 8
$\gamma_p = 3000$, $l_{ij} = 120$	Set 9	Set 10	Set 11	Set 12
$\gamma_p = 3000$, $l_{ij} = 150$	Set 13	Set 14	Set 15	Set 16

---

[2] https://github.com/swacil/Flocking.

4   Result and Discussion

In order to identify the impacts of factors N, l_{ij} and γ_p on the group performance, a series of tests were selected from the above experiments, in which only the object factor is an independent variable and other factors are fixed. Box-plots were used to show the distributions of distances with different populations N, sensor sensitivity l_{ij} and group force γ_p.

4.1   Effects of Time

Each experiment took 22 s and contained 336 sample points for each e-puck. In an attempt to simplify data analysis and comparison, every 28 data was averaged to represent the average distances, d_s, in every two seconds. As shown in Fig. 4(a), the median of the d_s held steadily around the initial value for the first 14 s that indicates the swarm was able to maintain the coherency as the initialisation within this period of time. However, distribution of the obtained results (size of boxes) increases over time hence it is evident that the maximum d_s increases, namely d_s had a greater chance to reach the maximum value of a test at the end of the experiment. During the experiments, the swarm spread apart as time goes on.

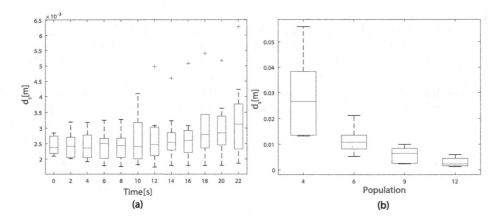

Fig. 4. (a) Box plots of average distance during experiments with $F_g = 3000$, $l_{ij} = 120$ using 12 robots within 22 s. (b) Box plots of average distance during experiments with $F_g = 2000$, $l_{ij} = 120$ using swarms with different population sizes.

4.2   Effects of Population

Figure 4(b) shows the swarm coherency, d_s, with $\gamma_p = 2000$, desired distance of $l_{ij} = 120$ and varying swarm size of $N \in \{4, 6, 9, 12\}$ robots. Considering the results of the experiments with the same goal force and the desired distance, the minimum and maximum d_s see declines as the population increase. It can be clearly observed that the median d_s decreases as population increases. All in all, the swarm with a larger population yields better group performance when other factors are the same.

4.3   Group Force

To compare group performance under different group forces, the data of the first and third row in Table 1 was chosen. The d_s with $l_{ij} = 120$ within 19 s are shown in Fig. 5. Comparing the d_s with $\gamma_p = \{2000, 3000\}$, there is a dramatic rise of the median and maximum d_s in all experiments, while the minimum d_s saw a decline apart from the swarm with 6 robots. Overall, the bigger force triggers larger d_s, resulting in worse coherency regardless of population size.

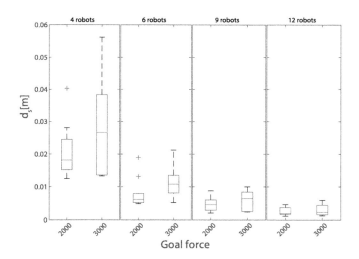

Fig. 5. Box-plots of average distance during experiments with $l_{ij} = 120$ and $N \in \{4, 6, 9, 12\}$ robots using different goal forces.

4.4   Desired Distance

Desired distance, l_{ij}, is the target distances between the robots, which is directly related to a specific sensor value in the simulation. We varied l_{ij} and kept $\gamma_p = 2000$ in the first and second row of experiments in Table 1. Figure 6 illustrates that a bigger desired distance leads to larger d_s for small population size. In contrast, bigger desired distance contributes to smaller d_s when the swarm has a larger population sizes.

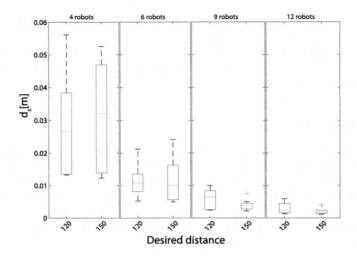

Fig. 6. Box plots of average distance during experiments with $F_g = 2000$ and $N \in \{4, 6, 9, 12\}$ robots using different desired distances.

4.5 Statistical Analysis

To compare the influence of the aforementioned factors, statistical analysis was made using the data in Table 1. According to the result of Kruskall-Wallis test, the data were normally distributed. As a common data analysis method, Analysis of Variance (ANOVA) test [22] was adopted to assess if the factors have a statistically significant effect on group behaviour. In the F-test, the three parameters: force, desired distance and population size are the factors and d_s is the result to represent the group behaviour. According to the properties of F-distributions, traditionally, $p = 0.05$ is chosen as a significance level [12]. When the p-value is less than the significance level, the null hypothesis is rejected and the corresponding factors have significant impacts on the result.

Results of the ANOVA are shown in Table 2. The number of robots plays the most significant role in coherency because p-value of the population size is far less than 0.5 ($p_p \approx 0.000$). Increasing the population size contributes to better group performance significantly. Though desired distance has an impact on the coherency of the swarm, the influence is less than the population size and force. Force has the least impact on coherency with $p_p \approx 0.691$. Compared with a swarm with fewer robots, large-sized swarms tend to have more coherent performance even under different goal forces and desired distances.

4.6 Discussion

This study showed that, an increase in population size of a swarm improve the collective performance. This is one of the main criteria of swarm robotics [19], which has been also reported in many studies [1, 10, 17, 18]. This shows that the

Table 2. Results of analysis of variance (ANOVA).

Factors	p-value	F-value
Force (F_g)	0.912	0.015
Desired distance (l_{ij})	0.691	0.381
Population (N)	0.000 ($<$0.001)	33.039

increased inter-robot interactions due to the higher population size resulted in improvement of collective behaviour of the swarm.

Also, it is interesting to note that the robots' ID is not necessary for all experiments during the decision-making process. Since the model here is decentralised and every controller of the robot is independent, each individual enables to make decisions with stochastic neighbours by sensor detection.

In some of the previous works [9,17], collective motion of a swarm relies on an extra observer, such as a camera or simulation platform. A portion of robots make decisions according to the specific robots' information which provides by the extra observer. Compared with these works, we provide a solution with less requirement of equipment and better tolerance for the change of neighbours when the accuracy of the group performance is less regulated.

The decision-making fully relies on sensors in this study, hence the sensitivity of sensor affects group performance. One source of weakness in this study which could have affected the group behaviour is sensor distribution. As mentioned in Sect. 3.1, e-puck has asymmetry sensor distribution which led to the unbalance of sensor values between the front and back. Also, this study did not consider an environment with obstacles and only focused on the flocking behaviour in an open space. Besides, the scope of this study was limited in terms of evaluation of group performance. Average distance is the only metric to describe the density of the swarm. The speed of group motion and group size were the paucities of the study.

5   Conclusion

This work was undertaken to implement a collective motion model using a realistic simulated robot swarm. We evaluated the influence of several factors on the swarm performance by quantitative experiments. The simulated robots were able to achieve flocking behaviour as a single, cohesive group. The results of the study demonstrated that the population size plays a significant role in flocking behaviour. The insights gained from this study may be of assistance to applying an abstract model to the real-world robotic applications. A natural progression of this work is to transfer the simulation to a real robotic scenario. A further study will investigate the application of this flocking mechanism in the inspection of farms for precision agriculture.

Acknowledgments. This paper was partially supported by the UK EPSRC RAIN (EP/R026084/1) and RNE (EP/P01366X/1) projects.

References

1. Arvin, F., Turgut, A.E., Krajník, T., Yue, S.: Investigation of cue-based aggregation in static and dynamic environments with a mobile robot swarm. Adaptive Behav. **24**(2), 102–118 (2016)
2. Ferrante, E., Turgut, A.E., Dorigo, M., Huepe, C.: Collective motion dynamics of active solids and active crystals. New J. Phys. **15**(9), 095011 (2013)
3. Flack, A., Nagy, M., Fiedler, W., Couzin, I.D., Wikelski, M.: From local collective behavior to global migratory patterns in white storks. Science **360**(6391), 911–914 (2018)
4. Gifford, C.M., et al.: A novel low-cost, limited-resource approach to autonomous multi-robot exploration and mapping. Robot. Auton. Syst. **58**(2), 186–202 (2010)
5. Grieve, B.D., et al.: The challenges posed by global broadacre crops in delivering smart agri-robotic solutions: a fundamental rethink is required. Global Food Secur. **23**, 116–124 (2019)
6. Grover, P., Bakshi, K., Theodorou, E.A.: A mean-field game model for homogeneous flocking. Chaos: Interdisc. J. Nonlinear Sci. **28**(6), 061103 (2018)
7. He, C., Feng, Z., Ren, Z.: A flocking algorithm for multi-agent systems with connectivity preservation under hybrid metric-topological interactions. PloS One **13**(2), e0192987 (2018)
8. Hein, A.M., Gil, M.A., Twomey, C.R., Couzin, I.D., Levin, S.A.: Conserved behavioral circuits govern high-speed decision-making in wild fish shoals. Proc. Natl. Acad. Sci. **115**(48), 12224–12228 (2018)
9. Hu, J., Bhowmick, P., Arvin, F., Lanzon, A., Lennox, B.: Cooperative control of heterogeneous connected vehicle platoons: an adaptive leader-following approach. IEEE Robot. Autom. Lett. **5**(2), 977–984 (2020)
10. Huang, X., Arvin, F., West, C., Watson, S., Lennox, B.: Exploration in extreme environments with swarm robotic system. In: IEEE International Conference on Mechatronics (ICM), vol. 1, pp. 193–198 (2019)
11. Jia, Y., Vicsek, T.: Modelling hierarchical flocking. arXiv preprint arXiv:1904.09584 (2019)
12. Laubscher, N.F., et al.: Normalizing the noncentral t and f distributions. Ann. Math. Stat. **31**(4), 1105–1112 (1960)
13. Liu, Z., West, C., Lennox, B., Arvin, F.: Local bearing estimation for a swarmof low-cost miniature robots. Sensors **20**(11) (2020)
14. Michel, O.: Cyberbotics ltd. webots^{TM}: professional mobile robot simulation. Int. J. Adv. Robot. Syst. **1**(1), 5 (2004)
15. Miletitch, R., Dorigo, M., Trianni, V.: Balancing exploitation of renewable resources by a robot swarm. Swarm Intell. **12**(4), 307–326 (2018)
16. Mondada, F., et al.: The e-puck, a robot designed for education in engineering. In: Proceedings of the 9th Conference on Autonomous Robot Systems and Competitions, vol. 1, pp. 59–65 (2009)
17. Na, S., et al.: Bio-inspired artificial pheromone system for swarm robotics applications. Adaptive Behav. 1–21 (2020)

18. Na, S., Raoufi, M., Turgut, A.E., Krajník, T., Arvin, F.: Extended artificial pheromone system for swarm robotic applications. In: The 2018 Conference on Artificial Life: A Hybrid of the European Conference on Artificial Life (ECAL) and the International Conference on the Synthesis and Simulation of Living Systems (ALIFE), pp. 608–615. MIT Press (2019)
19. Şahin, E.: Swarm robotics: from sources of inspiration to domains of application. In: Şahin, E., Spears, W.M. (eds.) Swarm Robotics. SR 2004. LNCS, vol. 3342, pp. 10–20. Springer, Heidelberg (2004). https://doi.org/10.1007/978-3-540-30552-1_2
20. Schranz, M., Caro, G.A.D., Schmickl, T., Elmenreich, W., Arvin, F.,Şekercioğlu, A.: Swarm intelligence and cyber-physical systems: concepts, challenges and future trends. Swarm Evol. Comput. (2020)
21. Shao, J., Wang, L., Yu, J.: Development of an artificial fish-like robot and its application in cooperative transportation. Control Eng. Practice **16**(5), 569–584 (2008)
22. Tabachnick, B.G., Fidell, L.S.: Experimental Designs Using ANOVA. Thomson/Brooks/Cole Belmont, CA (2007)
23. Tarapore, D., Floreano, D., Keller, L.: Influence of the level of polyandry and genetic architecture on division of labour. In: Proceedings of the Tenth International Conference on the Simulation and Synthesis of Living Systems, pp. 358–364. No. CONF, MIT Press (2006)
24. Tarapore, D., Timmis, J., Christensen, A.L.: Fault detection in a swarm of physical robots based on behavioral outlier detection. IEEE Trans. Robot. **35**(6), 1516–1522 (2019)
25. Thrun, M.C.: Projection-Based Clustering Through Self-organization and Swarm Intelligence: Combining Cluster Analysis with the Visualization of High-dimensional Data. Springer, Heidelberg (2018). https://doi.org/10.1007/978-3-658-20540-9
26. Torney, C.J., Hopcraft, J.G.C., Morrison, T.A., Couzin, I.D., Levin, S.A.: From single steps to mass migration: the problem of scale in the movement ecology of the serengeti wildebeest. Philos. Trans. Roy. Soc. B: Biol. Sci. **373**(1746), 20170012 (2018)
27. Turgut, A.E., Çelikkanat, H., Gökçe, F., Şahin, E.: Self-organized flocking in mobile robot swarms. Swarm Intell. **2**(2–4), 97–120 (2008)
28. Vicsek, T., Czirók, A., Ben-Jacob, E., Cohen, I., Shochet, O.: Novel type of phase transition in a system of self-driven particles. Phys. Rev. Lett. **75**(6), 1226 (1995)

Investigation of Cue-Based Aggregation Behaviour in Complex Environments

Shiyi Wang[1], Ali E. Turgut[2], Thomas Schmickl[3], Barry Lennox[1],
and Farshad Arvin[1(✉)]

[1] Swarm and Computational Intelligence Lab (SwaCIL),
Department of Electrical and Electronic Engineering,
The University of Manchester, Manchester, UK
{shiyi.wang,farshad.arvin}@manchester.ac.uk
[2] Department of Mechanical Engineering, Middle East Technical University,
Ankara, Turkey
[3] Artificial Life Lab, Institute of Biology, University of Graz, Graz, Austria

Abstract. Swarm robotics is mainly inspired by the collective behaviour
of social animals in nature. Among different behaviours such as foraging and flocking performed by social animals; aggregation behaviour is
often considered as the most basic and fundamental one. Aggregation
behaviour has been studied in different domains for over a decade. In
most of these studies, the settings are over-simplified that are quite
far from reality. In this paper, we investigate cue-based aggregation
behaviour using BEECLUST in a complex environment having two cues
–one being the local optimum and the other being the global optimum–
with an obstacle between the two cues. The robotic validation of the
BEECLUST strategy in a complex environment is the main motivation
of this paper. We measured the aggregation size on both cues with and
without the obstacle varying the number of robots. The simulations were
performed on a custom open-source simulation platform, *Bee-Ground*,
using *MONA* robots. The results showed that the aggregation behaviour
with BEECLUST strategy was able to overcome a certain degree of environmental complexities revealing the robustness of the method. We also
verified these results using our stock-flow model.

Keywords: Swarm robotics · Aggregation · BEECLUST

1 Introduction

Aggregation is a common behaviour observed in different species ranging from
amoeba to insects [11]. In aggregation individuals of a species gather in one area
with or without an environmental cue [16]. Aggregation is the basis and a pre-
requisite for many complex swarm tasks, e.g. flocking and foraging [10,27]. There
are two types of aggregation behaviour in nature: cue-based aggregation and self-
organised aggregation [11]. In cue-based aggregation, there is a cue present in

© ICST Institute for Computer Sciences, Social Informatics and Telecommunications Engineering 2021
Published by Springer Nature Switzerland AG 2021. All Rights Reserved
H. Gao et al. (Eds.): CollaborateCom 2020, LNICST 350, pp. 18–36, 2021.
https://doi.org/10.1007/978-3-030-67540-0_2

the environment where animals gather [8]. The cue represents the optimal location in terms of temperature, humidity or such. For example, young honeybees tend to gather in optimal areas with temperatures between 34 °C and 38 °C [17] while flies gather in hot and humid locations [14]. Hence, in cue-based aggregation environment plays a very important role and in most cases, environment is very complex having multiple cues, obstacles and predators that animals have to tackle. These make aggregation a very challenging task even for the animals. In self-organised aggregation, there is no cue in the environment. Animals gather in random locations without any particular external cue in the environment [15].

Researchers in swarm robotics study aggregation behaviour to better understand aggregation behaviour in biology [29] or to replicate it artificially with robots [8] since aggregation behaviour is the basis of most of the behaviours in swarm robotics [8,10,25,27,33]. Among different approaches, BEECLUST algorithm proposed by Schmickl et al. [29] that mimics the thermotaxis behaviour of honeybees is one of the simple yet a very elegant approach to cue-based aggregation in swarm robots.

Cue-based aggregation in swarm robotics can be traced back to the 1990s with Kube and Zhang's [22] pioneering experiment on robot aggregation around a light source. Since then, there have been many studies such as [18] proposing an aggregation scenario based on the infrared transmitter. In a recent study on the effects of environmental changes on the performance of swarm robotic aggregation [8] using real mobile robots, it has been shown that the BEECLUST algorithm is an efficient mechanism for cue-based aggregation with optimal performance in dynamic and complex environments. The study also proposed a probabilistic aggregation model to represent the influence of swarm parameters, e.g. population and arena size, on the aggregation size. Another study [30] proposed two macroscopic models based on the BEECLUST algorithm to predict the aggregation behaviour in a dynamic environment. Mathematical modelling is an efficient tool for predicting the behaviour of a swarm system. It is used to predict the collective behaviour of a group by analysing the random behaviour of the individuals [32]. In other studies [13,19,20], different models for predicting the macroscopic behaviour of swarms have been proposed. Another study [21] proposed a mathematical model which can account for peculiarities of energy transfer among swarm robots. In addition, modelling approaches from other research fields have also been used to study the modelling of swarm robotics problems. One such research study [23] used a chemical reaction network to model the interaction between individuals in robot swarms. In another study [30], *Stock-Flow* model was used, borrowed from macro-economics, to predict the performance of a cue-based swarm aggregation.

Most of the previous studies which are based on real-robot implementations used small swarm populations, and the duration of the experiments were very limited [4,8,25]. There are relatively few studies on the aggregation of swarm robots in a complex environment with multiple cues and obstacles. In this paper, we studied the aggregation of large number of robots using BEECLUST algorithm in an environment with two cues having different intensities and an obstacle. We investigated the effects of the environmental factors on aggregation. The robotic validation of the BEECLUST strategy in a complex environment is the main motivation of this paper. We also improved a mathematical model [30] and used it to predict the aggregation size on different settings. The simulation was based on the open-sourced robot MONA [3] and open-sourced simulation platform Bee-Ground [1].

The rest of this paper was structured as follows: in Sect. 2, we introduced the aggregation method. In Sect. 3, the stock-flow model was discussed. In Sect. 4, we presented the realisation of aggregation using Bee-Ground. In Sect. 5, we discussed the results of the experiments. In Sect. 6, we concluded the paper and discussed future research directions.

2 Aggregation Method

The aggregation method used in this study is based on BEECLUST [31]. It imitates the thermotaxis behaviour of young bees that aggregate at an optimal zone with temperature between 34 °C to 38 °C. Previous studies demonstrated that the BEECLUST algorithm successfully imitates the aggregation behaviour of bees on the optimal zone [6,8,9,28]. In most cases, the cue has been modelled by a light or sound source instead of a heat source due their simplicity in implementaion.

Based on the BEECLUST algorithm (depicted in Fig. 1), a robot moves forward in the environment until it encounters an object. It checks if the object is an obstacle or another robot.

- If it is an obstacle, the robot turns to avoid the obstacle and moves forward.
- If it is another robot, it stops and waits for a particular amount of time. The waiting time $w(t)$ in Eq. (1) depends on intensity of the cue. The stronger cue results in a longer waiting time.

When the waiting time finishes, the robot turns randomly and moves forward again. The waiting time, $w(t)$, is calculated by:

$$w(t) = w_{max} \cdot \frac{S(t)^2}{S(t)^2 + \mu}.$$

(1)

The waiting time, Eq. (1), was adopted from the previous studies [5]. $S(t)$ denotes the sensory reading. The sensor range is from 0 and 255. w_{max} is the maximum waiting time, and μ is a parameter which changes the steepness of the waiting curve. The constants w_{max} and μ were chosen empirically: $w_{max} = 60 \, s$ and $\mu = 5000$.

3 The Stock-Flow Model

The origin of the stock-flow model can be traced to work in the mid-twentieth century. It was used to study the dynamic relationship between the flow of

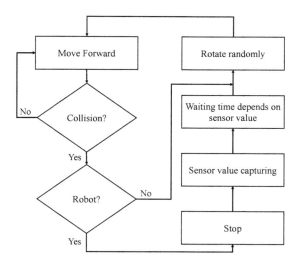

Fig. 1. Flowchart of the BEECLUST aggregation algorithms

income and expenditure and changes in national assets in the US economy [12]. In a recent work [30], this model was used to predict the aggregation of swarm robots in presence of two cues.

To model the aggregation behaviour in this study, we adopted the proposed model in [30] and used Vensim™ [2] software for the numerical modelling. In this model, 'stock' is the total number of robots in the arena.

These robots can be divided into three aggregation states, 1) the number of free robots $F(t)$, 2) the number of robots aggregated on the left source (weak source) at time t is $A_{left}(t)$, and 3) the number of robots aggregated on the right source at time t is $A_{right}(t)$. This is a constant value that does not change over time. The structure of our model is depicted in Fig. 2.

We have two sources hence two aggregations in our setting. The robots shift between each states at a certain rate. The rate at which robots join the aggregation on the left source at time t is $J_{left}(t)$. Meanwhile, the rate at which robots leave the aggregation on the left source at time t is $L_{left}(t)$.

Therefore, the rate of change the number of robots on the left source is:

$$\frac{A_{left}(t)}{dt} = J_{left}(t) - L_{left}(t). \tag{2}$$

The leave rate $L_{left}(t)$ can be described as:

$$L_{left}(t) = \frac{A_{left}(t)}{w_{left}(t)}. \tag{3}$$

$w_{left}(t)$ is the waiting time for a single robot on the left source at time t, which depends on the sensor value $S(t)$ obtained at the position where the robot stopped moving.

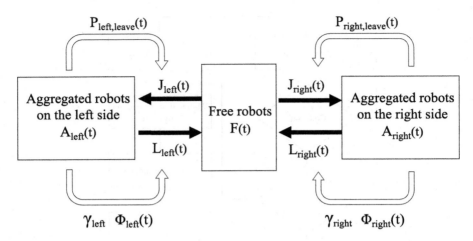

Fig. 2. Developed Stock-Flow model for aggregation

We modeled $L_{left}(t)$ with slight changes to the previous model proposed in [30]. The applied changes showed improvement in predicting the system, due to the following reasons:

1. After the waiting time is over, the robots randomly turn at an angle and continue moving forward. However, there is a high probability for at least one of them to stop immediately again. This occurs when both robots face each other again after a random rotation, or one of them face the other robots in the aggregation. Therefore, their position would not change at this moment and the robot can not leave the aggregation.
2. In more complicated situations in a large aggregation, the internal robots near the centre of the aggregation are not able to leave the group regardless of turning angles. This is because of the surrounding robots that are still in their waiting period.

In our model, we used a function $P_{leave}(t)$ to describe the probability of which robots can leave the aggregation which based on the size of the aggregation. In this paper, the function $P_{left,leave}(t)$ is the probability of which robots can leave the left aggregation. We fitted the function $P_{left,leave}(t)$ based on the observed results [4,7] from preliminary experiments. It is calculated as:

$$P_{left,leave}(t) = 0.25 * (-(\frac{\pi}{2})^{-1} \arctan(\frac{A_{left}(t)}{100} - 3.5) + 1) . \qquad (4)$$

The rate $L_{left}(t)$ in our model is described as:

$$L_{left}(t) = P_{left,leave}(t)\frac{A_{left}(t)}{w_{left}(t)}. \qquad (5)$$

Moreover, the joining rate $J_{left}(t)$ which is calculated from the number of robots which join the aggregation on the left hand side per second at time t is calculated by:

$$J_{left}(t) = \gamma_{left}F(t)\Phi_{left}(t)P_{detect}, \tag{6}$$

where γ_{left} is a fraction of a free robot appearing in the left hand side of the arena assuming that free robots are evenly distributed and $\Phi_{left}(t)$ is the probability of a robot meeting another robot in the arena. The area of the arena on the left hand side is 65.4% of the total arena, while the right-hand side area is 34.6% of the total arena. Thus, $\gamma_{left} = 65.4\%$.

$F(t)$ is the total number of free robots in the arena. P_{detect} is the probability of a robot to identify a robot-to-robot collision which is assumed:

$$P_{detect} = 0.5. \tag{7}$$

On the other hand, $\Phi_{left}(t)$ can be calculated through geometric considerations, mainly the ratio of the area covered by the robots to the total area of arena.

The area covered by a single robot is:

$$k_{aggr} = \pi r_{robot}^2. \tag{8}$$

The space covered by a free moving robot in time t is related to its speed and area, which is described by:

$$k_{mov}(t) = \pi r_{robot}^2 + 2r_{robot}vt, \tag{9}$$

where v represents the robots' speed ($v = 0.02$ m/s).

We assume that the free robots are evenly distributed across the arena, so the number of the free robots in the left side is $\gamma_{left}F(t)$.

Therefore,

$$\Phi_{left}(t) = \frac{\gamma_{left}F(t)k_{mov}(t) + A_{left}(t)k_{aggr}}{\gamma_{left}k_{arena}}, \tag{10}$$

where k_{arena} represents the total area of the arena.

The probabilities for the right-hand side of the arena are calculated similar to the left-hand side. Hence, the change rate of the free moving robots $F(t)$ in the arena can be described as:

$$F(t) = L_{left}(t) + L_{right}(t) - J_{left}(t) - J_{right}(t). \tag{11}$$

4   Experimental Setup

4.1   Simulation Platform

We developed a simulation platform, *Bee-Ground*, on Unity real-time development platform and using an open-source tool, Unity Machine Learning Agents.

Compare to other mainstream robotic simulation platforms such as ARGoS [26], Webot [24], Stage [34], Bee-Ground is able to simulate long-term operations of swarm robots in various complex and dynamic environments including obstacles and multiple sources. This is because it performs multi-layer and multi-scenario simulations simultaneously 40 times faster than the real-time without losing sampling resolution. Moreover, thanks to the built-in physics engines in Unity, our simulation platform, *Bee-Ground*, can realistically reproduce the operation of multiple robots in complex environments.

Figure 3 illustrates a randomly selected aggregation experiments with an obstacle between two sources.

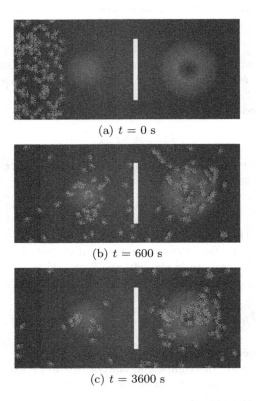

(a) $t = 0$ s

(b) $t = 600$ s

(c) $t = 3600$ s

Fig. 3. A sample simulation in Bee-Ground with swarm population of $N = 200$ robots in obstacle size with 60% of arena width. The right-hand side cue is the global optimal and the left hand side cue is a local optimum.

All experimental parameters, e.g. arena dimensions, obstacles' configuration and sources are defined by input array files. In this paper, we used a model of MONA robot [3] as shown in Fig. 4.

Fig. 4. Mona - an open-source robotic platform for swarm robotics.

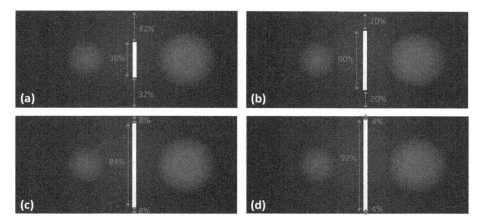

Fig. 5. Four different obstacle sizes in Bee-Ground with $\{36\%, 60\%, 84\%, 92\%\}$ of arena width.

4.2   Arena Setup

The arena size was 440 cm in length and 200 cm in width. It was surrounded by walls which stop the robots from moving out of the arena. An obstacle with different lengths were set up in the middle of the arena between the two sources. In this paper, we used five different obstacle lengths, $L_o \in \{0, 72, 120, 168, 184\}$ cm corresponding to no-obstacle, 36%, 60%, 84% and 92% of arena width, respectively (Fig. 5).

Two gradient-based heat sources were utilised as cues in this study (Fig. 3). The one on the left side, the local optimum, had a temperature of 125 unit (out of 255) at the centre point with a radius of 40 cm. The heat source on the right-hand side, the global optimal, had a temperature of 255 unit (out of 255) at the centre point with a radius of 60 cm.

4.3   Initialisation

All the robots were randomly placed on the left side of the arena to keep them away from the global optimal. They moved according to the BEECLUST algorithm during the simulations. Each simulation lasted for 3600 s and all the experiments were repeated 20 times. Table 1 shows the list of parameters and variables were used in this study.

Table 1. Parameter of Experiments

Parameter	Description	Range/Value
L_a	Length of arena	440 cm
W_a	Width of arena	200 cm
L_o	Length of obstacles	{72, 120, 168, 184} cm
W_o	Width of obstacles	8 cm
r_{cL}	Radius of the left cue	40 cm
r_{cR}	Radius of the right cue	60 cm
r_r	Radius of robot	3.25 cm
v_r	Robot forward speed	2 cm/s
ω_{turn}	Turning speed	1 rad/s
r_s	Radius of robot IR sensory system	8 cm
t_w	Waiting Time after collision	0 to 60 s

4.4   Experimental Configurations

We implemented aggregation with two different settings, without an obstacle and with a variable size obstacles.

Without Obstacles. In this setting, a group of $N \in \{100, 150, 200, 250, 300\}$ robots were deployed. At the beginning, the robots were randomly placed on the left side of the arena. We tracked the number of robots which are aggregated on both left and right-hand cues during the experiments.

With Obstacles. In this setting, $N = \{200, 300\}$ robots were used. We also set four different obstacle sizes—36% ($L_o = 72$ cm), 60% ($L_o = 120$ cm), 84% ($L_o = 168$ cm) and 92% ($L_o = 184$ cm)—sequentially. We observed the size of aggregations on both cues.

5   Results

In both configuration, the solid orange lines in the graphs show the median value of the number of robots that were aggregated on the main cue (global optimum on the right-hand side). The orange shades in each population indicate the lower quartile value to the upper quartile value. Also, the solid blue lines in the graphs show the median value of the number of robots that were aggregated on the weak cue (local optimum on the left side). The blue shades indicate the lower quartile value to the upper quartile value.

5.1   Without Obstacles

Figure 6 illustrates aggregation size on the main cue from experiments without obstacle. The five simulations showed a similar trend. The number of robots in the main aggregation increased to a stable value monotonically. Increasing the number of robots increased the size of aggregation. However, there was a significant difference in the time they reached the final number of aggregations. The group of $N = 300$ robots took approximately 910 s to reach 70% of the final aggregation size on the main cue (right-hand side). Meanwhile, the group of $N = 100$ robots took approximately 400 s to reach 70% of the final aggregation size on the main cue. The time had an approximate exponential relationship with the total swarm size.

It was not obvious, however worth mentioning that the ratio at which the final aggregation size to the total number of robots ($\frac{A_{right}}{N}$) increased slightly as the number of robots increased (Fig. 6).

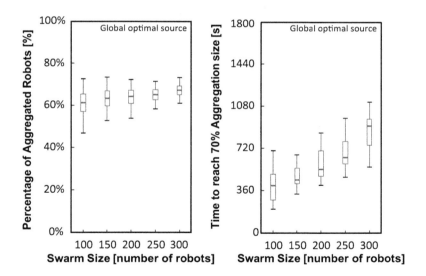

Fig. 6. Ratio of aggregated robots and the time to reach 70% aggregation size of $N = \{100, 150, 200, 250, 300\}$ robots without obstacles at global optimal source

Figure 7 shows the observed aggregation size on the main cue during 3600 s experiments with different population sizes N. Also, the predicted values from the our model were indicated with grey colour dashed lines.

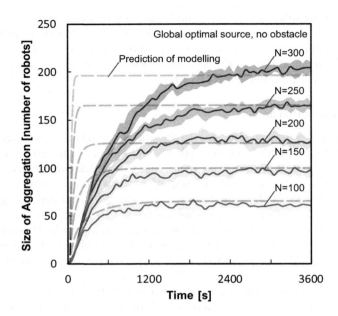

Fig. 7. Size of aggregation at the global optimal source with swarm population of $N = \{100, 150, 200, 250, 300\}$ robots without obstacles

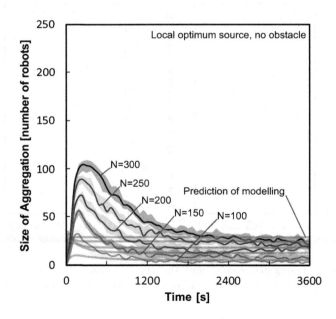

Fig. 8. Size of aggregation at the local optimum source with swarm population of $N = \{100, 150, 200, 250, 300\}$ robots without obstacles

Figure 8 illustrates the size of aggregation on the small cue (left hand size) during 3600 s. The results revealed that the situation in the local optimum source was more complex. In the beginning, the number of robots aggregated in this area reached a peak and then fell back to a stable value (minimum size) which was not shown on the right-hand side cue (in Fig. 7). There was a proportional relationship between the aggregation sizes and the number of robots in the first half time of the experiments. However, as the size of the aggregation stabilized, this proportional relationship gradually disappeared or became less obvious. The observed results show that in the simulation of $N \in \{200, 250, 300\}$ robots, the final sizes of the aggregation were almost the same. There was also no significant difference between $N \in \{100, 150\}$ robots.

5.2   With Obstacles

Figure 9 illustrates the size of aggregation with $N = 200$ robots in experiments with different obstacle sizes of $L_o \in \{36\%, 60\%, 84\%, 92\%\}$ of arena width. The grey dotted lines show the aggregation results in a no-obstacle environment with the same number of robots in each group (as the control). In the global optimal source, we found that the size of the obstacles affected the aggregation time; the larger the obstacle, the longer it took the robots to aggregate. In the arena divided by obstacles with 36% and 60% of the arena width, the final aggregation size reached a stable value that was almost the same as that in the no-obstacle arena. In arenas divided by 84% and 92% obstacles, the aggregation size increased very slowly and did not reach a stable value within the duration of 3600 s.

Figure 10 shows the aggregation size during 3600 s on the left cue (local optimum). It was found that the size of the obstacle affected the peak value of the aggregation size. Hence, in the experiments with the obstacles of 92% of the arena width, the number of aggregated robots was close to the maximum number of robots that could be aggregated in this area. Aggregation size and obstacle length had an inverse relationship, which the number of aggregated robots decreased as the obstacle size increased. In the arena divided by obstacles with 36% and 60% of the arena width, the final aggregation size was the same as in the no-obstacle area. However, in experiments with longer obstacles, 84% and 92% of the arena width, a stable value in the $N = 200$ robots was doubled and trebled as the others.

Figure 11 reveals the aggregation size for $N = 300$ robots with different obstacle sizes. The results were almost similar to the aggregation size observed with $N = 200$ robots.

Similarly, Fig. 12 shows the size of aggregation on the local optimum (left side cue) for experiment with $N = 300$ robots in presence of different obstacles lengths. Results from 36% and 60% showed a similar behaviour as shown in results with $N = 200$ robots.

However, in two extremely difficult situations with the obstacles lengths of 84% and 92% of the arena width, we did not expect that robots can successfully aggregate at the global optimal source during the $t = 3600$ s experiments.

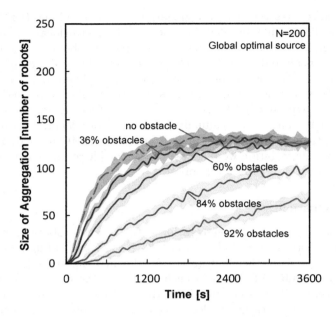

Fig. 9. Size of aggregation at the global optimal source with population of $N = 200$ robots in obstacle sizes with $\{36\%, 60\%, 84\%, 92\%\}$ of arena width

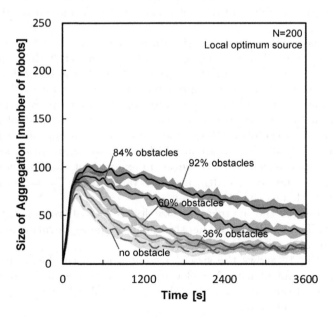

Fig. 10. Size of aggregation at the local optimum source with population of $N = 200$ robots in obstacle sizes with $\{36\%, 60\%, 84\%, 92\%\}$ of arena width.

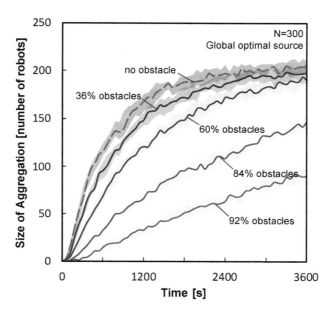

Fig. 11. Size of aggregation at the global optimal source with population of $N = 300$ robots in obstacle sizes with $\{36\%, 60\%, 84\%, 92\%\}$ of arena width.

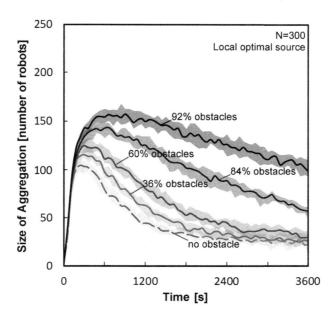

Fig. 12. Size of aggregation at the local optimum source with population of $N = 300$ robots in obstacle sizes with $\{36\%, 60\%, 84\%, 92\%\}$ of arena width.

Therefore, we just calculated the ratio in which robots overcame the obstacles. In experiments with $N = 200$ robots, 44.5% robots (89 of 200 robots) overcame the obstacles in 84% of the arena width. Meanwhile, this ratio in case of $N = 300$ robots is 39.7% (119 of 300 robots). The similar situation observed in the experiments with obstacles in 92% of the arena width in both groups. The ratios were 19% (38 of 200 robots) and 14.3% (43 of 300 robots) respectively.

6   Discussion

6.1   Arena Without Obstacles

Each group of our simulations showed that the global optimal source could eventually attract more robots to aggregate than the local optimum source. Although the capability of BEECLUST in finding the global optimum was demonstrated, the speed and efficiency of completing the aggregation were different. The total number of robots affected the aggregation speed in the global optimal source. The larger the number of robots, the slower the gathering speed of the robot. It is due to increase in number of inter-robot collisions.

In our stock-flow model, we discussed the $P_{leave}(t)$ function, the probability that a robot leaves an aggregation. It changed as the size of the aggregation changed. The larger the aggregation, the lower the probability that the robot would leave, and vice versa. Therefore, it was found that when there were two cues in the arena, the larger cue reduced the aggregation ability of the smaller cue because of the change of this probability because of the following reasons:

- As the size of aggregation in the bigger cue grows, free robots have a higher probability of entering larger aggregation. This probability increases continuously. As a result, more robots join and the number of free robots becomes smaller.
- Also, the rate at which robots enter the smaller cue becomes smaller. Each robot that leaves the smaller cue has a higher probability of joining the aggregation in the bigger cue. Hence, the aggregation size in the smaller cue decreases continuously results in the rate at which robots leave it increases.

Our model's prediction and simulation results of the aggregation size were close. The difference was at the beginning of the experiments, that was, the robots' initialisation position. In the model, we set the robots in random positions uniformly, however, we initialised the robots on the left hand side.

We found that the probability of a robot leaving the aggregation had a significant effect on the aggregation results in complex environments. We currently have only an approximate function $P_{leave}(t)$. There are relatively few related studies about this phenomenon.

6.2   Arena with Obstacles

It is shown that the length of the obstacle influenced the aggregation speed and ratio in the global optimal source on the right-hand side. The larger the obstacle, the slower the aggregation speed of the robot. It also affected the peak value of the number of aggregated robots in the local optimum on the left hand side.

It was unexpected that the presence of an obstacle, under certain circumstances, had a positive impact on the aggregation size. For example, in the group with $N = 200$ robots, the final aggregation performance in small obstacle experiments, $L_o \in \{36\%, 60\%\}$ were better than those in without obstacles. Our view is that the obstacle reduced the possibility for free robots on the right-hand side to leave the area.

On the other side, the number of robots was not always a positive factor for aggregation. In the experiments with 84% and 92% obstacles, the group with $N = 300$ robots performed worse than the one with $N = 200$ robots. That observed behaviour showed the higher the number of robots in the arena, the fewer the number of robots that can cross the passages (on top and bottom) made by the long obstacle. We observed the different stages of the experiment and found that the robots continually hovered near the passage, thereby reducing the possibility of other robots crossing the obstacle.

6.3   BEECLUST Algorithm in Complex Environment

The BEECLUST algorithm performed efficiently in the dual-cue environment with no obstacle, which was also reported in previous studies [8,31]. It coordinated individuals in the arena to find a global optimal source by following simple inter-robot interactions. Furthermore, when there were obstacles in the arena, the algorithm helped robots to successfully overcame the obstacle. Although the robots took longer to reach the peak of aggregation size in the arena with obstacle, the final aggregation size were similar to the no-obstacle configuration.

However, we have not observed if the algorithm can overcome a very complex situation e.g. a large obstacles during the specific experiment time, $t = 3600$ s. It still showed a trend that robots attempted to aggregate towards the global optimal source.

6.4   Simulation Platform: Bee-Ground

Our experiments proved that our simulation platform, Bee-Ground, is an efficient, flexible and realistic platform for the simulation of swarm robotic scenarios. It is an open-source platform [1] on GitHub. This platform is suitable for simulating more complicated environments such as a continuously changing environment and multiple cues.

7   Conclusion

A bio-inspired aggregation scenario, BEECLUST, was utilised to implement a bio-inspired swarm aggregation. The results showed that although the BEECLUST algorithm is a simple system, it was a successful algorithm in controlling a large swarm in a complex environment. The results demonstrated the feasibility of the aggregation in finding the global optimal source in the presence of multiple sources. It was illustrated that the robots could overcome a certain degree of environmental complexities, thanks to the robustness of swarm systems. Meanwhile, in a complex environment with limited access, the swarm population was not always a positive factor to the final aggregation size at the global optimal source. For the future work, we will investigate effects of multi-layer environmental complexity using real robots. Also, a probabilistic macroscopic model based on [32] will be proposed to predict the collective behaviour of the swarm with additional environmental factors.

Acknowledgment. This work was partially supported by the UK EPSRC RAIN (EP/R026084/1), RNE (EP/P01366X/1), and the Field of Excellence COLIBRI of the University of Graz projects.

References

1. Bee-Ground: Open-sourced simulation tool for aggregation of robotics swarms. https://github.com/wshiyi-cn/BeeGround. Accessed 15 Jan 2020
2. Vensim^{TM}. http://www.vensim.com. Accessed 15 Jan 2020
3. Arvin, F., Espinosa, J., Bird, B., West, A., Watson, S., Lennox, B.: Mona: an affordable open-source mobile robot for education and research. J. Intell. Robot. Syst. **94**(3–4), 761–775 (2019)
4. Arvin, F., Murray, J., Zhang, C., Yue, S.: Colias: an autonomous micro robot for swarm robotic applications. Int. J. Adv. Robot. Syst. **11**, 1 (2014)
5. Arvin, F., Turgut, A.E., Bellotto, N., Yue, S.: Comparison of different cue-based swarm aggregation strategies. In: Tan, Y., Shi, Y., Coello, C.A.C. (eds.) ICSI 2014. LNCS, vol. 8794, pp. 1–8. Springer, Cham (2014). https://doi.org/10.1007/978-3-319-11857-4_1
6. Arvin, F., et al.: ΦClust: pheromone-based aggregation for robotic swarms. In: IEEE/RSJ International Conference on Intelligent Robots and Systems (IROS), pp. 4288–4294 (2018)
7. Arvin, F., Turgut, A.E., Krajník, T., Yue, S.: Investigation of cue-based aggregation in static and dynamic environments with a mobile robot swarm. Adapt. Behav. **24**(2), 102–118 (2016)
8. Arvin, F., Turgut, A.E., Krajník, T., Yue, S.: Investigation of cue-based aggregation in static and dynamic environments with a mobile robot swarm. Adapt. Behav. **24**(2), 102–118 (2016). https://doi.org/10.1177/1059712316632851
9. Bodi, M., Thenius, R., Szopek, M., Schmickl, T., Crailsheim, K.: Interaction of robot swarms using the honeybee-inspired control algorithm beeclust. Math. Comput. Model. Dyn. Syst. **18**(1), 87–100 (2012)
10. Brambilla, M., Ferrante, E., Birattari, M., Dorigo, M.: Swarm robotics: a review from the swarm engineering perspective. Swarm Intell. **7**, 1–41 (2013). https://doi.org/10.1007/sd11721-012-0075-2
11. Camazine, S., Deneubourg, J.L., Franks, N.R., Sneyd, J., Bonabeau, E., Theraula, G.: Self-organization in Biological Systems. Princeton University Press, Princeton (2003)
12. Caverzasi, E., Godin, A.: Stock-flow consistent modeling through the ages. Levy Economics Institute of Bard College Working Paper, no. 745 (2013)

13. Correll, N., Martinoli, A.: Modeling self-organized aggregation in a swarm of minia-ture robots. In: IEEE International Conference on Robotics and Automation: Workshop on Collective Behaviors inspired by Biological and Biochemical Systems (2007)
14. Frank, D., Jouandet, G., Kearney, P., Macpherson, L., Gallio, M.: Temperature representation in the drosophila brain. Nature **519**, 358–361 (2015). https://doi.org/10.1038/nature14284
15. Gauci, M., Chen, J., Li, W., Dodd, T.J., Groß, R.: Self-organized aggregation without computation. Int. J. Robot. Res. **33**(8), 1145–1161 (2014)
16. Grünbaum, D., Okubo, A.: Modeling social animal aggregations. In: Levin, S.A. (ed.) Frontiers in Mathematical Biology. Lecture Notes in Biomathematics, vol. 100, pp. 296–325. Springer, Heidelberg (1994). https://doi.org/10.1007/978-3-642-50124-1_18
17. Heran, H.: Untersuchungen über den Temperatursinn der Honigbiene (Apis melli-fica) unter besonderer Berücksichtigung der Wahrnehmung strahlender Wärme. J. Comparat. Physiol. **34**, 179–206 (1952). https://doi.org/10.1007/BF00339537
18. Holland, O., Melhuish, C.: An interactive method for controlling group size in mul-tiple mobile robot systems. In: 8th International Conference on Advanced Robotics, pp. 201–206 (1997)
19. Hu, C., Arvin, F., Xiong, C., Yue, S.: Bio-inspired embedded vision system for autonomous micro-robots: the LGMD case. IEEE Trans. Cogn. Dev. Syst. **9**(3), 241–254 (2016)
20. Krajník, T., et al.: A practical multirobot localization system. J. Intell. Robot. Syst. **76**(3–4), 539–562 (2014)
21. Krestovnikov, K., Cherskikh, E., Ronzhin, A.: Mathematical model of a swarm robotic system with wireless bi-directional energy transfer. In: Kravets, A.G. (ed.) Robotics: Industry 4.0 Issues & New Intelligent Control Paradigms. SSDC, vol. 272, pp. 13–23. Springer, Cham (2020). https://doi.org/10.1007/978-3-030-37841-7_2
22. Kube, C.R., Zhang, H.: Collective robotics: from social insects to robots. Adapt. Behav. **2**(2), 189–218 (1993)
23. Mermoud, G., Matthey, L., Evans, W.C., Martinoli, A.: Aggregation-mediated col-lective perception and action in a group of miniature robots. In: International Conference on Autonomous Agents and Multiagent Systems, pp. 599–606 (2010)
24. Michel, O.: Cyberbotics LTD. webots™: professional mobile robot simulation. Int. J. Adva. Robot. Syst. **1**(1), 5 (2004)
25. Na, S., et al.: Bio-inspired artificial pheromone system for swarm robotics applica-tions. Adapt. Behav. 1–21 (2020). https://doi.org/10.1177/1059712320918936
26. Pinciroli, C., et al.: ARGoS: a modular, parallel, multi-engine simulator for multi-robot systems. Swarm Intell. **6**(4), 271–295 (2012)
27. Şahin, E.: Swarm robotics: from sources of inspiration to domains of application. In: Şahin, E., Spears, W.M. (eds.) SR 2004. LNCS, vol. 3342, pp. 10–20. Springer, Heidelberg (2005). https://doi.org/10.1007/978-3-540-30552-1_2
28. Schmickl, T.: How to engineer robotic organisms and swarms? In: Meng, Y., Jin, Y. (eds.) Bio-Inspired Self-Organizing Robotic Systems. Studies in Computational Intelligence, vol. 355, pp. 25–52. Springer, Heidelberg (2011). https://doi.org/10.1007/978-3-642-20760-0_2
29. Schmickl, T., Hamann, H.: BEECLUST: a swarm algorithm derived from honey-bees. In: Bio-inspired Computing and Communication Networks, pp. 95–137 (2011)

30. Schmickl, T., Hamann, H., Wörn, H., Crailsheim, K.: Two different approaches to a macroscopic model of a bio-inspired robotic swarm. Robot. Auton. Syst. **57**(9), 913–921 (2009)
31. Schmickl, T., et al.: Get in touch: cooperative decision making based on robot-to-robot collisions. Auton. Agent. Multi-Agent Syst. **18**(1), 133–155 (2009)
32. Soysal, O., Sahin, E.: Probabilistic aggregation strategies in swarm robotic systems. In: Proceedings 2005 IEEE Swarm Intelligence Symposium, SIS 2005, pp. 325–332. IEEE (2005)
33. Turgut, A.E., Çelikkanat, H., Gökçe, F., Şahin, E.: Self-organized flocking in mobile robot swarms. Swarm Intell. **2**(2–4), 97–120 (2008)
34. Vaughan, R.: Massively multi-robot simulation in stage. Swarm Intell. **2**(2–4), 189–208 (2008)

Towards Efficient and Privacy-Preserving Service QoS Prediction with Federated Learning

Yilei Zhang[1,2]([✉]), Xiao Zhang[1], and Xinyuan Li[1]

[1] Anhui Normal University, Wuhu, China
stonezyl@gmail.com, {zx95,bread}@mail.ahnu.edu.cn
[2] The Chinese University of Hong Kong, Hong Kong, China

Abstract. Cloud computing provides many service resources that enable large-scale cloud applications composed of services to be widely adopted in many crucial domains. Quality of Service (QoS) is often used as an indicator in service selection and composition to guarantee the quality of cloud applications. To facilitate QoS-based selection and composition, previous studies have employed collaborative filtering techniques to predict unknown QoS values as a supplement to limited user-perceived QoS data. However, Collaborative modeling approaches encounter privacy issues in the practice of QoS prediction. Users may be reluctant to collaborate through sharing data. As a result, addressing privacy threats has become a key effort towards making QoS prediction methods practical. In this paper, we leverage federated learning techniques and propose a privacy-preserving QoS prediction approach to address this challenge. We further propose several efficiency improvement techniques to significantly reduce system overhead so that the prediction model can provide results quickly and timely. We conduct experiments on a large-scale real-world QoS dataset to evaluate our approach, and the experimental results show that it can make fast and accurate predictions.

Keywords: QoS prediction · Cloud services · Privacy preservation · Federated learning · Communication efficiency

1 Introduction

Cloud computing provides many service resources that enable large-scale cloud applications, which provide services to millions of customers around the world on a 24/7 basis. With the rapid growth of service market, a large number of services are publicly accessible in cloud to provide various atomic functions, which makes Service-Oriented Architecture (SOA) has become the primary paradigm for constructing cloud applications. Desired services are discovered and composed by users (i.e. cloud applications) in a loosely-coupled way at both design time and runtime. Ensuring the performance of cloud applications especially those used in key areas such as government, health care, and finance, is a big challenge as

© ICST Institute for Computer Sciences, Social Informatics and Telecommunications Engineering 2021
Published by Springer Nature Switzerland AG 2021. All Rights Reserved
H. Gao et al. (Eds.): CollaborateCom 2020, LNICST 350, pp. 37–57, 2021.
https://doi.org/10.1007/978-3-030-67540-0_3

it is difficult to get details of the internal implementation of services provided by third parties.

In order to ensure the performance of applications, it is necessary to select high-quality services. Generally, effective service selection employs QoS as a primary indicator to distinguish a set of candidate services with equivalent or similar functions. QoS is widely used to describe the non-functional characteristics of services, such as response time, throughput, Availability, etc. The QoS values of services hosted by different providers may vary depending on the server side environment (e.g., workload, bandwidth, etc.). In addition, due to user-side factors (e.g., geographical location, device capabilities, etc.) and the network environment (e.g., bandwidth, congestion, etc.) between users and services, the observed QoS for each user is different from the others, even on the same service. Therefore, effective service selection is highly dependent on the user-perceived QoS values of services. However, it is impossible to get QoS values for all candidate services because each user invokes a very limited number of services and can only observe the QoS values for the corresponding service. It is also impractical to actively evaluate these QoS values due to the high cost of invoking a large number of services.

In recent literature, collaborative filtering is commonly employed by QoS prediction approaches [25,30,33] to solve the QoS prediction problem. A global QoS dataset is formed by collecting user-contributed local service usage records. By building a prediction model on the dataset, unknown QoS values can be estimated rather than measured by actual invocations. Unfortunately, the practical use of these approaches is severely hampered by privacy threat issues. In order to receive the predicted QoS values from the prediction model, users need to provide their local usage data, which may disclose their sensitive data and privacy. It is difficult for users to prevent others from inferring personal information or selling their data. Consequently, some users are concerned that their valuable data may be leaked and are not enthusiastic in participating in collaborative modeling. As a result, the limited historical training data leads to the low prediction accuracy of the model.

In order to allow more users to safely participate in the collaborative model construction, there is an urgent need to implement privacy-preserving mechanisms in the QoS prediction approach.

Different from the conventional collaborative filtering approaches which use the original data to build a prediction model, several privacy-preserving approaches leveraging encryption [12], obfuscation [35], anonymization [19], and other data conversion techniques for QoS prediction. While these methods can provide a certain degree of privacy protection, they still face the following limitations:

- **Overhead:** All users need to transfer local usage data to the central server. Large-scale online applications (i.e., users) typically run around the clock and support millions of customers, generating an incredible amount of usage data. Transferring large amounts of data over the network imposes a huge overhead on users and the central server.

– **Accuracy:** The use of obfuscated data techniques can lead to the loss of some useful information on the original QoS data, which will affect the accuracy of the model. It is difficult to determine the level of obfuscation to balance the trade-off between privacy preservation and prediction accuracy.
– **Efficiency:** Due to the large amount of data needed to train the predictive model, it puts forward a extremely high demand on the processing capacity of the server. It is a severe challenge for the central server to provide prediction results to all users in a timely manner at runtime. In addition, data obfuscation, data encryption and data decryption lead to extra computation time, which further delays the delivery of prediction results.

This paper proposes a highly efficient QoS prediction approach using federated learning techniques [26] to protect user privacy. Users and the server work together in a distributed and collaborative manner to train a global QoS prediction model. In this paradigm, users do not have to send a large amount of local data, but only a small number of parameters of the local model to the central server, thus achieving the goal of protecting user privacy. In order to improve the efficiency of the federated prediction model, we propose several techniques, including reducing the computational overhead, reducing the size of the transmission message, and reducing the number of communication rounds. Compared with the existing privacy-preserving QoS prediction approaches, our approach has several significant advantages: 1) Compared with QoS data, the data volume of model parameters is very small, which greatly reduces the transmission overhead. 2) The model is trained over real QoS data rather than obfuscated data, which ensures the accuracy of the model while protecting the privacy of users. 3) Since a lot of training and prediction work is distributed on the user side in the decentralized prediction framework, the workload of central server is cut down and the computing capacity on the server side is no longer a bottleneck. The proposed efficiency improvement techniques further reduce the processing time on the user side. We evaluate our approach on a real-world QoS dataset which has been widely employed for evaluating QoS prediction approaches in the literature. The experimental results show that our approach can still achieve high accuracy while preserving the privacy of users, and is more efficient than the existing approaches.

In summary, this paper makes the following contributions:

– We propose a novel decentralized framework for privacy-preserving QoS prediction by employing federated learning techniques.
– We further propose several effective techniques to significantly improve the efficiency of federated QoS prediction approach.
– We conduct extensive experiments on a large-scale real-world QoS dataset to evaluate the effectiveness and efficiency of our privacy-preserving prediction approach.

The rest of the paper is organized as follows. We describe the detailed privacy-preserving QoS prediction approach in Sect. 2. We conduct experiments and discuss the evaluation results in Sect. 3. We present some related works in Sect. 4. Finally, we conclude this paper in Sect. 5.

2   Efficient Privacy-Preserving QoS Prediction

In this section, we propose a novel privacy-preserving federated QoS prediction approach. We first introduce the QoS prediction problem of cloud services in Sect. 2.1. Then we describe the conventional matrix factorization and discuss its limitations on privacy-preserving issues in Sect. 2.2. The details of proposed privacy-preserving approach are described in Sect. 2.3. Finally, three techniques are proposed to enable efficient privacy-preserving QoS prediction in Sect. 2.4.

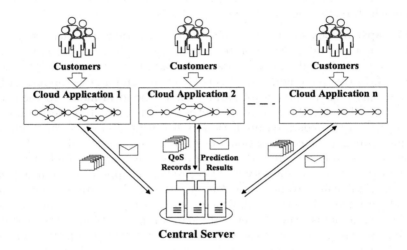

Fig. 1. QoS prediction problem

2.1   Problem Description

Figure 1 illustrates the QoS prediction problem we study in this paper. Many applications are deployed in the cloud to serve their customers on a 24/7 basis. In the workflow of an application, multiple services need to be invoked at runtime. By integrating response results in the workflow, applications can implement specific logic. Note that in this paper, we use users to refer to applications. Generally, a user can invoke a set of services, and a service can be invoked by different users. Since there exists many functionally equivalent or similar services, users can select services with better QoS performance. Each user records the QoS values observed for each service invocation, representing the personalized service performance experienced by that user at that time. By collecting data from all users with collaborative sharing mechanism [32], a global QoS matrix can be formed. In the matrix, each row represents a user, each column represents a service, and each entry represents the QoS value of the corresponding service observed by a user. An empty entry indicates that the corresponding user did not invoke the service before. Typically, a user invokes only a small number of services, while a large number of services are not invoked. Therefore, most of the entries in the QoS matrix are unknown. This paper investigates how to predict unknown QoS values more efficiently and accurately through data sharing mechanisms while protecting user privacy.

Formally, suppose there are m users and n services. $Q \in \mathbb{R}^{m \times n}$ represents the QoS matrix, with each entry q_{ij} representing the QoS value of service s_j observed by user u_i. $I_{ij} = 1$ indicates an observed entry, and $I_{ij} = 0$ indicates an empty entry. Given all the known entries $\{q_{ij}|I_{ij} = 1\}$, the empty entries $\{q_{ij}|I_{ij} = 0\}$ should be predicted representing unknown QoS values.

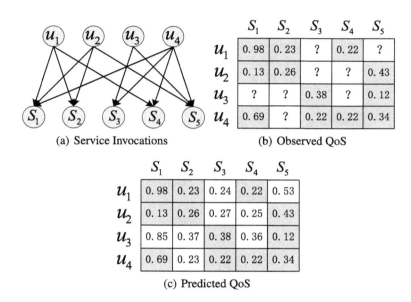

(a) Service Invocations   (b) Observed QoS

(c) Predicted QoS

Fig. 2. Matrix factorization

2.2 Conventional Matrix Factorization

In collaborative filtering [15], Matrix factorization [17] is widely used to build recommender systems. In cloud computing, some existing approaches [31] use it to collect data and build prediction models for QoS prediction. Matrix factorization derives user feature matrix and service feature matrix from the original QoS matrix. These two matrices are low-rank, high-quality, which represent latent factors of users and services, respectively. The premise behind a low-rank factor model is that the entries of Q have a large correlation in practice, therefore a small number of latent factors can represent the features more effectively. The degree to which user latent vector match the corresponding service latent vector determines what QoS value the user can observe. Figure 2 illustrates a toy example of matrix factorization in QoS prediction problem. As shown in Fig. 2(a), the invocation from a user to a service is represented by a solid line. The user records the QoS value it observed in the QoS matrix in the corresponding entry of this invocation as shown in Fig. 2(b). By decomposing the matrix, users and services are mapped to a low-dimensional joint latent factor space, so that the existing entries in the QoS matrix can be recaptured as the inner product of the latent vectors in the space. Finally, the unknown entries in the matrix can be predicted in a similar way using latent vectors as shown in Fig. 2(c).

We use $U \in \mathbb{R}^{l \times m}$ to denote the user latent matrix and each row represents the latent vector of a particular user. $S \in \mathbb{R}^{l \times n}$ denotes the service latent matrix and each row represents the latent vector of a particular service. l is the rank of the QoS matrix. Matrix factorization attempts to find the optimal U and S, so that Q can be well fitted by the inner product of U and S (i.e., $Q \approx U^T S$). It can be formalized to minimize the following loss function:

$$\mathcal{L} = \frac{1}{2} \sum_i \sum_j I_{ij}(q_{ij} - U_i^T S_j)^2 + \frac{\lambda}{2}(||U||^2 + ||S||^2), \tag{1}$$

where $I_{ij} = 1$ means q_{ij} is observed, and $I_{ij} = 0$ means q_{ij} is unknown. $|| \cdot ||$ is the Euclidean norm. The first term calculates the sum of squared errors between observed QoS values and the predicted ones by $U^T S$. We add two regularization terms in the last positions to avoid overfitting. λ controls the extent of regularization. The optimal U and S is calculated by an iterative process of gradient decent until convergence:

$$U_i \leftarrow U_i - \eta \frac{\partial \mathcal{L}}{\partial U_i}, \tag{2}$$

$$S_j \leftarrow S_j - \eta \frac{\partial \mathcal{L}}{\partial S_j}, \tag{3}$$

where η is the learning rate that controls the step size at each iteration. By calculate the inner product of user latent vectors and service latent vectors, the unknown QoS values can be predicted as follows:

$$p_{ij} = U_i^T S_j. \tag{4}$$

2.3   Privacy-Preserving Matrix Factorization

Despite the high prediction accuracy achieved by conventional matrix factorization models, it is difficult to apply them directly in practice. Users are concerned that their privacy will be compromised when providing data to third parties. Consequently, collaborative filtering approaches are difficult to work effectively. In this paper, we propose a privacy-preserving QoS prediction approach to address this challenge.

Data Transformation

Figure 3 illustrates the distributions of response time and throughput values in the real-world QoS dataset. Obviously, these two distributions have high skewness and large variance which is conflict with the probabilistic hypothesis of matrix factorization [17]. Moreover, different QoS attributes also have different value ranges (e.g., 0–7000 kbps for throughput, 0–100% for availability). This will affect the prediction accuracy of matrix factorization based approaches. To handle this issue, we conduct data transformation on the raw QoS data by employing Box-Cox transformation [18] before building prediction model. Box-Cox is a useful data transformation technique widely used to stabilize variance in data analysis. A Box-Cox transformation is a way to transform non-normal dependent variables into a normal shape as follows:

$$b(x) = \begin{cases} \dfrac{x^{\alpha} - 1}{\alpha}, & \text{if } \alpha \neq 0, \\ log(x), & \text{if } \alpha = 0, \end{cases} \tag{5}$$

where α controls the extent of the transformation. Let q_{max} and q_{min} be the upper and lower bounds of the QoS value respectively, which can be specified by the user. Since $b(x)$ is a monotonic non-decreasing function, $b(q_{max})$ and $b(q_{min})$ are the upper and lower bounds after data transformation. Then the QoS values can be mapped to the range $[0, 1]$ by the following transformation:

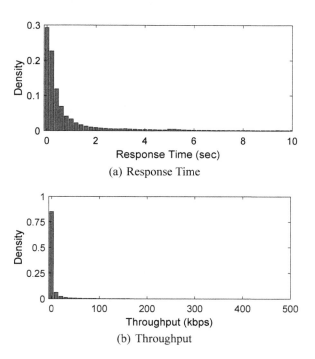

(a) Response Time

(b) Throughput

Fig. 3. QoS value distributions

$$\hat{q}_{ij} = \frac{b(q_{ij}) - b(q_{min})}{b(q_{max}) - b(q_{min})}. \tag{6}$$

Accordingly, the prediction result p_{ij} can be mapped to the range $[0, 1]$ by the logistic function, as simply done in [17]:

$$\hat{p}_{ij} = \frac{1}{1 + e^{-p_{ij}}}. \tag{7}$$

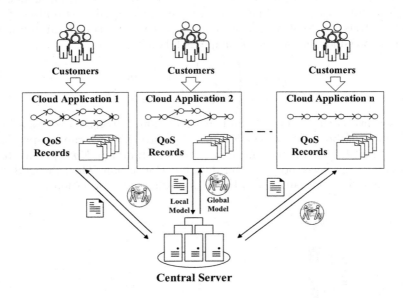

Fig. 4. Federated QoS prediction framework

Federated QoS Prediction

Figure 4 illustrates the federated QoS prediction framework. Users record the observed QoS data locally while the application is running, and this data does not need to be transmitted to the central server. Part of the matrix factorization process takes place locally at the user site and is integrated again at the central server through federated learning technologies. More specifically, user latent factors are learned on the user side and the service latent matrix is learned on the server side. The prediction model is continuously trained and updated in multi-rounds (i.e., time slices). In each round, the user receives the latest global service latent matrix from the central server. Each user then updates the local QoS matrix with the latest observed QoS values. The local matrix factorization process is then performed based on the global service latent matrix to obtain the local user latent vector and the updates to the global service matrix latent matrix. After that, each user simply sends the updated global service latent matrix to the central server and keeps the raw QoS data and user latent vector in local site. Then, the central server collects the updated from all users and combines them to determine the updates that should be made to the global service latent matrix in this round, and provides the updated global service latent matrix to the users in the next round. Each user can quickly predict unknown QoS values using local user latent vector and global service latent matrix.

More specifically, at the beginning of each round, the local QoS matrix $Q_i \in \mathbb{R}^{1 \times m}$ is updated with newly observed QoS data. Each user u_i get a local copy $S^i \in \mathbb{R}^{l \times n}$ of the global service latent matrix S from the central server. The local user latent vector $U_i \in \mathbb{R}^{l \times 1}$ are initialized with small random numbers. The local loss function is as follows:

$$\mathcal{L}_i = \frac{1}{2} \sum_j I_{ij}(\hat{q}_{ij} - \hat{p}_{ij})^2 + \frac{\lambda}{2}(\|U_i\|^2 + \|S^i\|^2), \tag{8}$$

and the global loss function is:

$$\mathcal{L} = \sum_i \mathcal{L}_i. \tag{9}$$

Instead of minimizing $\mathcal{L}$, the federated model relaxes to minimize the local loss function $\mathcal{L}_i$ at each user site. U_i and S^i are updated on local QoS matrix Q_i by the iterative process until converge:

$$U_i \leftarrow U_i - \eta((\hat{q}_{ij} - \hat{p}_{ij})\hat{p}'_{ij}S^i_j + \lambda U_i), \tag{10}$$

$$S^i_j \leftarrow S^i_j - \eta((\hat{q}_{ij} - \hat{p}_{ij})\hat{p}'_{ij}U_i + \lambda S^i_j), \tag{11}$$

where $\hat{p}'_{ij} \leftarrow \dfrac{e^{p_{ij}}}{(e^{p_{ij}} + 1)^2}$ is the derivative of $\hat{p}_{ij}$ and η is the learning rate. The user keeps U_i locally and send updated S^i to the central server. The central server then takes the average of updated S^i from all users and updates the global service latent matrix as follows:

$$S = \frac{1}{m} \sum_i S^i. \tag{12}$$

For each unobserved q_{ij}, the corresponding user u_i can conduct a quick local prediction as follows:

$$p_{ij} = U_i^T S^i_j. \tag{13}$$

Algorithm 1 shows the detailed process of federated QoS prediction. In each round, the local QoS data is stored and processed on the corresponding user side without exposing to others, thus protect the user from privacy threat.

Differential Privacy. In order to prevent the central server from inferring user privacy from gradients, we adopt differential privacy technology [7] to encrypt and decrypt gradient parameters transmitted between the user and the server.

Algorithm 1: Federated QoS Prediction

Input: QoS matrix Q, the current model: U, S, and the parameters η, λ.
Output: The updated model: U, S.

1 **repeat**
2   **foreach** *user u_i* **do**
3          $S^i \leftarrow S$;   `/* Local copy of global S */`
4   Update local QoS matrix Q_i with new QoS values;
5   Initialize U_i with small random numbers;
6   **repeat**   `/* Update local U_i and S^i */`
7              $p_{ij} \leftarrow U_i^T S_j^i$;
8              $\hat{p}_{ij} \leftarrow \frac{1}{1+e^{-p_{ij}}}$;
9              $\hat{p}'_{ij} \leftarrow \frac{e^{p_{ij}}}{(e^{p_{ij}}+1)^2}$;
10             $U_i \leftarrow U_i - \eta((\hat{q}_{ij} - \hat{p}_{ij})\hat{p}'_{ij}S_j^i + \lambda U_i)$;
11             $S_j^i \leftarrow S_j^i - \eta((\hat{q}_{ij} - \hat{p}_{ij})\hat{p}'_{ij}U_i + \lambda S_j^i)$;
12   **until** *converge*;
13   **end**
14     $S \leftarrow \frac{1}{m}\sum_i S^i$;   `/* Update global S */`
15 **until** *forever*;

For simplicity, we denote the gradient by g hereafter. A secret key comprising of two big prime numbers is generated and distributed to each user. Let x and y be the two prime numbers, $N = xy$ be the public parameter, and ϵ be the privacy budget. For each user u_i, the randomly sampled noise $Lap(\frac{\Delta f}{\epsilon})$ from Laplace distribution is add to the local gradients as follows:

$$g_{i,x} \equiv (g_i + Lap(\frac{\Delta f}{\epsilon})) \mod x, \tag{14}$$

$$g_{i,y} \equiv (g_i + Lap(\frac{\Delta f}{\epsilon})) \mod y, \tag{15}$$

where Δf denotes the sensitivity, and ϵ controls privacy budget. A smaller ϵ indicates a higher privacy level. The encrypted gradient is then calculated by:

$$E_i = y^{-1}yg_{i,x}^x + x^{-1}xg_{i,y}^y \mod N, \tag{16}$$

where y^{-1} and x^{-1} are the inverses of y and x, respectively. After receiving E_i from each user u_i, the central server performs global aggregation in the following way:

$$E = \sum_i E_i. \tag{17}$$

At the beginning of each round, each user u_i receives a local copy of the latest encrypted global gradients E and decrypts it by:

$$g \equiv E \mod x, \tag{18}$$

$$g \equiv E \mod y, \tag{19}$$

such that $g \approx \sum_i g_i$ is an unbiased estimate of the global gradient.

2.4   Efficient Federated Learning

While federated QoS prediction approach can protect user privacy, further improvements are needed to accommodate the diversity capacities of user devices. As cloud applications usually run on customers' devices, such as desktops, laptops, tablets, smart phones, and in-car systems, computing capacity and network bandwidth may become bottlenecks. In order to solve this problem, we propose several techniques to improve the model efficiency from three aspects: reducing computational overhead, reducing the amount of data transmission, and reducing the number of communications.

Reducing the Overhead of User-Side Computation in Each Round

As shown in Algorithm 1, each user needs to learn the potential user factor u_i and update the potential service matrix S during several rounds of iteration. The iterative process takes time from random initialization to convergence. Since the user latent vectors do not change much between two consecutive rounds, we initialized each round with the u_i from the previous round instead of random values to reduce learning time. Therefore, we only need to calculate the incremental updates in each round. For this purpose, we employ stochastic gradient descent (SGD) [16], a wildly-used online learning technique, to train the model. For each newly observed QoS value q_{ij} from user u_i in the current round, we have the pointwise loss function:

$$\Delta \mathcal{L}_{ij} = \frac{1}{2}(\hat{q}_{ij} - \hat{p}_{ij})^2 + \frac{\lambda}{2}(||U_i||^2 + ||S_j^i||^2), \tag{20}$$

such that $\mathcal{L}_i = \sum_j \Delta \mathcal{L}_{ij}$. Instead of minimizing $\mathcal{L}_i$, SGD relaxes to minimize the pointwise loss function $\Delta \mathcal{L}_{ij}$ over all newly observed QoS values. The corresponding updating equations are as follows:

$$U_i \leftarrow U_i - \eta((\hat{q}_{ij} - \hat{p}_{ij})\hat{p}'_{ij}S_j^i + \lambda U_i), \tag{21}$$

$$S_j^i \leftarrow S_j^i - \eta((\hat{q}_{ij} - \hat{p}_{ij})\hat{p}'_{ij}U_i + \lambda S_j^i), \tag{22}$$

where $\hat{p}'_{ij} = \dfrac{e^{p_{ij}}}{(e^{p_{ij}} + 1)^2}$ is the derivative of $\hat{p}_{ij}$ and η is the learning rate. The user then keeps U_i locally and only need to transfer the gradient descent $\frac{\partial \mathcal{L}_i}{\partial S^i}$ instead of S^i to the central server. The central server collects contributions from all users and takes advantage of the learning results to make a simple average to get a global gradient descent:

$$\frac{\partial \mathcal{L}}{\partial S_j} = \frac{1}{m} \sum_i \frac{\partial \mathcal{L}_i}{\partial S_j^i}. \tag{23}$$

Then the global service latent matrix can be updated as follows:

$$S_j \leftarrow S_j - \eta \frac{\partial \mathcal{L}}{\partial S_j}. \tag{24}$$

Reducing the Size of Transmitted Messages in Each Round

In practice, federated QoS prediction involves a massive number of users with various devices (e.g., desktops, laptops, smart phones, etc.), and communication on the Internet is many orders of magnitude slower than local computing [10]. The asymmetry of Internet speed (i.e., the uplink is usually much slower than the downlink) makes communication possible to become a bottleneck in the construction of a federated prediction model. It is necessary to reduce the communication cost, especially the uplink communication cost. We adopt three optimization measures to reduce the amount of data transmitted.

– Firstly, according to Eq. (22), only when user u_i observes an updated QoS value q_{ij} in the current round, will the corresponding service latent factor S_j^i be updated. Therefore, we can submit a subset of $\frac{\partial \mathcal{L}_i}{\partial S^i}$ (i.e., the set of updated $\frac{\partial \mathcal{L}_i}{\partial S_j^i}$) instead of the entire $\frac{\partial \mathcal{L}_i}{\partial S^i}$ to the central server without affecting accuracy. Accordingly, the central server aggregates the local updates by:

$$\frac{\partial \mathcal{L}}{\partial S_j} = \frac{1}{|N_j|} \sum_{u_i \in N_j} \frac{\partial \mathcal{L}_i}{\partial S_j^i}, \tag{25}$$

where N_j is the set of users update S_j^i in the current round.
– Secondly, instead of sending $\frac{\partial \mathcal{L}_i}{\partial S_j^i}$, each user only communicates $\frac{\partial \tilde{\mathcal{L}}_i}{\partial S_j^i}$ which is formed from a random subset of the values of $\frac{\partial \mathcal{L}_i}{\partial S_j^i}$. the central server aggregates the local updates to form a global update $\frac{\partial \tilde{\mathcal{L}}}{\partial S_j}$, which is an unbiased estimator of the true update:

$$\frac{\partial \mathcal{L}}{\partial S_j} = \mathbb{E}[\frac{\partial \tilde{\mathcal{L}}}{\partial S_j}]. \tag{26}$$

The mask can be randomized independently for each user in each round by a synchronized seed stored on both the user side and the server side. The percentage of omitted values is defined by the parameter r.
– Thirdly, we compress the transmitted data by quantizing the updates. Let $d \in \mathbb{R}^{1 \times l}$ denotes the vector of $\frac{\partial \mathcal{L}_i}{\partial S_j^i}$, $d_{max} = max(d_k)$ and $d_{min} = min(d_k)$. The quantized d is calculated as follows:

$$\bar{d}_k = \begin{cases} d', & \text{with probability } \dfrac{d - d_{min}}{d_{max} - d_{min}} \\ d'', & \text{with probability } \dfrac{d_{max} - d}{d_{max} - d_{min}} \end{cases}, \tag{27}$$

where $d' = \lceil \dfrac{d - d_{min}}{d_{max} - d_{min}} 2^b \rceil$ and $d'' = \lfloor \dfrac{d - d_{min}}{d_{max} - d_{min}} 2^b \rfloor$. It is easy to show that $\bar{d}$ is an unbiased estimator of d. Parameter b represents b-bit quantization, which controls the trade-off between accuracy and communication overhead. The error incurred by this compression was analyzed in [22].

Reducing the Number of Communication Rounds
Local service latent factors usually do not change much in one round. In order to reduce the number of communications for each user, a threshold δ can be set to control how often updates are transmitted to the central server. If the maximum value of the updates is greater than δ, the updates are transmitted. Otherwise, the updates are not transmitted and saved at the local site. Undelivered updates will be accumulated in the next round. In this way, large updates have influence on the model in time. The delay of small updates has little effect on the accuracy of the model, and small updates eventually produce an effect on the model in the form of cumulative updates. δ controls the tradeoff between efficiency and accuracy.

3 Experiments

In this section, we conduct several experiments on a real-world QoS dataset of Web services to evaluate our online prediction approach.

3.1 Dataset Description

We conduct experiments on a commonly used real-world dataset [31] to evaluate the QoS prediction approaches. This dataset contains two important QoS attributes, response time (RT) and throughput (TP), which characterize the non-functional quality of services. RT refers to the time it takes for a service to respond to a user request. TP refers to the data transfer rate during an invocation. Without losing generality, our approach can be easily extended to other QoS attributes.

Table 1. Statistics of QoS dataset

QoS	#Users	#Services	#Slices	#Records	Scale	Median
RT (sec)	142	4,532	64	27,393,508	0–20	0.377
TP (kbps)	142	4,532	64	25,643,690	0–1000	1.851

The QoS values are collected by invoking 4,532 services from 142 geographically distributed users on 64 consecutive time slices, each for 15 min. The users and the real-world services are distributed in 22 countries and 57 countries, respectively. The statistics of the dataset are summarized in Table 1. The range of response time is 0–20 s, and the range of throughput is 0–1000 kbps. The medians of response time and throughput are 0.377 s and 1.851 kbps, respectively. Note that there are some missing values in the dataset. Figure 3 shows the data distributions of response time and throughput, which are highly skewed. Figure 5 shows the data distributions closer to normal shape after data transformation by tuning α to -0.007 for RT and -0.005 for TP.

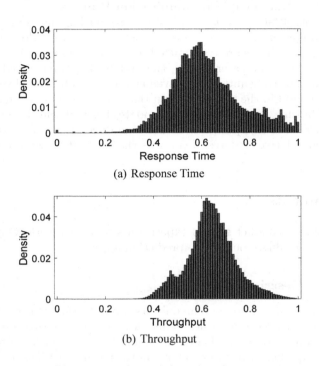

(a) Response Time

(b) Throughput

Fig. 5. Transformed data distributions

3.2 Evaluation Metric

We use MAE (Mean Absolute Error) to evaluate the accuracy of QoS prediction approaches. MAE is defined by:

$$MAE = \frac{1}{N} \sum_{I_{ij}=0} |q_{ij} - p_{ij}|, \tag{28}$$

where q_{ij} represents the real QoS value of an empty entry, p_{ij} is the predicted one by the model, and N represents the total number of predicted entries in the experiments. MAE is a widely used metric for evaluating the accuracy of a prediction model. Generally, the lower the MAE value, the better the performance of the model and the higher the prediction accuracy.

3.3   Prediction Accuracy

We compare the proposed privacy-preserving QoS prediction approach against other counterpart approaches in terms of accuracy. Note that some of these prediction approaches have no mechanism to protect privacy. The representative approaches in the literature are as follows:

– UIPCC [32]: This approach use historical QoS data of similar users and similar services to predict the QoS for current users. It is a collaborative filtering approach without any privacy-preserving mechanism.
– PMF [33]: This is a collaborative filtering approach employing matrix factorization techniques for QoS prediction, with no privacy-preserving mechanism.
– P-PMF [35]: This approach is similar to PMF, with the difference that it integrates a privacy-preserving mechanism through the use of data obfuscation techniques.
– **FMF:** This is the federated QoS prediction approach proposed in Sect. 2.3 with a privacy-preserving mechanism.
– **EFMF:** This approach further extends the FMF approach with efficiency improvement techniques proposed in Sect. 2.4.

We randomly designate some entries as unobserved ones and remove them from the QoS matrix. We vary the percentage of removed entries from 10% to 30% with a step of 5%. $l = 10$, $r = 20\%$, $b = 8$, and δ is tuned to optimal value in experiments for EFMF. We also optimize the parameters for counterpart approaches in experiments accordingly. We carry out the experiments 20 times with randomized initialization for each density and take the average accuracy as the final result. Table 2 shows the prediction accuracy. The results show that FMF and EFMF have higher accuracy than UIPCC and P-PMF. The predictive accuracy of FMF and EFMF is similar to that of PMF, whereas PMF does not consider privacy issues. Compared to approaches based on similar users and services, matrix factorization can retrieve more useful information in sparse data. Therefore, FMF and EFMF have a higher accuracy than UIPCC. P-PMF approach uses confused data, which will bring noise to the model. FMF and EFMFF use raw QoS data and are therefore more accurate than P-PMF. Generally, all approaches are more accurate as the data becomes more dense, since more training data means more information. The experimental results shows that the proposed privacy-preserving QoS prediction approach is effect.

Table 2. Prediction accuracy (w.r.t. MAE)

QoS	Approach	Data density				
		10%	15%	20%	25%	30%
RT	UIPCC	0.593	0.527	0.489	0.461	0.445
	PMF	0.506	0.485	0.468	0.451	0.439
	P-PMF	0.557	0.524	0.501	0.482	0.469
	FMF	0.517	0.492	0.475	0.458	0.446
	EFMF	0.522	0.500	0.481	0.462	0.454
TP	UIPCC	23.457	21.253	19.964	18.582	17.738
	PMF	16.580	15.307	14.681	14.178	13.957
	P-PMF	21.267	18.742	17.502	16.930	16.635
	FMF	17.270	16.028	15.145	14.354	14.067
	EFMF	17.714	16.351	15.296	14.397	14.105

3.4 Prediction Efficiency

We compare the convergence time of different approaches to evaluate the efficiency. Figure 6 shows that FMF and EFMF need less time to than other approaches to make prediction. Since the federated QoS prediction model is trained in a distributed manner, it includes both local models and a global model. Users process small amounts of data locally to update the local model. The central server, on the other hand, can update the global model by simply handling the updates collected for users. In contrast, other approaches require the central server to process large amounts of data and train the model over long periods of time. EFMF is more efficient than FMF due to the several techniques we used in Sect. 2.4. In addition, the convergence time EFMF in the first round is relatively long, and it is very short in the subsequent rounds. This is because EFMF only needs to incrementally update the model by using online learning techniques for new observations. These results demonstrate the efficiency of our approaches.

3.5 Scalability Analysis

We further analyze the scalability of our approach. We compare the median convergence time under different data volumes by varying the number of users and services. Due to the limited size of the real-world dataset, we generate large-scale synthetic datasets by repeating the data, which is reasonable for measuring convergence time. Figure 7 shows that the convergence time (log scale) of all approaches increases with the number of users and services. The increase of convergence time of FMF and EFMF is very gentle, while the convergence time of other approaches increases greatly. This is because the computing tasks are distributed to users in FMF and EFMF. In addition, EFMF employ online prediction techniques and make incremental updates on new data, while other approaches requires an iterative process in each time slice. These results demonstrate that EFMF can be easily extend to process large-scale data from more users and services.

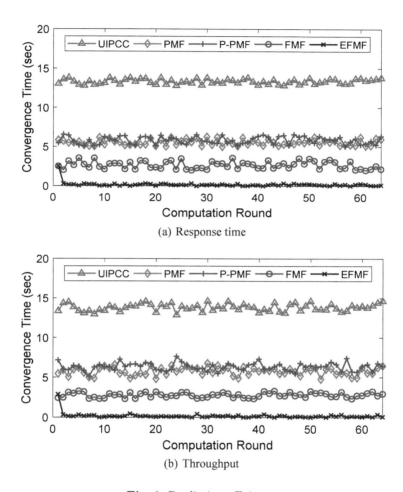

(a) Response time

(b) Throughput

Fig. 6. Prediction efficiency

4 Related Work

4.1 QoS Prediction

QoS refers to a series of non-functional characteristics of services (e.g., response time, throughput, availability, etc.) [5]. It has been widely employed to facilitate QoS-driven approaches in cloud and service computing [27]. QoS-aware service recommendation and selection help users select high-quality services from a set

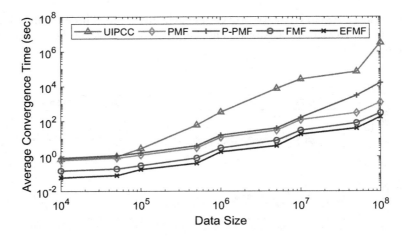

Fig. 7. Scalability result

of functional equivalent or similar service candidates. QoS-aware composition [11,24] aims to orchestrating a set of available services to provide complicated functions through a defined work flow. QoS-aware service adaption [6] help users replace current working services with corresponding service candidates at runtime to tolerate unacceptable QoS degradation or failures [36].

Accurate QoS value is important for QoS-aware techniques. Online testing [13,34] is proposed to obtain QoS values by active measurement. However, this approach incurs additional overhead and the observed QoS data is very sparse. In order to solve this problem, QoS prediction approaches [14,23,29–31] are proposed to use historical observed QoS data and make prediction for unobserved QoS data.

Collaborative filtering techniques [9,21] are widely used in social computing to predict unknown ratings on items (e.g., books, movies, etc.). They are introduced recently make QoS prediction for services. Matrix factorization is a representative model in collaborative filtering. The model derives user latent factors and service factors to represent the user-specific features and service-specific features, respectively. The unobserved QoS values are predicted by how the user-specific features apply to the corresponding service-specific features. In the literature, some of the state-of-the-art QoS prediction approaches such as PMF [17], CloudPred [30], WSPred [31], AMF [36], P-PMF [35], use matrix factorization techniques to provide accurate QoS prediction results. In contrast, this paper tries to protect user privacy while making QoS prediction.

4.2 Privacy Preservation

In the era of big data, privacy threat issues are particularly concerned by many studies. In cloud computing, many third parties provide services to enable service composition. Personal sensitive data usually flows across different parties.

Untrusted parties may expose these data and pose a threat to user privacy. To address this challenge, there appears some privacy-preserving approaches in service computing, such as privacy-preserving service selection [20,28], service recommendation [1,35] and service composition [2,4], etc. For example, Zhu et al. [35] propose a data obfuscation approach before sharing data across parties to prevent data leakage. Barhamgi et al. [2] analyze the privacy issue in composition of data services. Different from these studies, we focus on privacy-preserving issues for QoS prediction.

4.3   Federated Learning

Federated learning techniques have been used by major service providers [3] for train models over remote devices. In federated learning, a star network, where a central server is connected to a network of devices, is the main communication topology [26]. The decentralized training has been demonstrated to be effective when operating on networks with low bandwidth or high latency [8]. Under federated learning paradigm, users keep their data at local site without transferring it to the central server. In this paper, we conduct matrix factorization in a distributed manner by leveraging federated learning. Further more, we reduce computing cost and improve communication efficiency to make federated matrix factorization feasible.

5   Conclusion

In order to build and maintain high-quality cloud applications, privacy protection is the key to improve the accuracy of QoS prediction. In this paper, we propose a novel efficient and privacy-preserving QoS prediction approach for cloud services. We employ federated learning techniques to train the QoS prediction model in a decentralized manner. Users can keep their data at local site securely without fear of data leakage. We further make some improvements to significantly improve the efficiency of the QoS prediction model. The experimental results on a large-scale real-world QoS dataset demonstrate that our approach is effective and efficient, while effectively preventing privacy threats.

Acknowledgment. The work described in this paper was supported by the National Natural Science Foundation of China (61802003), and the Anhui Innovation Program for Overseas Students.

References

1. Badsha, S., et al.: Privacy preserving location-aware personalized web service recommendations. IEEE Trans. Serv. Comput. (2018)
2. Barhamgi, M., Perera, C., Yu, C.M., Benslimane, D., Camacho, D., Bonnet, C.: Privacy in data service composition. IEEE Trans. Serv. Comput. (2019)
3. Bonawitz, K., et al.: Towards federated learning at scale: system design. arXiv preprint arXiv:1902.01046 (2019)

4. Carminati, B., Ferrari, E., Tran, N.H.: A privacy-preserving framework for constrained choreographed service composition. In: 2015 IEEE International Conference on Web Services, pp. 297–304. IEEE (2015)
5. Chen, T., Bahsoon, R.: Self-adaptive and online QoS modeling for cloud-based software services. IEEE Trans. Softw. Eng. **43**(5), 453–475 (2016)
6. Chen, X., Wang, H., Ma, Y., Zheng, X., Guo, L.: Self-adaptive resource allocation for cloud-based software services based on iterative QoS prediction model. Future Gener. Comput. Syst. **105**, 287–296 (2020)
7. Dwork, C., McSherry, F., Nissim, K., Smith, A.: Calibrating noise to sensitivity in private data analysis. In: Halevi, S., Rabin, T. (eds.) TCC 2006. LNCS, vol. 3876, pp. 265–284. Springer, Heidelberg (2006). https://doi.org/10.1007/11681878_14
8. He, L., Bian, A., Jaggi, M.: COLA: decentralized linear learning. In: Advances in Neural Information Processing Systems, pp. 4536–4546 (2018)
9. He, X., Liao, L., Zhang, H., Nie, L., Hu, X., Chua, T.S.: Neural collaborative filtering. In: Proceedings of the 26th International Conference on World Wide Web, pp. 173–182 (2017)
10. Huang, J., et al.: An in-depth study of LTE: effect of network protocol and application behavior on performance. ACM SIGCOMM Comput. Commun. Rev. **43**(4), 363–374 (2013)
11. Li, J., Fan, G., Zhu, M., Yan, Y.: Pre-joined semantic indexing graph for QoS-aware service composition. In: 2019 IEEE International Conference on Web Services (ICWS), pp. 116–120. IEEE (2019)
12. Liu, A., et al.: Differential private collaborative web services QoS prediction. World Wide Web **22**(6), 2697–2720 (2019)
13. Liu, X., Sheu, R.K., Lo, W.T., Yuan, S.M.: Automatic cloud service testing and bottleneck detection system with scaling recommendation. Concurr. Comput.: Pract. Exp. **32**(1), e5161 (2020)
14. Luo, X., Wu, H., Yuan, H., Zhou, M.: Temporal pattern-aware QoS prediction via biased non-negative latent factorization of tensors. IEEE Trans. Cybern. **50**, 1798–1809 (2019)
15. Ma, H., Yang, H., Lyu, M.R., King, I.: SoRec: social recommendation using probabilistic matrix factorization. In: Proceedings of the 17th ACM Conference on Information and Knowledge Management, pp. 931–940 (2008)
16. Mandt, S., Hoffman, M.D., Blei, D.M.: Stochastic gradient descent as approximate Bayesian inference. J. Mach. Learn. Res. **18**(1), 4873–4907 (2017)
17. Mnih, A., Salakhutdinov, R.R.: Probabilistic matrix factorization. In: Advances in Neural Information Processing Systems, pp. 1257–1264 (2008)
18. Osborne, J.: Improving your data transformations: applying the box-cox transformation. Pract. Assess. Res. Eval. **15**(1), 12 (2010)
19. Qi, L., Xiang, H., Dou, W., Yang, C., Qin, Y., Zhang, X.: Privacy-preserving distributed service recommendation based on locality-sensitive hashing. In: 2017 IEEE International Conference on Web Services (ICWS), pp. 49–56. IEEE (2017)
20. Squicciarini, A., Carminati, B., Karumanchi, S.: A privacy-preserving approach for web service selection and provisioning. In: 2011 IEEE International Conference on Web Services, pp. 33–40. IEEE (2011)
21. Su, X., Khoshgoftaar, T.M.: A survey of collaborative filtering techniques. In: Advances in Artificial Intelligence 2009 (2009)
22. Suresh, A.T., Yu, F.X., Kumar, S., McMahan, H.B.: Distributed mean estimation with limited communication. In: Proceedings of the 34th International Conference on Machine Learning-Volume 70, pp. 3329–3337 (2017)

23. Wang, S., Zhao, Y., Huang, L., Xu, J., Hsu, C.H.: QoS prediction for service recommendations in mobile edge computing. J. Parallel Distrib. Comput. **127**, 134–144 (2019)
24. White, G., Palade, A., Clarke, S.: QoS prediction for reliable service composition in IoT. In: Braubach, L., et al. (eds.) ICSOC 2017. LNCS, vol. 10797, pp. 149–160. Springer, Cham (2018). https://doi.org/10.1007/978-3-319-91764-1_12
25. Xu, Y., Yin, J., Deng, S., Xiong, N.N., Huang, J.: Context-aware QoS prediction for web service recommendation and selection. Expert Syst. Appl. **53**, 75–86 (2016)
26. Yang, Q., Liu, Y., Chen, T., Tong, Y.: Federated machine learning: concept and applications. ACM Trans. Intell. Syst. Technol. (TIST) **10**(2), 1–19 (2019)
27. Zhang, Y., Lyu, M.R.: QoS Prediction in Cloud and Service Computing: Approaches and Applications. Springer, Singapore (2017). https://doi.org/10.1007/978-981-10-5278-1
28. Zhang, Y., Zhang, P., Luo, Y., Luo, J.: Efficient and privacy-preserving federated QoS prediction for cloud services. In: IEEE Conference on Web Services (ICWS) (2020)
29. Zhang, Y., Zhang, X., Zhang, P., Luo, J.: Credible and online QoS prediction for services in unreliable cloud environment. In: IEEE Conference on Services Computing (SCC) (2020)
30. Zhang, Y., Zheng, Z., Lyu, M.R.: Exploring latent features for memory-based QoS prediction in cloud computing. In: 2011 IEEE 30th International Symposium on Reliable Distributed Systems, pp. 1–10. IEEE (2011)
31. Zhang, Y., Zheng, Z., Lyu, M.R.: WSPred: a time-aware Personalized QoS Prediction Framework for Web services. In: 2011 IEEE 22nd International Symposium on Software Reliability Engineering, pp. 210–219. IEEE (2011)
32. Zheng, Z., Ma, H., Lyu, M.R., King, I.: WSRec: a collaborative filtering based web service recommender system. In: IEEE International Conference on Web Services, pp. 437–444. IEEE (2009)
33. Zheng, Z., Ma, H., Lyu, M.R., King, I.: Collaborative web service QoS prediction via neighborhood integrated matrix factorization. IEEE Trans. Serv. Comput. **6**(3), 289–299 (2012)
34. Zhong, H., Zhang, L., Khurshid, S.: TestSage: regression test selection for large-scale web service testing. In: 2019 12th IEEE Conference on Software Testing, Validation and Verification (ICST), pp. 430–440. IEEE (2019)
35. Zhu, J., He, P., Zheng, Z., Lyu, M.R.: A privacy-preserving QoS prediction framework for web service recommendation. In: 2015 IEEE International Conference on Web Services, pp. 241–248. IEEE (2015)
36. Zhu, J., He, P., Zheng, Z., Lyu, M.R.: Online QoS prediction for runtime service adaptation via adaptive matrix factorization. IEEE Trans. Parallel Distrib. Syst. **28**(10), 2911–2924 (2017)

A Reinforcement Learning Based Approach to Identify Resource Bottlenecks for Multiple Services Interactions in Cloud Computing Environments

Lingxiao Xu[1], Minxian Xu[2(✉)], Richard Semmes[3], Hui Li[3], Hong Mu[3], Shuangquan Gui[3], Wenhong Tian[1(✉)], Kui Wu[4], and Rajkumar Buyya[1,5]

[1] School of Software and Information Engineering, University of Electronic Science and Technology of China, Chengdu, China
tian_wenhong@uestc.edu.cn
[2] Shenzhen Institutes of Advanced Technology, Chinese Academy of Sciences, Shenzhen, China
mx.xu@siat.ac.cn
[3] Siemens Industry Software (Chengdu) Co., Ltd., Chengdu, China
[4] Department of Computer Science, University of Victoria, Victoria, Canada
[5] CLOUDS Lab, School of Computing and Information Systems, University of Melbourne, Melbourne, Australia

Abstract. Cloud service providers are provisioning resources including a variety of virtual machine instances to support customers that migrate their services to the cloud. From the customers' perspective, selecting the appropriate amount of resources is tightly coupled with performance and cost. By identifying the potential resource bottlenecks in the early stage of the service deployment process, resource planning can be significantly optimized. However, due to the unpredictable workloads and heterogeneous resources, it is difficult to identify resource bottlenecks that can degrade system performance. To support system non-functional requirements (NFR) in a better manner, we propose a reinforcement learning based approach to support the NFR management of system concerning the multiple services interactions scenario by identifying the potential resource bottleneck and optimizing the demanded resources. The proposed approach can predict the resource bottleneck for multiple services interactions, e.g. bottleneck in CPU or overloads in specific service, and provide guidance for resource planning. We modeled and simulated the proposed approach using an extended version of the CloudSim toolkit. Comprehensive evaluations with realistic use case from Siemens Digital Industries Software's MindSphere Solution on AliCloud show that our proposed approach can achieve high accuracy in terms of performance metrics, such as response time, queries per second (QPS), and resource usage.

© ICST Institute for Computer Sciences, Social Informatics and Telecommunications Engineering 2021
Published by Springer Nature Switzerland AG 2021. All Rights Reserved
H. Gao et al. (Eds.): CollaborateCom 2020, LNICST 350, pp. 58–74, 2021.
https://doi.org/10.1007/978-3-030-67540-0_4

Keywords: Cloud computing · Reinforcement learning · Service interactions · Non-functional requirement · Resource bottleneck

1 Introduction

The rapid development of cloud computing has made it be regarded as the fifth utility, like electricity, gas, and water [3]. Rather than assigning all tasks to a single local computer or a traditional computer cluster, cloud computing enables users to utilize computing or storage resources remotely, which provisions transparent and on-demand resources. In essence, the cloud is a networked computer paradigm based on virtualization techniques to improve resource usage. The pay-as-you-go model provided by cloud computing also helps the service providers and customers to start their business with minimal costs and eliminates the efforts to maintain the data centers [8].

Among all the benefits provided by cloud computing, some features are particularly attractive for customers. To be more specific, the first one is flexibility, which allows customers to acquire or release resources dynamically to fit their demands. The second is cost reduction, which means cloud computing service can convert capital expenditures to operational expenditures [19]. By utilizing the cloud, expensive infrastructures like servers and professionals will not be the main concern of customers anymore. And the third is the device and location independence, which supports customers to access computing resources anywhere and anytime, instead of using the specific interfaces to access the local machines and avoiding unavailability due to machine maintenance. Currently, the prominent IT companies such as Amazon, Google, Microsoft, and Alibaba, have established their own cloud data centers and become cloud providers.

Though cloud computing has significant advantages, based on the practice of utilizing resources of cloud computing, resource bottlenecks can still happen occasionally, like in CPU, memory, and bandwidth [20]. The reason is that when customers set up their services, they need to estimate how much of the resources will be used for their services and reserve specific amount of resources. However, due to the fluctuations of loads, overloads can occasionally happen and lead to bottlenecks. When bottlenecks exist, the system performance, such as non-functional requirement (NFR) can be significantly influenced [16], for instance, CPU bottleneck can cause insufficient computing power and reduced number of parallel processes.

As bottlenecks can greatly affect the performance and user experience of cloud services, it is important to identify the potential bottlenecks. However, different cloud services often provision various resource capacity, such as VM instances with different CPU, memory, and bandwidth, thus they may have bottlenecks under diverse load conditions. Besides, the bottlenecks of individual services are more difficult to be predicted based on divergent resource demands, running status and internal logic. If service providers can identify the potential bottleneck of each service and modify configurations, the NFR can be satisfied.

However, identifying the bottlenecks at the early stage is not easy. It's not feasible that arranging testers to perform a large number of stress tests to evaluate the bottlenecks when changing resource configurations. Apart from it, purchasing much more resources than required is not cost-effective and evaluation results may not be reproducible due to uncontrolled factors, such as network traffics.

To deal with the above challenges, using a simulation toolkit to simulate the real environment is a promising way. In this paper, we use CloudSim [4], which is a well-known cloud simulation toolkit that enables seamless modeling, simulation and experimentation of cloud computing. With CloudSim, we can model the resource provisioning of various cloud infrastructure configurations and generate reproducible results. Large-scale simulations can also be easily conducted. Our motivation is to predict the potential resource and service bottlenecks of a sample use case that is based on Siemens Digital Industries' MindSphere solution [13][1], which is a cloudbased Internet-of-Things (IoT) open operating system. We aim to generate various metrics to help plan hardware resources for the use case. We also make efforts to ensure our approach as generic as possible in order to extend it other cloud platforms with ease. To achieve these goals, we also have some challenges to address, including how to build a model for realistic scenarios, how to ensure that the system output results are consistent with the real test results, and how to make it a generic approach.

In this paper, we propose a data-driven framework to support the non-functional requirement, e.g. system performance, for multiple services interactions to identify the potential bottlenecks of service in the early stage. A policy gradient approach based on reinforcement learning is also proposed to predict the potential system bottleneck based on data collected from real test cases and guide companies to optimize the resource configuration. Additionally, the approach can also help to predict the response time and QPS when the system has reached the bottleneck.

Our key **contributions** are as follows:

- Presented a framework to identify the potential bottleneck for multiple services interacted in the cloud to optimize resource planning for companies.
- Proposed a policy gradient approach to model resource utilization and predict the system behaviors under different loads.
- Utilized CloudSim components to setup specific underlying IaaS infrastructure model, specific MindSphere service model, and realized performance test simulation.

The rest of the paper is organized as: we start by discussing the related work in Sect. 2, where we highlight the differences between our work and existing work. In Sect. 3, we introduce our proposed framework, named IRBS, for bottleneck identification. Then we discusses the modeling of multiple services interactions scenario and introduce our proposed reinforcement learning based approach for bottleneck prediction in Sect. 4. Afterward, the evaluation results based on the sample scenario are demonstrated in Sect. 5. Finally, conclusions and future work are given in Sect. 6.

---

[1] https://siemens.mindsphere.io/en.

2 Related Work

To utilize cloud service more efficiently and reduce the cost, it is important to model cloud services and scheduling process and optimize resources. Some work has been done to model services and optimize resources in Clouds. To decrease the task execution failure, Lattif et al. proposed the DCLCA (dynamic clustering league championship algorithm) scheduling technique for fault tolerance awareness to address cloud task execution which would reflect on the currently available resources and reduce the untimely failure of autonomous tasks [10]. Sekaran et al. presented a new meta-heuristic algorithm, named the dominant firefly algorithm, which can optimize load balancing of tasks among the multiple virtual machines in the Cloud server, thereby improving the response efficiency of Cloud servers that concomitantly enhances the accuracy of m-learning systems [15]. Cheng et al. introduced DRL-Cloud, a novel Deep Reinforcement Learning (DRL)-based RP and TS system, to minimize energy cost for large-scale CSPs with a very large number of servers that receive enormous numbers of user requests per day [5]. Nayak et al. used AHP (Analytic Hierarchy Process) as a decision-maker in the backfilling algorithm to choose the possible best lease from the given best-effort queue to schedule the deadline sensitive lease [12]. Priya et al. constructed a Fuzzy-based Multidimensional Resource Scheduling model to obtain resource scheduling efficiency in cloud infrastructure and increased utilization of Virtual Machines through effective and fair load balancing are then achieved by dynamically selecting a request from a class using Multi-dimensional Queuing Load Optimization algorithm [14]. These work can optimize resource usage, however, their objectives are not identifying the potential bottleneck of resources and services in the system.

In order to generate reproducible results and simulate a large-scale cloud environment, CloudSim is frequently used. For instance, Wickremasinghe et al. developed CloudAnalyst, which is a CloudSim based tool for simulating large-scale Cloud applications to study the behavior of such applications under various deployment configurations [17]. Jung et al. proposed a simulation tool that supports the MapReduce model, implemented on CloudSim [9]. Alla et al. presented Task Scheduling optimization using a novel approach based on Dynamic dispatch Queues (TSDQ) and hybrid meta-heuristic algorithms, which is based on CloudSim and showed a great advantage in terms of waiting time, queue length, makespan, cost, resource utilization, degree of imbalance, and load balancing [1]. Sharma et al. demonstrated a modified particle swarm optimization (MPSO) task scheduling algorithm in order to optimize execution time, transmission time, makespan, transmission cost, and load balancing of virtual machines and got the best cost as compared to original PSO on the CloudSim [2]. However, these works do not model service internal logic. Besides, they do not provide the model for the intended sample scenario and use cases which will be elaborated in Sect. 4.

Compared with other related work, our work mainly contributes to the current research area by providing an approach for identifying the resource and service bottlenecks for cloud system, it also models the multiple services interactions for the sample scenario. The proposed approach can be further applied to other related scenarios and service providers.

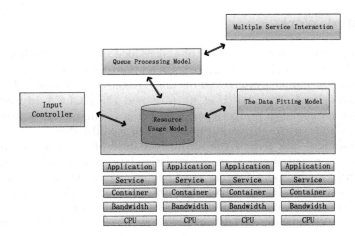

Fig. 1. Framework of IRBS

3 IRBS Framework

To make our proposed approach to be reusable, correct and universal, we propose a framework to identify resource and service bottlenecks for multiple services interactions in Cloud, named IRBS (Identifying Resource Bottlenecks Solution). IRBS can evaluate different kinds of cloud services to achieve their availability and scalability. IRBS aims to predict the bottlenecks of the cloud service and identify the metrics when the service has reached the bottleneck. From the companies' perspective, selecting the appropriate amount of resources is tightly coupled with performance and cost, and bottlenecks are the important metric to improve the rationality of resource allocation. We provide a high-level summary below and then discuss the details in the subsequent sections.

We aim to make generic design, thus IRBS should be applicable to cloud services of different specifications by changing configurations. We also target to fit the data to achieve high accuracy. Furthermore, when a new service is integrated into the system, our framework should also adapt it in a good manner.

The framework of IRBS is shown in Fig. 1, it contains five components: Input Controller, Resource Usage Model, Data Fitting Model, Queue Processing Model, Multiple Services Interactions Modelling.

The *Input Controller Component* handles the input workloads. The workloads can be generated by this component, then the concurrent loads can be processed by other components in the framework. The characteristic of loads can be defined in the input property files, including the number of loads, duration, ramp-up time etc., which will be loaded firstly. In our framework, we consider the online scheduling policy to fit the realistic scenario, and the loads are generated as per time slot, e.g. 1 s.

The *Resource Usage Model Component* is the key component of IRBS. It produces the resource pool containing the CPU pool and bandwidth pool, then simulates the real scenario to perform the distribution of loads. Based on this, results including real-time CPU usage, the real-time bandwidth usage, the initial response time can be obtained and modelled. More details will be provided in Sect. 4.

The *Data Fitting Model Component* supports to calculate the evaluation results. In the real scenario, the hardware, temperature, voltage, and other factors can have significant impacts on cloud service performance. As for the software, the scheduling policy and the adopted cloud service can also influence the NFR. This component provides the approach to train the prediction model based on the actual data.

Queue Processing Model controls two queues by monitoring the data in Resource Usage Model, which named the deferral queue and processing queue. It is responsible for managing the workloads in the system and collaborating with the components in the system to schedule the resources. The deferral queue stores the loads that are deferred when the resource pool is full, and the processing queue stores the loads under processing. The component notifies the Input Controller about the total loads in the system and then adjusts the generated loads dynamically based on resource usage.

The *Multiple Services Interactions Component* manages the interactions and collaborations between services. It can also represent the flow of workloads. Considering the collaboration of different services that if one service reaches its bottleneck, the other service will be affected, so IRBS designs the multiple service interactions to deal with the situation. By querying the Queue Processing Model, the component can obtain the previous service queue status and notify the next service, then achieve better resource optimization effects.

4 Problem Modelling and Proposed Solution

In this section, we will formulate our problem modeling for identifying the system bottlenecks by predicting the system resource usage. We then present our proposed approach based on policy gradient in reinforcement learning to solve the problem.

4.1 System Model

To support the understanding of our system model, the sample scenario is shown in Fig. 2, which mainly consists of 6 services collaborating together to support the scenario. The load generator is the service that generates loads for the system. SLB GW is used as the task scheduler (GW), which is responsible for maintaining the load balancing and performing task distribution. GW K8S is the service under test which is also the core component simulated in the whole system model, which will also refer to the simulation of the interaction between GW K8S and its backing database (Redis). SLB NFR is used to monitor various metrics in the entire system, and Mockservice is a black box component that generates as fixed processing delay component, e.g. 300 ms. Considering the

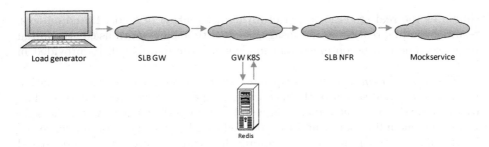

Fig. 2. The sample scenario for simulation

different functionalities of these services, the services consume different amount of resources, and there are interactions between the services. Our objective is to identify the bottlenecks in the system resulted from CPU usage and bandwidth usage. A more general system model is introduced as follows.

Assuming the whole observation time period is T, and at time slot t, $t \in \{1, 2, \ldots, T\}$, the number of load requests is L_t. We consider that there are M services collaborating together to support the sample scenario by providing services for the customers. These services can be deployed on public cloud, e.g. Alicloud, which provisions N ECS (Elastic Computing Service) VM nodes and I pods (containers). The pods are deployed on ECS. For ECS node E_n, $n \in \{1, 2, \ldots N\}$, it can have E_n^c cores. And for pod P_i, $i \in \{1, 2, \ldots, I\}$, it can have P_i^c cores. The cores represent the capacity to process the load requests coming into the system. The CPU utilization of each ECS and pod at time slot t can be represented as $E_n^u(t)$ and $P_i^u(t)$ respectively. For service S_m, $m \in \{1, 2 \ldots, M\}$, considering I_m pods are provided for it, thus the CPU utilization of service $S_m^u(t)$ can be represented:

$$S_m^u(t) = \frac{1}{I_m} \sum_{i=1}^{I_m} P_i^u(t) \tag{1}$$

CPU utilization is one of the key resource bottleneck that we would like to investigate.

Besides CPU utilization, we also consider the bandwidth resource bottleneck. Based on the obtained data, the bandwidth between different services can also experience bottlenecks due to the communications between services. For example, in Fig. 2 the requests are processed and distributed via the SLB GW service and then the requests are processed by GW K8S, where the bottleneck can exist due to the high volume of requests.

Considering services S_m and S_k, $m, k \in \{1, 2 \ldots, M\}$, are collaborating with each other. The bidirectional network traffics at time slot t can be represented as $B_{S_m, S_k}(t)$ and $B_{S_k, S_m}(t)$, where $B_{S_m, S_k}(t)$ represents the communication is from service S_m to S_k, and vice verse.

We use m^* to denote the id of the service with the highest CPU utilization, which can be calculated as:

Table 1. Performance metrics of Realistic Scenario

Case	Load	Pod CPU utilization	Throughput	Average response time	Bandwidth
1	1,500 vu	38.5 cores	10,000 calls/s	50 ms	300 Mbps
2	2,000 vu	46 cores	11,300 calls/s	75 ms	342 Mbps
3	3,000 vu	47 cores	11,600 calls/s	158 ms	352 Mbps

$$m^* = \arg\max_{m,t}\{S_m^u(t)\} \tag{2}$$

Compared with the predefined CPU utilization threshold T^c, when $S_{m^*}^u(t) \geq T^c$, it means the bottleneck exists in service m^*. Similarly, the communication with the highest bandwidth utilization between services can be represented as $B_{S_{m^*},S_{k^*}}(t)$, where the m^* and k^* can be calculated as:

$$\{m^*, k^*\} = \arg\max_{m,k,t}\{B_{S_m,S_k}(t)\} \tag{3}$$

Assuming that the bandwidth threshold from service m^* to k^* is T_{m^*,k^*}^b, the bandwidth bottleneck exists when $B_{S_{m^*},S_{k^*}}(t) \geq T_{m^*,k^*}^b$.

Some sample data for the Gateway service are shown in Table 1, which includes the information on load, pod utilization, throughput, average response time and bandwidth. Based on these data, we can define a given set of J input and output data as $D = \{(\boldsymbol{x}_i, y_1), (\boldsymbol{x_2}, y_2), \ldots, (\boldsymbol{x_J}, y_J)\}$, and $X = \{\boldsymbol{x_1}, \boldsymbol{x_2}, \ldots, \boldsymbol{x_J}\}$ and $Y = \{y_1, y_2, \ldots, y_J\}$ are the set of input and output data.

In our reinforcement learning based approach, we would like to predict some key metrics that can represent the system bottlenecks, e.g. CPU utilization and bandwidth. The predicted metrics can be calculated by $\hat{y}_i = f(\boldsymbol{x}_i)$, where $f(x)$ is the trained model by policy gradient approach. In our proposed approach, we aim to reduce the error between the predicted data and actual data, which can be evaluated by metrics such as Root Mean Square Error, Mean Relative Error, and R^2.

4.2   Policy Gradient Approach

Reinforcement Learning [11] is a data-driven approach for adaptively applying optimized control policies based on real-time feedback, which models the stochastic process under the framework of Markov Decision Process (MDP) [21]. Policy gradient [7] is one of the most common types of reinforcement learning algorithms. In the policy gradient approach, the optimal actions with model parameters can be learned directly. The control actions are selected as the ones that can maximize the system rewards. In this section, we exploit the policy gradient to seek the optimal policy.

In our problem, the approach will decide the amount of resources that should be allocated to each request load at each time slot. The key entities in our approach are defined as follows.

States: The system state $S(t)$ includes the status of (1) amount of loads $L(t)$, (2) CPU capacity for processing loads $W_c(t)$ (3) bandwidth capacity for processing loads $W_b(t)$, (4) CPU resource usage $U_c(t)$, and (5) bandwidth resource usage $U_b(t)$. The state space of $L(t)$ at time t is given as $L(t) < \overline{L} \in \mathbb{Z}^+$, where $\overline{L}$ is the maximum number of loads that the system can accommodate and $\mathbb{Z}^+$ is a set of non-negative integers. $W_c(t) \in \mathbb{R}^+$ and $W_b(t) \in \mathbb{R}^+$ represent the capacity to process loads based on CPU (per core) and bandwidth (per Mbps) respectively, where $\mathbb{R}^+$ is a set of non-negative real numbers. The CPU resource usage $U_c(t) \in [0,1]$ and bandwidth resource usage $U_b(t) \in [0,1]$ are obtained from system running status. Therefore, the system state at time t, $S(t)$, can be denoted by:

$$S(t) \triangleq [L(t), W_c(t), W_b(t), U_c(t), U_b(t)] \in \mathbb{S}, \tag{4}$$

where $\mathbb{S}$ stands for all possible states.

Actions: At the beginning of each time slot, the approach determines the actions to increase or decrease $W_c(t)$ and $W_b(t)$ to decide the CPU and bandwidth capacity to process loads. The next state $S(t+1)$ depends on the current state $S(t)$ by taking action $A(t) \triangleq [A_c(t), A_b(t)] \in \mathbb{A}$, where $A_c(t) \in (0, +\infty)$ and $A_b(t) \in (0, +\infty)$ are the corresponding actions for CPU capacity and bandwidth capacity, and $\mathbb{A}$ stands for all possible actions. Thus, the CPU and bandwidth capacity for processing loads at $t+1$ can be determined according to the following equations:

$$W_c(t+1) = A_c(t) \cdot W_c(t) \tag{5}$$

$$W_b(t+1) = A_b(t) \cdot W_b(t) \tag{6}$$

Rewards: At each unit time, the process is in a state $S(t)$, and we choose a possible action $A(t)$. The process randomly moves to the next state $S(t+1)$ at the next time slot, and gives the corresponding reward is $r(S(t), A(t))$. Our model concerns the Euclidean Distance. After adopting the action $A(t)$, the $U_c(t)$ and $U_b(t)$ that are received from the system, and the errors between the actual values and predicted values are accumulated. We set the cumulative errors of two consecutive actions, e.g.

$$r(t) = D(t-1) - D(t), \tag{7}$$

where $D(t-1)$ and $D(t)$ are the total cumulative errors in the current and previous time slots. The total cumulative errors at time slot t, is the summation of the cumulative errors of all actions from $t = 0$ to the current time slot. The positive reward values imply that actions are taken to decrease the total cumulative errors and the negative rewards imply an increase. With regard to the reward values, the actions can be taken to change the current state to other certain states in future time slots.

Objective Function: The approach chooses the actions based on policy π, which is defined as the mapping from the input state to a probability distribution over actions $\mathbb{A}$. We use parameters θ as policy parameters, and the policy distribution $\pi(A(t)|S(t);\theta)$ is learned by performing gradient descent on the policy parameters. In this research, we aim to maximize the reduction of cumulative errors, which represent the accuracy of resource utilization and bottleneck identification. The objective to maximize the reward values under the probability distribution $\pi(A(t)|S(t);\theta)$ can be denoted as:

$$J(\theta) = E_{\pi_\theta}[\sum_{t=0}^{T} \gamma^t r(t)] = E_{\pi_\theta}[R]. \tag{8}$$

Where γ is the discount factor, and R is defined as the sum of future discounted rewards.

In policy gradient approach, the gradient of the objective function in Eq. (8) can be given by:

$$\nabla_\theta J = \sum_{t=0}^{T} E_{\pi_\theta}[\nabla_\theta log(A(t)|S(t);\theta)R_t]. \tag{9}$$

The equation can update the policy parameters θ in the direction $\nabla_\theta log(A(t)|S(t);\theta)$ in order to increase the probability of action $A(t)$ at state $S(t)$ if the action can increase the cumulative reward values, and vice verse.

By utilizing actor-critic approach [7], the parameters update can be done by selecting actions to achieve the terminal state or taking K steps. Therefore, the parameters θ can be updated through the following equation:

$$\theta = \theta + \alpha \sum_{t} \nabla_\theta log(A(t)|S(t);\theta)V(S(t), A(t);\theta, \theta_v), \tag{10}$$

where α is the learning rate, $V(S(t), A(t);\theta, \theta_v)$ is the estimate of advantage function [11] comparing the state value function $V^{\pi_{\theta_v}}(S(t))$ of current state and the state after K steps.

The pseudocode of our proposed algorithm is shown in Algorithm 1. The algorithm is based on policy gradient to maximize the reward values, which controls by the actions. With the inputs of reachable states, possible actions, and predefined system parameters, the algorithm iterates over time slots from periods 1 to T. Line 3 explores the possible actions under policies and the expected reward of a state is calculated in line 4. The future discounted rewards calculation is performed by lines 5 to 10. And lines 12–14 update the parameters.

5 Performance Evaluations

In this section, we introduce our experimental environment, including the extension of CloudSim, environment settings, two scenarios under different loads, convergence analysis, performance evaluations, bottleneck prediction, and scalability analysis.

Algorithm 1. Policy gradient based reinforcement learning approach

---

Input: Reachable system states $S(t) \in \mathbb{S}$, observation time period T, possible actions $A(t) \in \mathbb{A}$, initialized parameters θ and θ_v, the maximum step K, discount factor γ, and learning rate α

Output: θ, θ_v

1: **for** t from 1 to T **do**
2:    **for** $S(t) \in \mathbb{S}$ **do**
3:        perform action $A(t)$ according to policy $\pi(A(t)|S(t);\theta)$
4:        obtain reward $R(t)$ and next state $S(t+1)$
5:        **if** $S(t)$ is terminal state **or** $T - t = K$ **then**
6:            $R = 0$
7:            **break**;
8:        **else**
9:            $R = V(S(t); \theta_v)$
10:        **end if**
11:        **for** $t^{'}$ from t to $t + K$ **do**
12:            $R \leftarrow R(t^{'} + \gamma R)$
13:            $\theta \leftarrow \theta + \alpha \nabla_{\theta} log(A(t^{'})|S(t^{'});\theta)(R - V(S(t^{'});\theta_v))$
14:            $\theta_v \leftarrow \theta_v + \frac{\partial (R - V(S(t^{'});\theta_v))^2}{\partial \theta_v}$
15:        **end for**
16:    **end for**
17: **end for**

5.1   Environment Settings

By following the simulation methodology of CloudSim 4.0, we have re-implemented some existing modules such as cloudlet, container, containerVM, and host. We add the queueing method and apply the resource scheduling approach to design our scheduling policy.

In our simulations, the time accuracy can be adjusted to make it more fine-grained. However, if the accuracy is improved, the simulation takes longer running time and consumes more memory resources. To balance the trade-offs between accuracy and running costs, we simulate and collect the CPU usage, bandwidth usage every second. And then, we draw the results based on the obtained metrics. We use the policy gradient algorithm introduced in Sect. 4.2 to train the resource usage parameters in our model, and the simulations are conducted for the situation under this configuration.

Based on the results, we can know when the bottleneck will exist under specific loads. Then we can adjust the configuration files according to the corresponding loads to optimize resource planning. In this way, the number of the loads when CPU resources or the bandwidth resources reach the bottleneck can be identified under this configuration. Additionally, we can use IRBS to predict some scenarios without evaluations under real tests, for example in what CPU configuration or bandwidth configuration the QPS will reach 20,000 calls/s, and optimize resource planning without changing hardware specifications, only via changing the cores or the maximum bandwidth.

Table 2. Infrastructure configuration

Case	Load (per sec.)	ECS number	Pod number	ECS cores	Pod cores	Bandwidth	Other service response time
1	1,500	9	18	8 cores	3 cores	350 Mbps	About 100 ms
2	2,000	9	18	8 cores	3 cores	350 Mbps	About 100 ms
3	3,000	9	18	8 cores	3 cores	350 Mbps	About 100 ms
4	3,020	20	40	8 cores	3 cores	650 Mbps	About 100 ms
5	4,000	16	32	8 cores	3 cores	800 Mbps	About 100 ms
6	4,000	20	40	8 cores	3 cores	610 Mbps	About 100 ms

In the scenario, GW K8S is the core component to be evaluated, as the whole data is confidential, only the key data is shown in Table 1. In the configurations, there are 9 ECS and 18 pods in total for the resource provisioning. Each ECS is equipped with 8 cores, each pod has 3 cores, and the bandwidth is configured 350 Mbps. The detailed configurations are shown in Table 2.

To evaluate the performance of our approach to verify system performance under different conditions, we firstly conduct experiments for underutilized scenario and overutilized scenario, which represent the system is running without bottlenecks and with bottlenecks. The discount factor γ in Algorithm 1 is configured as 0.99, and the learning rate α is set as 0.1. The following metrics are adopted to evaluate the performance, which are also key metrics that company would like to investigate:

- GW workload: number of input loads per second for GW service
- CPU usage: CPU usage base on percentage and number of cores at each time slot
- Outbound traffic: bandwidth usage in the outbound direction
- Inbound traffic: bandwidth usage in the inbound direction
- GW latency: response time of GW
- QPS: Queries per second that system can accommodate.

5.2 Underutilized Scenario

The underutilized scenario demonstrates the scenario when both the CPU and the inbound and outbound traffic have not reached the bottleneck, and all loads can respond in time. The results are demonstrated in Fig. 3. In this scenario, the load is about 1500 vu (virtualized users) per second at peak, and the loads grow gradually until 660 s and then keep stable until about 1200 s, and finally dropping to 0. During this period, the maximum CPU resource usage is about 70%, the used cores are about 38.5, the output bandwidth is about 300 Mbps, the inbound traffic is about 110 Mbps, the latency is about 50 ms, and the QPS is about 10,000. Compared with the actual data, the results conform to the system behavior, which illustrates that our approach can predict resource usage and system performance under the underutilized scenario.

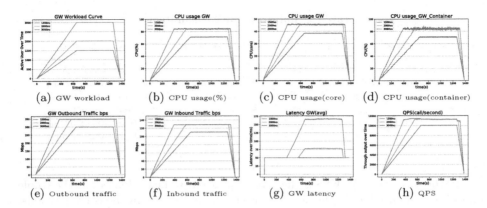

Fig. 3. The metrics under different load conditions

5.3 Overutilized Scenario

Considering the overutilized bandwidth as another example when the peak loads are 2000 and 3000 vu per second, which is also shown in Fig. 3. The increase and decrease patterns of loads are the same as the one with 1500 vu per second, which reaches the peak at time 660 s and starts to drop at time 1200 s. Based on the results, we can notice that when the outbound traffic reaches the peak and the saturation happens, the loads which arrive later cannot be processed by the service. Although CPU is not overutilized, the CPU utilization still grows up to be about 90% and the QPS also reaches 11,600 calls/s, which means bandwidth is the bottleneck rather than CPU in this case.

Given that resource competition will occur when resource usage approaches the bottleneck value, resulting in performance degradation, the average response time tends to rise in advance. Meanwhile, it is observed that when the bottleneck point is reached, the QPS remains at its peak, but the load is still rising. The results obtained by the simulation system are nearly the same as the actual test results. According to the results, it can be observed that when the load is close to 1,800 vu, it is the bottleneck in this configuration. Therefore, we can conclude that our approach supports identifying the system bottleneck.

Table 3. Regression models

Case	Method	Type	Regression formula
1	Single parameter regression	CPU	$\hat{U_c}(t) = 0.0324 * L(t) + 0.5691$
2	Single parameter regression	Bandwidth	$\hat{U_b}(t) = 0.1392 * L(t) - 1.0351$
3	Double parameters regression	CPU	$\hat{U_c}(t) = 0.01129 * L(t) + 0.0626 * t$
4	Double parameters regression	Bandwidth	$\hat{U_b}(t) = 0.0512 * L(t) + 0.2326$
5	Triple parameters regression	CPU	$\hat{U_c}(t) = 0.00009 * L(t) + 0.2185 * U_b(t) + 0.0118 * t$
6	Triple parameters regression	Bandwidth	$\hat{U_b}(t) = -0.00005 * L(t) + 4.5441 * U_c(t) - 0.052 * t$

Table 4. Performance comparison with baselines

Approach	RMSE (underutilized)	RMSE (overutilized)	RMSE	MAPE	R2
Single parameter regression	55.832	54.561	55.200	35.6%	−0.243
Double parameters regression	79.782	14.638	57.356	30.3%	−0.342
Triple parameters regression	8.822	10.132	9.499	4.0%	0.963
PSO	8.171	8.603	8.390	6.7%	0.971
IRBS	**6.306**	**4.466**	**5.464**	**3.5%**	**0.988**

5.4   Evaluation Metrics and Results

To show the accuracy of our proposed approach, we used a linear regression algorithm [6] and PSO (particle swarm optimization) algorithm [18] to compare the performance of our proposed algorithm. We adopt these two algorithms as baselines because there is no algorithm focusing on the same scenario we are going to apply and these two algorithms have been validated to be efficient to model resource usage in cloud systems. In the linear regression algorithm, we concern about the different number of parameters (single parameter, double parameter, and triple parameters) for training. The corresponding models based on linear regression are shown in the Table 3. For instance, the predicted CPU utilization $U_c(t)$ at time slot t with single parameter (loads number $L(t)$) regression can be represented as $0.0324 * L(t) + 0.5691$. PSO algorithm can be used as an automatic parameter training algorithm, in which specific parameters are trained and finally integrated into the system for evaluations. We evaluate the following metrics including Root Mean Square Error (RMSE), Mean Absolute Percentage Error (MAPE), and R^2, which have been widely used to evaluate the prediction accuracy:

$$RMSE = \sqrt{\frac{1}{n} \sum_{i=1}^{n} (y_i - \hat{y}_i)^2} \tag{11}$$

$$MAPE = \frac{100\%}{n} \sum_{i=1}^{n} \left| \frac{\hat{y}_i - y_i}{y_i} \right| \tag{12}$$

$$R^2 = 1 - \frac{\sum (y_i - \hat{y}_i)^2}{\sum (y_i - \bar{y})^2} \tag{13}$$

where n is the size of the dataset, y_i is the actual value, $\bar{y}$ is the mean value of actual data, and $\hat{y}_i$ is the predicted value based on our approach. The performance evaluation results are shown in the Table 4.

It can be noted that our approach has the highest value of RMSE for the underutilized scenario, the overutilized scenario, and the combined scenario. Our approach can also obtain the smallest MAPE value and the best R^2 value, which all illustrate that our approach can achieve the highest accuracy than other algorithms.

5.5   Bottleneck Prediction

Apart from identifying the bottleneck in specific configurations, IRBS can also predict the satisfaction of various metrics based on the parameters of the current configurations. Taking QPS as an example, Fig. 4 shows various situations when QPS reaches 20,000 calls/s, and the configurations of each situation are the same as the configurations in Table 2.

The first case is that the CPU and bandwidth are not overutilized. In this case, we can notice that when the number of loads reaches 3020 vu, the QPS is close to 20,000 calls/s. At this moment, the peak response time of GW service is 50 ms, and other services are also at the system condition that is not overutilized, the total response time of the other parts is about 100 ms, and the total response time is about 150 ms. Thus, we can conclude that the prediction results are accurate.

The second and third cases are about the condition when the CPU or bandwidth reaches the overutilization and becomes the bottleneck. In these cases, the CPU or bandwidth utilization is about 100% as overutilized. Under this situation, the peak response time of the GW service is about 100 ms. And the total response time is about 200 ms, which shows that the results are also accurate. Then the users can use these results to optimize their infrastructure configuration, e.g. adding more CPU or bandwidth resources to avoid the bottleneck.

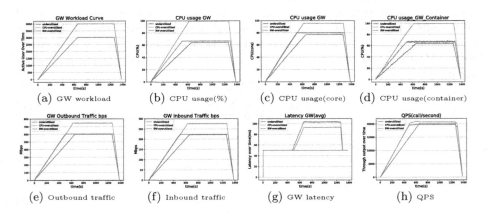

Fig. 4. The metrics of different loads when QPS is 20,000 calls/s.

6   Conclusions and Future Work

In this paper, we present a resource bottleneck identification approach, called IRBS, which can predict resource and service bottlenecks in multiple services interactions scenarios. We also proposed a policy gradient algorithm based on reinforcement learning to support the prediction of resource usage under different conditions. Based on the dataset provided, which is deployed on AliCloud, our simulated results based on the extension of CloudSim show that our approach can effectively identify the potential bottlenecks and achieve high accuracy for adopted metrics. The dominant advantage of the proposed approach is easy to operate and it provides a feasible solution for the customers to plan infrastructure resources.

Acknowledgments. This research is partially supported by Key-Area Research and Development Program of Guangdong Province (NO. 2020B010164003), the National Natural Science Foundation of China, with Grant ID 61672136 and 61828202, SIAT Innovation Program for Excellent Young Researchers. We thank teams in Siemens Industry Software Co., Ltd., China, for their discussion and comments on this work.

References

1. Ben Alla, H., Ben Alla, S., Touhafi, A., Ezzati, A.: A novel task scheduling approach based on dynamic queues and hybrid meta-heuristic algorithms for cloud computing environment. Clust. Comput. **21**(4), 1797–1820 (2018). https://doi.org/10.1007/s10586-018-2811-x
2. Ben Alla, H., Ben Alla, S., Ezzati, A., Mouhsen, A.: A novel architecture with dynamic queues based on fuzzy logic and particle swarm optimization algorithm for task scheduling in cloud computing. In: El-Azouzi, R., Menasché, D.S., Sabir, E., Pellegrini, F.D., Benjillali, M. (eds.) Advances in Ubiquitous Networking 2. LNEE, vol. 397, pp. 205–217. Springer, Singapore (2017). https://doi.org/10.1007/978-981-10-1627-1_16
3. Buyya, R., Yeo, C.S., Venugopal, S., Broberg, J., Brandic, I.: Cloud computing and emerging it platforms: vision, hype, and reality for delivering computing as the 5th utility. Futur. Gener. Comput. Syst. **25**(6), 599–616 (2009)
4. Calheiros, R.N., Ranjan, R., Beloglazov, A., De Rose, C.A., Buyya, R.: CloudSim: a toolkit for modeling and simulation of cloud computing environments and evaluation of resource provisioning algorithms. Softw.: Pract. Exp. **41**(1), 23–50 (2011)
5. Cheng, M., Li, J., Nazarian, S.: DRL-cloud: deep reinforcement learning-based resource provisioning and task scheduling for cloud service providers. In: Proceedings of the 2018 23rd Asia and South Pacific Design Automation Conference (ASP-DAC), pp. 129–134. IEEE (2018)
6. Fan, J.: Local linear regression smoothers and their minimax efficiencies. Ann. Stat. **21**(1), 196–216 (1993)
7. Funika, W., Koperek, P.: Evaluating the use of policy gradient optimization approach for automatic cloud resource provisioning. In: Wyrzykowski, R., Deelman, E., Dongarra, J., Karczewski, K. (eds.) PPAM 2019. LNCS, vol. 12043, pp. 467–478. Springer, Cham (2020). https://doi.org/10.1007/978-3-030-43229-4_40
8. Gao, H., Huang, W., Zou, Q., Yang, X.: A dynamic planning framework for QoS-based mobile service composition under cloud-edge hybrid environments. In: Wang, X., Gao, H., Iqbal, M., Min, G. (eds.) CollaborateCom 2019. LNICST, vol. 292, pp. 58–70. Springer, Cham (2019). https://doi.org/10.1007/978-3-030-30146-0_5
9. Jung, J., Kim, H.: MR-CloudSim: designing and implementing MapReduce computing model on CloudSim. In: Proceedings of the 2012 International Conference on ICT Convergence (ICTC), pp. 504–509. IEEE (2012)

10. Abdulhamid, S.M., Abd Latiff, M.S., Madni, S.H.H., Abdullahi, M.: Fault toler-ance aware scheduling technique for cloud computing environment using dynamic clustering algorithm. Neural Comput. Appl. **29**(1), 279–293 (2016). https://doi.org/10.1007/s00521-016-2448-8

11. Mousavi, S.S., Schukat, M., Howley, E.: Traffic light control using deep policy-gradient and value-function-based reinforcement learning. IET Intell. Transp. Syst. **11**(7), 417–423 (2017)

12. Nayak, S.C., Tripathy, C.: Deadline sensitive lease scheduling in cloud computing environment using AHP. J. King Saud Univ.-Comput. Inf. Sci. **30**(2), 152–163 (2018)

13. Petrik, D., Herzwurm, G.: iIoT ecosystem development through boundary resources: a Siemens MindSphere case study. In: Proceedings of the 2nd ACM SIG-SOFT International Workshop on Software-Intensive Business: Start-Ups, Plat-forms, and Ecosystems, pp. 1–6 (2019)

14. Priya, V., Kumar, C.S., Kannan, R.: Resource scheduling algorithm with load balancing for cloud service provisioning. Appl. Soft Comput. **76**, 416–424 (2019)

15. Sekaran, K., Khan, M.S., Patan, R., Gandomi, A.H., Krishna, P.V., Kallam, S.: Improving the response time of m-learning and cloud computing environments using a dominant firefly approach. IEEE Access **7**, 30203–30212 (2019)

16. Wang, Z., Wen, Y., Zhang, Y., Chen, J., Cao, B.: A resource usage prediction-based energy-aware scheduling algorithm for instance-intensive cloud workflows. In: Gao, H., Wang, X., Yin, Y., Iqbal, M. (eds.) CollaborateCom 2018. LNICST, vol. 268, pp. 626–642. Springer, Cham (2019). https://doi.org/10.1007/978-3-030-12981-1_44

17. Wickremasinghe, B., Calheiros, R.N., Buyya, R.: CloudAnalyst: a CloudSim-based visual modeller for analysing cloud computing environments and applications. In: Proceedings of the 2010 24th IEEE International Conference on Advanced Infor-mation Networking and Applications, pp. 446–452. IEEE (2010)

18. Wu, D., Jiang, N., Du, W., Tang, K., Cao, X.: Particle swarm optimization with moving particles on scale-free networks. IEEE Trans. Netw. Sci. Eng. **7**(1), 497–506 (2020)

19. Xu, M., Buyya, R.: Brownout approach for adaptive management of resources and applications in cloud computing systems: a taxonomy and future directions. ACM Comput. Surv. (CSUR) **52**(1), 1–27 (2019)

20. Xu, M., Tian, W., Buyya, R.: A survey on load balancing algorithms for virtual machines placement in cloud computing. Concurr. Comput.: Pract. Exp. **29**(12), e4123 (2017)

21. Xu, M., Toosi, A.N., Bahrani, B., Razzaghi, R., Singh, M.: Optimized renewable energy use in green cloud data centers. In: Yangui, S., Bouassida Rodriguez, I., Drira, K., Tari, Z. (eds.) ICSOC 2019. LNCS, vol. 11895, pp. 314–330. Springer, Cham (2019). https://doi.org/10.1007/978-3-030-33702-5_24

Differentially Private Location Preservation with Staircase Mechanism Under Temporal Correlations

Rong Fang, Jianmin Han[✉], Juan Yu, Xin Yao, Hao Peng, and Jianfeng Lu

College of Mathematics and Computer Science, Zhejiang Normal University,
Jinhua 321004, China
`hanjm@zjnu.cn`

Abstract. Location-Based Service (LBS) is one of basic services in collaborative applications. However, LBS applications may disclose user's location privacy, which receives considerable concerns. Many methods have been proposed to protect privacy in LBS. Planar Isotropic Mechanism (PIM) is a typical location privacy preservation method in the scenario of continuous location data release. However, the method is complicated, since it requires two convex hull transformations and one isotropic position transform. To solve the problem, we propose a Staircase Mechanism (SM) based location privacy preservation method for the scenario of continuous location data release. The proposed method replaces PIM with SM, whose implementation is simple and efficient. Furthermore, SM can achieve the same privacy budget with less noise addition, so it can maintain higher quality of services in LBS. Comprehensive experiments conducted on real location data demonstrate that the proposed method is efficient and can maintain high data utility compared with the method based on PIM.

Keywords: Differential privacy · Location privacy · Temporal correlation · Staircase mechanism

1 Introduction

LBS is a kind of basic services in collaborative applications. With the development of positioning technologies and popularity of mobile Internet and smartphones, location-based applications have permeated into our daily life, such as location-based points of interest searching, location-based games, location-based commerce and location-based social networks. To enable the location-based applications, users have to share their locations to service providers. However, the disclosure of users' locations could raise serious privacy concerns. Because locations could reflect users' religion and health conditions, and further expose them to attacks, e.g., unwanted location-based spams, even physical danger [1] etc.

In order to protect the location privacy of users in LBS, various location privacy preservation technologies have been proposed. These technologies can be classified into

© ICST Institute for Computer Sciences, Social Informatics and Telecommunications Engineering 2021
Published by Springer Nature Switzerland AG 2021. All Rights Reserved
H. Gao et al. (Eds.): CollaborateCom 2020, LNICST 350, pp. 75–92, 2021.
https://doi.org/10.1007/978-3-030-67540-0_5

three categories, i.e., Private Information Retrieval (PIR) [2], location generalization [3], and location perturbation [4]. PIR is based on cryptography and can provide provable privacy preservation. However, it tends to be computationally expensive and not practical because that PIR need design different query for different query types. Location generalization hides a user's exact location in an area so that attackers cannot infer the exact location of the user with high probability. However, location generalization technologies rely on syntactic privacy models such as k-anonymity, or ad-hoc uncertainty models, and could not provide rigorous privacy. Location perturbation disturbs the sensitive location by adding random noises, so as to protect the sensitive location privacy. Differential privacy based on location privacy preservation are realized via location perturbation. Differential privacy [5] has been widely recognized as a leading privacy preservation method in both industrial and academic community, as it provides a formal and provable privacy guarantee and it can preserve privacy against attackers with arbitrary background knowledge. The idea is that the presence or absence of one single individual in a database shall not change significantly the probability of any outcome of an aggregate function.

Traditional location preservation methods only consider static scenarios or perturb the location at single timestamps without considering the temporal correlations of a moving user, and hence are vulnerable to various inference attacks. Therefore, Xiao et al. [6] considered the privacy leakage issue caused by the temporal correlation between locations, and proposed a systematic framework to protect location privacy for continuous location sharing scenarios. They first modeled the temporal correlation between locations with the Markov chain, and then introduced the concept of δ-location set to hide the real location, finally they proposed a Planar Isotropic Mechanism (PIM) to achieve δ-location set based differential privacy. However, the PIM is computational expensive as it requires two convex hull transformations and one isotropic position transform. Furthermore, the isotropic position transformation relies on an invertible matrix T which is too strict to be obtained.

To solve the problems of PIM, we propose a Staircase Mechanism (SM) based location perturbation method for location privacy preservation in continuous location sharing scenario. Main contributions are summarized as follows:

First, we propose a new perturbation mechanism, i.e., SM. Compared with the PIM, it could achieve the same privacy budget with less noise addition, and thereby maintaining higher quality of location-based services. Second, we verify theoretically and experimentally that SM has lower time complexity and better data utility compared with PIM.

The rest of the paper is organized as follows. Section 2 discusses related work. Section 3 introduces the basics of differential privacy and the relevant definitions. Section 4 proposes differentially private location preservation method with staircase mechanism under temporal correlations. Section 5 evaluates the performance of the proposed method with extensive experiments. Section 6 concludes the paper.

2   Related Work

Differential privacy is an effective privacy preservation method on location or trajectory data. Machanavajjhala et al. [7] first introduced differential privacy into location privacy preservation. They proposed differential privacy mechanism to securely release commuting patterns of users without compromising individuals' privacy. Andres et al. [8] proposed a geo-indistinguishability method based on differential privacy to make sure that attackers can hardly identify the difference between the exact location and the approximate locations within a circular region of radius r. Bordenabe et al. [9] devised an optimal geo-indistinguishable mechanism to minimize the LBS service quality loss. Niu et al. [10] investigated the long-term observation attacks on geo-indistinguishable mechanisms and proposed a three-phase differential location privacy framework, i.e., Eclipse, which combines geo-indistinguishability and k-anonymity. Gursoy et al. [11] proposed DP-Star, a framework, a methodical framework for publishing trajectory data with differential privacy guarantee as well as high utility preservation.

The methods mentioned above do not consider data correlations, which will lead in new privacy problems. Different types of correlations should adopt different privacy preservation method. To user-to-user correlation, Zhu et al. [12] defined the correlated differential privacy and the correlated sensitivity, and proposed a correlated data release mechanism to preserve privacy. Liu et al. [13] demonstrated the vulnerability of traditional differential privacy mechanisms under data dependence, and proposed a generalized dependent differential privacy framework, which introduced dependence coefficients to measure the sensitivity of different queries under probabilistic dependence between tuples. Yang et al. [14] concentrated on the privacy leakage caused by probabilistic correlations between tuples, and modeled the correlations by a Gaussian Markov Random Field and proposed Bayesian differential privacy (BDP). The other type is the temporal correlations among single user's data at different timestamp. Xiao et al. [6] investigated adversaries with knowledge of temporal correlations of single user, and proposed PIM for continuous location sharing scenarios. Cao et al. [15, 16] quantified the privacy loss of differential privacy mechanisms caused by temporal correlations in the context of continuous aggregate release.

3 Preliminaries

In this section, we introduce some preliminary definitions. The associated symbols are shown in Table 1. We use bold lowercase letters for vectors, such as $\mathbf{a}$, and bold capital letters for matrices, such as $\mathbf{A}$.

Definition 1 (δ-location set) [6]. δ-location set is a set containing minimum number of locations that have prior probability sum not less than $1 - \delta$,

$$\Delta X_t = min\left\{ s_i \mid \sum_{s_i} p_t^-[i] \geq 1 - \delta \right\} \tag{1}$$

where p_t^- is the prior probability vector of a user's location at timestamp t, the size of ΔX_t is related to the size of δ.

For example, S denote the domain of space. If we divide the space S into $\{s_1, s_2, s_3, s_4, s_5, s_6\}$, then $p_t^- = [0.1, 0.5, 0.05, 0.3, 0.03, 0.02]$ corresponds to $[s_1, s_2, s_3, s_4, s_5, s_6]$, where p_t^- represents the probability that a user appears in each

Table 1. Summary of notations.

Symbols	Symbolic meaning
s_i	The grid number of the user's real location after area grid
$u^* = (x^*, y^*)$	User's real location coordinates
$u_t^* = (x_t^*, y_t^*)$	User's real location coordinates at timestamp t
$z_t = (x_t, y_t)$	User's location after the disturbance at timestamp t
$\Pr(z_t \vert u_t^* = s_i)$	The probability of releasing z_t given $u_t^* = s_i$
p_t^-	The prior probability vector at timestamp t
$p_t^-[i]$	The prior probability of the i th location of p_t^-
p_t^+	The posterior probability vector at timestamp t
$p_t^+[i]$	The posterior probability of the i th location of p_t^+
ΔX	δ-location set
ΔX^1	The set of horizontal coordinates of the points in the δ-location set
ΔX^2	The set of vertical coordinates of points in the δ-location set
$\Vert \cdot \Vert_p$	L_p norm
M	Transfer matrix
m_{ij}	The probability of a user moving from grid i to grid j

cell. When $\delta = 0.1$, $\Delta X_t = \{s_2, s_4, s_1\}$; $\delta = 0.05$, $\Delta X_t = \{s_2, s_4, s_1, s_3\}$. In special case, when $\delta = 0$, δ-location set contains all the possible locations, where s_i represents the i th cell of the grid partition.

In practice, there is a small probability that the real location is not in δ-location set. We denote this phenomenon as drift. If it happens, we handle it with a surrogate approach. The two concepts are defined as follows.

Definition 2 (Drift) [6]. A real location which is not in δ-location set is defined Drift.

Definition 3 (Surrogate) [6]. Surrogate $\overline{u} = (\overline{x}, \overline{y})$ is the closest location in ΔX to the real location $u^* = (x_t^*, y_t^*)$.

$$\overline{u} = (\overline{x}, \overline{y}) = \operatorname*{argmin}_{(x_s, y_s) \in \Delta X} dist\big((x_s, y_s), (x_t^*, y_t^*)\big) \tag{2}$$

where $dist(\cdot)$ is the Euclidean distance.

Definition 4 (Differential privacy on δ-location set) [6]. Given a random mechanism $f : x \to z$. At timestamp t, f satisfies ε-differential privacy on δ-location set ΔX_t, if for any output z_t and any two locations x_1 and x_2, the following holds.

$$\frac{\Pr(f(x_1) = z_t)}{\Pr(f(x_2) = z_t)} \leq e^{\varepsilon} \tag{3}$$

where ε is privacy budget.

Differential privacy is achieved by adding random noise. The magnitude of the noise is influenced by the sensitivity of a query function. As we focus on protecting the user's real two-dimensional location coordinates, the sensitivity on two-dimensional is defined as follows.

Definition 5 (Sensitivity on δ-location set). Given a query function $f : \Delta X \rightarrow R$, where ΔX is δ-location set and R is the return result of the query function. The sensitivity of any two adjacent locations in ΔX is as follows.

$$\Delta f = \max_{u_1^*, u_2^* \in \Delta X} \left\| u_1^* - u_2^* \right\|_1 = \max(\Delta_1, \Delta_2) \tag{4}$$

where $\Delta X^1 = \{x^* | u^* = (x^*, y^*) \in \Delta X\}$ represents the set of x coordinates (longitudes), $\Delta X^2 = \{y^* | u^* = (x^*, y^*) \in \Delta X\}$ is the set of y coordinates (latitudes), u^* is the real location, $\Delta_1 = \max_{a,b \in \Delta X^1} |a - b|$, and $\Delta_2 = \max_{a,b \in \Delta X^2} |a - b|$.

Definition 6 (Distance). The distance between the real location $u^* = (x^*, y^*)$ and the perturbed location $z = (x, y)$ is defined as.

$$dis(u^*, z) = \left\| u^* - z \right\|_2. \tag{5}$$

$dis(u^*, z)$ can be used as an error measurement between the real location and the perturbed location.

Definition 7 (Staircase mechanism) [18]. Given a multidimensional query function $f : D \rightarrow R^d$, the Staircase mechanism is defined as:

$$M(D) = f(D) + Staircase(\Delta, \varepsilon, \gamma)^d \tag{6}$$

where $Staircase(\Delta, \varepsilon, \gamma)^d$ is taken to be random variable subjecting to Staircase distribution. Δ is sensitivity, ε is privacy budget and $\gamma \in [0, 1]$.

4   Location Release Model

4.1   Framework

We focus on the scenario of continuous location data release, in which users need to send their real-time locations to servers frequently to obtain the corresponding services during their moving. We assume that LBS providers are untrusted, and users' locations should not be directly released to the providers to protect their location privacy. The user's real location under each timestamp is treated as a sensitive location, so the real location is only visible to the user himself.

In order to protect users' location privacy under continuous timestamps, we propose a location preservation framework. If we want to release a location of a timestamp, we should construct a δ-location set for it. And then, we adopt a differential privacy mechanism to release a perturbed location.

Specifically, the framework steps are as follows.

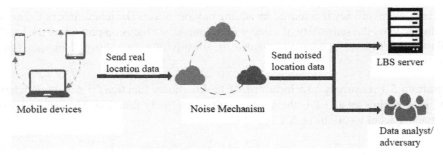

Fig. 1. Location release framework.

Step 1. Grid partition: partition the map area with grid where the data set is located.
Step 2. Calculate the transition matrix M between locations. Each $m_{ij} \in M$ represents the probability of the user moving from grid i to grid j.
Step 3. Calculate the prior probability vector p_t^- under each timestamp, where the prior probability refers to the probability before releasing the perturbed location. $\mathcal{G}(\mathcal{G} \in \{1, 2, \ldots, i, \ldots\}$ is the number of grid cells. T_t is the total number of locations at timestamp t. ω_t^i is the number of locations in grid cell i. Then the prior probability vector p_t^- at timestamp t is calculated as follows:

$$\begin{cases} p_1^- = \left[\frac{\omega_1^1}{T_1}, \frac{\omega_1^2}{T_1}, \ldots, \frac{\omega_1^i}{T_1}, \ldots \right] & t = 1 \\ p_t^- = p_{t-1}^+ M & t \geq 2 \end{cases} \tag{7}$$

where p_{t-1}^+ represents the posterior probability of the timestamp $t - 1$.
Step 4. Calculate the posterior probability vector p_{t-1}^+ under each timestamp,

$$p_{t-1}^+[i] = \Pr\left(u_{t-1}^* = s_i | z_{t-1}\right) = \frac{\Pr\left(z_{t-1} | u_{t-1}^* = s_i\right) p_{t-1}^-[i]}{\sum_j \Pr\left(z_{t-1} | u_{t-1}^* = s_j\right) p_{t-1}^-[j]} \tag{8}$$

where $p_{t-1}^+[i]$ is the ith element in p_{t-1}^+. $\Pr\left(z_{t-1} | u_{t-1}^* = s_i\right)$ represents emission probability and z_{t-1} represents the perturbed location of $t - 1$ timestamp.

To make it easier to understand, Table 2 shows the flow of alternate calculation of prior probability vector and posterior probability vector. The posterior probability vector p_t^+ can be obtained according to the emission probability and the prior probability at timestamp t.

Table 2. Prior probability and posterior probability under each timestamp.

timestamp 1	timestamp 2	timestamp 3	...	timestamp $t-1$	timestamp t
p_1^-	p_1^+	p_2^+	...	p_{t-2}^+	p_{t-1}^+
↓(B)	↓(M)	↓(M)	...	↓(M)	↓(M)
p_1^+	p_2^-	p_3^-	...	p_{t-1}^-	p_t^-
	↓(B)	↓(B)	...	↓(B)	↓(B)
	p_2^+	p_3^+	...	p_{t-1}^+	p_t^+

4.2 Protecting Location with Staircase Mechanism

Algorithm 1 is the implementation process of the proposed framework. Step 1–3 calculates the prior probability vector at timestamp t. Steps 4–10 is the location releasing process. Specifically, when the user's location needs to be released at timestamp t, δ-location set ΔX_t is constructed and the real location is hidden in the δ-location set. If the real location $u_t^* = (x_t^*, y_t^*)$ is not in ΔX_t, we choose a surrogate location $\bar{u} = (\bar{x}, \bar{y})$ in ΔX_t. Then, we generate the perturbed location z_t through adding noise to the real location or the surrogate location. Finally, we release the z_t. Steps 11 calculates the posterior probability vector p_t^+, which is subsequently used to compute the prior probability for the next timestamp $t+1$. When the location at timestamp $t+1$ needs to be released, we repeat the above process. The process of generating δ-location set and the process of noise generation are shown in details in Algorithm 2 and Algorithm 3, separately.

Algorithm 1. Location_perturbation

Input: $\varepsilon, \delta, M, p_1^-, u_t^* = (x_t^*, y_t^*), \Delta_1, \Delta_2, \gamma \in [0,1]$

Output: $z_t = (x_t, y_t)(t = 1, \ldots, T)$

1. $p_1^- = \left[\frac{\omega_1^1}{\mathcal{T}_1}, \frac{\omega_1^2}{\mathcal{T}_1}, \ldots, \frac{\omega_1^i}{\mathcal{T}_1}, \ldots\right]$

2. for t in $(2, \ldots, T)$:

3. $p_t^- = p_{t-1}^+ M$ //Markov transition

4.   Construct ΔX_t (Algorithm 2);

5.   if $(x_t^*, y_t^*) \notin \Delta X_t$ then

6.   Compute the surrogate $(\bar{x}, \bar{y})$ according to equation (2)

7. $(x_t^*, y_t^*) = (\bar{x}, \bar{y})$

8.   end if

9. $(\alpha_t, \beta_t) \leftarrow Staircase\_noise\ (\varepsilon, \Delta_1, \Delta_2, \gamma \in [0,1]))$(Algorithm 3)

10. $z_t = (x_t^* + \alpha_t, y_t^* + \beta_t)$

11.   Compute the posterior probability vector p_t^+ according to equation (8)

12. end for

13. return $z_t = (x_t, y_t)(t = 1, \ldots, T)$

Algorithm 2 is the process of generating δ-location set, where δ-location set is a set containing minimum number of locations that have prior probability sum no less than $1 - \delta$. Algorithm 2 input the prior probability vector $\boldsymbol{p}_t^-$ for timestamp t. The prior probability vector $\boldsymbol{p}_t^-$ is obtained by the posterior probability vector $\boldsymbol{p}_{t-1}^+$ of the timestamp $t - 1$ and the Markov transition matrix $\boldsymbol{M}$, i.e., $\boldsymbol{p}_t^- = \boldsymbol{p}_{t-1}^+ \boldsymbol{M}$, where $\boldsymbol{p}_{t-1}^+$ is calculated by Eq. (8).

Algorithm 2. Generate_δ-location set

Input : D: location set, $p_t^-[i]$: the probability of each location, δ

Output : δ-location set ΔX_t

1.$sort(D, p_t^-[i])$ //Sort by probability in descending order

2.$\Delta X_t = \emptyset$

3.$p = 0$

4.$for\ x\ in\ D$:

5. $p += p_t^-[i]$

6. $\Delta X_t = \Delta X_t + \{x\}$ // Add the ith location to location set D.

7.   if $p \geq \delta$:

8.   *break*

9.   end if

10. end for

11.*return* ΔX_t

Algorithm 3 is the process of generating random noise following the Staircase distribution for real locations. In Algorithm 3 S determines the noise symbol, G determines the interval in which the noise is located $[G\Delta, (G+1)\Delta)$, B determines the interval in which the noise is located $[G\Delta, (G+\gamma)\Delta)$ and $[(G+\gamma)\Delta, (G+1)\Delta)$ (Δ is sensitivity), and U contributes to the uniform appearance interval.

Algorithm 3. Staircase_noise

Input: $\varepsilon, \Delta_1, \Delta_2, \gamma \in [0,1]$

Output: (α_t, β_t)

1. Generate a $r.v.$ S with $\Pr[S = 1] = \Pr[S = -1] = 1/2$.

2. Generate a geometric $r.v.$ G with $\Pr[G = i] = (1 - b)b^i$ for integer $i \geq 0$, where $b = e^{-\varepsilon}$.

3. Generate a $r.v.$ U uniformly distributed in $[0,1]$.

4. Generate a binary $r.v.$ B with $\Pr[B = 0] = \gamma/(\gamma + (1 - \gamma)b)$ and $\Pr[B = 1] = (1 - \gamma)b/(\gamma + (1 - \gamma)b)$

5. $\alpha_t \leftarrow S((1 - B)((G + \gamma U)\Delta_1) + B((G + \gamma + (1 - \gamma)U)\Delta_1)$

6. $\beta_t \leftarrow S((1 - B)((G + \gamma U)\Delta_2) + B((G + \gamma + (1 - \gamma)U)\Delta_2)$

7. *return* (α_t, β_t)

The time complexity of Algorithm 1 is $O(Tn^2)$, where n is the number of the grids, T is the number of timestamps. Step 1 is original prior probability. Step 2–2 is T-cycle about timestamps. Where step 3 calculates the prior probability at each timestamp, it takes $O(n)$. Step 4 generates δ-location set ΔX by Algorithm 2, it takes $O(nlogn)$. Steps 9–10 add random noise by Algorithm 3, it takes $O(1)$. Step 11 computes the posterior probability vector p_t^+ by Bayesian inference, it takes $O(n^2)$. So, the time complexity is $T(O(n) + O(nlogn) + O(1) + O(n^2)) = O(Tn^2)$.

4.3 Privacy Analysis

In this subsection, we focus on analyzing the privacy level of the proposed methods, and prove that it satisfies differential privacy. As the proof depends on the definition of Adversarial Privacy, we firstly introduce the definition, then formalize a theorem and prove it.

Definition 8 (Adversarial Privacy) [20]. A mechanism satisfies the ε-adversarial privacy for any location $s_i \in S$, if and only if any output z and any adversaries knowing the real location in ΔX, the following holds:

$$\frac{\Pr(u_t^* = s_i | z_t)}{\Pr(u_t^* = s_i)} \leq e^{\varepsilon} \tag{8}$$

$\Pr(u_t^* = s_i)$ is the prior probability, $\Pr(u_t^* = s_i | z_t)$ is the posterior probability for adversaries, z_t is the location of the release, $u_t^* = s_i$ is the user's real location.

Theorem 1. At any timestamp t, Algorithm 1 satisfies ε- differential privacy on δ-location set.

Proof. Algorithm 1 include construct ΔX_t and add noise on releasing locations corresponding to Algorithm 2 and Algorithm 3 respectively. Where Algorithm 2 include Algorithm 3, because Algorithm 2 need to add noise during construct ΔX_t. Therefore, prove Algorithm 1 satisfies ε- differential privacy, which is equivalent to prove Algorithm 2 satisfies ε- differential privacy.

If Algorithm 2 satisfies ε -differential privacy on 0-location set, then it also satisfies ε-differential privacy on δ-location set, since 0-location set contains all possible locations in δ-location set. Therefore, we only need to prove Algorithm 2 satisfies ε-differential privacy on 0-location set.

If $u_t^* = (x_t^*, y_t^*) \in \Delta X_t$, then the location z_t of perturbed by Algorithm 3 is released, where $z_t = (x_t^* + \alpha_t, y_t^* + \beta_t)$, (α_t, β_t) is a noise following Staircase distribution. Staircase distribution satisfies the differential privacy has been proved in [18], then

$$\frac{\Pr(u_t^* = s_i | z_t)}{\Pr(u_t^* = s_i)} \leq e^{\varepsilon_t}. \tag{10}$$

If $u_t^* = (x_t^*, y_t^*) \notin \Delta X_t$, releases $\widetilde{v}_t = (\widetilde{x}_t, \widetilde{y}_t)$ for u_t^* in ΔX_t. Then

$$\frac{\Pr(u_t^* = s_i | z_t)}{\Pr(u_t^* = s_i)} = \frac{\sum_k \Pr(u_t^* = s_i | \widetilde{x}_t = s_k) \Pr(\widetilde{x}_t = s_k | z_t)}{\sum_k \Pr(u_t^* = s_i | \widetilde{x}_t = s_k) \Pr(\widetilde{x}_t = s_k)} \leq e^{\varepsilon_t} \tag{11}$$

It has been proved in [6] that differential privacy for continuous location data release is equivalent to adversarial privacy. Therefore, Algorithm 1 satisfies the ε- differential privacy on 0-location set. That is, theorem 1 is proved.

5   Experimental Evaluation

5.1   Experimental Settings

In this section, we evaluate the performance of the proposed SM based location perturbation algorithm on Geolife data. We compare it with the LM and PIM in terms of the
size of ΔX, drift ratio and distance, the precision and recall on k NN query. All of the
algorithms are implemented in Python on a machine with AMD Ryzen 5 2500U with
Radeon Vega Mobile Gfx 8 CPUs 2.0 GHz and 8192 MB RAM, running Windows 10.

Geolife Data [21] . Geolife data was collected from 182 users over a period of more than
five years (from April 2007 to August 2012). The data has a series of tuples containing
latitude, longitude, and timestamps. We select the trajectories within the third ring of
Beijing to calculate the Markov transition matrix with the map is divided into 0.34 ×
0.34 km^2 cells.

Evaluation Metrics. We use size of ΔX, drift ratio and distance as experimental
metrics.

Size of ΔX [6]. Because our privacy definition is based on ΔX, we evaluated the size
of ΔX to understand how ΔX grows or changes. The smaller size of ΔX means the
better experimental results, when ε is the same.

Drift Ratio [19] . If δ is not properly valued, the real location may be filtered out with
a small probability. We denote it as drift. $Driftratio = \frac{l_t}{S_t}$ is drift ratio at timestamp t,
where l_t is the number of locations that drift ratio at timestamp t, S_t is the total number
of locations in ΔX at timestamp t. The smaller drift ratio means the better experimental
results, when ε is the same.

Distance. The distance means Euclidean distance between the real location and the
released location, which is used to measure data utility. The smaller distance means the
better experimental results, when ε is the same.

5.2   Performance Over Time

To evaluate the performance of the release mechanism when a user moves under a
continuous timestamp. We randomly selected a test track containing 500 timestamps
from the Geolife data. We tested LM, PIM and SM at each timestamp with $\varepsilon = 1$ and
$\delta = 0.01$. Each method runs 20 times and releases the average.

(1)   Track Release
     Figure 2(a) shows the original trajectory. Figure 2(b), 2(c), and 2(d) show the release
     locations under each timestamp. We can see that the released locations of SM are
     closer to real locations, compare with LM and PIM. It means that SM has less
     perturbation at each timestamp with $\varepsilon = 1$ and $\delta = 0.01$, compare with LM and
     PIM.

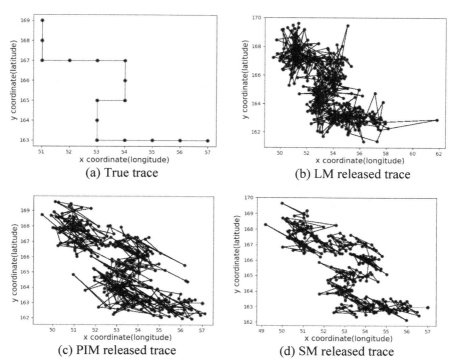

Fig. 2. Track Release: (a) True trace; (b) LM released trace; (c) PIM released trace; (d) SM released trace. ($\varepsilon = 1$ and $\delta = 0.01$)

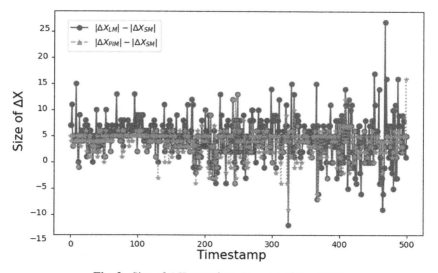

Fig. 3. Size of ΔX over time. ($\varepsilon = 1$ and $\delta = 0.01$)

(2) Size of ΔX

Figure 3 shows the change of size of ΔX over time. $|\Delta X_{LM}| - |\Delta X_{SM}|$ represents the difference between size of ΔX on the LM and size of ΔX on the SM. $|\Delta X_{PIM}| - |\Delta X_{SM}|$ shows the difference between Size of ΔX on PIM and size of ΔX on the SM. Figure 3 as a whole, the difference between size of ΔX is greater than 0. The size of ΔX generated by SM is less than that generated by LM and PIM. It shows that SM achieve the same privacy budget with less noise addition, and we can get high data utility in most case.

(3) Drift ratio

Figure 4 shows the change of drift ratio over time. $|Driftratio_{LM}| - |Driftratio_{SM}|$ represents the difference between drift ratio on the LM and drift ratio on the SM. $|Driftratio_{PIM}| - |Driftratio_{SM}|$ shows the difference between drift ratio on PIM and drift ratio on the SM. The orange line is closer to 0. It means that SM has a smaller drift ratio than LM and PIM in most case.

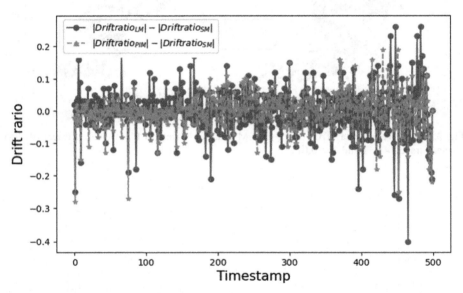

Fig. 4. Drift ratio over time. ($\varepsilon = 1$ and $\delta = 0.01$)

(4) Distance

Figure 5 shows the change in distance over time. It shows that SM's distance is less than LM's and PIM's distance. SM releases locations more accurately under each timestamp. It has the small noise, so its distance is closer to 0. SM has better data utility compared with LM and PIM under $\varepsilon = 1$ and $\delta = 0.01$. The emission probability and prior probability are more accurate at timestamp t, we can calculate the posterior probability by Eq. (8) at timestamp t. Therefore, in most case, the posterior probability distribution obtained by SM is more accurate than that obtained by LM and SM.

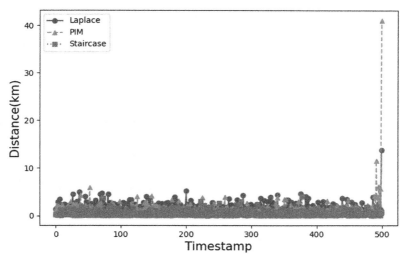

Fig. 5. Distance over time. ($\varepsilon = 1$ and $\delta = 0.01$)

5.3 Impact of Parameters

Different trajectories may have some influence on the accuracy of experiment results. We select 100 trajectories from 100 users, each containing 500 timestamps, to evaluate the overall performance of method and the impact of different parameters. In this section, Size of ΔX, Drift ratio and Distance represent the average of 100 trajectories from 100 users under 500 timestamps.

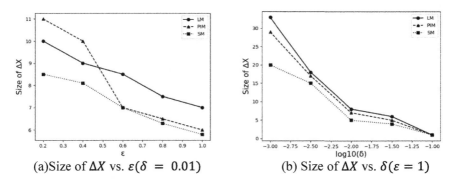

(a)Size of ΔX vs. $\varepsilon(\delta = 0.01)$ (b) Size of ΔX vs. $\delta(\varepsilon = 1)$

Fig. 6. Impact of ε and δ on size of ΔX.

Size of ΔX vs. ε.

Figure 6(a) demonstrates the variation of size of ΔX with the change of ε while fixing δ. It shows that when $\delta = 0.01$, the sizes of ΔX generated by the three mechanisms all decrease with the increasing of ε. Because a larger ε implies lower privacy, and a large

ε requires a small ΔX for hiding the real location. In addition, Fig. 6(a) also shows that the size of ΔX generated by SM is the smallest for each ε. That is to say that SM has the lowest data perturbation and the best data utility compared with LM and PIM under the same privacy budget ε.

Size of ΔX vs. δ.

Figure 6(b) demonstrates the variation of size of ΔX with the change of δ while fixing ε. It shows that when $\varepsilon = 1$, the sizes of ΔX generated by the three mechanisms all decrease with the increasing of δ. With δ increasing, the size of ΔX become smaller according to definition 1. In practice, δ can't be too large because ΔX should contain at least one location (real location). Therefore, we set $\delta = 0.01$ as the default value. In addition, Fig. 6(b) also shows that the size of ΔX generated by SM is smaller than LM and PIM under the same δ. It shows that SM provide higher data utility than LM and PIM in most case.

Drift Ratio vs. ε.

Figure 7(a) demonstrates the variation of drift ratio with the change of ε while fixing δ. It shows that when $\delta = 0.01$, with ε increasing, the drift ratio becomes smaller, because a large ε implies less location perturbation. In addition, Fig. 7(a) also shows that drift ratio of SM is smaller than LM and PIM under the same ε. It shows that SM provides higher data utility than LM and PIM in most case.

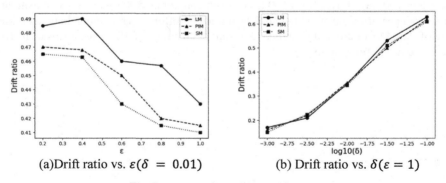

(a)Drift ratio vs. $\varepsilon(\delta = 0.01)$ (b) Drift ratio vs. $\delta(\varepsilon = 1)$

Fig. 7. Impact of ε and δ on drift ratio.

Drift Ratio vs. δ.

Figure 7(b) demonstrates the variation of drift ratio with the change of δ while fixing ε. It shows that when $\varepsilon = 1$, the drift ratio increases with δ increasing, because with δ increasing, the size of ΔX decreases. In addition, Fig. 7(b) also shows that drift ratio of SM is smaller than LM and PIM under the same ε. It shows that SM provides higher data utility than LM and PIM in most case.

Distance vs. ε.

Figure 8(a) demonstrates the variation of distance with the change of ε while fixing δ.

It shows that when $\delta = 0.01$, the distance decreases with ε increasing, because with ε increasing, the size of ΔX is smaller. At this time, the location in ΔX is closer, so the distance is smaller. In addition, Fig. 8(a) shows that the distance of SM is smaller than LM and PIM under the same ε. It shows that SM provide higher data utility than LM and PIM in most case.

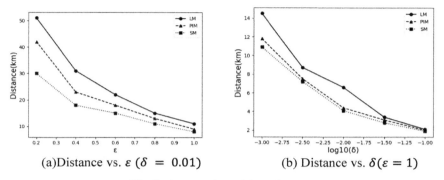

(a)Distance vs. ε ($\delta = 0.01$)   (b) Distance vs. $\delta(\varepsilon = 1)$

Fig. 8. Impact of ε and δ on distance.

Distance vs. δ.
Figure 8(b) demonstrates the variation of distance with the change of δ while fixing ε. It shows that when $\varepsilon = 1$, the distance decreases with δ increasing. Because with δ increasing, the size of ΔX is smaller. The rest of the location in ΔX is closer. Both the real location and the release location are in ΔX, so the distance is smaller. In addition, SM has a smaller distance and better data utility than LM and PIM under same δ. It shows that SM provide higher data utility than LM and PIM in most case.

5.4   Utility for Location Based Queries

In order to verify the utility of released locations, we used k NN query on 100 trajectories which has 500 timestamps to obtain the query precision and recall under each timestamp. We use k NN on the original trajectories and k'NN on the release trajectories, when $\varepsilon = 1$, $\delta = 0.01$.

Figure 9(a) shows that when $k = k'$(Precision = Recall), the F1 increases with $k(k')$ increasing. Because the scope of the query will be expanded, the neighboring locations will be queried over a larger area. In addition, SM has better F1 than LM and PIM. It shows that SM provide more accurate query results in most case.

Figure 9(b) shows that when $k = 5$, SM's F1 is bigger than LM and PIM. It shows that SM is more accurate than LM and PIM.

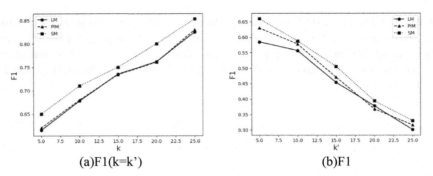

(a)F1(k=k')   (b)F1

Fig. 9. k NN results: (a) F1 under $k = k'$; (b) F1 vs k'.

5.5 Running Time

Figure 10 shows the running time of LM, PIM, and SM with continuous timestamps. It shows that when $\delta = 0.01$ and $\varepsilon = 1$, the running time for three mechanisms all increase with the large timestamp. Because if we need to calculate the prior probability and the posterior probability timestamp t, we need to calculate the prior probability and the posterior probability of the preceding timestamp $t - 1$. Compared with the LM and PIM, SM has a shorter running time.

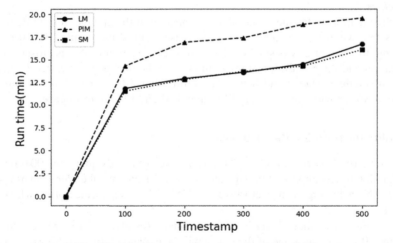

Fig. 10. Run time over time. ($\varepsilon = 1$ and $\delta = 0.01$)

6 Conclusion

A differential privacy location preservation method based on Staircase mechanism is proposed to solve the problem of location privacy preservation under temporal correlations. Considering the temporal correlation between locations, this method protects the privacy of the user's real location in each timestamp, making it difficult for the adversary to infer the user's real location from the released location information. Experiments show that SM has good utility while achieving the effect of location differential privacy.

However, this method does not take into account the privacy preservation level of the user's location, and the default location posted by the user is a strong sensitive location. The further work will consider the appropriate privacy level allocation of users' location, and more privacy budget should be allocated to the highly sensitive location, so that it can be fully protected. For weakly sensitive locations, less privacy budget is allocated to improve the availability of location data on the premise of satisfying differential privacy preservation.

Acknowledgment. The authors would also like to appreciate the anonymous reviewers for their valuable suggestions, which lead to a substantial improvement of this paper. This research has been funded by the National Natural Science Foundation of China (Grant No. 61672468, 61702148).

References

1. Primault, V., Boutet, A., Mokhtar, S.B., Brunie, L.: The long road to computational location privacy: a survey. IEEE Commun, Surv. Tutorials **21**(3), 2772–2793 (2019). https://doi.org/10.1109/COMST.2018.2873950
2. Tan, Z., Wang, C., Yan, C., Zhou, M., Jiang, C.: Protecting privacy of location-based services in road networks. IEEE Trans. Intell. Transport. Syst. 14 (2020). https://doi.org/10.1109/TITS.2020.2992232
3. Chatzikokolakis, K., Elsalamouny, E., Palamidessi, C., Pazii, A.: Methods for location privacy: a comparative overview. Found. Trends® Privacy Secur. **1**(4), 199–257 (2017). https://doi.org/10.1561/3300000017
4. Wei, J., Lin, Y., Yao, X., Zhang, J.: Differential privacy-based location protection in spatial crowdsourcing. IEEE Trans. Serv. Comput. 1 (2019). https://doi.org/10.1109/TSC.2019.2920643
5. Dwork, C.: Differential privacy. In: Bugliesi, M., Preneel, B., Sassone, V., Wegener, I. (eds.) ICALP 2006. LNCS, vol. 4052, pp. 1–12. Springer, Heidelberg (2006). https://doi.org/10.1007/11787006_1
6. Xiao, Y., Xiong, L.: Protecting locations with differential privacy under temporal correlations. In: Proceedings of the 22Nd ACM SIGSAC Conference on Computer and Communications Security, New York, NY, USA, pp. 1298–1309 (2015). https://doi.org/10.1145/2810103.2813640
7. Machanavajjhala, A., Kifer, D., Abowd, J., Gehrke, J., Vilhuber, L.: Privacy: theory meets practice on the map. In: 2008 IEEE 24th International Conference on Data Engineering, pp. 277–286, April 2008. https://doi.org/10.1109/ICDE.2008.4497436
8. Andrés, M.E., Bordenabe, N.E., Chatzikokolakis, K., Palamidessi, C.: Geo-indistinguishability: differential privacy for location-based systems. In: Proceedings of the 2013 ACM SIGSAC Conference on Computer & Communications Security, pp. 901–914, New York, NY, USA (2013). https://doi.org/10.1145/2508859.2516735
9. Bordenabe, N.E., Chatzikokolakis, K., Palamidessi, C.: Optimal geo-indistinguishable mechanisms for location privacy. In: Proceedings of the 2014 ACM SIGSAC Conference on Computer and Communications Security - CCS 2014, Scottsdale, Arizona, USA, pp. 251–262 (2014). https://doi.org/10.1145/2660267.2660345

10. Niu, B., Chen, Y., Wang, Z., Li, F., Wang, B., Li, H.: Eclipse: preserving differential location privacy against long-term observation attacks. IEEE Trans. Mob. Comput. 1 (2020). https://doi.org/10.1109/TMC.2020.3000730

11. Gursoy, M.E., Liu, L., Truex, S., Yu, L.: Differentially private and utility preserving publication of trajectory data. IEEE Trans. Mob. Comput. **18**(10), 2315–2329 (2019). https://doi.org/10.1109/TMC.2018.2874008

12. Zhu, T., Xiong, P., Li, G., Zhou, W.: Correlated differential privacy: hiding information in non-IID data set. IEEE Trans. Inf. Forensics Secur. **10**(2), 229–242 (2015). https://doi.org/10.1109/TIFS.2014.2368363

13. Liu, C., Chakraborty, S., Mittal, P.: Dependence Makes You Vulnerable: Differential Privacy Under Dependent Tuples. In: 23rd Annual Network and Distributed System Security Symposium, NDSS 2016, p. 15. CA, USA, San Diego (Feb. 2016)

14. Yang, B., Sato, I., Nakagawa, H.: Bayesian differential privacy on correlated data. In: Proceedings of the 2015 ACM SIGMOD International Conference on Management of Data, New York, NY, USA, May 2015, pp. 747–762 (2015). https://doi.org/10.1145/2723372.2747643

15. Cao, Y., Yoshikawa, M., Xiao, Y., Xiong, L.: Quantifying differential privacy in continuous data release under temporal correlations. IEEE Trans. Knowl. Data Eng. **31**(7), 1281–1295 (Jul. 2019). https://doi.org/10.1109/TKDE.2018.2824328

16. Cao, Y., Yoshikawa, M., Xiao, Y., Xiong, L.: Quantifying differential privacy under temporal correlations. In: 2017 IEEE 33rd International Conference on Data Engineering (ICDE), San Diego, CA, USA, April 2017, pp. 821–832 (2017). https://doi.org/10.1109/ICDE.2017.132

17. Dwork, C., Naor, M., Pitassi, T., et al.: Differential privacy under continual observation. Stoc 715–724 (2010)

18. Geng, Q., Kairouz, P., Oh, S., Viswanath, P.: The staircase mechanism in differential privacy. IEEE J. Sel. Topics Sig. Process. **9**(7), 1176–1184 (2015). https://doi.org/10.1109/JSTSP.2015.2425831

19. Geng, Q., Viswanath, P.: The optimal mechanism in differential privacy. In: International Symposium on Information Theory, pp. 2371–2375 (2014)

20. Chen, R., Fung, B.C.M., Desai, B.C.: Differentially private trajectory data publication. arXiv Preprint, arXiv: 1112.2020 (2011)

21. Yu, Z., Zhang, L., Xie, X., Ma, W.-Y.: Mining interesting locations and travel sequences from GPS trajectories. In: Proceedings of International Conference on World Wild Web (WWW 2009), Madrid, Spain, pp. 791–800. ACM Press (2009)

Resource Management

Mobile Edge Server Placement Based on Bionic Swarm Intelligent Optimization Algorithm

Feiyan Guo, Bing Tang[(✉)], Linyao Kang, and Li Zhang

School of Computer Science and Engineering,
Hunan University of Science and Technology,
Xiangtan 411201, China
btang@hnust.edu.cn

Abstract. By offloading computing tasks from mobile devices to edge servers with sufficient computing resources, network congestion and data propagation delays can be effectively reduced. The placement of edge servers is the core of task offloading and is a multi-objective optimization problem with multiple resource constraints. An optimization model of edge server placement has been established in this paper by minimizing both access delay and workload difference as the optimization goal. Then, based on Glowworm Swarm algorithm, it proposes a mobile edge server placement approach called GSOESP to achieve a multi-objective optimization goal. In this study, we use the improved Glowworm Swarm Optimization (GSO) algorithm to find the optimal places as the clustering center which is the edge server placement address, and every base station in edge server's neighbor list is allocated to the edge server. After many iterations, we gradually approach the optimal target. So, the optimal placement scheme is obtained to achieve the goals of minimizing the distance for users to access the edge server and balancing the workload. The GSOESP algorithm is similar to a fast clustering algorithm with good time performance. Experimental results using Shanghai Telecom's real dataset show that the proposed approach achieves an optimal balance between low latency and workload balancing, while guaranteeing service quality, which outperforms several existing representative approaches.

Keywords: Mobile edge computing · Edge server placement · Artificial firefly algorithm · Performance optimization

1 Introduction

With the development of mobile communication technologies, mobile applications have become increasingly diverse and complex, and the demand for computing resources, storage resources, and various network resources is also increasing. In order to deal with the needs of users in the network fastly, Mobile Edge

© ICST Institute for Computer Sciences, Social Informatics and Telecommunications Engineering 2021
Published by Springer Nature Switzerland AG 2021. All Rights Reserved
H. Gao et al. (Eds.): CollaborateCom 2020, LNICST 350, pp. 95–111, 2021.
https://doi.org/10.1007/978-3-030-67540-0_6

Computing (MEC). Compared with centralized cloud computing, MEC deploys computing and storage resources in a distributed manner near the user's network edge (such as mobile base stations, wireless hotspots or edge routers). The computing tasks in the mobile application can be offloaded nearby to effectively reduce the communication overhead and network delay of the application. At the same time, the processing method near to the user also greatly eases the processing pressure of the core network and data center.

In order to solve the needs of low latency and high bandwidth, the deployment of mobile edge servers will be the next key task. On the premise of meeting user's demands and goals, a specific method should be adopted for the placement of edge servers. Selecting a suitable geographic location within a certain geographic area to deploy edge servers and achieving the goal of minimizing network delay and balancing access resources are the current problems that need to be solved in edge server deployment. To further illustrate the problem of mobile edge server placement, we take Fig. 1 as an instance to illustrate the placement of edge server. In the wireless network environment, mobile users access multiple wireless base stations through various wireless terminals to obtain various services. Due to the low latency and high bandwidth requirements of some applications, the existing base stations cannot meet the needs of mobile users well. Therefore, edge server is deployed near the user's position to improve user's access performance. In Fig. 1, the edge server is located next to the base station. Therefore, it is necessary to find a suitable location from multiple base stations' positions for edge server's deployment and achieve the goal of minimizing the access delay of each mobile user and balancing the workload of each edge server. As shown in Fig. 1, an edge server can be responsible for processing multiple users' requests forwarded by multiple base stations.

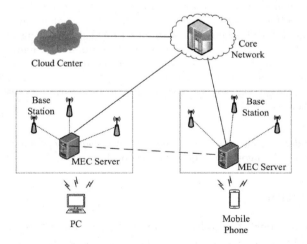

Fig. 1. Edge server placement system model.

The existing edge computing related research on the placement of edge servers has the following problems: 1) The research basis of most studies is that the relevant edge equipment has been assumed to be set in a certain location, and then the access delay is studied, and the problems related to load offload are not considered; 2) There is almost no research on the mobile edge server placement model in the existing research; 3) Most of the existing studies do not have real experimental data and environment for mobile edge computing.

The main contributions of this paper are listed as follows:

1. Express the placement problem of edge server as a constrained multi-objective optimization problem.
2. Use a bionic swarm intelligence optimization algorithm to find the best place of edge server, minimizing access delay and balance workload between edge servers.
3. Based on a real dataset Shanghai Telecom to evaluate our approach and the experimental results show that our method exceeds other methods.

The remainder of the paper as follows: Sect. 2 describes some researches related to the placement of cloudlet and edge cloud. Section 3 introduces the edge server's placement system model and problem definition. Section 4 proposes the edge server placement method. Section 5 presents the experimental results and discussions. Finally, Sect. 6 summarizes the characteristics of the proposed approach and introduces future work.

2   Related Work

At present, there are many scholars who have carried out research on mobile edge services, but most of the existing researches focuse on offloading mobile user workloads to the micro-cloud or remote cloud to achieve low latency for mobile devices, and these studies assumed that devices such as micro-clouds have been placed [11,18,21]. Few researchers concerned about the deployment of micro-clouds and offloading mobile user's workloads to edge servers and improving mobile user access performance [7,20]. Similar to mobile cloud and micro-cloud, there are also problems with the placement of edge servers in MEC environments [5,12].

In mobile edge computing environment, there are few related studies on the deployment of mobile edge servers, but in recent years, there have been some related researches on the deployment of cloudlets and edge clouds [3,8,13].

Cloudlets are often described as small and mobile data centers located at the edge of the enterprise and can be used by nearby devices in a network to relieve network congestion [4,14,17]. There are many similarities between the placement of cloudlets and edge servers [6,16,19]. Jia et al. [6] provided a task offloading model for multiple mobile users. Xu et al. [19] also studied the deployment problem of cloudlets on multiple wireless access points. Although these studies are effective, Jia et al. [6] focused only on the workload balance of cloudlets, and others [2,10,15,16,19] only considered the access delay.

There have several studies on edge cloud placement in recent years [1,9,11, 20,21]. Jia et al. [6] proposed two different heuristic algorithms for the edge cloud placement problem. Xu et al. [19] studied the placement of edge clouds in a large-scale wireless metropolitan area network. They placed edge clouds in strategic locations and used heuristic algorithms to allocate mobile users to minimize the average communication delay between mobile devices .

Inspired by above, we comprehensively consider the two factors of access delay and workload balancing during the placement of the edge server in MEC environment.

3 System Model and Problem Definition

Related symbols are described in Table 1.

Table 1. Symbols.

Symbol	Description
G	The network of MEC
K	The number of MEC servers
N	The number of base stations in MEC
S	MEC servers in G
B	Base stations in G
E	The links between base stations and edge servers
W_b	Workload size of b and b comes from B
W_s	Workload size of s and s comes from S
P_b	Place of base station b and b comes from B
P_s	Place of edge servers and s comes from S
E_s	Base stations allocated to s and s comes from S

3.1 System Model

In the environment of MEC, we define the problem of edge server placement as an undirected graph $G = (N, E)$. There are lots of mobile users, base stations and edge servers to be deployed in this network. $N = B \cup S$, and B is the set of base stations, S is the set of edge servers in N. E is the links between base stations and edge servers in MEC.

All base station's address can be used as the location of edge server, and every base station must be assigned to an edge server. If the number of edge servers is K, K edge servers' locations need to be found in B, and all base stations are allocated to the K edge servers. The base station can directly access the edge server which it belongs to, and can offload tasks to the edge server.

Each edge server is responsible for processing network requests forwarded by all base stations in this edge server's coverage.

As shown in Fig. 1, taking the left MEC server as an example, it serves base stations surrounding itself, and the relations are represented by the links between it and the base stations. Because the relationship among base stations, and the relationship between base stations and mobile edge devices are not focused in this paper, so we did not show these in Fig. 1.

3.2   Problem Definition

Suppose that in the undirected network G, $S = \{s_1, s_2, ..., s_k\}$ is a set of edge servers, each edge server is isomorphic, and the calculation capacity is limited and the same; $B = \{b_1, b_2, ..., b_n\}$ is a set of all base stations, responsible for forwarding service requests of mobile network users within its coverage. After the edge server is placed, each edge server is responsible for processing mobile network user service requests forwarded by all base stations within its coverage.

In our study, there are two main problems we need to solve: one is the placement of K edge servers; the other is the allocation of N base stations. We suppose that $D = \{p(s_1), ..., p(s_j), ..., p(s_k)\}$ is an edge server placement plan, $C = \{E_{s_1}, ..., E_{s_j}, ..., E_{s_k}\}$ is the base station's allocation scheme, so (D, C) is the solution of edge server placement. The workload of edge server is the user requests forwarded by base stations in the edge server's domain, $w(s_j) = \sum_{b_i \in c_j} w(b_i)$. The purpose of edge server placement is to minimize user access latency and minimize edge server's workload differences. Therefore, we formulate the problem of edge server placement as a multi-objective optimization problem. Assuming that each edge server has the same computing ability, so the edge server workload variance can be used to detect edge server's workload differences. Assuming that $Wl(D, C)$ represents the workload balance between edge servers, the formula is as follows:

$$Wl(D, C) = \frac{\sum\limits_{j=1}^{k} (w(s_j) - \bar{w}(s))^2}{k} \tag{1}$$

$$\bar{w}(s) = \frac{\sum\limits_{j=1}^{k} w(s_j)}{k} \tag{2}$$

where $\bar{w}(s)$ represents the average value of edge server's workload. The smaller the value of $Wl(D, C)$, the workload more balanced among edge servers. Edge server is placed at the location of base station in network N, so the transmission path between a base station and its edge server is composed of the links in E. Let $\upsilon(b, s)$ be the access delay between a base station b and the linked edge server s, it is the value of the shortest path between base station b at location $p(b)$ and edge server s at location $p(s)$ in E. Let $Ad(D, C)$ be the access delay between all the base stations and edge servers in an edge server placement solution, it can be formulated as follows:

$$Ad(D,C) = \frac{\sum\limits_{s_j \in S} \sum\limits_{b_i \in E_{s_j}} v(b_i, s_j)}{N} \tag{3}$$

When seeking the optimal solution of the edge server placement, the following constraints need to be considered:

1) When all mobile edge servers are placed, every base station will be allocated to the corresponding MEC server. There will be no crossover base station between edge servers: $E_{s_i} \cap E_{s_j} = \phi$.

$$\bigcup_{s \in S} E_s = B \tag{4}$$

2) The edge server will process all user requests forwarded by base stations within its field.

$$W_s = \sum_{b \in E_s} W_b \tag{5}$$

From Eq. (1)–(5), edge server placement problem can be described as Eq. (6), where (D, C) is an m-dimensional vector of decision variables and $Wl(D, C)$, $Ad(D, C)$ represent functions that are defined for constraints. Then, we can transform Eq. (1)–(5) into (6) as a multi-objective optimization problem (denoted as Problem 1).

$$\min Ad(D,C), \min Wl(D,C)$$
$$s.t. \begin{cases} E_{s_i} \cap E_{s_j} = \phi \\ \bigcup\limits_{s \in S} E_s = B \\ W_s = \sum\limits_{b \in E_s} W_b \end{cases} \tag{6}$$

In order to obtain the Pareto or weakly Pareto optimal solution of Problem 1, we convert the multi-objective optimization problem of edge servers' placement into a single-objective optimization problem using the weighting method. The weighted objective optimization function is established as follows:

$$F(D,C) = \lambda Ad_{nor}(D,C) + (1-\lambda)Wl_{nor}(D,C) \tag{7}$$

where λ is the weighting coefficient between $[0, 1]$, $AD_{nor}(D,C)$ and $WL_{nor}(D,C)$ present normalized values $AD(D,C)$ and $WL(D,C)$. Problem 1 is converted into the following single-objective optimization problem (denoted as Problem 2), as shown in Eq. (8).

$$\min (\lambda Ad_{nor}(D,C) + (1-\lambda)Wl_{nor}(D,C))$$
$$s.t. \begin{cases} E_{s_i} \cap E_{s_j} = \phi \\ \bigcup\limits_{s \in S} E_s = B \\ W_s = \sum\limits_{b \in E_s} W_b \end{cases} \tag{8}$$

4   Optimization Approach

The swarm intelligence optimization algorithm is a bionic simulation evolution algorithm. This type of algorithm treats all possible solution sets of the problem as a solution space, starting from a subset that represents the possible solution of the problem, by applying some operator operation to the subset to generate a new solution set, and gradually make the population evolve to the state containing the optimal solution or near optimal solution. In the evolution process, only the information of the objective function is needed, and the optimal solution can be found without being restricted by the continuous or differentiable search space. Glowworm swarm algorithm (GSO) is a typical swarm intelligence optimization algorithm. Inspired by this, we use glowworm swarm algorithm to solve the edge server placement problem. This algorithm is named GSOESP (**G**lowworm **S**warm **O**ptimized **E**dge **S**erver **P**lacement).

The GSOESP algorithm is mainly used to solve the problems of edge server location selection and base station allocation. First, we set the base station space as $B = \{b_1, b_2, ..., b_i, ..., b_n\}$, and b_i is a two-dimensional vector containing the address and workload. The vector $S = \{s_1, s_2, ..., s_k\}$ represents k edge servers, is also the k fireflies which are more attractive, s_j and b_i are equal sized vectors. Taking the k base stations where k fireflies are located as the first locations of the k edge servers and find the most attractive neighbors in these k fireflies' neighbors to move towards them. During the search process, it uses the maximum attraction force iterates for the iteration criteria until the positions of the k fireflies no longer change, so k cluster centers are obtained, that is, where the k edge servers are located. In the traditional artificial firefly algorithm, the firefly uses a random selection mechanism with a greedy strategy when it moves to an individual with a higher brightness in its perceptual range. This mechanism easily causes problems such as excessive time or reduced convergence accuracy. Based on the research content, this paper proposes a local optimal selection strategy, which solves the problem that the decision domain in the artificial firefly algorithm needs to be dynamically adjusted due to the density of neighbors.

The brightness and attraction formulas of GSOESP are shown in Eqs. (9) and (10).

$$I = I_0 \times e^{-\gamma d_{ij}} \tag{9}$$

$$\beta = \beta_0 \times e^{-\gamma d_{ij}^2} \tag{10}$$

In Eqs. (9) and (10), I_0 is the maximum brightness of fireflies (at their own positions), γ is the medium light intensity absorption factor, β_0 is the maximum attraction of fireflies (at their own positions), and d_{ij} is the distance between fireflies i and j, the maximum brightness of fireflies I_0 is shown in Eq. (11).

$$I_0 = \frac{1}{1 + F(D, C)} \tag{11}$$

In GSOESP, firefly i is attracted to firefly j, and its displacement is shown in Eq. (12).

$$x_i^{t+1} = x_i^t + \beta_0 \times e^{-\gamma d_{ij}^2} \times (x_j^t - x_i^t) + \alpha \times (rand - 0.5) \tag{12}$$

In Eq. (12), x_i^t, x_j^t are the positions of fireflies i and j, α is the step factor, and rand is a random factor that follows the distribution of [0, 1]. The interval [0, 1] follows a uniform distribution. The first part on the right side of Eq. (12) represents the current position of fireflies, the second part represents the amount of position change caused by being attracted by other fireflies, and reflects the global optimization ability of the algorithm, and the third part represents the local random search movement of fireflies, and reflects the local optimization ability of the algorithm. Therefore, the firefly algorithm has both a good global optimization ability and a certain local optimization ability, which is suitable for solving the problem of edge server placement.

Choose the k edge server locations and every base station in edge server's neighbour list $NL(F_i)$ is allocated to the edge server S_i. The nearest N neighbour base stations of the firefly F_i to form the neighbour list $NL(F_i)$ as shown in Eq. (13).

$$NL(F_i) = \{b_j, b_j \in MinN(d(F_i, b_j))\} \tag{13}$$

The N neighbour base stations are composed as shown in Eq. (14).

$$N = \frac{n}{10 * k} \tag{14}$$

In the process of base station allocation, the base station relationship matrix Tt in the GSOESP are introduced. The base station relationship matrix Tt is defined as Eq. (15), where τ_{ij} records the allocation mapping relationship between base station b_i and edge server s_j, n is the number of all base stations, k is the number of all edge servers and the value of $\tau_i j$ is defined as Eq. (16).

$$T_t = \begin{bmatrix} \tau_{11} & \cdots & \tau_{1k} \\ \cdots & \tau_{ij} & \cdots \\ \tau_{n1} & \cdots & \tau_{nk} \end{bmatrix} \tag{15}$$

$$\tau_{ij} = \begin{cases} 0 \ (b_i \notin E_{s_j}) \\ 1 \ (b_i \in E_{s_j}) \end{cases} \tag{16}$$

When the base station allocation is completed, the sum of the base stations corresponding to row i with element 1 on the j-th column of the matrix is the total workload of the edge server s_j, so $w_{s_j} = \sum w_{b_i}$, if $\tau_{ij} = 1$. Because firefly considers the distance between the base stations and the workload size at the same time when searching for the optimal location, the edge server location obtained by the GSOESP algorithm can not only meet the requirements of delay minimization, but also improve the workload balancing degree, resulting in an optimal edge server placement scheme (D, C). Therefore, the whole algorithm can be described as the following Algorithm 1.

5 Experiments and Analysis

To validate our proposed approach, we adopted a real base station dataset. Our experiments primarily consisted of two parts: 1) we compared our approach with other known placement approaches; 2) we determined the number of edge servers K that are placed using our approach.

5.1 Datasets

These experiments use a real base station dataset of Shanghai Telecom, which contains 562,914 records accessed by 6,262 mobile users on 2,786 base stations. In these base stations, there are 2400 valid base stations. The dataset contains the detailed time of each base station visited by service requester. Using Google map tools, we make a distribution map of the Shanghai base station in Fig. 2. Table 2 describes some base stations' workloads which is calculated by all the

---

Algorithm 1. GSOESP

Input: B, k, n, I_o, β_o, γ, α, $IterNum$
Output: $Wl(D,C)$, $Ad(D,C)$, $F(D,C)$, optimal allocation scheme (D,C)
1: $i \leftarrow 1, j \leftarrow 1$
2: $w_a \leftarrow 0$
3: $v_a \leftarrow 0$
4: select k base stations from B as the center of clusters
5: calculate distance from stations to k and allocate other stations into the nearest cluster
6: **for** i to $IterNum$ **do**
7:     calculate I_0 according to Eq. (11)
8:     calculate I and β according to Eq. (9) and (10)
9:     low brightness fireflies move to high brightness fireflies
10:     calculate x_i^{t+1} according to Eq. (12)
11:     **if** fireflies's positions changed **then**
12:         recalculate I of fireflies
13:         record new I, P_s and P_b
14:     **end if**
15: **end for**
16: calculate $\bar{w}(s)$ according to Eq. (2)
17: **while** $j \leq k$ **do**
18:     $w_a = w_a + w(s_j) - \bar{w}(s)$
19: **end while**
20: **for** j to k **do**
21:     **for** i to $|E_{s_j}|$ **do**
22:         $v_a = v_a + v(b_i, s_j)$
23:     **end for**
24: **end for**
25: calculate $Wl(D,C)$ according to Eq. (1)
26: calculate $Ad(D,C)$ according to Eq. (3)
27: calculate $F(D,C)$ according to Eq. (7)

Fig. 2. Shanghai base station distribution map.

Table 2. Workload of each base station in Shanghai.

Base station ID	The number of users	Workload (min)
12	354	24958
23	147	10655
345	28	824
400	1242	61972
543	311	12566
726	67	1331
1227	1	89
2428	3	78
139	440	9639
106	261	14026

visited time of the base station. From Table 2 we can find that some base stations are overloaded, some base stations are not fully utilized, and there are serious imbalances among base stations' workloads.

5.2 Baseline Approaches

- **Top-K.** This approach considers workload mainly, choose the k largest workload as the edge servers' locations to minimize workload differences, and every base station is allocated to the nearest edge server.
- **Random.** Select k edge servers' addresses randomly from n base stations, and each base station selects the nearest edge server to be allocated.

- **K-means.** K-means only considers access delay and does not consider edge server's workload balancing. It divides all base stations into multiple clusters according to the principle of minimizing user's access delay, with the multiple cluster centers as edge servers' addresses, and base stations' allocation adopt the nearest distribution principle.

Our approach adopts the glowworm swarm intelligence optimization algorithm to find the optimal placement solution of MEC server.

5.3   Evaluation Metric

Definition 1. Access Delay.

We use the time that the base station forwards the user's request to the edge server as the access delay. The longer the distance between the base station and the edge server, the bigger the access delay. Due to the limitation of this dataset, we regard the shortest distance between the base station and the MEC server as the access delay. In our experiment, we compare the performance of the average distance between all base stations and MEC servers.

$$Ad = \frac{\sum\limits_{j=1}^{K} \sum\limits_{i=1}^{n} \upsilon(i,j)\tau_{i,j}}{n} \tag{17}$$

Definition 2. Workload Balance

We used Wl to represent the workload balance of edge servers and compared the workload of each edge server with the average workload, the calculation formula as follows:

$$Wl(D,C) = \frac{\sum\limits_{j=1}^{k} (w(s_j) - \bar{w}(s))^2 - (n-k) \times \bar{w}(s)^2}{k} \tag{18}$$

Definition 3. Comprehensive Performance

Comprehensive performance value means considering the two attributes of average distance and the standard deviation of the workload simultaneously, we set the weight ratio of each attribute λ as 0.5.

5.4   Comparison Results with the Number of Base Stations

When the placement ratio R of the edge server is 0.1 and the number of base stations n is increased from 300 to 2400, as shown in Fig. 3. The average access delay in GSOESP is the smallest, followed by K-means, Top-K and Random. With the number of base station n increases, the average access delay has shown a descending trend in GSOESP approach.

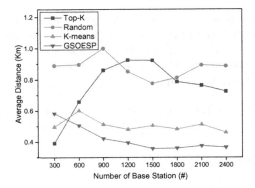

Fig. 3. Base station access average distance under different placement approaches when $R = 0.1$.

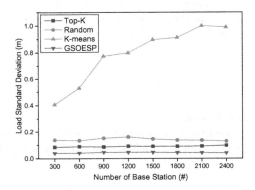

Fig. 4. Load standard deviation under different placement methods when $R = 0.1$.

When the number of base stations becomes larger and the number of edge servers gradually increases, the standard deviation of the workload of each edge server changes differently in these four approaches, as shown in Fig. 4. In K-means approach, the workload standard deviation gradually increases when the number of base stations becomes larger. The other three approaches change only a little when the number of base stations changes. The load standard deviation of GSOESP is the smallest. Top-K approach uses the top K base stations with the largest workload as the edge server address, so the standard deviation of this approach's workload is the second. The K-means obtains the worst result.

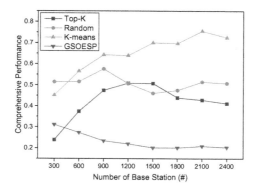

Fig. 5. Comprehensive performance value under different placement methods when $R = 0.1$.

Considering the two attributes of average distance and the standard deviation of the workload together, we set the weight ratio of each attribute λ to 0.5. We found that the average distance of the base station access and the standard deviation of the workload in GSOESP approach is the smallest, and the comprehensive performance of GSOESP approach outperforms others, as shown in Fig. 5.

According to Eq. (10), we take $\lambda = 0.5$, and the value of $pw(l)$ in the four different placement approaches is shown in Table 3.

Table 3. Comprehensive performance value under different placement methods with $R = 0.1$.

n	Top-K	Random	K-means	GSOESP
300	0.2385	0.5148	0.4518	0.3110
600	0.3740	0.5158	0.5656	0.2731
900	0.4743	0.5760	0.6434	0.2335
1200	0.5085	0.5069	0.6395	0.2196
1500	0.5072	0.4611	0.7007	0.2012
1800	0.4394	0.4751	0.6974	0.2012
2100	0.4283	0.5149	0.7560	0.2085
2400	0.4117	0.5082	0.7253	0.2028

5.5   Comparison Results with the Number of Edge Servers

The number of base stations n is 2400, and the number of edge server starts from 100 to 500 with the step size of 100. The average access delay and workload standard deviation of the four approaches under different edge server numbers are evaluated. As shown in Fig. 6, as the number of edge servers increases, the average access distance of base station decreases, that is the more edge servers are deployed, the more the nearest edge server can be selected by the base station, so the average access distance gradually decreases. Among the four approaches, GSOESP obtains the minimal average access distance, and Random approach obtains the worst result.

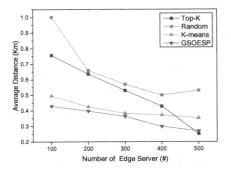

Fig. 6. Base station access average distance under different placement approaches when K changes.

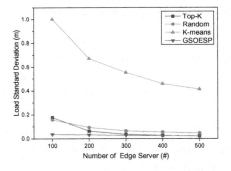

Fig. 7. Workload standard deviation under different placement approach when K changes.

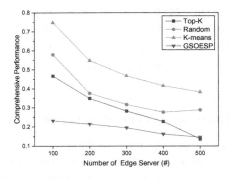

Fig. 8. Comprehensive performance value under different placement approaches.

The standard deviation of the workload of the edge server also gradually decreases as the number of edge servers increases. The workload standard deviation of GSOESP is the smallest, followed by Top-K, Random, and K-means, as shown in Fig. 7.

According to Eq. (3), we take $\lambda = 0.5$, among the four placement approaches, $pw(a)$ is the comprehensive performance parameter shown in Fig. 8.

The $pw(a)$ is shown in Table 4. By Comparing the performance of the edge server placement among the four approaches, GSOESP outperforms others, as shown in Fig. 8.

Table 4. Comprehensive performance value under different placement methods when K changes.

K	Top-K	Random	K-means	GSOESP
100	0.4663	0.5786	0.7473	0.2317
200	0.3489	0.3755	0.5476	0.2137
300	0.2819	0.3162	0.4671	0.1942
400	0.2272	0.2763	0.4151	0.1617
500	0.1365	0.2877	0.3824	0.1455

6   Conclusion

Mobile edge computing is an important technology for extending remote computing resources. Through the placement of edge servers to improve the access quality of mobile users, this paper studied the placement problem of edge servers. In mobile edge computing, the placement problem is considered as a multi-objective optimization problem and an optimal deployment scheme is proposed. The main approach is the combination of improved Glowworm Swarm optimization algorithm and adaptive algorithm to comprehensively evaluate the performance of edge server workload balancing and communication delay under four different approaches. We have designed three experiments using real Shanghai Telecom's base station dataset and the experiment results show that the overall performance of GSOESP approach is acceptable and GSOESP approach achieves the best performance compared to the other approaches.

Currently in this paper, we assume that all edge servers to be placed are the same, but in actual network environment, edge servers are heterogeneous. Therefore, in future work, we will continue to study the placement problem of heterogeneous edge servers.

Acknowledgment. The authors would like to thank all anonymous reviewers for their invaluable comments. This work is supported by the Scientific Research Fund of Hunan Provincial Education Department under grant no. 18A186, the Natural Science Foundation of Hunan Province under grant no. 2018JJ2135, as well as the National Natural Science Foundation of China under grant no. 61602169.

References

1. Ahmed, E., Akhunzada, A., Whaiduzzaman, M., Gani, A., Hamid, S.H.A., Buyya, R.: Network-centric performance analysis of runtime application migration in mobile cloud computing. Simul. Model. Pract. Theory **50**, 42–56 (2015)
2. Chen, X., Jiao, L., Li, W., Fu, X.: Efficient multi-user computation offloading for mobile-edge cloud computing. IEEE/ACM Trans. Netw. **24**(5), 2795–2808 (2016)
3. Chun, B., Ihm, S., Maniatis, P., Naik, M., Patti, A.: CloneCloud: elastic execution between mobile device and cloud. In: Kirsch, C.M., Heiser, G. (eds.) European Conference on Computer Systems, Proceedings of the Sixth European conference on Computer systems, EuroSys 2011, Salzburg, Austria, 10–13 April 2011, pp. 301–314. ACM (2011)
4. Clinch, S., Harkes, J., Friday, A., Davies, N., Satyanarayanan, M.: How close is close enough? Understanding the role of cloudlets in supporting display appropriation by mobile users. In: Giordano, S., Langheinrich, M., Schmidt, A. (eds.) 2012 IEEE International Conference on Pervasive Computing and Communications, Lugano, Switzerland, 19–23 March 2012, pp. 122–127. IEEE Computer Society (2012)
5. Hoang, D.T., Lee, C., Niyato, D., Wang, P.: A survey of mobile cloud computing: architecture, applications, and approaches. Wirel. Commun. Mob. Comput. **13**(18), 1587–1611 (2013)
6. Jia, M., Cao, J., Liang, W.: Optimal cloudlet placement and user to cloudlet allocation in wireless metropolitan area networks. IEEE Trans. Cloud Comput. **5**(4), 725–737 (2017)
7. Krishnanand, K.N., Ghose, D.: Glowworm swarm optimization for simultaneous capture of multiple local optima of multimodal functions. Swarm Intell. **3**(2), 87–124 (2009)
8. Lee, H., Lee, J.: Task offloading in heterogeneous mobile cloud computing: modeling, analysis, and cloudlet deployment. IEEE Access **6**, 14908–14925 (2018)
9. Li, H., Dong, M., Liao, X., Jin, H.: Deduplication-based energy efficient storage system in cloud environment. Comput. J. **58**(6), 1373–1383 (2015)
10. Li, H., Dong, M., Ota, K., Guo, M.: Pricing and repurchasing for big data processing in multi-clouds. IEEE Trans. Emerg. Top. Comput. **4**(2), 266–277 (2016)
11. Liang, T., Li, Y.: A location-aware service deployment algorithm based on K-means for cloudlets. Mob. Inf. Syst. **2017**, 8342859:1–8342859:10 (2017)
12. Mach, P., Becvar, Z.: Mobile edge computing: a survey on architecture and computation offloading. IEEE Commun. Surv. Tutor. **19**(3), 1628–1656 (2017)
13. Peng, K., Leung, V.C.M., Xu, X., Zheng, L., Wang, J., Huang, Q.: A survey on mobile edge computing: focusing on service adoption and provision. Wirel. Commun. Mob. Comput. **2018**, 8267838:1–8267838:16 (2018)
14. Satyanarayanan, M., Bahl, P., Cáceres, R., Davies, N.: The case for VM-based cloudlets in mobile computing. IEEE Pervasive Comput. **8**(4), 14–23 (2009)
15. Tao, M., Ota, K., Dong, M.: Foud: integrating fog and cloud for 5G-enabled V2G networks. IEEE Netw. **31**(2), 8–13 (2017)
16. Varghese, B., Reaño, C., Silla, F.: Accelerator virtualization in fog computing: moving from the cloud to the edge. IEEE Cloud Comput. **5**(6), 28–37 (2018)
17. Wolbach, A., Harkes, J., Chellappa, S., Satyanarayanan, M.: Transient customization of mobile computing infrastructure. In: Cáceres, R., Cox, L.P. (eds.) Proceedings of the First Workshop on Virtualization in Mobile Computing, Breckenridge, CO, USA, 17 June 2008, pp. 37–41. ACM (2008)

18. Xiang, H., et al.: An adaptive cloudlet placement method for mobile applications over GPS big data. In: 2016 IEEE Global Communications Conference, GLOBE-COM 2016, Washington, DC, USA, 4–8 December 2016, pp. 1–6. IEEE (2016)
19. Xu, Z., Liang, W., Xu, W., Jia, M., Guo, S.: Efficient algorithms for capacitated cloudlet placements. IEEE Trans. Parallel Distrib. Syst. **27**(10), 2866–2880 (2016)
20. Yao, H., Bai, C., Xiong, M., Zeng, D., Fu, Z.: Heterogeneous cloudlet deployment and user-cloudlet association toward cost effective fog computing. Concurr. Comput. Pract. Exp. **29**(16), e3975 (2017)
21. Zhao, J., Ou, S., Hu, L., Ding, Y., Xu, G.: A heuristic placement selection approach of partitions of mobile applications in mobile cloud computing model based on community collaboration. Clust. Comput. **20**(4), 3131–3146 (2017)

A MOEAD-Based Approach to Solving the Staff Scheduling Problem

Feng Hong, Hao Chen, Bin Cao$^{(\boxtimes)}$, and Jing Fan

Zhejiang University of Technology, Hangzhou, China
{hongfeng,chenhao,bincao,fanjing}@zjut.edu.cn

Abstract. Due to the impact on increase of the utilization efficiency of the staff and decrease of operating cost of enterprises, the staff scheduling problem has attracted the interests of many scholars. Actually, the staff scheduling problem can be considered to be how to assign the right staff to the right shift on the right time period based on constraints, meanwhile the objectives should be optimized. Hence, designing an algorithm to satisfy all the requirements mentioned above is challenging. First, there are prohibitive combinations of assigning the staff to shifts from a sheer numbers perspective; Next, there are potential conflicts among optimization objectives, which means objectives may not reach the optimization at the same time and the optimal schedule can not be found; Finally, rare work about the fairness of optimization objectives has been studied. The existing works usually focus on optimization objectives in total, ignoring the fairness of them. A schedule with best optimization objectives can not provide the highest fairness. Hence, we propose an approach based on multi-objective evolutionary algorithm based on decomposition (MOEAD) to solve the staff scheduling problem in the fairness aspect. A series of experiments are performed and prove that the proposed method can effectively find the schedule with fairness.

Keywords: Multi-objective optimization · MOEAD · Staff scheduling · NP-hard

1 Introduction

The staff scheduling problem has drawn significant attention during the last few decades, and it can be considered to be how to assign the right staff to the right shifts on the right time period [1]. A schedule with high quality contributes to improve the utilization efficiency of the staff and reduce the operating cost of the enterprises [2,3].

The staff scheduling problem and some variants are well known as NP-hard [4] and attract the interests from both industry and academia [5,6]. However, existing works usually take into account the optimization of the objectives such as minimizing the personnel costs, while ignoring the fairness of these objectives.

© ICST Institute for Computer Sciences, Social Informatics and Telecommunications Engineering 2021
Published by Springer Nature Switzerland AG 2021. All Rights Reserved
H. Gao et al. (Eds.): CollaborateCom 2020, LNICST 350, pp. 112–131, 2021.
https://doi.org/10.1007/978-3-030-67540-0_7

A schedule with minimal cost does not guarantee the highest fairness, which increases the complexity of the staff scheduling problem due to the resulting trade-off between efficiency and fairness. Actually, the staff scheduling in the fairness aspect is a multi-objective optimization problem.

In this paper, what we need to take into account is: (a) the constraints that the scheduler must be considered to make a schedule, and (b) the evaluation metrics used to measure a schedule. We divided the constraints into two categories: (1) *Hard constraint*. The schedule is invalid if any one of them is broken, and (2) *Soft constraint*. It is desirable to meet the soft constraint. Hence, we make a strategy to guarantee that the hard constraint would not be broken. Each schedule will be tested for whether this schedule violates hard constraints. If the schedule violated hard constraints, the shifts of violating them will be substituted according to the strategy. Then this schedule performed by the operation of substitution will be treated as a potential schedule option. As for the soft constraint, the number of violating it is set to an evaluation metric *Shifting_Satisfaction*, which gets higher as the number of violating soft constraints is less. Besides, there are other evaluated metrics, i.e., average workload coverage *Ave_Coverage* and coverage fairness *Fairness_Coverage*. The staff deployed is asked for working as effectively as possible, *Ave_Coverage* is used to measure whether the number of the staff can cover the work demand at different times of each day. But due to constraints for scheduling, the number of the staff on the shifts of different times is varied, and the workload coverage may be different, the *Fairness_Coverage* is introduced to be an evaluation metric. Based on these metrics, the schedules will be compared with each other. The evaluation result of the schedule is worse by others, we say that this schedule is dominated by others. For example, if two schedules s_i and s_j satisfy the hard constraint, yet s_i would result in less shifting coverage, less workload coverage, and less coverage fairness than s_j, we say that s_i dominates s_j. Then, we substitute s_j and return those schedules that are *not* dominated by others as the results of our approach.

One simple method to find the optimal schedule is to compute all the combinations of shifts for the staff. However, generating the optimal schedule in this way needs prohibitive expensive as it encounters a large number of combinations of shifts assigned to the staff.

What is more, there are potential conflicts among evaluation metrics, which means that evaluation results may not reach the optimization value at the same time and the optimal schedule may not exist. Hence, finding a set of (near)optimal schedules is an acceptable way, and we propose an approach based on multi-objective evolutionary algorithm based on decomposition (MOEAD [7]) to address our issue. Our approach can be divided into three steps. In the first step, we encode the schedule as an individual, and a series of individuals constitute a population. In the second step, we perform the operations of mutation and crossover to generate a series of new schedules. During these steps, we perform our substitution strategy to ensure that each schedule is valid. Then these schedules after the strategy will be treated as potential schedule options to be compared with those schedules in the original population. In the third step, the newly generated schedules will be measured by evaluation metrics and their evaluation results will be compared with original schedules, where the schedules with better evaluation results will be selected to be made up to a new population. These three steps will be performed consecutively until the number of repeating reaches the pre-set value. Then the schedules of the last population are the set of (near)optimal schedules.

The experimental analysis also shows the performance of our approach. In general, the contribution of this paper can be summarized as follows:

1. Previous multi-objective evolutionary algorithms (MOEAs) rarely take coverage fairness as an optimization objective used to measure the schedules. In our work, we model the fairness coverage based on the real-world scene and use it to measure the schedules, which makes the evaluation metrics more reasonable.
2. We adopt MOEAD to solve the proposed problem by converting the multiple objective into several singe-objectives. Every single objective problem is optimized, and the corresponding (near)optimal schedules are found and treated as the solutions to our problem.
3. We execute a series of extensive experimental evaluations to show the performance of MOEAD on our problem. Compared with three other MOEAs (the nominated sorting genetic algorithm II (NSGA-II [8]), the reference vector guided evolutionary algorithm (RVEA [9]) and the preference-inspired coevolutionary algorithms [10]), MOEAD has been proven that its mechanism of decomposition can effectively address our problem.

The rest of the paper is organized as follows: Sect. 2 introduces the related work of our algorithm; Sect. 3 introduces the preliminaries for helping readers to understand our algorithm; Sect. 4 introduces the main contents of the algorithm; Sect. 5 describes the experiment contents; Finally, Sect. 6 summarizes the full text.

2 Related Work

Current algorithms to address the staff shifting problem can be classified into three categories. The first category is the heuristic algorithm. Leksakul et al. [11] proposed a scheduling algorithm based on genetic algorithms to minimize the objective functions based on the overtime cost of nurses. Their algorithm was able to reduce 12% of the staffing cost and 13% of overtime costs. Compared with the classical search algorithms based on genetic algorithm [12], the new memetic algorithms [13] can find better results in terms of global optimization and produce more number of optimization results under the condition of balanced development and exploration in the search space. However, these algorithms do not measure the satisfaction of soft constraints, which determines the quality of the final results. Then the bee colony algorithm [14] is proposed to satisfy the above problem, but the fairness of workforce satisfaction between different days in the schedules remains an issue. Our algorithm introduces the idea of workforce satisfaction standard deviation, and it's treated as a metric to evaluate the schedules.

The second category to address the staff shifting problem is based on mathematical programming. Various mathematical models and integer programming techniques are proposed to solve the staff shifting problem. Howell [15] proposed a cyclical model to create a schedule within a duration. Due to a large number of infeasible solutions in a cyclic shifting, a method of modeling scheduling problems with mixed integer programming is studied. Pour et al. [16] proposed a hybrid framework for staff shifting problem, they implemented the global constraint programming model to find the (near)optimal solutions. Compared with constraint programming, this model can produce better solutions by taking the requirements of staff into consideration. Hamid et al. [17] proposed a multi-objective nurse scheduling mathematical model, which takes the decision-making style of nurses into account. Then the data envelopment analysis method is used to sort the obtained Pareto solutions. Although this category of the method has high effectiveness, a large amount of computation leads to low efficiency and high responding time. However, our algorithm can generate (near)optimal solutions with less responding time, which is considered as improving the exceeding speed of the algorithm based on a bit reduction of effectiveness.

The third category is the multi-objective evolutionary algorithm. According to the evolutionary mechanism, they can be divided into MOEAs based on domination [18,19], MOEAs based on decomposition [7,20], and MOEAs based on indicators [21–23]. The main idea of MOEAs based on decomposition is to decompose multi-objective optimization problems into several scalar quantum problems for optimization. Each sub-problem is optimized by using information from its several neighboring sub-problems. And in this paper, the MOEAD is the most classical algorithm among the MOEAs based on decomposition. Though multi-objective evolutionary algorithms have natural advantages in solving the staff shifting problems, it is usually an algorithm that has the best effect in solving a specific problem due to the different constraints in the problem. Our algorithm performs the evolution based on the uniformly distributed weight vector, which makes that the solutions of the algorithm will converge to Pareto Font more quickly. Besides, MOEAD has been proven that it will perform well in dealing with similar problems [7].

3 Preliminaries

This section presents a set of preliminaries that are important to set the stage for understanding our algorithm and its vision. In particular, we will introduce our preliminaries from following aspects: basic concepts, scheduling objectives and our problem definition.

3.1 Basic Concepts

Shifts and scheduling constraints are two basic concepts for staff scheduling problems. In our work, we present them as follows:

Shifts. Each day is divided into two shifts, namely the day shift (*Day_shift*) and the night shift (*Night_shift*). Every day, each shift requires a certain number of staffs to finish the job and each staff can work at most one shift, i.e., he can take this day off if both shifts on this day have been assigned with enough workers. Note that different shifts on different days may require different numbers of staffs.

Scheduling Constraints. When assigning staffs to shifts, constraints like staffs' preferences, business demands or legal regulations should be taken into account. Usually, there are two types of constraints, i.e., hard constraints and soft constraints.

- **Hard constraints.** The entire schedule is invalid if any constraint of this type fails. In our work, we mainly consider following two hard constraints: (1) H_{osod}. *Each staff can only be assigned to at most one shift per day;* and (2) H_{mwd}. *Each staff can work at most k consecutive working days.* The second hard constraint actually indicates two different cases of how a staff can take holiday. Specifically, a staff must take at least one day off if he had worked k consecutive days. In the other case, it is also allowable if a staff rest a few days within a period of k consecutive days.
- **Soft constraints.** Violating these constraints will not make the schedule invalid, but it is desirable to meet them. In our work, we merely consider one soft constraint: *If one staff was assigned to the night shift (Night_shift) on one day, assigning a day shift on the next day is not allowed.*

3.2   Schedule Evaluation

Given a group of staffs N and a scheduling horizon H, it is possible to find more than one valid schedule that can satisfy the aforementioned hard and soft constraints. Hence, to determine the optimal schedule, we propose three evaluation metrics, i.e., *shifting satisfaction, average workload coverage,* and *coverage fairness*.

Shifting Satisfaction. For a valid schedule, though the soft constraint on shifts rotation for a staff could be violated due to reasons like insufficient workforce or the impact of hard constraints, it is better to reduce the number of violations as many as possible. Because the employee satisfaction matters a lot in real scenarios. Hence, we propose the metric of shifting satisfaction to measure the degree of soft constraint violation, and it can be computed as follows:

$$Shift\_Satisfaction = 1 - \frac{k}{|N|(|H| - 1)} \times 100\% \qquad (1)$$

where the $|N|$ denotes the number of total staffs, the $|H|$ is the number of days in a scheduling horizon, k represents the number of violations on the soft constraint.

Since each staff can only be assigned at most one shift per day and each shift assignment after the first day corresponds to a judgment that whether the assigned shift follows the soft constraint, i.e., whether the day shift is arranged after the night shift for a staff, the number of soft constraints that are needed to be tested for each staff should be $|H| - 1$. Then for $|N|$ staffs, the total number of judgments on the soft constraint is $|N|(|H| - 1)$. Hence, the shifting satisfaction will finally be represented as $1 - \frac{k}{|N|(|H|-1)}$. Obviously, the value of shifting satisfaction ranges from 0 to 1, and it will reach the maximum value of 1 if there is no soft constraint violated.

Average Workload Coverage. One of the critical issues that a schedule must follow is to have the staff deployed as effectively as possible, and this usually means whether the number of the staff at work can cover the work demand of different times of the day. Here the work demand in our paper refers to the workload of different shifts on different days, e.g., the minimal number of required staffs to finish the job for the corresponding shift. Considering that the workload in real scenarios may vary by different shifts, we use the average workload coverage for the past $|H|$ days to reflect the overall coverage level, and it can be computed as follows:

$$Ave\_Coverage = \frac{1}{2|H|} \times \sum_{j=1}^{|H|} (Coverage_{H_j}^{Day\_shift} + Coverage_{H_j}^{Night\_shift}) \quad (2)$$

where $Coverage_{H_j}^{Day\_shift}$ and $Coverage_{H_j}^{Night\_shift}$ respectively denotes the coverage of the day shift and night shift on day H_j. Specifically, the day/night shift coverage equals to the ratio of the number of assigned staffs to the number of required staffs. Note that, since two shifts are involved on each day, the average workload coverage for the scheduling horizon of H days should be divided by 2.

Coverage Fairness. Due to the existence of hard and soft constraints for scheduling, the workload coverage for different shifts on different days may vary a lot. For example, suppose the workload for the day shift and night shift is 30 and 20 respectively for one day, and 40 staffs in total are available for scheduling, how many would really be put in place? If we choose to have schedules that fully cover the day shift, then the maximal number of staffs that can be arranged for the night shift would be only 10, and the average coverage of this day is 0.75 $((30/30 + 10/20)/2)$. Compared with the 100% coverage of the day shift, the night shift coverage is merely 50%. The coverage deviation of these two shifts are too huge to be applied in real scenarios, in fact, this deviation can cause the problem of overstaffing and understaffing. To avoid above problems and achieve the fairness in staff scheduling, we calculate the standard deviation for the coverage of all the shifts as follows:

$$Coverage\_Fairness = \left\{ \frac{1}{2|H|} \times \sum_{j=1}^{|H|} \left[(Coverage_{H_j}^{Day\_shift} - Ave\_Coverage)^2 \right. \right.$$
$$\left. \left. + (Coverage_{H_j}^{Night\_shift} - Ave\_Coverage)^2 \right] \right\}^{1/2} \quad (3)$$

Take the above example again, the coverage fairness is $[(1 - 0.75)^2 + (0.5 - 0.75)^2]^{1/2}/2 = 0.18$. Now suppose the day shift and night shift are now assigned with 21 and 16 staffs, though the average coverage is still 0.75, the coverage deviation becomes smaller, i.e., the day shift coverage is 0.7 while it is 0.8 for the coverage of night shift, hence the coverage fariness is 0.04. Moreover, three staffs in this schedule can take off this day, which then can provide some room for future staffing.

3.3   Problem Definition

Definition 1. Given a group of staffs N and a scheduling horizon H, the shifts *Day_shift* and *Night_shift* are assigned to staffs, along with shifting satisfaction *Shift_Satisfaction*, average workload coverage *Ave_Coverage* and coverage fairness *Coverage_Fairness*, the goal is to find a set of schedules S, where each schedule $\forall s \in S$, the following hold:

1. Following hard constraints must be satisfied,
   - Each staff can only be assigned to at most one shift per day;
   - Each staff can work at most k consecutive working days;
2. argmax *Shift_Satisfaction*;
3. argmax *Ave_Coverage*;
4. argmin *Coverage_Fairness*;
5. s is in the set of Pareto solutions.

There are many staffs remaining to be assigned to shifts, it's difficult from a sheer number perspective with so many different possible combinations of schedules to optimize all the evaluated metrics. Besides, there may be potential conflicts among the evaluated metrics, which means that all the evaluated metrics may not reach the optimization at the same time. Under this situation, finding the (near)optimal schedules is acceptable way. The evaluated results of different metrics are hard to be compared with each other for selecting (near)optimal schedules, which makes finding (near)optimal schedules challenging. Actually, some variants of staff scheduling problem have been proven NP-hard [4].

3.4   Problem Definition

Definition 2. Given a group of employees E and a scheduling horizon H, where each employee $e \in E$ has the performance P_e, the shifts *day_Shift* and *Night_shift* are assigned to employees, along with shifting satisfaction *Shift_Satisfaction*, average workload coverage *Ave_Coverage* and coverage fairness *Coverage_Fairness*, our algorithm finds the optimal schedule, the following hold:

1. Following hard constraints must be satisfied,
   - Each staff can only be assigned to at most one shift per day;
   - Each staff can work at most k consecutive working days;
2. argmax *Shift_Satisfaction*;
3. argmax *Ave_Coverage*;
4. argmin *Coverage_Fairness*;

Scheduler often face the situation about assigning lots of employees to the shifts, it's a difficult job from a sheer numbers perspective. Since so many possibilities and combinations of potential options need to be taken into account while maximizing the Coverage_Fairness, minimizing the Ave_Coverage and the Shifting_Satisfaction. Besides, these optimization goals may potentially conflict with each other, which means that they may not reach the optimal value at the same time. What's more, not only are there countless possibilities for scheduling options, but juggling these and deciding on the right combination is very challenging because of all the employees involved in the process. For example, the workload of day shift on Monday is 2, the performances of employees A, B and C are 1, 1 and 2. Employee A and B can work together to satisfy the demand of day shift, employee C can finish the job alone. Although different employees can be assembled to perform the job, how to choose right combination remains to be a challenging problem.

4   MOEAD-Based Shifting

A naive way to find the optimal schedule is to compute all the combinations about assigning staffs to shifts. Though the solution looks simple, if all the combinations are taken into consideration, the computation is so prohibitive that we can't afford to support it. Actually, the staff scheduling problem is a multi-objective optimization problem. We adopt MOEAD to convert our problem into several single-objective optimization problems based on the idea of decomposition. And the original evaluation metrics are transformed into a certain metric called mapping distance by the Tchebyshev function [24], each single objective optimization problem will be optimized by this metric, we will get a set of (near)optimal schedules as the solutions to our problem.

In this section, we first overview MOEAD, then modeling our problem based on MOEAD. The detail of modeling can be divided into three steps: (a) encoding the individual and population, (b) the operations of mutation and crossover, and (c) the selection of a set of (near)optimal schedules.

4.1   Overview of MOEAD

MOEAD adopts the mechanism of decomposition to convert the multi-objective optimization problem into multiple single objective optimization sub problems and works concurrently to solve these sub problems. Each sub problem is optimized with the help of the information gained from its neighborhood [25].

MOEAD uses the mechanisms inspired by biological evolution, such as individual, population, generation, mutation, crossover, fitness function and selection. An individual represents a solution to the problem, the population consists of a series of individuals and denotes a set of solutions. During the process of MOEAD, the initialized population is called the first generation. Then by the operation of mutation and crossover, the individuals in initialized population will be generated new ones. All the individuals will get scored by the fitness function, where some individuals with higher score will be selected to make up of a new population. This new population is considered as the second generation. MOEAD will repeat the above operations on the previous generation to generate the next until the number of generations reaches the pre-set value. The individuals in the final generation are considered as the (near)optimal solutions.

However, the fitness function is composed of different evaluation metrics, and the evaluated results of different solutions are hard to be compared with each other. For example, the *Shifting_Satisfaction* and *AVE_Coverage* of solution A are 0.7 and 0.5, and those of solution B are 0.5 and 0.7. weighing up with these two solutions, B has a higher AVE_Coverage, but A owns better Shifting_Satisfaction. It's a difficult job to compare the solutions from different evaluation metrics. MOEAD adopts idea of mechanism to use the Tchebyshev function [24] to solve this situation. The function is as follows:

The Tchebyshev function is shown as follows:

$$C(x \mid w) = \max_{i \in \{1,...,m\}} w_i \mid f_i(x) - P_i \mid \tag{4}$$

where x represents the set of solutions, the $f_i(x)$ is the computing functions of metrics and m is the number of metrics, w_i denotes the weight on the metric $f_i(x)$, the $C(x|w)$ is the mapping distance between schedules x and weight vector w, P_i represents the best evaluated result of the metric $f_i(x)$ among all the solutions.

The weight vector represents a sub problems, and is uniformly distributed in the space of solution, each weight vector corresponds to each individual The original three evaluation metrics are transformed into only one metric called mapping distance, as Fig. 1 shows. Suppose that there are two evaluation metrics F1 and F2, a schedule is positioned as Fig. 1 shows. The schedule is mapped to the vector w_1, and the mapping point on the w_1 is c, the distance between the origin and c is called mapping distance. If this schedule is comared with others, these schedules will be mapped to the weight vector w_1 corresponds to the schedule and be computed to the mapping distances, which are used to be compared with the schedule'. Thus, the mechanism of decomposition contributes to selecting the feasible schedules.

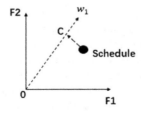

Fig. 1. The diagram of schedule mapping to the weight vector

Original evaluated results of different metrics are converted into a mapping distance on the weight vector. The solution gets better as the mapping distance is shorter. Then the solutions can be compared with each other by their mapping distances, and the solutions with shorter mapping distance will be selected to make up of the next generation.

4.2   Encoding the Individual and Population

In our problem, the individual denotes a schedule, and the population represents a set of schedules. Then we will introduce the process of encoding the individual and population as the following steps: (1) Encoding the shifts of schedules; (2) Splicing the shift sequence of each day; (3) Setting the constrained control of schedules.

Encoding the Shifts of Schedules. Schedule is fulfilled with shifts, dates and staffs as shown in Fig. 2(a). Note that the staffs can not be assigned to two different shifts on one day in the schedule of our problem as the hard constraint H_{osod} asks. In MOEAD, the individuals are asked for being transformed into a vector. Hence, both the individual and the population need to be encoded. We use the example of schedule as Fig. 2(a) to explain the operations of encoding.

Timetable T	Staff 1	Staff 2
Day 1	Night shift	Night shift
Day 2	Day shift	Rest-day

Timetable T	Staff 1	Staff 2
Day 1	2	2
Day 2	1	0

(a) the example of schedule   (b) the example of schedule where the shifts are encodeed

Fig. 2. The diagram of encoding the shifts

The shifts consist of *Rest-day*, *Day shift* and *Night shift*. We use the number 0 to represent the *Rest-day*, number 1 and 2 denote the *Day shift* and *Night shift*. Then the example of schedule is shown in Fig. 2(b) after the shifts are encoded.

Splicing the Shift Sequence of Each Day. The schedule needs to be transformed into a vector. A schedule can be treated as the combination of shift sequence on each staff, e.g., the shift sequence of staff 1 is 21, and staff 2 has the shift sequence 20. Then the shift sequence of each staff is spliced according to the staffs' number, the quantities of staffs and days in the schedule are unique. Hence, we can know that each number on the sequence represents this shift is performed by which staff on which day, and the information of the schedule will not be lost. As Fig. 3 shows, the schedule is transformed into a vector, where the shift sequence of each day is connected according to the staffs' number. By the mentioned above operations, the new schedule T' is generated. The number of days in the original schedule T is 2, every two numbers on the new schedule T' is a shift sequence and represents the shifts performed by a staff. For example, the first two numbers 2 and 1 denote the shift sequence of staff 1, and the shift sequence of staff 2 is the next two numbers 20. Note that the shift sequence of each staff is arranged according to the day, and the shift sequence 12 makes no sense. The shift sequence (21) of staff 1 represents that the staff 1 performs the night shift on Day 1 and the day shift on Day 2. All the individuals are encoded into a vector based on the above transformation.

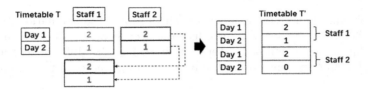

Fig. 3. The diagram of splicing the shift sequence of each day

The population is composed of individuals, each individual is a vector, and the population is a set of vectors.

Setting the Constrained Control of Schedules. For our problem, there are several constraints, which can be divided into two categories: hard constraints and soft constraints. The hard constraints must be followed. However, the original operations of mutation, crossover and selection in MOEAD can not provide the guarantee that each individual will follow the hard constraints. Hence, we need to set a optimization strategy to ensure that the hard constraints will be followed.

There are two hard constraints in our problem: (1)H_{osod}, each staff can only be assigned to one shift per day; (2)H_{mwd}, the maximal number of consecutive working days do not exceed k days. First, during the process of encoding the individuals, the hard constraint H_{osod} has bee set to be followed. All the schedules used in MOEAD will not appear that one staff is assigned to two different shifts on one day. As for the hard constraint H_{mwd}, we traverse each individual, where the examination is performed in a sliding window of k+1 consecutive numbers and its step is one number. More in detail, the rest-day is encoded to the number 0, we multiply these numbers and get a value. When the value equals to 0, some these numbers is 0, which means the the assignment of these k consecutive shifts obeys the hard constraint H_{mwd}. Otherwise, the last one of these k + 1 consecutive numbers will be replaced with the number 0. We repeat above operations until the lower bound of the window reaches the last number of this staff. The sliding window will take the first number as the upper bound to continue the examination.

For example, as shown in Fig. 4. Suppose that the maximal consecutive working days is 1, and the population consists of three staffs and each staff is assigned to shift on three consecutive days. The length of sliding window is 2, and the upper and lower bound is the first two numbers as Fig. 4(a) shows. The value of multiplying first two numbers is 4, which do not equal to 0. The night shift represented by the lower bound of sliding window break the hard constraint H_{mwd}.

Then this night shift will be reset to the rest-day and the window will slides down one number as Fig. 4(b). When the lower bound of sliding window reaches the last shift performed by staff 1, it will continue the examination from the first shift of next staff (staff 2) as Fig. 4(c) shows. All the shifts of schedule T1 are examined as Fig. 4(d) shows, the sliding window will start the examination from the first shift of next schedule until all the schedules are performed the examination.

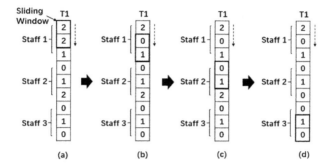

Fig. 4. The diagram of examining the individuals

Compared with abandoning the schedules which break the hard constraint H_{mwd}, substituting the shifts breaking H_{mwd} is more reasonable. Because the number of individuals in the population is fixed. If the schedules which do not obey the hard constraints H_{mwd} are abandoned, we need to supplement the same number of schedules to keep the number of individuals in the population constant. This way works inefficiently and costs a lot of extra computations, and our substitution strategy is more feasible.

4.3 The Operations of Mutation and Crossover

The operations of mutation and crossover are performed on individuals of the previous population to generate new schedules. The mutation can be considered that some of a schedule is replaced with the part of another one according to the mutation point, note that the mutation point is randomly generated. For example, as shown in Fig. 5, there are three encoded schedules T1, T2 and T3. We generate randomly the mutation point of T1, the corresponding shift in the position of mutation point is 0, and the shifts of T2 and T3 with the same position are 1 and 2, which means the shift 0 of the position of mutation point on T1 can be substituted by the shifts 1 and 2. Which shift will be used to substitute the shift 0 depends on the polynomial of MOEAD. The shift 0 is an input to the polynomial, the computed result is the shift 2, then the shift 0 with the position of mutation point on T1 can be substituted by the shift 2.

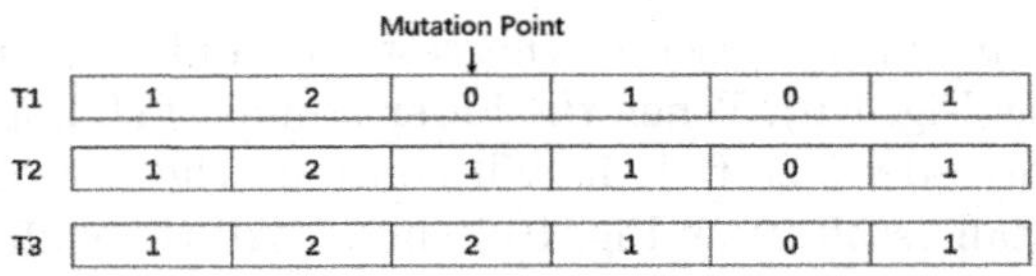

Mutation Point

| T1 | 1 | 2 | 0 | 1 | 0 | 1 |

| T2 | 1 | 2 | 1 | 1 | 0 | 1 |

| T3 | 1 | 2 | 2 | 1 | 0 | 1 |

Fig. 5. The example operation for mutation

As for the crossover, MOEAD adopts the simulated binary crossover strategy to achieve it. It can be treated as exchanging some shift sequences of two schedules to generate new schedules according to the crossover point, note that the crossover point is randomly generated.

Fig. 6. The example operation of crossover

For example, there are two schedules T1 and T2, we randomly generate the crossover point on these two schedules. The schedules will be divided int shift sequences according to the crossover point. Thus, we get four shift sequences 12, 12, 0, and 1. Then we splice these shift sequences which are not in the same schedule as Fig. 6 shows. We get two new schedules T1′ and T2′.

4.4 The Selection of a Set of (Near)optimal Schedules

The input to this section is the new schedules generated by the operations of mutation and crossover, and they will be compared with the schedules in the previous population. As the first subsection in this section shows, the Chebyshev function converts the evaluated metrics into a mapping distance between weighted vectors and three evaluated metrics, and this mapping distance is used to be a metric for schedules to compared with each others.

This selection can be divided into two steps: (1) Converting the evaluated metrics into the mapping distance; (2) Comparison between original schedules and new generated ones.

Converting the Evaluation Metrics into the Mapping Distance. Each schedule can be considered as a point in the solution space, and the mapping distance can be understood as an Euclidean distance between the point and origin point of the solution space. Note that this Euclidean distance needs to be converted into a distance on the corresponding weight vector by the Chebyshev function.

As Fig. 7 shows, suppose that there is a schedule which is represented by the point X in the coordinate system. The weight vector w responds to point

X. First, we connect the point X and the origin o. Then take one point A on w, vectors $\overrightarrow{OX}$ is perpendicular to $\overrightarrow{XA}$. The distance of OX can be computed based on Euclidean distance function, and the angle between the vector $\overrightarrow{OX}$ and weight vector w can be computed. Then we can get the length of OA, which is the mapping distance. The evaluated results of three metrics are converted into a distance OA.

Similarly, the schedule will be compared with its neighboring schedules, they will be mapped to the weight vector w and are generated to the mapping distances. These mapping distances will be compared with that of x. If none of them has shorter mapping distance than x, x will be keep. Otherwise, any of the schedules that have shorter mapping distance can be selected to substitute x.

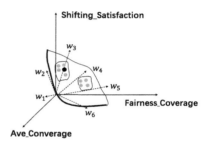

Fig. 7. The diagram of one schedule mapping to weight vector w

Fig. 8. The diagram of selecting the schedules

Comparison Between Original Schedules and New Generated Ones.
Combined with Fig. 8, we explain the detail of replacing. Suppose that a schedule (black point) corresponds to w_3, and the number of weight vector in the domain of w_3 is 4, the schedule will be selected randomly from four schedules (four gray point around black point) as a parent, and will be performed the operation of mutation and crossover with black point to generate a new schedule. Then we perform the operation of replacing. Three schedules (a selected gray point, black point and a new one) will be computed the mapping distance by Chebyshev function. Then the mapping distance of new schedule will be compared with parents'. If one of parents has longer mapping distance than the new one's, this parent will be replaced the new schedule. If both of them are longer than the new one's, choose any one of parents to be replaced. If the mapping distance of new schedule is longer than each parent's, the new schedule will be abandoned.

The operation of replacement will be performed in the order of weight vectors on each generation. When all the operations of replacement were finished, the next generation would have been generated.

5 Experiments

This section describes the experiments used to test the ideas outlined in the previous section. All experiments are performed based on a mixture of real and synthetic data sets. The real part comes from Hangzhou center of China Telecom, containing the call arrivals of each day from June 2019 to November 2019. The synthetic data set is the staff of each month during the above time period.

There are a series of measures to evaluate the quality of Pareto Fronts (PFs), and the hypervolume (HV) [26] is used into the following experiments. As for a three-dimension problem, HV is treated as the volume surrounded by each point on the PF and a point. HV is adopted to compare the performance of PFs. The larger the value of HV and the better the PF is.

In our work, in order to simulate the real scene, the schedule horizon of the experiment is set to one month. Besides, we assume that the performance of each staff is the same, and the number of staff is fixed. Each experiment will start from the same set of the initial population, and be executed 20 times. Our model is compared with four very effective algorithms for solving staff scheduling problems, including the nondominated sorting genetic algorithm II (NSGA-II [8]), the reference vector guided evolutionary algorithm (RVEA [9]) and the preference-inspired coevolutionary algorithms [10]. NSGA-II relies on its elite selection strategy, fast non-dominant sorting, and crowded distance calculation to find the (near)optimal schedules, these characteristics guarantee the generated schedules effectiveness and efficiency. RVEA introduces the idea of the reference vector, and achieve the search of (near)optimal schedules based on angle constrained local search. As for PICEAg, it finds the (near)optimal schedules based on co-evolutionary of candidate solutions and optimization functions. These three MOEAs achieve better performance than others in practice. All experiments are evaluated on a laptop with Intel(R) Core(TM) CPU i3-2310M 2.10 GHz processor and 6 GB RAM with Windows 10.

5.1 Parameter Setting

It's significant to choose proper parameters for evolutionary algorithms. In our work, we determine parameters, the rate of mutation $rate_m$, the rate of crossover $rate_c$ and the size of population though experiments on a instance. In order to find the optimal value for $rate_m$ and $rate_c$ for all the algorithms, a large number of experiments are constructed. The maximal number of generations is set to 1000, first, and $rate_m$, $rate_c$ are increased from 0.1 to 1 in the step of 0.1. For each combination of $rate_m$ and $rate_c$, the average HV (AVE.HV) which is the average value of 10 HVs over 20 independent runs. For each algorithm, we choose the most suitable $rate_m$ and $rate_c$ based on the value of HVs, and the detail value is as Table 1 shows.

With the above parameters, experiments are further conducted to test the effect of population size. The population size is set to 100, 200 and 400, and 20 independent runs are conducted. The AVE.HV are reported in Table 2. Compared with other algorithms, the values of AVE.HV for MOEAD is better than

Table 1. The rates of mutation and crossover used in each algorithm

Parameters	MOEAD	NSGA-II	RVEA	PICEAg
Rate_m	0.1	0.2	0.9	0.7
Rate_c	0.7	0.8	0.8	0.8

others in general, and its HV gets better as the population size increases. The number of solutions in a larger population is more, which strengthens the ability to search for feasible schedules.

Table 2. The AVE.HV generated by varying the size of population in each algorithm

Population size	MOEAD	NSGA-II	RVEA	PICEAg
100	**3.362**	3.32	3.29	3.346
200	**3.384**	3.341	3.326	3.351
400	**3.413**	3.351	3.348	3.361

5.2   Overall Evaluation on Each Data Set

The scheduler needs to choose a feasible schedule from a set of metric schedules. Here we suppose that according to the actual requirement of *Shifting_Satisfaction*, *Ave_Coverage* and *Fairness_Coverage*, the scheduling scheme which makes *Shifting_Satisfaction*, *Ave_Coverage* be maximum, and *Fairness_Coverage* be minimum will be chosen. 20 independent runs of each algorithm are conducted with the data set of each month, respectively. Table 3 shows the AVE_HV.

Table 3. The AVE.HV produced by adopting different algorithms on each data set

Dataset	MOEAD	NSGA-II	RVEA	PICEAg
June	**7.1831**	6.6484	6.7101	6.8251
July	**6.2350**	6.1819	5.5967	5.8449
August	6.6285	**7.3530**	7.2048	6.5801
September	**6.1173**	2.705	2.6505	2.4207
October	**6.3404**	5.953	5.7377	5.829
November	**5.6613**	3.1141	3.3386	3.389

With the AVE.HVs of data set of each month, MOEAD performs better than other algorithms in general. As for the performance of August, it's caused by the weight vectors. Whatever the shape of PF is, MOEAD will use a fixed

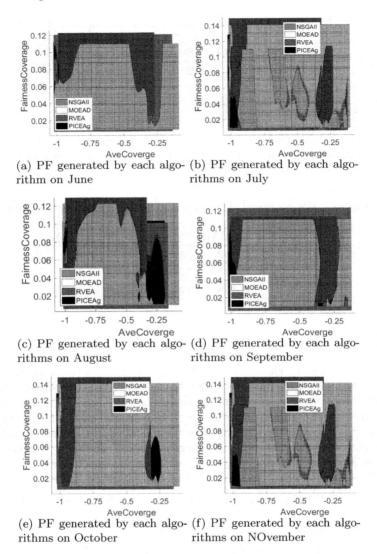

(a) PF generated by each algorithm on June

(b) PF generated by each algorithms on July

(c) PF generated by each algorithms on August

(d) PF generated by each algorithms on September

(e) PF generated by each algorithms on October

(f) PF generated by each algorithms on NOvember

Fig. 9. The PFs of each algorithm on the data sets of 6 months

set of weight vectors and all the sub-problems are assigned the same amount of computation in each generation, which causes that some of PF is harder to be converged and the reduction of AVE.HV.

Besides, we select the best result of these algorithms for comparison, and Fig. 9 lists the comparison of PFs of all the algorithms. For facilitating observing the differences among PFs, we transform the Shifting_Coverage into the coverage of colors and present the PFs in a two-dimension chart. In this two-dimension (Fairness_ Coverage and Ave_Coverage) aspect, a solution with lower Shifting_Coverage will be covered by others with a higher value. The colors shown in the figure represent that the corresponding solutions have higherShifting_Coverage. Then we can see the distribution of each PF directly and the proportion of its solutions on the approximate PF for our problem. Then as Fig. 9 shows, the color representing MOEAD has the largest area among all the algorithms, and the distribution of solutions towards argmax Ave_Converage and argmin Fairness_Coverage is superior to other algorithms in general. The decomposition mechanism of MOEAD transforms our problem into several single objective optimization problems, which explores the space of solutions. As for others, the non-dominant schedules of NSGA-II increases exponentially, which makes trouble in distinguishing the quality of schedules. RVEA is performed in the local search to find the (near)optimal solutions based on the reference vector, which reduces the space of solutions. And PICEAg is more suitable for dealing with the high-dimension problem, solving our problem does not make the co-evolutionary works well.

5.3 Evaluation of Varying Maximal Number of Consecutive Working Days

We add more choices for the scheduler about the maximal number of consecutive days based on the real-world scenes. Hence, we add the experiments about the maximal number of consecutive working days, which are set from 3 to 7. The parameters for each algorithm are set to a suitable situation. Each experiment is performed as the number of evolution is set to 5000, and runs 10 times. All the data in the table is the average of experimental results. According to the data in Table 4.

Table 4. The AVE.HV generated by varying the maximal number of consecutive working days of each algorithm

Working days	MOEAD	NSGA-II	RVEA	PICEAg
3	3.211	3.215	3.17	2.991
4	**2.865**	2.736	2.754	3.341
5	**2.8398**	2.1629	2.7732	2.8327
6	**3.3419**	3.3391	3.3389	3.0839
7	**3.3843**	3.1225	3.3120	3.3277

The days need to be assigned shift for the staff increases as the maximal number of consecutive working days decreases. For example, the maximal number of consecutive working days is set to 3, then for a month, the hard constraint H_mwd must be met, there are 10 days to rest and 20 days needs to be assigned to shifts at least. And the value is set to 7, there are 4 days to rest and 26 days needs to be assigned to shifts at least. When the number of working days increasing, the space of solutions will grow up. Similarly, the decomposition mechanism of MOEAD works well in this situation. As for other algorithms, their working mechanism results in the worse schedules than MOEAD's.

6   Conclusion

In this paper, we adopt the MOEAD to address the staff scheduling problem, where MOEAD can optimize multiple objectives simultaneously and provides a set of feasible schedules instead of only one optimal schedule. The numerical experiment and a case study on 6 data sets from the real-world scene shows that MOEAD can find PFs of high quality than other classical MOEAs for this problem. Besides, we perform the experiments of varying the maximal number of consecutive working days, which verifies the applicability of our approach in the real-world scene. What's more, a series of experiments prove the effectiveness of the decomposition mechanism of MOEAD.

Acknowledgment. This research was partially supported by: National Key R&D Program of China (2018YFB1402800), and the Fundamental Research Funds for the Provincial Universities of Zhejiang (RF-A2020007).

References

1. Strandmark, P., Yi, Q., Curtois, T.: First-order linear programming in a column generation based heuristic approach to the nurse rostering problem. Comput. Oper. Res. **120**, 104945 (2020)
2. Akbarzadeh, B., Moslehi, G., Reisi-Nafchi, M., Maenhout, B.: A diving heuristic for planning and scheduling surgical cases in the operating room department with nurse re-rostering. J. Schedul. **23**(2), 265–288 (2020). https://doi.org/10.1007/s10951-020-00639-6
3. Maenhout, B., Vanhoucke, M.: An exact algorithm for an integrated project staffing problem with a homogeneous workforce. Kluwer Academic Publishers (2016)
4. Augustine, L., Faer, M., Kavountzis, A., Patel, R.: A brief study of the nurse scheduling problem (NSP). University of Pittsburgh Medical Center (2009)
5. Solos, I.P., Tassopoulos, I.X., Beligiannis, G.N.: A generic two-phase stochastic variable neighborhood approach for effectively solving the nurse rostering problem (2013)
6. Zhou, Y., Liu, J., Zhang, Y., Gan, X.: A multi-objective evolutionary algorithm for multi-period dynamic emergency resource scheduling problems. Transp. Res. Part E Logs Transp. Rev. **99**, 77–95 (2017)
7. Zhang, Q., Li, H.: MOEA/D: a multiobjective evolutionary algorithm based on decomposition. IEEE Trans. Evol. Comput. **11**(6), 712–731 (2007)
8. Deb, K., Pratap, A., Agarwal, S., Meyarivan, T.: A fast and elitist multiobjective genetic algorithm: NSGA-II. IEEE Trans. Evol. Comput. **6**(2), 182–197 (2002)
9. Cheng, R., Jin, Y., Olhofer, M., Sendhoff, B.: A reference vector guided evolutionary algorithm for many-objective optimization. IEEE Trans. Evol. Comput. **20**(5), 773–791 (2016)
10. Wang, R., Purshouse, R.C., Fleming, P.J.: Preference-inspired coevolutionary algorithms for many-objective optimization. IEEE Trans. Evol. Comput. **17**(4), 474–494 (2013)

11. Leksakul, K., Phetsawat, S.: Nurse scheduling using genetic algorithm,". Mathematical Problems in Engineering **2014**(pt.21), 1–16 (2014)
12. Aickelin, U., Dowsland, K.A.: An indirect genetic algorithm for a nurse-scheduling problem. Comput. Oper. Res. **31**(5), 761–778 (2004)
13. Lü, Z., Hao, J.-K., Glover, F.: Neighborhood analysis: a case study on curriculum-based course timetabling. J. Heurist. **17**(2), 97–118 (2011)
14. Todorovic, N., Petrovic, S.: Bee colony optimization algorithm for nurse rostering. IEEE Trans. Syst. Man Cybern.: Syst. **43**(2), 467–473 (2012)
15. Howell, J.P.: Cyclical scheduling of nursing personnel. Hospitals **40**(2), 77 (1966)
16. Pour, S.M., Drake, J.H., Ejlertsen, L.S., Rasmussen, K.M., Burke, E.K.: A hybrid constraint programming/mixed integer programming framework for the preventive signaling maintenance crew scheduling problem. Eur. J. Oper. Res. **269**(1), 341–352 (2018)
17. Hamid, M., Tavakkoli-Moghaddam, R., Golpaygani, F., Vahedi-Nouri, B.: A multi-objective model for a nurse scheduling problem by emphasizing human factors. Proc. Inst. Mech. Eng. Part H: J. Eng. Med. **234**(2), 179–199 (2020)
18. Zhang, Z., Qin, H., Yao, L., Liu, Y., Jiang, Z., Feng, Z., Ouyang, S.: Improved multi-objective moth-flame optimization algorithm based on r-domination for cascade reservoirs operation. J. Hydrol. **581**, 124431 (2020)
19. Yu, X., Yao, X., Wang, Y., Zhu, L., Filev, D.: Domination-based ordinal regression for expensive multi-objective optimization. In: 2019 IEEE Symposium Series on Computational Intelligence (SSCI), pp. 2058–2065. IEEE (2019)
20. Mashwani, W.K., Salhi, A., Yeniay, O., Jan, M.A., Khanum, R.A.: Hybrid adaptive evolutionary algorithm based on decomposition. Appl. Soft Comput. **57**, 363–378 (2017)
21. Tian, Y., Cheng, R., Zhang, X., Cheng, F., Jin, Y.: An indicator-based multiobjective evolutionary algorithm with reference point adaptation for better versatility. IEEE Trans. Evol. Comput. **22**(4), 609–622 (2017)
22. Jiang, S., Zhang, J., Ong, Y.-S., Zhang, A.N., Tan, P.S.: A simple and fast hypervolume indicator-based multiobjective evolutionary algorithm. IEEE Trans. Cybern. **45**(10), 2202–2213 (2014)
23. Luo, J., Yang, Y., Li, X., Liu, Q., Chen, M., Gao, K.: A decomposition-based multi-objective evolutionary algorithm with quality indicator. Swarm Evol. Comput. **39**, 339–355 (2018)
24. Apostol, T.M.: Introduction to Analytic Number Theory. Springer, Heidelberg (2013). https://doi.org/10.1007/978-1-4757-5579-4
25. Omran, S.M., El-Behaidy, W.H., Youssif, A.A.A.: Decomposition based multi-objectives evolutionary algorithms challenges and circumvention. In: Arai, K., Kapoor, S., Bhatia, R. (eds.) SAI 2020. AISC, vol. 1229, pp. 82–93. Springer, Cham (2020). https://doi.org/10.1007/978-3-030-52246-9_6
26. Zitzler, E., Thiele, L., Laumanns, M., Fonseca, C.M., Fonseca, V.G.D.: Performance assessment of multiobjective optimizers: an analysis and review. IEEE Trans. Evol. Comput. **7**(2), 117–132 (2003)

A Deep Reinforcement Learning Based Resource Autonomic Provisioning Approach for Cloud Services

Qing Zong[1], Xiangwei Zheng[1(✉)], Yi Wei[1(✉)], and Hongfeng Sun[2]

[1] School of Information Science and Engineering, Shandong Normal University, Jinan, China
zongqing_sdnu@163.com, xwzhengcn@163.com, weiyi@sdnu.edu.cn
[2] School of Data and Computer Science, Shandong Women's University, Jinan, China
nameshf@163.com

Abstract. Resource elastic scheduling is a key feature of cloud services. The elastic makes cloud services have the ability to flexibly increase or decrease resources to satisfy user needs, and dynamically allocate resources for cloud services on demand. The amount of resources to be configured is determined at runtime based on the changes in workload to flexibly respond to the fluctuating demands of cloud services. Appropriate resources need to be configured in advance. In this article, we propose a dynamic resource provisioning framework based on the MAPE loop, and use a two-tier elastic resource configuration for collaborative work. In order to implement the proposed framework, we propose an elastic resource scheduling algorithm based on a combination of the autonomic computing and deep reinforcement learning (DRL) to reduce task rejection rate of the virtual machine (VM) and increase utilization to obtain as much profit as possible. In this paper, Experimental results using actual Google cluster tracking results show that the proposed policy reduces the total cost about 17%–58% and increases the profit by up to not less than 9%, reduces the service level agreement (SLA) violations to less than 0.4% to better guarantee the quality of service (QoS).

Keywords: Cloud computing · DRL · Resource scheduling · Autonomic computing

1 Introduction

Cloud computing has emerged as the most popular commerce computing model in today's computer industry. Cloud computing can provide users with computing resources such as computing power, storage, and bandwidth as needed. Cloud applications are software products that run on cloud environments. Generally, each cloud application consists of one or more cloud services. According to different service patterns that users need, cloud computing can be divided into three types: Infrastructure as a Service (IaaS), Platform as a Service (PaaS) and Software as a Service (SaaS). In this paper, we consider a common three-tier cloud environment which involves three roles: IaaS vendors, SaaS

H. Gao et al. (Eds.): CollaborateCom 2020, LNICST 350, pp. 132–153, 2021.
https://doi.org/10.1007/978-3-030-67540-0_8

providers, and users. IaaS vendors provide their customers access to various computing resources, such as storage, servers and networking. SaaS providers rent several VM instances from IaaS vendors to construct their cloud services, and users purchase cloud services from SaaS providers.

The elastic characteristics of cloud computing enable SaaS providers to adapt to the changes in the workload of their cloud services by dynamically configuring and reallocating resources. Ideally, the available resources are as close as possible to current needs at each time point. However, user requests are potentially uncertain. It is not easy to determine the appropriate amount of resources for cloud services during execution. On the one hand, if the number of resources provided by the SaaS provider is greater than the demand of user requests, it will cause the waste of resources and unnecessary monetary cost. On the other hand, if the number of resources provided by the SaaS provider is less than the amount of user requests, the under-provisioned situation may result in lower revenue and missing potential customers.

In order to deal with the above resource supply problem of cloud applications, dynamic resource supply is used. Dynamic resource supply is an effective method, and its basic idea is to supply resources based on changes in the workload of cloud applications. The goal is to automate the dynamic configuration of resources by minimizing the cost of renting resources from IaaS providers and meeting users' SLA. For SaaS providers, they concern how to maximize their profits and guarantee customer satisfaction during the execution of their cloud applications.

In this paper, we combine the concepts of autonomous computing and deep reinforcement learning (DRL), and propose a cloud elastic resource scheduling algorithm. IBM has proposed a MAPE (Monitoring, Analysis, Planning, and Execution) control loop [1], which contains four stages: monitoring (M), analysis (A), planning (P), and execution (E). We use a two-layer MAPE control loop to better allocate resources for SaaS providers. The first MAPE control loop is in charge of renting appropriate virtual machines from the IaaS provider. The second MAPE control loop is responsible for resource sharing and coordination among cloud services. DRL is an important field of machine learning. It can simultaneously give play to the representational advantages of deep learning and the decision-making advantages of reinforcement learning. In the first MAPE control loop, we use DRL as the decision-making agents that interact repeatedly with the cloud environment and make resource adjustment decisions automatically. The main contributions of this paper can be summarized as follows:

- We design a dynamic resource supplying system, which consists of two sets of the MAPE loop to adjust the supplied resources for cloud services or balance resources among cloud services.
- We use DRL as the decision-making agents to improve the performance of the resource dynamic provisioning. At the same time, multiple cloud services learn collaboratively, which effectively accelerates the learning speed.
- For the elastic scaling of cloud resources, we adopt a vertical and horizontal hybrid method to adjust cloud resources elastically, which has a higher utilization rate and a lower SLA violation rate.
- We use real workload tracking to conduct experiments under different metrics to evaluate the performance of our approach.

The remainder of this paper is organized as follows. Section 2 reviews the related work. Section 3 formulates the problem and proposes a formal description. The necessary theoretical background is presented in Sect. 4. In Sect. 5, we propose a dynamic resource supplying framework and the corresponding elastic resource scheduling algorithm. Section 6 presents an evaluation and discusses the experimental results. Section 7 concludes our work and provides future research directions.

2   Related Works

At present, many scholars have done a lot of research on the resource provisioning in the cloud environment [1–3]. Traditional resource provisioning techniques employ predefined policies to guide application scaling, and manually specify scaling rules after an application is deployed. These rules specify a pair of thresholds and change the number of VMs after triggering the expansion condition [4, 5]. However, the threshold approach adds or reduces fixed numbers of VM instances at a certain time, which may not provide suitable resources in time when workload changes. In order to scale the amount of resource properly, it is necessary to predict workloads of cloud services for better performance. Prediction algorithm can allocate resources in advance, but the fluctuated amount of user requests will bring uncertain error [6–8].

With the development of the theory of machine learning, machine learning methods become more and more attractive in the field of cloud computing. Mazidi et al. [9] propose a new method based on the K-nearest neighbor algorithm that is used to analyze and label virtual machines, and statistical methods are used to make expansion decisions. Wei et al. [10] propose a cloud application auto scaling approach based on Q-learning method to help SaaS providers make optimal resource allocation decisions in a dynamic and stochastic cloud environment. Ghobaei et al. [11] propose a cloud service hybrid resource supply method based on the concepts of autonomous computing and reinforcement learning (RL), and an autonomous resource supply framework inspired by the cloud layer model. Li et al. [12] present the resource controllers based on statistical machine learning which execute on different VMs in cloud environments to achieve application service level objects (SLOs) under fluctuating time-varying workloads and unpredictable variations of system situations. Gill et al. [13] propose a based particle swarm optimization resource scheduling approach which is used to execute workloads effectively on available resources. Salah et al. [14] present an analytical model based on Markov chains to predict the number of or VM instances needed to satisfy a given SLO performance requirement. In our paper, we use a shared resource pool to reduce prediction errors, and DRL-based agents that interact repeatedly with the cloud environment to make better decisions.

3   Problem Statement

We consider a common three-tier cloud environment which involves three roles: IaaS vendors, SaaS providers, and users. The user submits tasks to the SaaS provider for execution. The SaaS provider rents a certain number of VMs from the IaaS provider and deploys cloud services to complete the requests submitted by users. IaaS providers provide almost unlimited resources in the form of virtual machines. The number of user requests is continuous and fluctuating. The main goal of SaaS providers is to maximize profits. Therefore, SaaS providers need to rent as few resources as possible and respond flexibly to workload fluctuations in order to minimize the cost of VM rental and the penalty caused by SLA violations. However, deciding the suitable amount of VMs for cloud services during execution is quite difficult.

Notions and variables used throughout the paper are defined in Table 1. In the cloud environment, the IaaS provider provides M types of VMs. The price of each VM type m is $VPrice_m$. The SaaS provider provides I types of cloud services. Each cloud service is composed of different types of VMs and the service fee is $CPrice_i$. Cloud services serve user requests continuously. Each user request Req_u^r contains the arrival time AT_u^r, the running time $RTime_u^r$ and the deadline time DL_u^r.

An SLA violation occurs when the SaaS provider fails to guarantee the predefined user SLA. FT_u^r and DL_u^r are the finishing time and deadline time for the user request Req_u^r. If the finishing time is greater than the deadline, an SLA violation will occur. The SLA is defined as:

$$SLA(Req_u^r) = \begin{cases} Yes(1) \ FT_u^r - DL_u^r > 0 \\ No(0) \ otherwise \end{cases} \tag{1}$$

The Total cost is the cost consumed by the SaaS provider to satisfy all cloud services. It includes the VM Costs and the Compensation, and it can be described as:

$$Total \ Cost = VM \ Costs + Compensation \tag{2}$$

VM cost is the total cost of all VMs rented from the IaaS provider:

$$VM \ Costs = \sum_{n=1}^{N} VM \ cost_n \tag{3}$$

For $VM \ cost_n$, it depends on the VM price $VPrice_m$ and the initiation VM price $init \ price_m$ of type m, and the duration of time for which the VM is on $VTime_i$, which is calculated as:

$$VM \ Cost_n = (VPrice_m \times VTime_n) + init \ price_m \quad \forall n \in N; m \in M \tag{4}$$

$Compensation$ is the total penalty cost for all user requests, and it is expressed as:

$$Compensation = \sum_{u=1}^{U} \sum_{r=1}^{R_u} Penalty(Req_u^r) \tag{5}$$

For each user request Req_u^r, the $Penalty$ is similar to related work [15] and is modeled as a Linear function can be described as:

$$Penalty(Req_u^r) = \lambda_u^r \times \frac{FT_u^r - DL_u^r}{\Delta t} \tag{6}$$

where λ_u^r is the penalty rate for the failed request which depends on the request type, and Δt is a fixed time interval.

The purpose of this paper is to enable SaaS providers to minimize the payment for using resources from IaaS provider, and to guarantee the QoS requirements of users.

Table 1. Notations and definitions.

Notation	Definition
U	Number of users
Use_u^r	The uth user
C	Total number of requests for all users
Ru	The number of requests belongs to u^{th} user
Req_u^r	r^{th} request of the uth user
$RTime_u^r$	The running time of the rth request of the uth user
AT_u^r	The time at which the rth request belongs to u^{th} user arrival system
ST_u^r	The start time of the user request Req_u^r
FT_u^r	The finish time of the user request Req_u^r
DL_u^r	The deadline time of the user request Req_u^r
$Cloud_i$	i^{th} cloud service offered by SaaS provider
$CPrice_i$	i^{th} cloud service of cloud service charges
$NumVM_i(\Delta t)$	Number of VMs allocated to the $Cloud_i$ at the Δt^{th} interval
$Num_i(\Delta t)$	Number of CPU resources allocated to the VM of the $Cloud_i$ at the Δt^{th} interval
$Remain_i(\Delta t)$	Number of remaining available CPU resources for the VM of the $Cloud_i$ at the Δt^{th} interval
$NumReq_i(\Delta t)$	Number of requests for the $Cloud_i$ at the Δt^{th} interval
$AvgReq_i$	Average (Ave) number of requests for the $Cloud_i$
$NumRun_i(\Delta t)$	The number of requests being executed by the $Cloud_i$ at the Δt^{th} interval
$NumWait_i(\Delta t)$	The number of requests blocked by the $Cloud_i$ at the Δt^{th} interval
$Utili_i(\Delta t)$	The CPU Utilization of the $Cloud_i$ at the Δt^{th} interval
N	Number of initiated VMs
M	Number of VM types
I	The total number of cloud services offered by the SaaS provider
$VPrice_m$	Price of VM type m
$init\ price_m$	The initiation VM price of type m
$VTime_i$	The duration of time for the nth VM is on
$RTime_u^r$	The running time of the rth request of the uth user

4   Theoretical Background

This section exposes a brief overview of the autonomic computing and deep reinforcement learning techniques.

4.1 Autonomic Computing

To achieve autonomic computing, IBM has suggested a reference model for auto-nomic control loops, which is called the MAPE (Monitor, Analysis, Plan, Execute, Knowledge) loop and is depicted in Fig. 1. The intelligent agent perceives its environment through sensors and uses the information to control effectors carry out changes to the managed element [16].

Fig. 1. The MAPE control loop

In the cloud environment, the MAPE loop can be used to manage cloud resources dynamically and automatically. The managed element represents cloud resources. Monitor collects CPU, RAM and other related information from managed element through sensors. The analysis phase analyzes the data received in the monitor phase in accordance with the prescribed knowledge. The plan stage processes the analyzed data and makes a certain action plan. In the execute phase, the action requirements are reflected to the effectors to execute on the managed element. Knowledge is used to control the shared data in the MAPE-K cycle, constrain each stage and adjust the configuration according to the indicators reported by the sensors in the connected environment.

4.2 Deep Reinforcement Learning

Deep Reinforcement Learning (DRL) [17] is an important field of machine learning. It has the representational advantages of deep learning and the decision-making advantages of reinforcement learning, providing the possibility to solve more complex large-scale decision-making control tasks. The general framework of deep reinforcement learning is shown in Fig. 2. The agent uses a deep neural network to represent the value function, strategy function or model, and the output of deep learning is the agent's action a. Next, the agent obtains feedback reward r by performing action a in the environment, and takes r as the parameter of the loss function. Subsequently, the loss function is derived based on the stochastic gradient algorithm. Finally, the weight parameters in the deep neural network are optimized after the training of the network model.

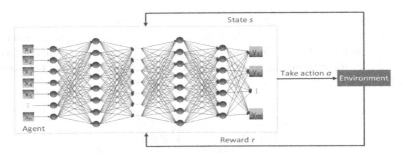

Fig. 2. Reinforcement learning with DNN

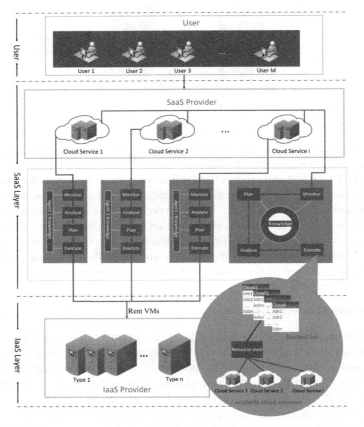

Fig. 3. Resource provisioning framework based on the control MAPE loop.

5   Framework and Algorithm

In this section, we propose a dynamic resource provisioning framework based on the MAPE loop, and describe in detail the elastic resource scheduling algorithm used to control the MAPE loop.

5.1   The Dynamic Resource Provisioning Framework

In order to better rent and manage VM resources, we propose a dynamic resource provisioning framework for the SaaS provider, which is shown in Fig. 3. With the support of the MAPE loop, each cloud service uses predictive technology and DRL technology to rent the appropriate number of VMs from IaaS provider. All this happens in parallel and cloud services do not interfere with each other. When all cloud services complete the MAPE loop, another MAPE cycle will be started to adjust the resource waste caused by prediction errors. The sensors collect resource waste of cloud service and user request blocking. The MAPE loop implements appropriate strategies to share resources between cloud services. The effector creates a resource pool based on the decision of the MAPE loop, and transfers blocking tasks to resources for execution. After deploying the SaaS provider with various cloud services in the dynamic resource provisioning framework, the MAPE loop runs until there are no cloud services running in the SaaS provider.

5.2   Elastic Resource Scheduling Algorithm

In this section, we will explain our elastic resource scheduling algorithm for cloud services in more detail. The algorithm consists of two groups of control MAPE loops. The first group of MAPE includes Monitoring Phase A, Analysis Phase A, Planning Phase A, Execution Phase A. It adjusts the number of VMs in advance by predictive technology and DRL algorithm to achieve horizontal scaling. Because new machines generally take between 5 and 10 min to be operational, we set it loop every five minutes. The second group of MAPE includes Monitoring Phase B, Analysis Phase B, Planning Phase B, Execution Phase B. It loops every one minute, and achieve vertical scaling by resource sharing, collaborative work and self-adapting management. The proposed algorithm (see Algorithm 1) will run until the SaaS provider does not receive a new user request.

Algorithm 1: Pseudo code for Autonomic Resource Provisioning

1: **begin**
2: **for** $k := 0$ to K **do** // k is a time of counter and the number of one-minute intervals is K,
3:    **if** $k \equiv 0$ mod 5 **then**
4:        **for** (every cloud service $Cloud_i$ in SaaS Provider at the interval Δt) **do**
5:            Monitoring Phase A ($Cloud_i$)
6:            Analysis Phase A ($Cloud_i$)
7:            Planning Phase A ($Cloud_i$)
8:            Execution Phase A ($Cloud_i$)
9:        **end for**
10:    **end if**
11:    Monitoring Phase B
12:    Analysis Phase B
13:    Planning Phase B
14:    Execution Phase B
15: **end for**

---

(1) Monitoring Phase

In monitoring phase A, for each cloud service $Cloud_i$ provided by the SaaS provider, the monitor will collect the number of VMs $NumVM_i(\Delta t)$ that the SaaS provider rents from the IaaS provider, and the user's request for the cloud service Cloudi, the number of requests $NumReq_i(\Delta t)$ and the number of requests that the cloud service Cloudi has not started $NumWait_i(\Delta t)$. At the same time, the monitor will also detect the CPU utilization of the VM of the cloud service Cloudi $Utili_i(\Delta t)$. In the monitoring phase B, for each cloud service Cloudi provided by the SaaS provider, the monitor collects the remaining resources of the VM leased by each cloud service $NumSur_i(\Delta t)$ and the number of requests for the cloud service Cloudi that have not started $NumWait_i(\Delta t)$. All the information obtained in the monitoring phase is stored in the database and will be used in other phases.

(2) Analysis Phase

The analysis phase A obtains the current number of user requests $NumReq_i(\Delta t)$ for the cloud service $Cloud_i$ from the monitoring phase A, in order to predict the number of requests $NumReq_i(\Delta(t+1))$ for cloud service $Cloud_i$ in the next interval $\Delta(t+1)$. In this paper, we use the Autoregressive Integrated Moving Average Model (ARIMA) [18] to predict future demands for cloud service $Cloud_i$, as shown in Algorithm 2.

ARIMA (p, d, q) model first uses the difference method for the historical data X_t of the non-stationary user request, and performs the smoothing preprocessing on the sequence to obtain a new stationary sequence $\{Z_1, Z_2, ..., Z_{t-d}\}$. The difference method is expressed as:

$$\Delta y_{X_t} = y_{X_{t+1}} - y_{X_t} \quad t \in Z \tag{7}$$

We then fit the $ARMA(p, q)$ model, and restore the original d-time difference to get the predicted data Y_t of X_t. Among them, the general expression of $ARMA(p, q)$ is:

$$X_t = \varphi_1 X_{t-1} + \cdots + \varphi_1 X_{t-p} + \varepsilon_t - \theta_1 \varepsilon_{t-1} - \cdots - \theta_q \varepsilon_{t-q} \quad t \in Z \qquad (8)$$

In the formula, the first half is the autoregressive part, the non-negative integer p is the autoregressive order, $\varphi_1, \ldots, \varphi_p$ is the autoregressive coefficient. The second half is the moving average part, and the non-negative integer q is the moving average order, $\theta_1, \ldots, \theta_q$ is the moving average coefficient. X_t is the related sequence of user request data, ε_t is the sequence of independent and identically distributed random variables, and satisfies $V\,ar\varepsilon_t = \sigma_\varepsilon^2 > 0$.

Algorithm 2: Pseudo code for Analysis Phase A ($Cloud_i$)

1: **Input:** The number of requests for cloud service $Cloud_i$

2: **Output**: Prediction value $(Cloud_i)//Y_t$

3: **begin**

4: $\{X_1, X_2, \ldots, X_t\} \leftarrow$ History $(Cloud_i)$ // History of the number of request for service $Cloud_i$

5: $\{Z_1, Z_2, \ldots, Z_{t\text{-}d}\} \leftarrow$ make the tranquilization of $\{X_1, X_2, \ldots, X_t\}$ using Eq.(7)

6: prediction value$(Cloud_i) \leftarrow$ Calculate X_t based on $\{Z_1, Z_2, \ldots, Z_{t\text{-}d}\}$ using Eq.(8)

7: **return** prediction value $(Cloud_i)$

8: **end**

Analysis phase B selects cloud services according to certain criteria for resource transfer in subsequent stages.

(3) Planning Phase

In planning phase A, we use a DRL-based method to make decisions to achieve horizontal scaling of cloud services and divide the entire process into three states: normal-resources, lack-resources, and waste-resources. According to algorithm 3, firstly, we obtain the predicted value of the number of tasks requested in the next interval from the analysis stage and calculate the cloud service $Cloud_i$ resource at time t to meet the CPU load requested by all tasks at next time (line 2), and then determine the current state based on it (line 3–5). We randomly choose an action with the probability of ε, otherwise choose the action with the largest Q value in the network (line 8–9). Then we perform the action a_t, boot users to request access to cloud services and observe reward r_t (line10–12). We store pair (S_k, A_k, R_k, S_{k+1}) as knowledge D (line 13), and take some pairs out of the knowledge D to optimize the parameters of the network θ.

Algorithm 3: Pseudo code for Planning Phase A($Cloud_i$)

Input: Prediction value (Y_t),

        The number of requests blocked ($NumWait_i(\Delta\ t)$)

        The number of requests being executed ($NumRun_i(\Delta\ t)$)

        The number of CPU resources allocated to the VM ($Num_i(\Delta\ t)$)

        The time of counter k

Output: selected action ($Cloud_i$) = a_k

Initialization: Network parameters; $U_{upper-threshold}$=1.0; $U_{lower-threshold}$=0.25

1: **begin**

2: Calculate the utilization ($Utili_i(\Delta)$) of the latest VM using Eq.9

3: **if** $Utili_i(\Delta\ t) > U_{upper-threshold}$ **then** predict state = lack-resources

4: **else if** $Utili_i(\Delta\ t) < U_{upper-threshold}$ **then** predict state = waste-resources

5: **else** predict state = normal- resources

6: **end if**

7: s_k = predict state

8: with probability ε select a random action a_k

9: otherwise select $a_k = argmax_a Q(\phi(s_t),a,\theta)$

10: executes action a_t

11: boot users to request access to cloud services

12: observe reward r_t ,and observe state transition at next decision time with a new state s_{k+1}

13: store transition (s_k, a_k, r_k, s_{k+1}) in D

14: sample random minibatch of transitions (s_k, a_k, r_k, s_{k+1}) from D

15: set $target_k = \begin{cases} r_k & \text{if episode terminates at step } k+1 \\ r_k + \gamma \max_a \widehat{Q}\left(\phi_{k+1}, a', \theta'\right) & \text{otherwise} \end{cases}$

16: perform a gradient descent step on $(target_k - Q(s_k, a_k; \theta))^2$ with respect to the network parameters θ

17: **end**

In planning phase B (see Algorithm 4), we first build a resource pool that consists of all unused resources of cloud services (lines 2–5), and then blocked requests of all cloud services are transferred to the resource pool to execution (lines 6–8). We calculate the time from the current moment to the next planning stage A and the average number of task requests per minute to estimate the resources to be reserved for the next planning stage A, release a certain number of VMs (line 9).

Algorithm 4. Pseudo code for Planning Phase B

Input: The number of requests blocked $NumWait_i(\Delta t)$

 Number of remaining available CPU resources $Remain_i(\Delta t)$

 Average number of requests for cloud service $Cloud_i$ $AvgReq_i$

 The time of counter k

Output: Task migration.

1: **begin**

2: set a resource pool

3: **for** (every cloud service $Cloud_i$ in SaaS provider at the interval Δt) **do**

4:   $Remain_i(\Delta t)$ join the resource pool

5: **end for**

6: **for** (every cloud service $Cloud_i$ in SaaS provider at the interval Δt) **do**

7:   Transfer The number of requests blocked $NumWait_i(\Delta t)$ to the resource pool

8: **end for**

9: **if** $k \equiv p$ mod 5 **then** reserve $(4 - p) \times AvgReq_i$ free resources to free the remaining empty VMs.

10: **end if**

11: **end**

---

(4) Execution Phase

In the execution phase A, the VM manager component executes the actions determined in the planning phase A. The VM manager creates a new VM for the cloud service $Cloud_i$ based on Table 2 or releases the lowest CPU utilization for scale-out action or scale-in action.

 In the execution phase B, based on the determined behavior in the planning phase B, the resource manager performs task migration, and releases the VM according to the judgment criteria of the planning phase.

6 Evaluation

In this section, we present an experimental evaluation of the approach proposed in the previous section. First, we describe the simulation settings and performance indicators, and then introduce and discuss the experimental results.

6.1 Experimental Setup

We simulate 5 sets of cloud services to work together, and take 5 data segments with the same length from the load tracking of the Google data center [19] to evaluate our algorithm. The use of real load tracking makes the results more real and reliable, and can better reflect the randomness and volatility of user requests. The time interval is 5 min, the data segment lasts for more than 16 h, including 200 intervals. The selected workload data segments are shown in Fig. 4, corresponding to cloud service 0 cloud service 1 cloud service 2 cloud service 3 cloud service 4 respectively. In addition, we

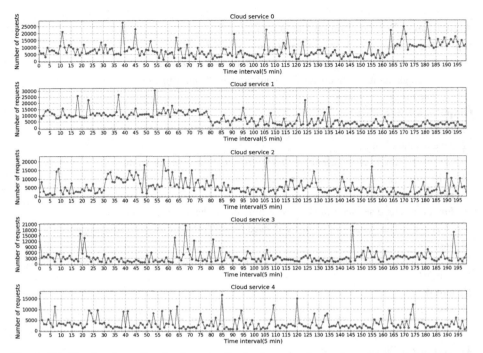

Fig. 4. Workload patterns for cloud service 0, cloud service 1, cloud service 2, cloud service 3 and cloud service 4.

Table 2. Different settings of VMs

VM type	Extra large	Large	Medium	Small
Core	8	4	2	1
CPU (MIPS)	20000	10000	5000	2500
RAM (GB)	150	75	37.5	17
Disk (GB)	16000	8500	4100	1600
VM price ($/h)	28	14	7	3.75
VM initiation price ($)	0.27	0.13	0.06	0.025

assume that the general scheduling strategy is first come first service strategy. There are four types of VMs offered by an IaaS provider: small, medium, large, and extra-large and they have different capacities and costs, as shown in Table 2.

For comparison, we conducted a group of experiments with our policy (Proposed). The contrast polices are as follow:

1. Constant resource policy (Constant): this policy is allocated with a static, constant number of cloud service resources.
2. Basic threshold-based policy (Threshold) [5]: this policy sets upper and lower bounds on the performance trigger adaptations to determines the horizontal scaling of cloud service resources.
3. RL controller-based policy (RL controller) [12]: this policy uses the workload prediction method and RL algorithm.

We applied the following metrics for a comparison of our policy with other strategies:

Utilization: The utilization of the cloud service $Cloud_i$ at the Δt^{th} interval is defined as the ratio of the resources occupied by user requests to the total CPU resources offered by VMs, at the Δt^{th} interval, and is defined as:

$$Utili_i(\Delta t) = \frac{Allocated(\Delta t)}{Total(\Delta t)} \tag{9}$$

Wasted Resources: The wasted resources of the cloud service $Cloud_i$ at the Δt^{th} interval is defined as the difference between the resources occupied by user requests and the total resources offered by VMs, at the Δt^{th} interval, and it can be described as:

$$Remain_i(\Delta t) = Allocated(\Delta t) - Total(\Delta t) \tag{10}$$

SLA Violation: The percentage SLA violation of the cloud service $Cloud_i$ at the Δt^{th} interval for user request Req_u^r of cloud service $Cloud_i$ at the Δt^{th} interval, which is calculated as:

$$SLAV_i(\Delta t) = \frac{1}{C} \sum_{c=1}^{C} SLA\left(Req_u^r\right) \tag{11}$$

Load Level: The load level of the cloud service $Cloud_i$ at the Δt^{th} interval is defined as ratio of the amount of resources required by the cloud service $Cloud_i$ to meet all requests to the amount of resources owned by the cloud service $Cloud_i$ at the moment, at the Δt^{th} interval, which is calculated as:

$$Load_i(\Delta t) = \frac{Need(\Delta t)}{Total(\Delta t)} \tag{12}$$

Total Cost: This metric is defined as the total cost of all cloud services $Cloud_i$ incurred by the SaaS provider at the Δt^{th} interval, and is expressed as:

$$Total\ Cost_i(\Delta t) = VM\ Cost_i(\Delta t) + Penalty\ Cost_i(\Delta t) \tag{13}$$

Profit: This metric is defined as the profit gained by the SaaS provider to serve all requests of cloud services, and is expressed as:

$$Profit_i(\Delta t) = Income_i(\Delta t) - Total\ Cost_i(\Delta t) \tag{14}$$

$$Income_i(\Delta t) = \sum_{u=1}^{U} \sum_{r=1}^{R} CPrice_i \times RTime_u^r \tag{15}$$

$CPrice_i$ is the service fee for the user request Req_u^r on the cloud service $Cloud_i$ and $RTime_u^k$ is the running time r^{th} request of the uth user.

6.2 Experimental Results

Figure 5 shows the CPU utilizations of the four approaches for 5 workloads at each interval. The average CPU utilizations under 5 workloads for constant resource-based policy, basic threshold-based policy, RL controller-based policy, and the proposed policy are 54.56%, 85.17%, 78.14%, and 88.23%, respectively, as shown in Table 3. From the results, we observe that basic threshold-based policy, RL controller-based policy and

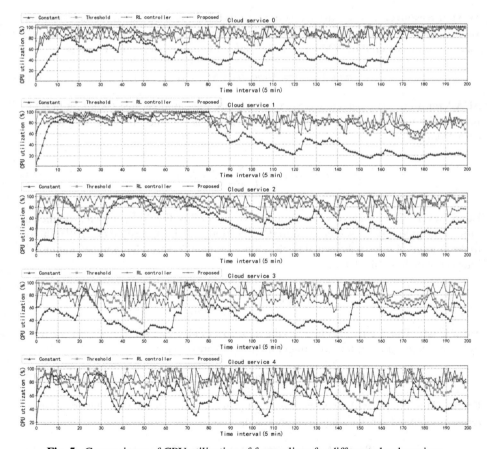

Fig. 5. Comparisons of CPU utilization of four polices for different cloud services.

Table 3. The average CPU utilization of four polices for different cloud services

	Constant	Threshold	RL controller	Proposed
Cloud service 0	58.62	**90.81**	81.83	90.68
Cloud service 1	56.07	**85.76**	80.95	84.23
Cloud service 2	53.34	86.87	77.49	**93.57**
Cloud service 3	48.57	81.41	73.26	**86.37**
Cloud service 4	56.20	80.99	77.18	**86.31**
Average	54.56	85.17	78.14	**88.23**

the proposed policy are able to utilize resources more fully, and the constant resource-based policy wastes more resources in 5 workloads for most intervals. This is because constant resource-based policy statically allocates a lower number of VMs. If the constant resource-based policy allocates enough VMs statically, the utilization will always be 100%, and there will be more waste of resources.

Figure 6 shows the waste of resources of the four polices for 5 workloads at each interval. The average resources remaining under 5 workloads for constant resource-based policy, basic threshold-based policy, RL controller-based policy, and the proposed policy are 30535.59, 5512.49, 9849.35 and 4315.68, respectively, as shown in Table 4. Our proposed policy outperforms other policies by reducing about 22%–86% of the resources wasting in 5 workloads.

The under-provisioning of resources causes SLA violations, and subsequently, leads to lower profit and fewer users. Compare with constant resource-based policy and basic threshold-based policy, RL controller-based policy has prediction-based controller, it is able to prepare sufficient resources in advance. Moreover, the pro-posed policy maintains a resource pool to make up for the prediction error of the prediction-based controller to reduce SLA violation rate. The proposed policy can reduce the SLA violations to less than 0.5%, as shown in Table 5.

Figure 7 shows the CPU load of the four polices for 5 workloads at each interval. The average CPU load under 5 workloads for constant resource-based policy, basic threshold-based policy, RL controller-based policy, and the proposed policy are 55.29%, 89.44%, 78.61%, and 88.27%, respectively, as shown in Table 6. The average value of the standard deviation (SD) about CPU load under 5 workloads for constant resource-based policy, basic threshold-based policy, RL controller-based policy, and the proposed policy are 23.31, 24.55, 11.63, and 8.46, respectively, as shown in Table 6. The smaller the standard deviation, the less volatile the CPU load. The smaller the data fluctuation, the better the solution to the multi-goal optimization problem of reducing SLA violation rate and increasing CPU utilization. The proposed policy gets a much higher standard deviation, so the proposed policy effectively reduced the SLA violation rate and increased the CPU utilization.

Our method is that multiple cloud services work together, we need to consider all cloud service costs and compare their sums. As mentioned above, the total cost depends on the cost of SLA violation and the cost of rented VMs. Our approach wastes the

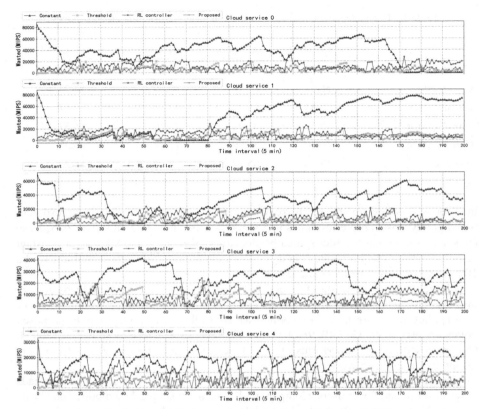

Fig. 6. Comparisons of wasted resources of four polices for different cloud services.

Table 4. The average wasted resources of four polices for different cloud services

	Constant	Threshold	RL controller	Proposed
Cloud service 0	37243.41	5103.99	11706.33	**5074.52**
Cloud service 1	39537.41	**6221.43**	10774.91	7122.11
Cloud service 2	32663.21	5221.75	10315.04	**2055.44**
Cloud service 3	25715.65	5872.27	9169.47	**3648.96**
Cloud service 4	17518.29	5143.03	7280.98	**3677.36**
Average	30535.59	5512.49	9849.35	**4315.68**

Table 5. The average SLA violations of four polices for different cloud services

	Constant	Threshold	RL controller	Proposed
Cloud service 0	13.30	23.38	3.26	**0.00**
Cloud service 1	14.16	14.49	2.72	**0.00**
Cloud service 2	7.57	16.11	1.66	**0.31**
Cloud service 3	1.01	13.96	3.61	**0.40**
Cloud service 4	**0.00**	11.55	6.65	0.30
Average	7.21	15.90	3.58	**0.20**

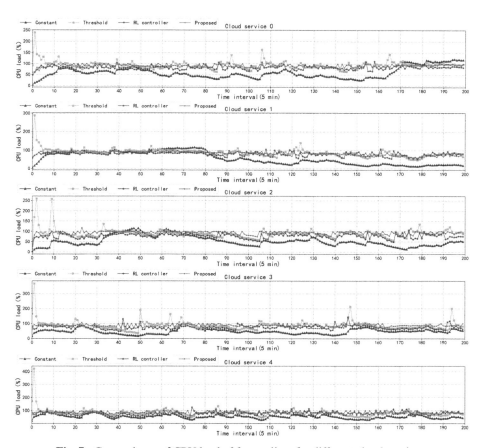

Fig. 7. Comparisons of CPU load of four polices for different cloud services.

Table 6. The average CPU load and the standard deviation of CPU load of four polices for different cloud services

	Constant (Ave/SD)	Threshold (Ave/SD)	RL controller (Ave/SD)	Proposed (Ave/SD)
Cloud service 0	60.23/25.77	**94.89**/17.50	82.12/8.77	90.68/**6.94**
Cloud service 1	57.45/33.67	**88.84**/21.20	81.15/10.33	84.23/**9.64**
Cloud service 2	53.92/24.88	90.91/23.04	77.63/11.81	**93.65/8.00**
Cloud service 3	48.62/17.81	**87.59**/31.41	74.12/13.40	86.41/**8.89**
Cloud service 4	56.20/14.43	84.97/29.65	78.04/13.88	**86.35/8.86**
Average	55.29/23.31	**89.44**/24.56	78.61/11.64	88.27/**8.46**

least amount of resources and has the lowest rate of violations, as shown in Table 4 and Table 5. Figure 8 shows the total cost of the four approaches for 5 workloads at all interval. The proposed policy outperforms other approaches by saving cost about 17%–58% in 5 workloads, as shown in Table 7.

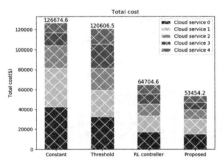

Fig. 8. The sum of total cost at five cloud service for different polices.

Table 7. The total cost of four polices for different cloud services.

	Constant	Threshold	RL controller	Proposed
Cloud service 0	42591.21	32392.45	17124.12	**15111.93**
Cloud service 1	38700.50	26726.88	16034.35	**14783.79**
Cloud service 2	22593.87	22145.12	12178.67	**9819.02**
Cloud service 3	12788.99	23848.74	10057.89	**7204.37**
Cloud service 4	10000.00	15493.29	9309.60	**6535.13**
Sum	126674.57	120606.47	64704.62	**53454.25**

Figure 9 shows the profit of the four approaches for 5 workloads at all interval, and we observe that basic RL controller-based policy and the proposed policy are able to make considerable profits, as shown in Table 8. Table 5 shows the proposed policy less on SLA violations than what RL controller-based policy and guarantee the quality of cloud service to increased customer retention. Therefore, the proposed policy can make the most profit in the long term.

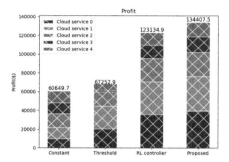

Fig. 9. The sum of profit at five cloud service for different polices.

Table 8. The profit of four polices for different cloud services.

	Constant	Threshold	RL controller	Proposed
Cloud service 0	9978.23	20750.97	35982.77	**39780.51**
Cloud service 1	11931.88	23939.15	34598.04	**36656.28**
Cloud service 2	14750.14	15169.65	25162.92	**25904.66**
Cloud service 3	11484.88	376.59	14196.51	**16590.64**
Cloud service 4	12504.54	7016.54	13194.70	**15475.38**
Sum	60649.68	67252.91	123134.94	**134407.46**

Based on the above results, we compared the CPU utilization, resources remaining, SLA violations, CPU load, total cost and profit of four approaches under 5 workloads. We observed that the proposed approach increases the CPU utilization by up to 3%–34%, the profit by up to not less than 9% and reduces the waste of resources by up to about 22%–86%, the SLA violations to less than 0.4%, the total cost saving about 17%–58% compared with the other polices.

7 Conclusion

In this study, we proposed an elastic resource scheduling algorithm based on a combination of the autonomic computing and DRL. In order to implemented the proposed approach, we presented a resource provisioning framework that supported the MAPE control loop, and used a two-tier elastic resource configuration. Our framework could adjust the number of VMs in advance by workload prediction and DRL algorithm to achieve horizontal scaling, and achieve vertical scaling by resource sharing, collaborative work and self-adapting management. Experimental results using actual Google cluster tracking results showed that the proposed approach could increase the resource utilization and decrease the total cost, while avoiding SLA violations. In the future, we plan to consider the type of request based on the proposed method to configure a more suitable cloud service. In addition, some emergency situations such as the sharp fluctuations of user requests and VM failures will be considered.

Acknowledgements. The authors are grateful for the support of the National Natural Science Foundation of China (61373149), the Natural Science Foundation of Shandong Province (ZR2020QF026) and the project of Shandong Normal University (2018Z29).

References

1. Singh, S., Chana, I.: Cloud resource provisioning: survey, status and future research directions. Knowl. Inf. Syst. **49**(3), 1005–1069 (2016)
2. Yousafzai, A., Gani, A., Noor, R.M., et al.: Cloud resource allocation schemes: review, taxonomy, and opportunities. Knowl. Inf. Syst. **50**(2), 347–381 (2017)
3. Suresh, A., Varatharajan, R.: Competent resource provisioning and distribution techniques for cloud computing environment. Cluster Comput. 1–8 (2019)
4. Chieu, T.C., Mohindra, A., Karve, A.A., et al.: Dynamic scaling of web applications in a virtualized cloud computing environment. In: 2009 IEEE International Conference on e-Business Engineering, pp. 281–286. IEEE (2009)
5. Yang, J., Liu, C., Shang, Y., et al.: A cost-aware auto-scaling approach using the workload prediction in service clouds. Inf. Syst. Front. **16**(1), 7–18 (2014)
6. Roy, N., Dubey, A., Gokhale, A.: Efficient autoscaling in the cloud using predictive models for workload forecasting. In: 2011 IEEE 4th International Conference on Cloud Computing, pp. 500–507. IEEE (2011)
7. Weingärtner, R., Bräscher, G.B., Westphall, C.B.: Cloud resource management: a survey on forecasting and profiling models. J. Netw. Comput. Appl. **47**, 99–106 (2015)
8. Nikravesh, A.Y., Ajila, S.A., Lung, C.H.: An autonomic prediction suite for cloud resource provisioning. J. Cloud Comput. **6**(1), 3 (2017)
9. Mazidi, A., Golsorkhtabaramiri, M., Tabari, M.Y.: Autonomic resource provisioning for multi-layer cloud applications with K-nearest neighbor resource scaling and priority-based resource allocation. Softw. Pract. Exp. (2020)
10. Wei, Y., Kudenko, D., Liu, S., et al.: A reinforcement learning based auto-scaling approach for SaaS providers in dynamic cloud environment. Math. Probl. Eng. **2019** (2019)
11. Ghobaei-Arani, M., Jabbehdari, S., Pourmina, M.A.: An autonomic resource provisioning approach for service-based cloud applications: a hybrid approach. Future Gener. Comput. Syst. **78**, 191–210 (2018)

12. Li, Q., Hao, Q., Xiao, L., et al.: An integrated approach to automatic management of virtualized resources in cloud environments. Comput. J. **54**(6), 905–919 (2011)
13. Gill, S.S., Buyya, R., Chana, I., et al.: BULLET: particle swarm optimization based scheduling technique for provisioned cloud resources. J. Netw. Syst. Manag. **26**(2), 361–400 (2018)
14. Salah, K., Elbadawi, K., Boutaba, R.: An analytical model for estimating cloud resources of elastic services. J. Netw. Syst. Manag. **24**(2), 285–308 (2016)
15. Wu, L., Garg, S.K., Buyya, R.: SLA-based resource allocation for software as a service provider (SaaS) in cloud computing environments. In: 2011 11th IEEE/ACM International Symposium on Cluster, Cloud and Grid Computing, pp. 195–204. IEEE (2011)
16. Huebscher, M.C., Mccann, J.A.: A survey of autonomic computing—degrees, models, and applications. ACM Comput. Surv. **40**(3), 1–28 (2008)
17. Mnih, V., Kavukcuoglu, K., Silver, D., et al.: Human-level control through deep reinforcement learning. Nature **518**(7540), 529–533 (2015)
18. Sowell, F.: Modeling long-run behavior with the fractional ARIMA model. J. Monet. Econ. **29**(2), 277–302 (1992)
19. Google Cluster-Usage Traces. http://code.google.com/p/googleclusterdata

End-to-End QoS Aggregation and Container Allocation for Complex Microservice Flows

Min Zhou, Yingbo Wu$^{(\boxtimes)}$, and Jie Wu

School of Big Data and Software Engineering, Chongqing University,
Chongqing, China
1395234171@qq.com, wyb@cqu.edu.cn, 201824131016@cqu.edu.cn

Abstract. Microservice is increasingly seen as a rapidly developing architectural style that uses containerization technology to deploy, update, and scale independently and quickly. A complex microservice flow that is composed of a set of microservices can be characterized by a complex request execution path spanning multiple microservices. It is essential to aggregate quality of service (QoS) of individual microservice to provide overall QoS metrics for a complex microservice flow. Besides, leveraging the cost and performance of a complex microservice flow to find an optimal end-to-end container allocation solution with QoS guarantee is also a challenge. In this paper, we define an end-to-end QoS aggregation model for the complex microservice flow, and formulate the end-to-end container allocation problem of microservice flow as a non-linear optimization problem, and propose an ONSGA2-DE algorithm to solve this problem. We comprehensively evaluate our modeling method and optimization algorithms on the open-source microservice benchmark Sock Shop. The results of experiments show that our method can effectively assist in the QoS management and container allocation of complex microservice flow.

Keywords: Microservice flow · Container allocation · Optimization · Quality of service

1 Introduction

Microservice architecture is a development model that has become more and more popular in recent years. Microservice-based applications typically involve the execution and interoperability of multiple microservices, each of which can be implemented, deployed, and updated independently without compromising application integrity [1,2]. In recent years, especially due to the proliferation of cloud technologies, the focus has been on applying workflow techniques on distributed environments. Motivated by this observation, we envision workflows that are composed of microservices, i.e., services which move away from the more

© ICST Institute for Computer Sciences, Social Informatics and Telecommunications Engineering 2021
Published by Springer Nature Switzerland AG 2021. All Rights Reserved
H. Gao et al. (Eds.): CollaborateCom 2020, LNICST 350, pp. 154–168, 2021.
https://doi.org/10.1007/978-3-030-67540-0_9

traditional large monolithic back-end applications by splitting the functionality into a set of smaller more manageable services [3]. Microservice flow, containing a set of microservices with data dependency between each other, is processed by multiple microservices that correspond to the tasks in the workflow. Microservice flow is a new manifestation of application independence. This independence in microservice flow management dramatically improves the scalability, portability, renewability, and availability of applications. Nevertheless, the price is high-cost overhead and complex resource usage dynamics [4]. Blindly allocating resources to the application to meet increasing loads is not cost-effective. Cloud providers would like to automatic elasticity management. Minimize the cloud resources allocated to an application to reduce operational cost while ensuring that the chosen cloud resources can support the service level agreement (SLA) agreed by the client. Currently, there is a lack of research on microservice flow in the field of microservices. Therefore, end-to-end quality of service (QoS) modeling [5] and container allocation optimization are one of the most critical concerns for complex microservice flows.

Different from traditional monolithic applications, end-to-end complex microservice flows bring two new challenges to QoS aggregation and container resource allocation. Firstly, the microservice flow is characterized by a complex request execution path spanning multiple microservices [6], forming a directed acyclic graph (DAG). Secondly, microservices run as containers hosted on virtual machines (VMs) clusters in a cloud environment, and application performance is often degraded in unpredictable ways, such as [8,9]. Traditional cloud providers support simple and relatively fast resource allocation for applications. It is not a simple task to run a specific application based on QoS requirements to determine an optimal and cost-effective container allocation strategy. Container allocation is a typical nonlinear optimization problem and has been proved to be an NP-hard problem [7]. An effective container resource allocation method not only satisfies the load requirements of the cluster but also ensures other QoS indicators, such as the reliability of the cluster. So far, minimal efforts have been directly devoted to QoS management and container allocation optimization for microservice flows.

Aiming at the above challenges, we define an end-to-end QoS aggregation model method for the complex microservice flow and represent the container allocation problem of microservice flow as a nonlinear optimization problem which minimizes the communication cost and the load balancing between containers under a user-specified response time constraint, and propose an ONSGA2-DE algorithm to solve this problem. In particular, we make the following contributions.

(a) End-to-end QoS aggregation can provide an overall QoS metric evaluation method for complex microservice flows. To the best of our knowledge, this is the first attempt at designing an end-to-end QoS aggregation model for the complex microservice flow.
(b) End-to-end container allocation provides an overall configuration optimization solution for complex microservice flows that meet SLAs. We represent

the container allocation problem of microservice flow as a nonlinear optimization problem and solve effectively by the proposed ONSGA2-DE algorithm.

(c) We demonstrate the feasibility of using the proposed QoS aggregation model and container allocation strategy to make effective container scaling and achieve container allocation optimization of microservice flow.

The rest of this paper is organized as follows. In Sect. 2, we introduce our system model. In Sect. 3, we propose a heuristic algorithm for container allocation optimization. Next, in Sect. 4, performance evaluations of our solution are presented, and Sect. 5 discusses related research. Section 6 concludes the article.

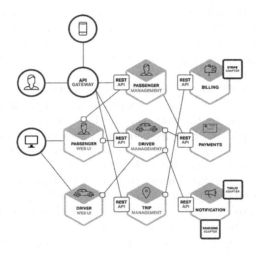

Fig. 1. Microservice composition and deployment [10].

2   System Model

2.1   Microservice Flow

As shown in Fig. 1, the microservice architecture [10] defines a framework that structures an application structure as a set of loosely coupled, collaborative, and interconnected services, each implementing a set of related functions. A workflow is defined as "the automation of a business process, in whole or part, during which documents, information, or tasks are passed from one participant to another for action, according to a set of procedural rules" by the workflow management coalition (WFMC) [11]. In particular, each workflow is implemented as a tightly coupled set of tasks and has its workflow execution plan that specifies how to run the workflow on a distributed computation infrastructure [12].

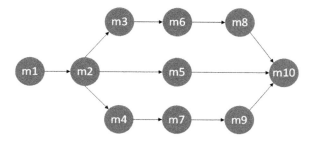

Fig. 2. Microservice flow with ten microservices.

Definition 1. *Microservice Flow.*

The microservice flow is the set of microservices and the interoperability between them. These interoperability relationships are established when microservices consume results generated by other microservices. According to the definition of cloud workflow, a microservice flow $W = (M, E)$ is modeled as a DAG where the set of vertices $M = \{m_1, m_2, \cdots, m_n\}$ represents microservice ms_i, and E is a set of directed edges that represent data or control dependencies between the microservices. A dependency e_{ij} is a precedence constraint of the form (m_i, m_j), where, $m_i, m_j \in M$, and $m_i \neq m_j$. It implies that microservice ms_j (child microservice) can only be started after the microservice ms_i (parent microservice) has completed execution and transferred reliant data to ms_j. Thus, a child microservice cannot execute until all its parent microservices have finished execution and satisfied the required dependencies. For a complex microservices flow, we abstract it, as shown in Fig. 2. Each microservice ms_i takes responsibility to complete a specific subfunction for requests to the microservice flow. Each microservice is deployed and scaled independently without impacting the rest of the application.

Each microservice is characterized as a tuple $(msreq_i, res_i, QoS_i, set_i)$, where $msreq_i$ is the number of microservice id i requests required to satisfy one application workload; res_i represents the computational resources consumed to satisfy one request of the microservice id i; QoS_i represents the QoS parameters for an instance of the microservice id i; set_i is the set of consumer microservices for the microservice ms_i. Each microservice is executed in the system encapsulated in one or more containers $cont_k$. Similar to existing work, it is guaranteed that each request for microservice ms_i would be assigned to one container.

2.2 QoS Aggregation Model

In a workflow management environment, a modeller will create an abstract workflow description, which is used for the integration of available services into the workflow. In the case of a workflow consisting of services only, the result of the integration process can be seen as a service composition [13]. Since the motivation for this work is to use the composition of microservice in the workflow domain,

Table 1. QoS aggregation model for four composition modes.

Parameters	Sequential	Parallel	Conditional	Loop
Response time (Q_{RT})	$\sum_{i=1}^{n} q_{RT}\|ms_i\|$	$max(q_{RT}\|ms_1\|, q_{RT}\|ms_2\|, \cdots, q_{RT}\|ms_n\|)$	$\sum_{i=1}^{n} pr_i*$ $q_{RT}\|ms_i\|$	$h*$ $q_{RT}\|ms_i\|$
Reliability (Q_{Rel})	$\sum_{i=1}^{n} q_{Rel}\|ms_i\|$	$max(q_{Rel}\|ms_1\|, q_{Rel}\|ms_2\|, \cdots, q_{Rel}\|ms_n\|)$	$\sum_{i=1}^{n} pr_i*$ $q_{Rel}\|ms_i\|$	$h*$ $q_{Rel}\|ms_i\|$
Availability (Q_{Ava})	$\sum_{i=1}^{n} q_{Ava}\|ms_i\|$	$max(q_{Ava}\|ms_1\|, q_{Ava}\|ms_2\|, \cdots, q_{Ava}\|ms_n\|)$	$\sum_{i=1}^{n} pr_i*$ $q_{Ava}\|ms_i\|$	$h*$ $q_{Ava}\|ms_i\|$

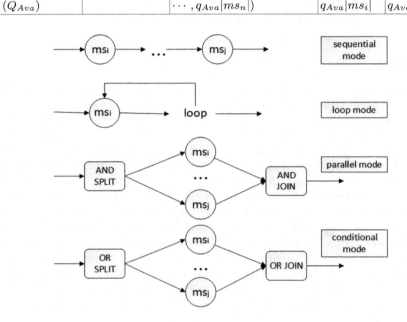

Fig. 3. Different patterns of the composition model.

the composition patterns are based on structural elements as used in workflow descriptions. Services in a microservice flow can be composed based on four basic modes: sequential mode, loop mode, conditional mode, and parallel mode, as shown in Fig. 3. A complete microservice flow can be easily decomposed into these four modes. Applying this procedure, the aggregation can be performed on the basis of each composition pattern. [13] has presented a composition model that can serve as a basis for aggregation of service properties regarding QoS dimensions. Inspired by it, this paper uses QoS aggregation to model the QoS of the entire microservice flow. There are five aggregation operations for developing an aggregation formula, including summation, average, multiplication, min, and max. Here, we choose QoS_i as $(q_{RT_i}, q_{Rel_i}, q_{Ava_i})$. These three indicators are the most representative of QoS. Based on them, the aggregation method of microservice flow is given. Table 1 presents the response time, the reliability, and the availability calculation formulas based on the four basic relations of sequential, parallel, conditional, and loop, respectively, where n is the number of microservice.

3 Solution: Container Allocation Optimization

3.1 Optimization Model

Container allocation mechanisms of microservice flow are based not only on the resource capacity but also consider the nature of the workload. Each microservice of the microservice flow has multiple running instances, and each microservice instance is encapsulated in one container. The scalability of microservice flow means that the container allocation can be appropriately scaled according to the resource usage rate, ensuring that some microservice instances would not become the performance bottleneck of system. In this paper, we focus on the allocation of container with QoS guarantee, so the communication cost and load balancing are formulated as optimization objectivities, and the QoS of response time is modeled as the QoS constraint. If some microservice flow applications have high QoS requirements for the availability and reliability, the optimization model can also include the QoS of availability and reliability as constraints. Table 2 presents the key notations.

The first optimization objective is the communication cost. The communication cost among microservices is related to two key factors: the network distance between containers assigned to two interoperable microservices, and the number of requests between the two microservices. Consider that consumer microservices and provider microservices may run multiple container instances simultaneously. This paper uses the average network distance of all the container pairs between consumer microservice and provider microservice to calculate the communication cost between two microservices. The communication cost for node j to communicate with other nodes is formalized in (1).

$$commCost_j = \sum_{i=1}^{n} \frac{x(j,i)}{scale_i} \sum_{ms_k \in set_i \wedge m \in list(ms_k)} dist(j,m) \tag{1}$$

For the second optimization objective, we consider that the cluster needs to be load balanced. The cluster is balanced when the use of the nodes is uniform across the cluster, i.e., the consumption of computational resources is uniformly distributed among the VMs [14]. A high $scale_i$ of a microservice is penalized when the container is underused, and a low $scale_i$ is penalized when the number of microservice requests is very high. With $scale_i$ changes, we can achieve automatic elasticity management. We use the standard deviation of computing resource usage to evaluate the balance of the cluster. The resources consumed by each node are formalized in (2).

$$nodeUsage_j = \sum_{i=1}^{m} \frac{ureq \times msreq_i \times res_i}{scale_i} \times x(j,i) \tag{2}$$

Table 2. Summary of the system model.

Parameter	Description
$ureq$	Number of user workloads
$msreq_i$	ms_i requests number needed for each workload
res_i	Computational resources required for a ms_i request
QoS_i	QoS parameters for a ms_i instance
set_i	The set of consumer microservice for ms_i
$scale_i$	Number of containers for ms_i
$dist(i,j)$	Network distance between vm_i and vm_j
$x(j,i)$	The number of container assigned on vm_j for microservice ms_i
$list(ms_i)$	List of containers to which ms_i is allocated
Q_{RT}	Response time for the entire microservice flow
Q_{Rel}	Reliability for the entire microservice flow
Q_{Ava}	Availability for the entire microservice flow
T	The response time for the entire microservice flow specified in the service level agreement
pr_i	Conditional probability
h	Loop times
ms_i	Microservice with id.i
$cont_k$	Container with id.k
vm_j	Virtual machine with id.j

To summarize, our aim can be defined as a multi-objective optimization problem, where the solution establishes the allocation of containers to the VMs, by minimizing (i) the communication cost and (ii) the load imbalance between containers. We formulate the optimization problem as follows: Given a microservice flow instance and a set of available VMs, we wish to find an allocation $A(t) : ms_i(cont_k) \rightarrow vm_j, vm_j \in VMs$, by minimizing

$$\sum_{j \in N} commCost_j, \qquad (3)$$

$$\sigma(nodeUsage_1, nodeUsage_2, \cdots, nodeUsage_j), \qquad (4)$$

subject to the constraint

$$Q_{RT} < T, \qquad (5)$$

$$\sum_{j \in N} x(j,i) = scale_i. \qquad (6)$$

This optimization problem is NP-complete because all the possible container allocation strategies should be evaluated to find the optimal solution. A meta-heuristic method must be employed to address the problem [15].

Table 3. ONSGA2-DE algorithm execution parameters.

Parameter	Value
populationSize	200
generationNumber	200
Mutation probability	0.25
Scale factor F	2
Control ratio k	rand(0,1)
P_{cmax}	1
P_{cmin}	0

3.2 Proposed Container Allocation Algorithm

The container allocation problem is represented as a multi-objective optimization problem. The Non-dominated Sorting Genetic Algorithm-II (NSGA2), which is considered a standard approach for multi-objective optimization problems in resource management problem for container allocation. NSGA2 [16] has high computational efficiency and high-quality solution set when processing the low-dimensional multi-objective optimization problem. In Differential Evolution (DE) Algorithm, the most novel feature is its mutation operation. In the initial stage of the algorithm iteration, the individual differences of the population are considerable. Such a mutation operation will make the algorithm have robust global search ability. By the end of the iteration, the population tends to converge, and the differences between individuals decrease, which makes the algorithm have strong local search capabilities. In order to further improve the diversity of the population and prevent the algorithm from converging to a local optimum, we propose a heuristic ONSGA2-DE algorithm in this paper. The algorithm introduces opposition-based learning in the iterative process of the population, and a crossover and mutation strategy of DE algorithm is applied as the operator. The ONSGA2-DE algorithm increases the diversity of the population and prevents local convergence.

Opposition-Based Learning. Apply the opposition-based learning to the evolution process, calculate the reverse individual X^* for the individual X in each generation of the evolution process. During opposition-based learning, the population is evolved towards better fitness value by the specific evolutionary strategy. And the better individuals are selected from both the original population and the opposite population to produce the next generation. Similarly, an opposite point in the multi-dimensional case can be defined as follows [17]: Let $P(x_1, x_2, \cdots, x_D)$ be a point in D- dimensional space, where, $x_1, x_2, \cdots, x_D$ are real numbers and $x_j \in [a_j, b_j], j = 1, 2 \cdots, D$. The opposite point of P^* is denoted by $P^*(x_1{}^*, x_2{}^*, \cdots, x_D{}^*)$ where

$$x_j{}^* = a_j + b_j - x_j \tag{7}$$

Algorithm 1. ONSGA2-DE

1: **INPUT:** $populationSize, generationNumber,$
2: $mutationPro, P_{cmax}, P_{cmin}, F, k$
3: **OUTPUT:** Pareto optimal
4: Generate randomly P_t of size $populationSize$, $t = 0$
5: **for** $t < generationNumber$ **do**
6:   **if** $rand(0,1) > P_c$ **then**
7:     **for** $i < populationSize$ **do**
8:       **for** $child_i, father_1, father_2, father_3$ are selected by Tournament
9:       $child_i{}^{new} = father_1 + F(father_2 - father_3)$
10:       **if** $random() < mutationPro$ **then**
11:         $mutation(child_i{}^{new})$
12:       **end if**
13:       $Q_t = Q_t \cup child_i{}^{new}$
14:     **end for**
15:   **else**
16:     **for** $i < populationSize$ **do**
17:       $child_i{}^{new} = k(a_i + b_i) - child_i$
18:       $Q_t = Q_t \cup child_i{}^{new}$
19:     **end for**
20:   **end if**
21:   $R_t = Q_t \cup P_t$
22:   $fitness = calculateFitness(R_t)$
23:   $fronts = fastNonDominatedSort(R_t, fitness)$
24:   $distances = calculateCrowdingDistances(R_t, fronts)$
25:   $P_{t+1} = SelectTopN(R_t, fronts, distances)$
26:   $t = t + 1$
27: **end for**
28: $Solution = fronts[1]$

However, considering that in the later stage of the algorithm, the population P has formed a certain rule and is close to the optimal solution region. It is of little significance to solve the reverse population in the later stage of the algorithm and the running speed of the algorithm is slowed down. In every generation of evolution, its opposite individual X^* is calculated by a certain probability of opposition, and in the process of evolution, P_c(probability) is linear decline. Therefore, the following method is designed in this paper:

$$P_c = P_{cmax} - \frac{t}{populationSize}(P_{cmax} - P_{cmin}) \qquad (8)$$

In each algorithm iteration, a set of new individuals, with the same size as the population, is generated (lines 6 to 20 in Algorithm 1). Including the crossover and mutation of the DE algorithm to generate a new population (lines 7 to 14). And generate a new population through opposition-based learning (lines 16 to 19). These new individuals are joined with the individuals in the parent population (line 21), obtaining a set that is double the size of the population. Only half of the individuals need to be selected for the offspring. The front levels and crowding distances of the new set are calculated and applied to order the solution set (lines 22 to 25). The best half of the solutions are selected as individuals of the offspring, and the remainder are rejected. The time complexity of the algorithm is related to the population P and the number of optimization objectives. The time complexity is $O(MN^2)$, N represents the population size, and M represents the number of optimization objectives. Table 3 summarizes the parameters for the execution of the ONSGA2-DE algorithm.

4   Experiment

For container allocation of microservice flows, we use Sock Shop. It is an open-source microservices benchmark specifically for container platforms and a microservices demo application. The Sock Shop simulates an e-commerce site, with the specific goal of helping to demonstrate and test existing microservices and cloud-native technologies. The parameter values for the microservices stack was estimated from [7].

In order to study the results of our method under different experimental conditions, we set up several different experimental configurations. Workload changes |workloads| = [10, 20, 30, 40] reflect the flexible management of the application. The capacity of the cluster is the number of available |VMs| = [100, 200] and their computational capacities $cap = 800$. Network distance among the nodes is $dist(i, j) = [1.0, 4.0]$. The simulation $q_{RT}(ms_i) = 0.458, 0.891, 0.514, 0.17, 0.523, 0.364, 0.683, 0.371, 0.236, 0.537$. Using the QoS aggregation formula, we calculate the Q_{RT} of the microservice flow instance. And it is less than the response time specified in the service level agreement. The comparisons of the three algorithms are performed in two aspects: communication cost, and load balancing of containers. Among them, we need to normalize two optimization objectives. This $Solution(x)$ was calculated as the weighted average of the normalized values of the two objectives in (9). $max_{commCost}, min_{commCost}$ are the maximum and minimum values of the communication cost, respectively. $max_{nodeUsage}, min_{nodeUsage}$ are the maximum and minimum values of the standard deviation of the resource usages of the nodes.

$$Solution(x) = \frac{1}{2} \times \frac{commCost(x) - min_{commCost}}{max_{commCost} - min_{commCost}} + \frac{1}{2} \times \frac{nodeUsage(x) - min_{nodeUsage}}{max_{nodeUsage} - min_{nodeUsage}} \tag{9}$$

As shown in Fig. 4(a), (b), experiments with ten workloads, and 100 nodes are performed. We can observe the change of the iterative process of two optimization objectives under constraints. Figure 4(a) is an iterative process of communication cost for the three algorithms. Communication cost is measured in terms of the average network distance and request number between all interacting microservices. We want two microservice instances with interaction to be distributed on the same node or adjacent nodes. This can reduce communication cost. We

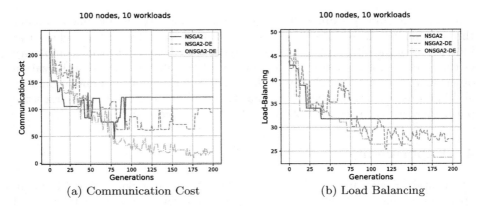

(a) Communication Cost   (b) Load Balancing

Fig. 4. Values of the two objective functions throughout the algorithm generations for the experiment with 10 workloads and 100 VMs.

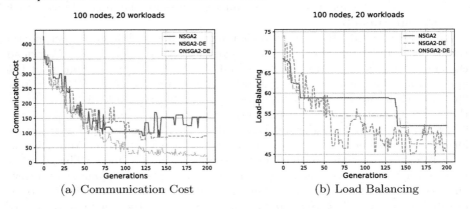

(a) Communication Cost   (b) Load Balancing

Fig. 5. Values of the two objective functions throughout the algorithm generations for the experiment with 20 workloads and 100 VMs.

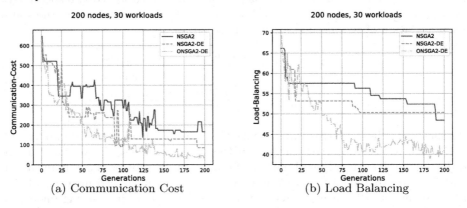

(a) Communication Cost   (b) Load Balancing

Fig. 6. Values of the two objective functions throughout the algorithm generations for the experiment with 30 workloads and 200 VMs.

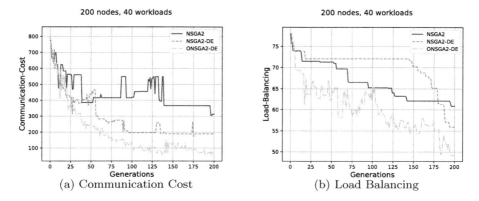

(a) Communication Cost   (b) Load Balancing

Fig. 7. Values of the two objective functions throughout the algorithm generations for the experiment with 40 workloads and 200 VMs.

clearly see that communication cost decreases as iteration progress. This also means that the distribution of nodes is ideal. Figure 4(b) shows the iterative process of the standard deviation. The standard deviation value represents the degree of balance of the cluster load. We can see that the standard deviation value gradually decreases with the process of iteration, which also means that the load between the containers is approaching equilibrium. From Fig. 4(a), (b), we can see that the NSGA2 algorithm performs worst, and the ONSGA2-DE algorithm performs best. Our proposed algorithm performs better than the contrasted algorithms in terms of convergence speed and optimal value. Because the ONSGA2-DE algorithm uses reverse learning, it has optimized the diversity and convergence of the population. The crossover and mutation stages of the ONSGA2-DE algorithm are combined with the DE algorithm. Then adds DE operator to generate new individuals, which improves the generation stage. So it is better than NSGA2-DE.

Figure 5 shows the comparative results of the two objective functions for the three algorithms with 20 workloads and 100 nodes. Increased user requests and the values of the two objective functions increased correspondingly. We increase the number of workloads to 30 and 40, and the number of nodes to 200 and observe the changes in the two objective values again in Fig. 6, Fig. 7. In general, we can see that the ONSGA2-DE algorithm works best. The ONSGA2-DE algorithm can find an optimal solution, and the value of the optimization objective is always better than the other two algorithms.

5   Related Work

QoS management for cloud applications has been an important research topic for many years [18]. Due to many complex interactions and performance changes caused by cloud computing, significant challenges are presented in modeling system performance, identifying key resource bottlenecks, and effective management [19,20]. From web services to cloud services, QoS management research focuses on QoS requirements specification, and QoS service selection [21] that is based on QoS aggregation technology [13]. End-to-end QoS aggregation model can provide an overall QoS metric evaluation method for complex microservice flows. Cloud providers would like to minimize the cloud resources allocated to an application to reduce operational cost while ensuring that the chosen cloud resources can support the SLA.

Since entering the era of cloud computing, resource allocation methods in the cloud have been widely studied. However, more research on cloud application mainly focuses on resource allocation of VMs to achieve performance-oriented load balancing or energy-oriented load integration [22]. With the development of container technology, some practical container allocation solutions are proposed, but there are still some important problems to be solved in container-based microservice management. Container allocation will improve the overall performance and resource utilization of the system. This characteristic of container allocation helps to reduce energy consumption and achieve better system improvements and user satisfaction. Some research focused on the optimization of resource management in the field of microservice applications [9,23]. According to some of the surveys released, it included a wide range of techniques for addressing resource management issues, such as scheduling [27], provisioning [24], and distribution [25]. We can describe the container allocation problem as a multi-objective optimization problem, which can be solved by evolutionary algorithms [26]. Also, genetic algorithms are often used in resource management of cloud environments. The adequacy of genetic algorithms, especially NSGA2, has been widely proven [16]. Guerrero et al. [7] used genetic algorithms to optimize container allocation in cloud structures, which enhanced system provisioning, system performance, system failure, and network overhead. Bao et al. [5] proposed a performance modeling and workflow scheduling method for microservice-based applications in the cloud, and a heuristic scheduling scheme was used for an application containing multiple microservices. Gribaudo et al. [9] proposed a performance evaluation method for applications based on large-scale distributed microservices.

To the best of our knowledge, the proposed method differs from the existing efforts in three aspects: (i) Our work is oriented end-to-end complex microservice flows. (ii) We propose an end-to-end QoS aggregation model and container allocation method for the complex microservice flow. (iii) We prove that the heuristic ONSGA2-DE algorithm implements a container allocation strategy and automatic elasticity management.

6   Conclusion

Microservice architecture styles are becoming increasingly popular in both academia and industry. QoS management and container allocation are two critical issues for microservice flows. In this paper, we have proposed a QoS aggregation model based on microservice flow, formulate a microservice flow container allocation problem to minimize communication cost and load imbalance between containers, and solve the problem by using ONSGA2-DE algorithm. Our work on QoS management and container resource optimization has been validated and evaluated through experimental and simulation results. Experimental results show that our model can find an optimal decision scheme to guarantee the QoS of microservice flow.

Future work will be pursued in several directions. Firstly, we assumed that VMs had the same capacity, but in reality, there were different types of VMs. In future work, we will consider the effects of changes in the capacity and type of VMs on the experimental results. Secondly, this paper focuses on the QoS management and container resource allocation of a single flow consisting of multiple microservices. Therefore, microservices can be maintained separately, scaled, or discarded as necessary. The problem of allocation for multiple complex microservice flows is the direction of our future research.

Acknowledgment. This work was supported in part by National Key Research and Development Project under grant 2019YFB1706101, in part by the Science-Technology Foundation of Chongqing, China under grant cstc2019jscxmbdx0083.

References

1. Dragoni, N., et al.: Microservices: yesterday, today, and tomorrow. Present and Ulterior Software Engineering, pp. 195–216. Springer, Cham (2017). https://doi.org/10.1007/978-3-319-67425-4_12
2. Jamshidi, P., Pahl, C., Mendonca, N.C.: Microservices: the journey so far and challenges ahead. IEEE Softw. **35**(3), 24–35 (2018)
3. Gil, Y., Deelman, E., Ellisman, M., Fahringer, T., Fox, G., Gannon, D.: Examining the challenges of scientific workflows. Computer **40**(12), 24–32 (2007)
4. Fazio, M., Celesti, A., Ranjan, R., Liu, C., Chen, L., Villari, M.: Open issues in scheduling microservices in the cloud. IEEE Cloud Comput. **3**(5), 81–88 (2016)
5. Bao, L., Wu, C., Bu, X.: Performance modeling and workflow scheduling of microservice-based applications in clouds. IEEE Trans. Parallel Distrib. Syst. **30**(9), 2114–2129 (2019)
6. Rahman, J., Lama, P.: Predicting the end-to-end tail latency of containerized microservices in the cloud. In: IEEE International Conference on Cloud Engineering, pp. 200–210 (2019)
7. Guerrero, C., Lera, I., Juiz, C.: Genetic algorithm for multi-objective optimization of container allocation in cloud architecture. J. Grid Comput. **16**(1), 113–135 (2018). https://doi.org/10.1007/s10723-017-9419-x
8. Barakat, S.: Monitoring and analysis of microservices performance. J. Comput. Sci. Control Syst. **5**(10), 19–22 (2017)
9. Gribaudo, M., Iacono, M., Manini, D.: Performance evaluation of massively distributed microservices based applications. In: European Council for Modelling and Simulation (ECMS), pp. 598–604 (2017)
10. Pattern: Microservice architecture (2019). http://microservices.io/patterns/microservices.html
11. Wiley, J.: Workflow Handbook (2019). http://pl.wikipedia.org/wiki/Workflow

12. Liu, L., Zhang, M., Lin, Y., Qin, L.: A survey on workflow management and scheduling in cloud computing. In: 14th IEEE/ACM International Symposium on Cluster, Cloud and Grid Computing. IEEE (2014)

13. Jaeger, M.C., Rojec-Goldmann, G., Muhl, G.: QoS aggregation for web service composition using workflow patterns. In: 8th IEEE International Conference on Enterprise Distributed Object Computing, pp. 149–159 (2004)

14. Rusek, M., Dwornicki, G., Orłowski, A.: A decentralized system for load balancing of containerized microservices in the cloud. In: Świątek, J., Tomczak, J.M. (eds.) ICSS 2016. AISC, vol. 539, pp. 142–152. Springer, Cham (2017). https://doi.org/10.1007/978-3-319-48944-5_14

15. Wei, G., Vasilakos, A.V., Zheng, Y., Xiong, N.: A game-theoretic method of fair resource allocation for cloud computing services. J. Supercomput. **54**(2), 252–269 (2010). https://doi.org/10.1007/s11227-009-0318-1

16. Deb, K., Agrawal, S., Pratap, A., Meyarivan, T.: A fast and elitist multiobjective genetic algorithm: NSGA-II. IEEE Trans. Evol. Comput. **6**(2), 182–197 (2002)

17. Tizhoosh, H.R.: Opposition-based learning: a new scheme for machine intelligence. In: International Conference on Computational Intelligence for Modelling, Control and Automation, pp. 695–701. IEEE (2005)

18. Ghosh, Q., Longo, F., Naik, V.K., Trivedi, K.S.: Modeling and performance analysis of large scale IaaS clouds. Future Gener. Comput. Syst. **29**(5), 1216–1234 (2013)

19. Vakilinia, Q., Ali, M.M., Qiu, D.: Modeling of the resource allocation in cloud computing centers. Comput. Netw. **91**, 453–470 (2015)

20. Jindal, A., Podolskiy, V., Gerndt, M.: Performance modeling for cloud microservice applications. In: 10th ACM/SPEC International Conference on Performance Engineering (ICPE 2019), pp. 25–32 (2019)

21. Liangzhao, Z., Benatallah, B., Ngu, A.H.H., Dumas, M., Kalagnanam, J., Chang, H.: QoS-aware middleware for web services composition. IEEE Trans. Softw. Eng. **30**, 311–327 (2004)

22. Yuan, H., Li, C., Du, M.: Optimal virtual machine resources scheduling based on improved particle swarm optimization in cloud computing. J. Softw. **9**(3), 705–708 (2014)

23. Amaral, M., Polo, J., Carrera, D., Mohomed, I., Unuvar, M., Steinder, M.: Performance evaluation of microservices architectures using containers. In: IEEE 14th International Symposium on Network Computing and Applications (NCA), pp. 27–34. IEEE (2015)

24. Khazaei, H., Barna, C., Beigi-Mohammadi, N., Litoiu, M.: Efficiency analysis of provisioning microservices. In: IEEE International Conference on Cloud Computing Technology and Science (CloudCom), pp. 261–268. IEEE (2016)

25. Akhter, N., Othman, M.: Energy aware resource allocation of cloud data center: review and open issues. Clust. Comput. **19**(3), 1163–1182 (2016). https://doi.org/10.1007/s10586-016-0579-4

26. Lucken, C., Baran, B., Brizuela, C.: A survey on multi-objective evolutionary algorithms for many-objective problems. Comput. Optim. Appl. **58**(3), 707–756 (2014). https://doi.org/10.1007/s10589-014-9644-1

27. Singh, S., Chana, I.: A survey on resource scheduling in cloud computing: issues and challenges. J. Grid Comput. **14**(2), 217–264 (2016). https://doi.org/10.1007/s10723-015-9359-2

A DQN-Based Approach for Online Service Placement in Mobile Edge Computing

Xiaogan Jie[1], Tong Liu[1,2,3,4(✉)], Honghao Gao[1], Chenhong Cao[1,3], Peng Wang[1], and Weiqin Tong[1,3,4]

[1] School of Computer Engineering and Science, Shanghai University, Shanghai, China
{jiegan,tong_liu,gaohonghao,caoch,pengwang,wqtong}@shu.edu.cn
[2] Shanghai Key Laboratory of Data Science, Shanghai, China
[3] Shanghai Institute for Advanced Communication and Data Science, Shanghai University, Shanghai, China
[4] Shanghai Engineering Research Center of Intelligent Computing System, Shanghai, China

Abstract. Due to the development of 5G networks, computation intensive applications on mobile devices have emerged, such as augmented reality and video stream analysis. Mobile edge computing is put forward as a new computing paradigm, to meet the low-latency requirements of applications, by moving services from the cloud to the network edge like base stations. Due to the limited storage space and computing capacity of an edge server, service placement is an important issue, determining which services are deployed at edge to serve corresponding tasks. The problem becomes particularly complicated, with considering the stochastic arrivals of tasks, the additional latency incurred by service migration, and the time spent for waiting in queues for processing at edge. Benefiting from reinforcement learning, we propose a deep Q network based approach, by formulating service placement as a Markov decision process. Real-time service placement strategies are output, to minimize the total latency of arrived tasks in a long term. Extensive simulation results demonstrate that our approach works effectively.

Keywords: Mobile edge computing · Service placement · Deep reinforcement learning

1 Introduction

With the emergence of 5G communication technology and the proliferation of mobile smart devices, many new computation and data intensive applications with low latency requirements have come forth to mobile devices, such as online interactive games, augmented reality and video stream analysis [2,9,19]. Unfortunately, traditional cloud computing cannot meet the requirements of the applications, due to the long data transmission time. To address the issue, mobile

H. Gao et al. (Eds.): CollaborateCom 2020, LNICST 350, pp. 169–183, 2021.
https://doi.org/10.1007/978-3-030-67540-0_10

edge computing [7, 10] is put forward to provide services to the applications, by deploying computing resources at the edge of Internet. A mobile edge computing system enabled by 5G networks is shown in Fig. 1, in which base stations endowed with servers act as edge nodes.

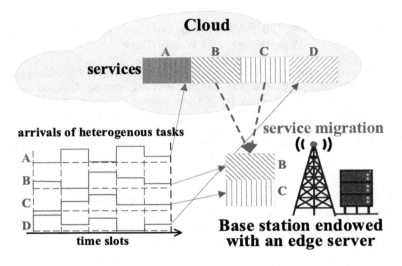

Fig. 1. An illustration of the 5G-enabled mobile edge computing system, in which base stations endowed with servers act as edge nodes.

To reduce the response time, computation tasks generated by different applications on mobile devices should be offloaded to an edge server for processing, instead of offloaded to the cloud. However, to serve the tasks of a particular application, the corresponding service should be placed on the edge server in advance, including installing required softwares and caching related databases/libraries. Unlike the cloud, both storage space and computing capacity of an edge server are extremely limited. Thus, service placement is an important issue in mobile edge computing, i.e., determining which services should be deployed at the edge. Particularly, the problem becomes complicated with the consideration of heterogenous tasks randomly arriving to the base station in real time. An online optimal service placement strategy is desired, to achieve the minimal response time of all tasks.

Some existing works have paid attention to the service placement in mobile edge computing. Most of the works focus on designing service placement policies to improve the quality of service. In [3, 6, 15, 18], the authors consider a stationary environment and propose several offline service placement algorithms to determine which services are selected and where they are placed. And these works achieve provable close-to-optimal performance. In addition, some works [1, 4, 12, 14, 21] consider the varying demands of computation requests arriving to a mobile edge computing system in real time. In [1, 8, 12, 17], the authors propose online service placement schemes, assuming the future dynamics like user mobility can be predicted in advance or make a milder assumption that user mobility follows a Markov chain. However, it is hard to obtain the future dynamics as prior knowledge in practice. Some other works [4,5,13,14,21] solving the service placement problem without a priori knowledge of future dynamics.

In this work, to make optimal service placement decisions in real time is non-trivial due to the following challenges. *Firstly*, the tasks generated by various applications request differential computing resources, and the arrivals of tasks randomly vary over time. Dynamic service placement strategy is desired, adapting to the changing demands of tasks, since only the tasks served by placed services can be executed at the edge. *Secondly*, different services request distinguishing storage for caching the corresponding softwares and databases. Thus, the placements of different services are interactive. Moreover, the set of services to be placed on the edge server is extremely constrained by its storage space. *Thirdly*, additional latency is incurred for migrating a service from the cloud to the edge server, including the time spent for transmitting and operating. This latency can be omitted if the service has been placed in the previous time slot. Thus, the service placement strategies in two successive time slots are coupled with each other. It is hard to derive the optimal strategy without any future information.

To tackle the difficulties, we propose an online learning-based approach for the base station to make real-time service placement decisions, aiming to minimize the total latency of arrived tasks in a long term. Specially, we first analyze the switching latency, serving latency, and offloading latency incurred by a service placement decision respectively, and formulate the service placement problem as a non-linear optimization with coupled constraints. Due to the superiority of reinforcement learning in making controls in dynamic and random environments, we model service placement as a Markov decision process and adopt the deep Q network (DQN) method, in which the reward achieved by a service placement decision is approximated by a deep neural network. The main contributions of this work are summarized as follows:

- We formulate the online service placement in a 5G-enabled mobile edge computing system as a Markov decision process. We consider stochastic arrivals of heterogenous tasks which are served by different services, and the limited storage space and computing capacity of the edge server.
- We propose a deep reinforcement learning based approach to obtain the optimal service placement strategy, with the objective of minimizing the total latency of tasks in a long term.
- Extensive simulations are conducted to evaluate the performance of our proposed approach, compared with several baselines. The simulation results confirm that our approach performs better than baselines.

The rest of this paper is organized as follows. Section 2 reviews related works. In Sect. 3, we describe the system model and problem formulation. Then, we propose our approach in Sect. 4. The results of extensive simulations are shown in Sect. 5, followed by the conclusion in Sect. 6.

2   Related Work

Considering the limited resources of the edge server, some exiting works [1, 3–6, 8, 12–15, 17, 18, 21] have studied on service placement in mobile edge computing. We roughly classify related works into two categories, according to whether the dynamic nature of an edge computing system is considered.

With considering a stationary environment, several offline service placement algorithms are proposed, to determine which services are selected and where they are placed. Specially, both [15] and [6] focus on the joint service placement and request scheduling problem, aiming to maximize the number of computing requests processed by edge servers. A near-optimal algorithm with a constant approximation factor is provided by each of them, respectively. Ascigil et al. [3] propose a set of uncoordinated strategies for solving the service placement problem, with considering multiple services with different latency deadlines. However, the above works do not take the operation cost incurred by service migration into account. Differently, Wang et al. [18] jointly optimize the service activation cost and service placement cost of an edge cloud provider, which provisions a social VR application based on edge clouds. The authors convert the cost optimization into a graph cut problem and employ a max-flow algorithm.

Considering the varying demands of computation requests arriving to a mobile edge computing system, services placed on each edge server should be adapted over time. To tackle the challenge, several works [1, 12] propose online service placement schemes, assuming the future dynamics like user mobility can be predicted in advance. Some other works [8, 17] make a milder assumption that user mobility follows a Markov chain. However, all the works consider future dynamics as prior knowledge, which is hard to obtain in practice.

Besides, there are several existing works solving the service placement problem without a priori knowledge of future dynamics. Xu et al. [21] adopt the Lyapunov optimization to solve the service placement problem as a queue control problem, which lacks taking service migration cost into consideration. Differently, Farhadi et al. [4] takes operation cost incurred by service placement into consideration. To balance the cost and the performance of serving requests, a two-time-scale framework for joint service placement and request scheduling is proposed. By taking advantage of set function optimization, a greedy-based service placement algorithm based on shadow request scheduling is proposed, which achieved constant-factor approximation. Similarly, performance-cost tradeoff is pursued in [14] with in view of the operational cost of service migration. Considering the long-term cost budget constraint, the authors design an online service placement algorithm based on Markov approximation and Lyapunov optimization. In addition, a time-efficiency distributed scheme is also provided based on the best response update technique. In [5], Gao et al. jointly study the access network selection and service placement, with the objective to balance the access, switching and communication delay. An efficient online framework is proposed for decomposing the long-term optimization and an iteration-based algorithm is designed to derive a computation-efficient solution. Ouyang et al. [13] propose a multi-armed bandit and Thompson-sampling based online learning algorithm to make optimal service placement strategy, which aims to optimize the completion latency of tasks and overhead caused by service migration.

Different from the above works, we jointly consider the switching latency incurred by service migration and the waiting time spent by tasks for processing at the edge. As the power of reinforcement learning has been witnessed in complex control under stochastic and dynamic environments, we propose a DQN-based online service placement approach without knowing future dynamics of the mobile edge computing system.

3   System Model and Problem Formulation

3.1   System Model

In this work, we consider a 5G-enabled mobile edge computing system, in which base stations endowed with servers act as edge nodes. Numerous heterogenous computation-intensive tasks arrive to the base stations in real time. For the convenience of modeling an online system, we discretize time into a set of equal-interval time slots, denoted by $\mathcal{T} = \{1, 2, \cdots, T\}$. The interval of each time slot is represented as Δt.

To execute the tasks of a certain type on an edge server, the corresponding service should be placed firstly. Specially, a service is an abstraction of an application, such as video stream analysis, interactive gaming, and augmented reality. To run a particular service, an amount of associated data should be cached by the edge server, including the software and database required by the application. Only the services cached on the edge server can be employed to serve the corresponding tasks.

We consider a library of L services are provided by the mobile edge computing system, denoted by $\mathcal{L} = \{1, 2, \cdots, L\}$. We assume different services have various amounts of associated data and require different CPU frequencies to process tasks. Specially, we use s_l and f_l to denoted the associated data size and the required CPU frequency of service $l \in \mathcal{L}$, respectively. Initially, all the services are stored on a remote cloud, as shown in Fig. 1. Each edge server can manually migrate one or more services from the cloud to its locality in an online manner. However, which services can be placed is restricted by the limited storage space and computing ability of an edge server. Generally, we denote the maximal storage space and CPU frequency of the edge server on a base station as s_{max} and f_{max}, respectively.

In each time slot $t \in \mathcal{T}$, we assume that the arrival of tasks served by service l follows a Poisson process at rate n_l^t. It means the average number of tasks served by l arriving in time slot t equals to n_l^t. The overall computation demand arriving to the base station in time slot t is noted as $\mathbf{n}^t = [n_1^t, n_2^t, \cdots, n_L^t]$. Besides, we assume the required CPU cycles of each task served by service l follows an exponential distribution with mean c_l. Thus, the computing time of such a task also follows an exponential distribution with mean c_l/f_l, if it is processed on the edge server.

According to the real-time arrivals of heterogenous tasks and their computation requirements, the base station needs to make the service placement decision online. In other words, the base station should determine which services to be placed on the edge server in the current time slot. To be specific, the services have been placed in the last time slot can be kept or removed according to the current decision. On the other hand, the unplaced services need to be migrated from the cloud.

We employ a binary variable $I_l^t \in \{0,1\}$ to indicate whether service l is determined to be placed on the edge server or not in time slot t. The service placement decision in t can be represented by vector $\mathbf{I}^t = [I_1^t, I_2^t, \cdots, I_L^t]$. Specially, if service l is placed on the edge server in time slot t, then $I_l^t = 1$; Otherwise, $I_l^t = 0$. Due to the limited storage space and computing capacity, there exist

$$\sum_{l=1}^{L} I_l^t \cdot s_l \leq s_{max}, \forall t \tag{1}$$

$$\sum_{l=1}^{L} I_l^t \cdot f_l \leq f_{max}, \forall t \tag{2}$$

3.2 Service Placement

In this work, we consider the *switching latency*, the *serving latency*, and the *offloading latency* incurred by different service placement decisions, which are described in detail as follows.

Switching Latency. To migrate a particular service from the remote cloud to the edge server, a certain amount of time is spent for transmitting its associated data, which is referred as the switching latency in our work. Specially, for service l, we assume the switching latency in time slot t is ψ_l^t, which depends on the data size of service l and the network condition between the base station and the cloud in time slot t.

According to the service placement decision in the last time slot $\mathbf{I}^{t-1}$, we can derive the overall switching latency incurred by a current service placement decision $\mathbf{I}^t$ as

$$\Psi^t = \sum_{l=1}^{L} \psi_l^t \cdot \mathbb{1}_{\{I_l^t - I_l^{t-1} = 1\}}, \tag{3}$$

where $\mathbb{1}_{\{.\}}$ is an indicator function. If the condition in the brace is satisfied, then the value of the indicator function equals to one; otherwise, it equals to zero. Note that only the services unplaced in the last time slot but needed in the current time slot will incur the switching latency.

Serving Latency. Note that only the tasks served by the services placed on the edge server can be processed on the base station, while the other tasks have to be offloaded to the remote cloud for execution. We firstly analyze the latency incurred for completing a task on the edge server, which is referred as the serving latency. The serving latency of a task is defined as the period from when the task arrives to the base station to when the task is completed, which consists of the waiting time and the computing time. We employ a specific M/M/1 queue to model the execution process of a task served by service l on the edge server, as both the arriving interval and the computing time of the task obey exponential distributions. And hence, the average serving time of the task (known as sojourn time in queuing theory) is

$$\omega_l^t = \frac{1}{f_l/c_l - n_l^t/\Delta t} = \frac{c_l \Delta t}{f_l \Delta t - n_l^t c_l}. \tag{4}$$

To make Eq. (4) sensible, we guarantee that serving rate f_l/c_l is greater than arriving rate $n_l^t/\Delta t$, by offloading extra tasks served by service l to the cloud.

Therefore, the overall serving latency incurred by a service placement decision $\mathbf{I}^t$ in time slot t can be obtained, i.e.,

$$\Omega^t = \sum_{l=1}^{L} I_l^t \cdot n_l^t \cdot \omega_l^t. \tag{5}$$

Offloading Latency. We consider the cloud has all services and adequate computing resource, and hence the executing time of a task offloaded to the cloud can be ignored. For a task whose required service l is not placed on the edge server, we assume the expected time spent for transmitting its input data to the cloud (referred as the offloading latency) is ϕ_l^t. Obviously, it depends on the input data size and the network condition between the base station and the cloud in t. Therefore, the overall offloading latency incurred by a service placement decision $\mathbf{I}^t$ in time slot t can be obtained as

$$\Phi^t = \sum_{l=1}^{L} (1 - I_l^t) \cdot n_l^t \cdot \phi_l^t. \tag{6}$$

3.3   Problem Formulation

In this work, we focus on the online service placement problem in a mobile edge computing system, where heterogenous computation-intensive tasks stochastically arrive in real time. We try to find an optimal service placement strategy for the base station, aiming to minimize the total latency of all tasks over time, i.e.,

$$\min_{\mathbf{I}^t} \sum_{t=1}^{T} \Psi^t + \Omega^t + \Phi^t$$

$$\text{s.t.} \sum_{l=1}^{L} I_l^t s_l \leq s_{max}, \forall t,$$

$$\sum_{l=1}^{L} I_l^t f_l \leq f_{max}, \forall t.$$

Without complete future information, e.g., stochastic task arrivals and dynamic network conditions, it is difficult to obtain the optimal service placement strategy. Moreover, as the switching latency not only depends on the current service placement decision but also the previous service placement, there exists a trade-off between migrating unplaced services or offloading tasks to the cloud, in terms of the total latency incurred. Specially, service migration introduces certain switching latency, while the execution latency of tasks can be reduced.

To this end, we propose an online learning-based approach for the problem via adopting reinforcement learning [22], which is applicable for decision making in stochastic and dynamic environments without any future information in prior.

4   Approach Design

To adopt reinforcement learning, we firstly reformulate the online service placement problem as a Markov Decision Process (MDP). Then, we propose a DQN-based online service placement approach.

4.1   Reformulated MDP Problem

A typical MDP model mainly consists of state space $\mathcal{S}$, action space $\mathcal{A}$, and reward function R. In what follows, we formally model the online service placement problem as a MDP, by defining each component mentioned above.

State Space. Each state $s^t \in \mathcal{S}$ represents the information observed from the mobile edge computing system in time slot t. Specially, state s^t contains the service placement in the last time slot, the arrivals of heterogenous tasks in the current time slot, and the current network condition, i.e., $s^t \triangleq (\mathbf{I}^{t-1}, \mathbf{n}^t, B^t)$. Here, B^t represents the real-time bandwidth of the channel between the base station and the cloud.

Action Space. In each time slot t, the base station needs to make a service placement decision according to the current state, determining which services placed on the edge server, i.e., $a^t \triangleq \mathbf{I}^t$, and hence $\mathcal{A} = \{0, 1\} \times_L$.

Reward Function. After taking action a^t under state s^t, the base station will receive a reward r^t defined as the total latency incurred in time slot t, i.e.,

$$r^t = -(\Psi^t + \Omega^t + \Phi^t). \tag{7}$$

Obviously, minimizing the overall latency is equivalent to maximizing the reward.

4.2   DQN-based Approach Design

Given the MDP model defined above, we can adopt reinforcement learning to obtain the optimal service placement policy. A feasible model-free reinforcement learning solution is Q-learning [20], which is widely used for making controls in a dynamic environment without any prior knowledge. In Q-learning, Q-function

$Q(s^t, a^t)$ is introduced to represent the estimated cumulative reward achieved by taking action a^t under state s^t.

$$Q(s^t, a^t) \triangleq \mathbb{E}\left[\sum_{\tau=t}^{T} \gamma^{\tau-t} \cdot r^\tau\right], \tag{8}$$

where $\gamma \in [0, 1]$ is a discount factor to indicate the degree of future rewards we look into. An optimal action a^t can be obtained by maximizing the Q-function.

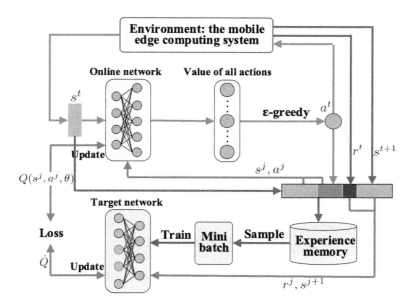

Fig. 2. The workflow and training process of our proposed DQN-based approach.

However, Q-learning maintains the Q-function value of each state-action pair in a Q-table, which will enlarge explosively with the increase of the state and action space. In our MDP model, the number of state-action pairs is infinite, as the number of arrived tasks and the network condition vary continuously. Therefore, Q-learning cannot be used to solve the online service placement problem. To tackle this challenge, the DQN solution [11] has been developed, in which the Q-function of each state-action pair is estimated by a deep neural network (DNN). In this work, we design a DQN-based approach for the online service placement problem.

Approach Workflow. The workflow of our proposed approach is presented in Fig. 2. In each time slot t, the base station firstly observes the current state s^t of the mobile edge computing system, and input it to a DNN with parameters θ which is called the *online network*. Then, the network outputs the Q-function value of each action. At last, the base station selects an action a^t according to the ϵ-greedy algorithm [16]. To be specific, the action with the maximal value $Q(s^t, a^t; \theta)$ is picked by the edge server with probability $1 - \epsilon$, which is referred as *exploitation*. On the other hand, the edge server also has the chance to randomly chooses an arbitrary action with probability ϵ to prevent being trapped in the local optimum. This way is referred as *exploitation*. Parameter $\epsilon \in (0, 1)$ is tunable to balance the tradeoff between exploitation and exploration. After taking action a^t, a reward r^t is returned by the environment and the state transforms to s^{t+1}. The process is recorded by a tuple (s^t, a^t, r^t, s^{t+1}), referred as an *experience*, which is stored in the memory of the edge server.

Network Training. We train the parameters of the online network based on historical experiences. Firstly, we randomly sample a mini batch of experiences from the memory, and input them to the online network and another DNN which has the same structure with the online network but different parameters θ', referred as the *target network*, as shown in Fig. 2. The target network is used to estimate the ground truth of Q-function values, to help train the parameters of the online network.

We take one sampled experience (s^j, a^j, r^j, s^{j+1}) as an example to illustrate the training process. Specially, we can obtain the Q-function value estimated by the online network given s^j and a^j, i.e., $Q(s^j, a^j; \theta)$. Then, s^{j+1} is input to the target network and the target Q-function value $\hat{Q}$ can be obtained according to the Bellman equation [22], i.e.,

$$\hat{Q} = r^t + \gamma \cdot \max_a Q(s^{j+1}, a; \theta'). \tag{9}$$

In order to reduce the difference between $Q(s^j, a^j; \theta)$ and $\hat{Q}$, we define the loss function to train the online network as follows,

$$Loss = (\hat{Q} - Q(s^j, a^j; \theta))^2. \tag{10}$$

We leverage the gradient descent algorithm to minimize the loss function. And hence, the parameters θ can be updated as

$$\theta \leftarrow \theta + \eta(Q(s^j, a^j; \theta) - \hat{Q})\nabla Q(s^j, a^j; \theta), \tag{11}$$

where η is the learning rate.

To maintain the stability of the training process, we update the parameters of the online network and the target network asynchronously. To be specific, the parameters of the online network are updated immediately after each training process. Whereas, the target network copy the parameters of the online network after a certain time slots.

5   Performance Evaluation

5.1   Simulation Setup

We conduct extensive simulations to evaluate the performance achieved by our proposed approach, compared with four baselines listed as follows.

- *Cloud Processing Only Approach*: All tasks are offloaded to the remote cloud directly, which has all services and adequate computing resources.
- *Stochastic Approach*: Each service is randomly determined to be placed on the edge server or not, with considering the storage space and CPU frequency constraints.
- *Service-prior Greedy Approach*: As many as possible services are placed on the edge server, under the storage space and CPU frequency limitations.
- *Task-prior Greedy Approach*: Services are placed on the edge server one by one. The service which can serve the most tasks in the current time slot is selected each time, unless the resource constrains cannot be satisfied.

We evaluate the performance of different approaches according to the their achieved rewards, i.e., the total latency of all tasks.

The default values of parameters in our simulations are set as follows. There are four types of services. We set the associated data size s_l and the required CPU frequency f_l of each service are within $[30,40]$ GB and $[2,3]$ GHz, respectively. The maximal storage space and CPU frequency of the edge server are set as $s_{max} = 100$ GB and $f_{max} = 5$ GHz, respectively. Besides, the required CPU cycles of each task served by service l follows an exponential distribution with mean $c_l \in [0.02, 0.03]$ GHz. The switching latency of migrating service l from the cloud to the base station is $\psi_l^t \in [0.1, 0.4]$ seconds, and the latency for offloading a task served by service l to the cloud is $\phi_l^t \in [0.5, 1]$ seconds. We set the number of tasks served by l arriving in time slot t, i.e., n_l^t, is uniformly distributed in $[50,100]$. The interval of each time slot is set as 1 s. In addition, we set each episode in the training phase of our approach equals to 250 time slots, in which a whole interacting process between the base station and the environment is completed. In default, the parameters of the target network are updated once when the parameters of the online network have been updated 40 times (referred as *update frequency* in our evaluation results). Besides, the batch size sampled for training is set as 64. Discount factor γ to 0.9, and learning rate η is 0.005.

5.2   Evaluation Results

We firstly study the impact of the batch size sampled for training and the update frequency of the target network to the convergence speed of the training process in our approach, as shown in Fig. 3 and Fig. 4 respectively. Figure 3 plots the rewards achieved by our approach varying with the episode, when the batch sizes are 64, 128, and 256, respectively. We can find that when the batch size equals to 64, the training process of our approach can converge faster than when the batch size is 128 or 256. This is because the smaller batch size, the steeper direction of the gradient descent. In other words, the parameters of the online network can be updated more quickly, according to Eq. (11). Moreover, a larger batch size consumes more time for training. Hence, we set the default batch size as 64 in the training phase. Figure 4 presents the rewards obtained by our approach changing with the episode, when the update frequencies are 40, 80, and 120, respectively. Apparently, the larger update frequency, the slower convergence speed of the training process. In order to make a tradeoff between maintaining the stability of the training process and achieving faster convergence, we set the update frequency as 40.

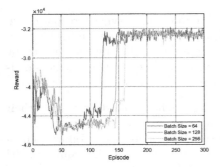
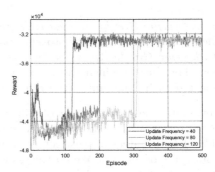

Fig. 3. Convergence v.s. batch size.   **Fig. 4.** Convergence v.s. update frequency.

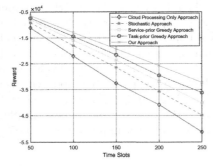
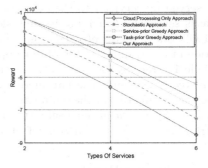

Fig. 5. Reward v.s. number of time slots.   **Fig. 6.** Reward v.s. number of services.

Next, we evaluate the performance of our proposed approach, compared with the four baselines, under different parameter settings. As shown in Fig. 5, we plot the rewards achieved by different approaches when the number of time slots varies from 50 to 250. The more time slots, the larger total latency achieved by each approach, since tasks successively arrive to the base station over time. We can find that the cloud processing only approach achieves the worst performance, due to the long data transmission delay to the cloud. Therefore, it is highly recommended to extend some services to the network edge. Furthermore, it can be found that the task-prior greedy approach achieves better performance than the other baselines, as more tasks can be served on the edge server. However, it obtains larger total latency than our approach, due to the increase of the switching latency introduced by frequent service migration. Obviously, our approach achieves the best performance under different settings of the number of time slots. Specially, when there are 250 time slots, the total latency achieved by our approach is 10.91%, 19.82%, 28.11%, 36.86% lower than the four baselines, respectively. Figure 6 presents the rewards achieved by the five approaches, when the number of service belongs to $\{2, 4, 6\}$, respectively. When there are only two services, the performance of our approach is as same as the service-prior greedy approach and the task-prior greedy approach, because the two services can be placed on the edge server at the same time. With the increase of the number of services, our approach significantly outperforms the baselines, which indicates our approach is suitable to complex service placement scenarios. Specially, when there are six services, the total latency obtained by our approach is 14.96%, 21.17%, 28.48%, 36.47% lower than the four baselines, respectively.

6   Conclusion

In this work, we focus on the online service placement problem in a 5G-enabled mobile edge computing system, in which numerous tasks randomly arrive to the edge (i.e., base station) in real time. The base station should make service placement decisions in real time, it is necessary to find an optimal service placement policy to minimize the total completion latency. However, which services are placed on the base station should be dynamically adjusted according to the computation demands. Moreover, the additional latency incurred by service migration and the time spent for waiting in queues should be considered. To overcome the challenges, we formulate the service placement problem by an MDP model and propose an DQN-based online service placement approach to obtain the optimal service placement policy. Extensive simulations are conducted to validate the performance of our proposed approach. The simulation results show that our approach achieves higher reward compared with baseline methods, no matter how the number of time slots and different types of services.

Acknowledgement. This research is supported by NSFC (No. 61802245), the Shanghai Sailing Program (No. 18YF1408200), and STSCM (No. 19511121000). This work is also supported by the Open Project Program of Shanghai Key Laboratory of Data Science (No. 2020090600002).

References

1. Aissioui, A., Ksentini, A., Gueroui, A.M., Taleb, T.: On enabling 5G automotive systems using follow me edge-cloud concept. IEEE Trans. Veh. Technol. **67**(6), 5302–5316 (2018)
2. Al-Shuwaili, A., Simeone, O.: Energy-efficient resource allocation for mobile edge computing-based augmented reality applications. IEEE Wirel. Commun. Lett. **6**(3), 398–401 (2017)

3. Ascigil, O., Phan, T.K., Tasiopoulos, A.G., Sourlas, V., Psaras, I., Pavlou, G.: On uncoordinated service placement in edge-clouds. In: 2017 IEEE International Conference on Cloud Computing Technology and Science (CloudCom), pp. 41–48. IEEE (2017)

4. Farhadi, V., et al.: Service placement and request scheduling for data-intensive applications in edge clouds. In: IEEE INFOCOM 2019-IEEE Conference on Computer Communications, pp. 1279–1287. IEEE (2019)

5. Gao, B., Zhou, Z., Liu, F., Xu, F.: Winning at the starting line: joint network selection and service placement for mobile edge computing. In: IEEE INFOCOM 2019-IEEE Conference on Computer Communications, pp. 1459–1467. IEEE (2019)

6. He, T., Khamfroush, H., Wang, S., La Porta, T., Stein, S.: It's hard to share: joint service placement and request scheduling in edge clouds with sharable and non-sharable resources. In: 2018 IEEE 38th International Conference on Distributed Computing Systems (ICDCS), pp. 365–375. IEEE (2018)

7. Hu, Y.C., Patel, M., Sabella, D., Sprecher, N., Young, V.: Mobile edge computing-a key technology towards 5G. ETSI White Paper 11(11), 1–16 (2015)

8. Ksentini, A., Taleb, T., Chen, M.: A Markov decision process-based service migration procedure for follow me cloud. In: 2014 IEEE International Conference on Communications (ICC), pp. 1350–1354. IEEE (2014)

9. Liu, J., Zhong, L., Wickramasuriya, J., Vasudevan, V.: uWave: accelerometer-based personalized gesture recognition and its applications. Pervasive Mobile Comput. 5(6), 657–675 (2009)

10. Mao, Y., You, C., Zhang, J., Huang, K., Letaief, K.B.: A survey on mobile edge computing: the communication perspective. IEEE Commun. Surv. Tutor. 19(4), 2322–2358 (2017)

11. Mnih, V., et al.: Human-level control through deep reinforcement learning. Nature 518(7540), 529–533 (2015)

12. Nadembega, A., Hafid, A.S., Brisebois, R.: Mobility prediction model-based service migration procedure for follow me cloud to support QoS and QoE. In: 2016 IEEE International Conference on Communications (ICC), pp. 1–6. IEEE (2016)

13. Ouyang, T., Li, R., Chen, X., Zhou, Z., Tang, X.: Adaptive user-managed service placement for mobile edge computing: an online learning approach. In: IEEE INFOCOM 2019-IEEE Conference on Computer Communications, pp. 1468–1476. IEEE (2019)

14. Ouyang, T., Zhou, Z., Chen, X.: Follow me at the edge: mobility-aware dynamic service placement for mobile edge computing. IEEE J. Sel. Areas Commun. 36(10), 2333–2345 (2018)

15. Poularakis, K., Llorca, J., Tulino, A.M., Taylor, I., Tassiulas, L.: Joint service placement and request routing in multi-cell mobile edge computing networks. In: IEEE INFOCOM 2019-IEEE Conference on Computer Communications, pp. 10–18. IEEE (2019)

16. Sutton, R., Barto, A.: Introduction to Reinforcement Learning. MIT Press, Cambridge (1998)

17. Taleb, T., Ksentini, A., Frangoudis, P.: Follow-me cloud: when cloud services follow mobile users. IEEE Trans. Cloud Comput. PP, 1 (2016)

18. Wang, L., Jiao, L., He, T., Li, J., Mühlhäuser, M.: Service entity placement for social virtual reality applications in edge computing. In: IEEE INFOCOM 2018-IEEE Conference on Computer Communications, pp. 468–476. IEEE (2018)

19. Wang, S., Dey, S.: Adaptive mobile cloud computing to enable rich mobile multimedia applications. IEEE Trans. Multimed. 15(4), 870–883 (2013)

20. Watkins, C.J., Dayan, P.: Q-learning. Mach. Learn. **8**(3–4), 279–292 (1992). https://doi.org/10.1007/BF00992698
21. Xu, J., Chen, L., Zhou, P.: Joint service caching and task offloading for mobile edge computing in dense networks. In: IEEE INFOCOM 2018-IEEE Conference on Computer Communications, pp. 207–215. IEEE (2018)
22. Zeng, D., Gu, L., Pan, S., Cai, J., Guo, S.: Resource management at the network edge: a deep reinforcement learning approach. IEEE Netw. **33**(3), 26–33 (2019)

A Hybrid Collaborative Virtual Environment with Heterogeneous Representations for Architectural Planning

Krishna Bharadwaj[✉] and Andrew E. Johnson

University of Illinois at Chicago, Chicago, USA
{kbhara5,ajohnson}@uic.edu

Abstract. We developed a collaborative virtual environment for architectural planning that facilitates groups of people to work together using three different interfaces. These interfaces enable users to interact with the scene with varying levels of immersion and different interaction modalities. We conducted a user study to gauge the general usability of the system and to understand how the different interfaces affect the group work. In this paper we present the architecture of the system along with its different interfaces. We also present the user study results and the insights we gained from the study.

Keywords: Group work · Collaborative environments

1 Introduction

Improvements in network connectivity and advances in web technology have made it possible for people from across the globe to partner up with one another for working on large projects through remote collaboration. We see that more and more project teams, from both industry as well as academia, are becoming geographically distributed. As a result, we see an increase in efforts being made to come up with collaborative environments that support such large distributed groups to work together. Initially such collaborative work was limited to working on textual data but off late attempts are being made to support collaboration over data of graphical nature, such as geo-data [7] for mapping events, spatial 3D data [2,8] for sculpting of 3D models, architectural planning [1] and so on. A few of these emphasize on remote collaboration while others are more geared towards collocated collaboration.

We have developed a hybrid collaborative virtual environment for architectural planning, in that it supports both collocated as well as remote collaboration. The system consists of three different interfaces, namely a 2D interface, a 3D interface, and an immersive VR (Virtual Reality) interface, all interacting with a virtual environment for architectural planning. All three interfaces are

H. Gao et al. (Eds.): CollaborateCom 2020, LNICST 350, pp. 184–199, 2021.
https://doi.org/10.1007/978-3-030-67540-0_11

synchronised using a centralized server to enable real time remote collaboration between them. Further, the 2D and 3D interfaces are developed as custom applications on top of SAGE2 (Scalable Amplified Group Environment) [3] leveraging the affordances of SAGE2 for collocated collaboration. This allows more than one person to simultaneously interact with, and control, both of these interfaces. The VR interface has been developed, using Unity 3D game engine, for HMDs (Head Mounted Display). The 2D interface presents a monochromatic top view of the scene over a grid layout, mimicking a floor plan. The 3D interface presents an in-person perspective of the scene with a limited field of view, giving the users a driver-seat view of the scene. The VR interface presents a fully immersive 360° view through the HMD to the user interacting with it. All the three interfaces are complete applications in their own right, in the sense that they can be individually used to create architectural designs. All functionalities concerning architectural design creation such as the creation of walls, creation and manipulation of furniture, and so on have been implemented in all the three different interfaces.

The motivation for this amalgamation of interfaces with different perspectives and viewpoints is threefold. First, in a collaboration consisting of multiple sites and users, a practical issue is the availability of hardware and other resources at every site. For example, if the required hardware has a large physical footprint, such as a CAVE (Cave Automatic Virtual Environment), chances are that not every site will have access to it or some user might be away from the hardware (say, working from home) and hence may not have access to it. In such cases, a system that is flexible in its hardware and software requirements for allowing a user to participate in the collaborative session is desirable. Second, each representation i.e 2D, 3D, and immersive VR, has its own affordances and limitations with respect to interaction with data. For example, while it can be very easy for a user to quickly place furniture in a floor plan using a 2D interface (or even just to draw on an actual floor plan), some level of immersion in a 3D scene is preferable to assess whether there is enough room to "walk around" in the space being designed. Further, while the immersive VR gives the user a chance to interact with the space at full scale, it prevents them from participating in face to face interaction with other collocated members of the team, whereas a 3D interface supporting collocated collaboration could provide a middle ground in such a case. Hence, a system that supports multiple ways of interacting with data can help the users by compensating for the limitations of one interface with the affordances of another. Third, project team sizes can have a wide range with the teams being split into smaller groups of collocated people, spread geographically. A system that can support both collocated as well as remote collaboration can enable more people from such teams to take part in the collaboration. Hence, by making each interface complete with all the functionalities and designing the synchronization server to handle multiple instances of each interface, we hope this model will afford flexibility to the users at each site, in choosing an interface that is best suited for them to participate in the collaboration.

2 Related Work

2.1 Distributed Collaborative Virtual Environments

Recent years have seen an increase in distributed collaborative systems that allow multiple users to interact with, and in some cases edit, data in virtual environments. Okuya et al. [4] discusses such a system that allows real time collaboration between users interacting with a wall-sized display and a CAVE-like system to edit CAD data. Even though the system combines two different VR platforms, it presents the same representation of the CAD data through both of these platforms, to the users. Our proposed system, in contrast, provides heterogeneous representations to users at different sites. Pick et al. [5] present a system to combine IVR (Immersive Virtual Reality) systems such as the CAVE with a lightweight web based counterpart that offers the same functionality and perspectives as the IVR system but in a reduced capacity to facilitate the integration of IVR in the factory planning process. The lightweight application provides a slice of the view that is presented to the IVR users. While our system presents users with heterogeneous representations, it ensures that every user is able to fully interact with the virtual environment, albeit with different interaction modalities.

2.2 Collaborative Environments with Asymmetric Viewpoints

A few collaborative virtual environments have incorporated asymmetric views or perspectives of the virtual world. CALVIN [6] is one of the earliest systems to present this approach where two sets of users interact with a virtual environment, one set from an in-person perspective, while the other interacts "from above". This second set of users is presented with a miniaturized version of the virtual environment to create the effect of interacting from above with a scaled model of the environment that the first set interacts with. Avatars are employed to facilitate co-presence. All instances are run in CAVEs. DollhouseVR [10] is another system that deals with asymmetric viewpoints in the form of a table top surface presenting a top view of the virtual environment while a head mounted display provides an in person view of the same. Although they make use of different technologies to present the two different viewpoints, both of those viewpoints show the same representation of the virtual world. Moreover, the system is intended for collocated collaboration. MacroScope [9] is a mixed reality application that aids in collocated collaboration by presenting a VR user with a first person perspective of an actual physical scale model that other team members in the room interact with. Our proposed system, in comparison, combines different perspectives with multiple representations with the aim of complementing the limitations of each representation or perspective with the affordances of a different representation or perspective.

3 System

3.1 Functional Requirements for an Architectural Design Application

All the three different interfaces should incorporate all the functional requirements for editing the architectural design. Here we briefly describe these requirements to help the reader understand the implementation details of the different interfaces that are explained below. We decided to have a basic set of four types of elements as the building blocks of our design space. They are walls, doors, windows, and furniture. Walls have a fixed height and thickness, but variable lengths and can be placed anywhere within the scene. So it would suffice for any interface facilitating creation of walls to provide the users a way to specify the start and end points of the wall. Doors and windows are functionally very similar to each other, in the sense that they are both units that are placed within the walls, have fixed shapes and sizes, take on the orientation of the wall within which they are placed, and can be placed anywhere throughout the length of the wall as long as they stay entirely within the bounds of the wall length. To facilitate creation of doors and windows, an interface must ensure that once created, a door or a window must only take on a point along the length of a preexisting wall as its valid stationary position, and have its y-axis rotation in line with that of the wall. Furniture objects such as couches, tables, chairs, and so on all have fixed shapes and sizes. Interaction with furniture thus gets limited to changing the position and y-axis rotation of a piece of furniture.

3.2 Architecture

SAGE2 exposes an API for creating custom applications. We developed the 2D and the 3D interfaces as SAGE2 apps to leverage the multi-user interaction capabilities that SAGE2 offers. This makes it possible for more than one user to simultaneously interact with these two interfaces, thus enabling collocated collaboration. SAGE2 allows multiple users to interact with applications on a large display using their personal devices such as laptops. Multiple users can simultaneously access its web interface to connect to the large display and interact with it. By making the 2D and the 3D interfaces be custom applications on top of SAGE2 we let SAGE2 handle multi-user interaction for 2D and 3D interfaces. As shown in Fig. 1, we developed a synchronization server that relays messages between all the instances of the three different interfaces and keeps them updated and in sync. Every action of each client such as moving a piece of furniture, deleting a wall and so on is conveyed, in real time, to all the participating clients. Further, the pointer location of the users at the 2D interface, the camera location and orientation of the 3D interface, and the head location and orientation of the VR users are all conveyed to all the clients in real time as well. Thus any action performed at any site is immediately replicated at all the sites. The position (and orientation) information of different users are used to animate their corresponding virtual representations at all the different sites.

This allows users to know where "within" the scene, each user is and what they are doing at any given point in time.

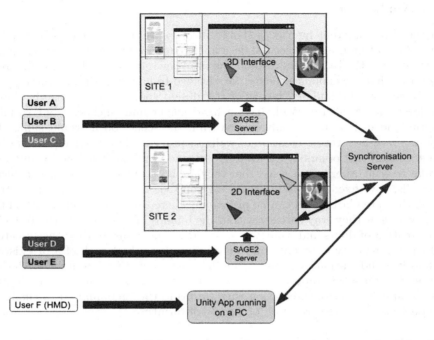

Fig. 1. Architecture of the collaborative setup showing one instance each of the 2D, 3D, and VR interfaces

3.3   2D Interface

The 2D interface has been implemented as a SAGE2 custom application in order to leverage the affordances provided by the SAGE2 platform such as scalable resolution and simultaneous multi-user real time interaction. The 2D interface as shown in Fig. 2, has a grid layout with each grid block representing 1 foot. These grid lines along with the rulers on the border depicting the foot units of the lines are meant to serve as guides to the users for placement of objects within the scene. A menu button has been provided on the top left corner but can be moved around (to avoid occlusion of the grid space) anywhere within the layout of the interface. The menu contains options to create walls, doors, windows, furniture objects, and flags. A contextual "help" text appears on the bottom left corner of the layout, to guide the users with information about the possible next steps that the users can take when interacting with either the objects in the scene or during one of special modes. The layout can be zoomed in and out enabling the users to work at a scale that is comfortable to them. We did not provide panning functionality in the 2D interface since the screen real estate in a SAGE2 environment is sufficiently large to present the layout in its entirety.

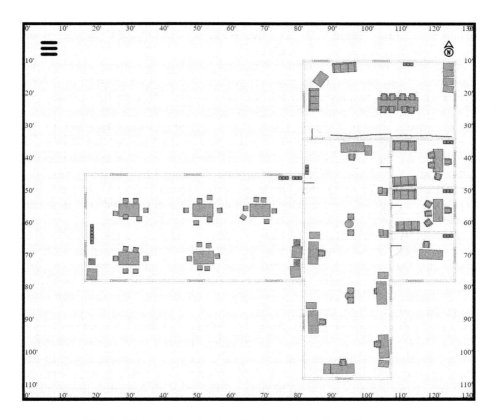

Fig. 2. 2D interface showing the floor plan of an office space.

In the 2D interface by default, users will be in "Selection" mode. In this mode, a user can "select" objects in the scene by clicking on them, for continued interaction with the selected object. For example, in case of furniture, the selection allows the users to move around, rotate, or delete the piece of furniture from the scene. When a user is in a different mode such as the "Wall Creation" mode, the user can get back to the "Selection" mode by clicking on the "Selection" mode icon from the menu. To create a wall, users can click on the wall icon from the menu. This makes the user enter the "Wall Creation" mode. This is indicated in the help text as "click to start wall". Now, a click anywhere on the grid initiates a new wall. The point of click is mapped to the nearest grid point and that grid point is used as the actual starting point of the wall, to get the effect of snapping to the grid. The current mouse position becomes the ending point of the wall, resulting in a "rubber band" wall fixed on one end and moving with the mouse pointer on the other end. A second click fixes the ending point of the wall to the location of the click (mapped to the nearest grid point), thereby completing the wall creation. To facilitate quick creation of adjoining walls, this end point of the wall is also treated as the starting point for a new wall. A user can easily break this chain by using a designated key to remove the current wall (rubber band wall). The user remains in the "Wall Creation" mode and can continue to create new walls.

To create a door or a window in the 2D interface, users can click on the respective icon in the menu and a corresponding item gets attached to the user's pointer and starts to move with the pointer. A user will now be in "Door or Window" mode. When a user hovers the pointer on a wall, the attached door or window orients itself in line with the wall as a way of providing feedback to the user that is a potential "drop" point for the door or window. At such a location the user could then make a single click to fix the door or window at that point on the wall. A click anywhere else other than on a wall has no effect, and the new instance of door or window continues to be attached to the user's pointer. If the user decides not to place the item on any wall, a designated key press can be used to remove the attached item from the pointer, bringing the user out of "Door or Window" mode. To create a piece of furniture in the 2D interface, users can click on the respective icon in the menu and a new instance of the chosen piece of furniture gets attached to the pointer and moves with it. A click anywhere on the layout will "drop" the new piece of furniture at that point.

To change the position of a previously placed door or window, users can "pull" the instance from its location on the wall by performing a mouse down and a slight drag. This action results in the instance of door or window getting attached to the mouse pointer, thereby bringing the user to "Door or Window" mode. Now the user can place it at a new position on a wall anywhere within the layout or discard it, as explained above. To move a piece of furniture, users can perform a mouse down and a slight drag on the piece of furniture in question. This attaches the piece of furniture to the mouse pointer. Now the user simply "drops" it off at a new position with a click of the mouse at the desired position. To change the orientation (rotate) of a piece of furniture, users can "select" it as explained above and using designated keys on the keyboard, can rotate the piece of furniture along its y-axis. To remove an object (wall, door, window, or furniture) from the scene, users can "select" it as explained above and then use a designated key press to remove it from the scene.

Flags are special objects that can be used as points of reference within the scene. They can be used to draw different users' attention to a part of the scene. Creation of flags follow the same process as pieces of furniture explained above. The flags are shown in the 2D interface as colored circles as shown in Fig. 5a. The color of a flag is chosen randomly by the interface from a predefined set of colors. A flag's color is shared across all sites thus allowing it to act as a point of reference within the scene. Any user from any site can refer to a flag by its color and the other users will be able to unambiguously and accurately infer where within the scene the flag is.

Fig. 3. Two users interact with the 3D interface

3.4   3D and VR Interfaces

The 3D interface presents the users with an in-person perspective of the scene, rendered on a large display in a rectangular window as shown in Fig. 3, whereas the VR interface presents the same in-person perspective in a fully immersive head mounted display as shown in Fig. 4. The scene itself consists of a floor laid with grid lines for aiding the users with placement of objects. A menu, identical to that of the 2D interface both in appearance as well as functionality, is provided in the 3D as well as the VR interfaces. Both of these interfaces also have "Selection", "Wall Creation", and "Door or Window" modes similar to the 2D interface, albeit with interaction metaphors that are more appropriate to 3D interaction. For example, unlike the 2D interface, newly created pieces of furniture do not follow the pointer, and are placed in front of the camera and the VR user respectively. Users can then "pick" them up and move them around.

The 3D interface allows the users to navigate through the scene by mapping the mouse scroll to forward and backward movements of the camera within the scene and designated keys for turning left and right. The VR interface allows the user to freely look around and walk within the scene as well as teleport to any point within the scene. Flags are presented as tall (40ft) colored pillars within the scene in the 3D and the VR interfaces as shown in Fig. 5b and c. This is to help users to see the flags despite being behind walls and other structures that might occlude part of their view. Additionally, flags in the VR interface act as teleportation targets. This allows the VR user to easily reach a flag despite being anywhere within the scene, since the flags are tall and can be seen even from a distance and even when the user is behind any structure or objects in the scene.

Fig. 4. VR View of a part of the designed space

The 3D interface has an overview map that allows the users to quickly and easily get an idea of the entire scene as well as where within the scene, the other participants are. This makes up for the comparatively slower navigation of the 3D interface within the scene (the "drive through the scene" metaphor of the 3D interface is relatively slower than the VR users' ability to teleport instantly to any part of the scene or the 2D interface users' ability to see the whole scene at once and "be" at any point within the scene by simply moving the mouse pointer over that location on the grid). In this way, when a location or an object within the scene is referred to by the other users, giving some description of the location or a reference point such as a flag, whereas the VR users and 2D users "navigate" to the point as part of the collaborative exchange, the 3D users can make sense of the reference with the help of the overview and thus still be able to meaningfully participate in such a collaborative exchange.

3.5   Representations of Different Users Within the Scene

In a typical collaborative virtual environment co-presence of different users is achieved by representing users as avatars. This makes sense when all the users are fully immersed in the environment and also have the same interaction affordances. However, in a hybrid system like ours, different users have different levels of immersion and affordances of different interfaces impose differences in how they interact with the scene. To achieve meaningful co-presence and to facilitate effective communication in such a case, any representation of a user should reflect these differences. Keeping this in mind we created representations as follows: The 2D interface serves as a top view and this is reinforced in the way the VR user as well as the 3D interface's camera within the scene are depicted in the 2D interface as shown in Fig. 5a. The VR user's representation shows a human icon wearing a VR headset, from the top view. The camera of the 3D interface is shown as a straight line with bent edges indicating a "window" in to the scene with a limited field of view. At the 3D and VR interfaces, the mouse pointers of users from the 2D interface are represented using 3D arrows in the scene and they continuously move about within the scene (similar to the "God-like" interaction technique [11]) following the pointer movements of users within the grid of the 2D interface. The VR user is represented as an avatar (Fig. 5b) in the 3D interface, with two rays attached, that reflect where the VR user is pointing the controllers. The camera of the 3D interface is represented as a rectangular

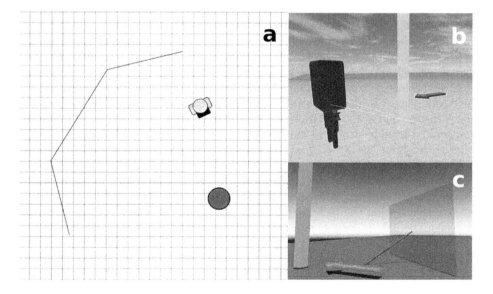

Fig. 5. a. 2D interface showing 3D user and VR user and a flag. b. 3D interface showing 2D user's pointer as an arrow and VR user as an avatar with the same flag shown as a column. c. VR interface showing 2D user's pointer as an arrow and 3D user as a 3D window with a pointer coming out of it and the flag shown as a column

window in the VR interface (Fig. 5c). Whenever a user at the 3D interface has their pointer on the interface window, a ray is shown as coming out of the window, enabling others in the scene to identify where the user is pointing to. That is, the pointer location of users are used to cast rays from the camera into the scene, and this is shown to other users in the scene.

4   Evaluation

To get an idea of the general usability of our system and to gain insights into how the differences in representations across the interfaces affect the group work, we conducted a group user study. In the study we asked a group of participants to use all three of the interfaces to collaborate in designing an office space.

4.1   Method

The study consisted of 5 trials (and two mock trials prior to the actual study to catch any interaction issues with the interfaces), with each trial involving 4 participants. One participant interacts with the 2D interface, and one interacts with the VR interface through an HMD, and two participants interact with the 3D interface. Even though both 2D and 3D interfaces are capable of handling multiple users in a collocated collaborative manner, due to lack of availability of participants we decided to limit collocated collaboration to only one of those interfaces in the study. Each interface was situated in a different room within our lab to reflect a real remote collaboration scenario. 3D interface was presented on a 23.8 × 6.7 foot (37.3 megapixel) large display running a SAGE2 instance. The display was situated in a 41 × 24 foot room. 2D interface was presented on a 13.4 × 5.7 foot (12.6 megapixel) large display running a SAGE2 instance. The display was situated in a 41 × 28 foot room. The participants were seated in front of these large screen displays and interacted with them using laptops. In a third room, the VR interface was presented using an HTC vive HMD to the user. The VR user had a 10 × 10 foot space to walk around. All three locations were connected through an audio conference and participants could speak to each other throughout the study. The participants were given a brief practice session at the beginning of each trial to introduce them to different functionalities of the interface they were going to work with and to familiarize them with co-presence, tele-pointing, and communicating with each other. After undergoing the practice session the participants were given a set of high level requirements to design the office space such as, "A conference room that can host 8 people", "An open floor area to seat 4 employees", and so on. We set a time limit of 75 min to give the participants enough time to work, however they could finish earlier. Every session was audio and video recorded. Additionally, we recorded the head orientation of the users at the 3D interface. Also, every action of each user was logged capturing details such as their location in the scene at the time of the action and the object/s in the scene that the user interacted with, in taking that action. A brief survey was administered to the participants at the end of the session.

4.2   Results

We asked the participants to fill out a survey at the end of the user study session. The survey contained a set of questions aimed at getting an idea of the general usability of our system including its affordances for group communication and collaboration. Table 1 shows the results we obtained from the survey. While the results indicate that the participants were fairly satisfied with its usability, relatively lower scores were reported for the 3D interface on navigation and object manipulation metrics. We had also asked descriptive questions towards understanding any issues the participants might have had in interacting with the system. Some of the answers we obtained helped us understand those lower scores. On a few occasions when the view had too many closely placed objects, the 3D interface made it difficult to accurately select objects. The navigation difficulty was also reported when the view had too many objects. This is mainly due to the object picking algorithm that we implemented at the 3D interface and has been fixed since the running of the user study. The different representations of different interfaces to achieve co-presence did not impede the collaboration as we can see from the results. In fact, we noted through the video recordings of the sessions that the users very quickly became accustomed to how others perceived the space and how they interacted with the scene.

Table 2 gives a summary of the interactions in each trial along with duration of the trials. This data shows the participants interacted quite a fair amount with the system and thus further supports the subjective scores from the Table 1.

Table 1. Average scores of metrics for usability, co-presence awareness, and ease of collaboration (on a scale of 1–5)

Metric	2D Interface	3D Interface	VR Interface
Scene Navigation	4.3	2.7	3.2
Object Creation	4.4	4.1	4.4
Object manipulation	4.2	2.6	3.2
Locating others users within the scene	5.0	4.1	3.8
Tell where other users were looking or pointing	4.6	3.9	3.6
Tell what objects others were interacting with	4.2	3.5	3.4
Tell what interactions others were performing	4.0	3.5	3.0
Draw other users' attention	4.8	4.3	4.4
Communicate	4.3	4.3	3.8
Convey Ideas	4.6	4.7	4.2
Collaborate	4.3	4.3	3.6
Complete the task	4.6	5.0	5.0

Table 2. Total duration and summary of objects created and edits in the scene

Trial	Time to completion (in min)	Objects created	Total edits
1	34.5	135	118
2	52.8	261	174
3	38.2	185	146
4	29.5	138	90
5	68.0	286	139

The chart in Fig. 6 shows a breakup of all the objects that were created by different participants (we combine the numbers of both the collocated participants at the 3D interface, as the same affordances apply to both of them) of all the trials. As can be seen in this chart, except for the fifth trial, in every other trial, the walls were mostly created by the 2D interface. Through the recording we observed that, at the beginning of each the participants briefly discussed how to proceed with the task, and in the first four trials, the users felt that large wall creation was easier for the participant interacting with the 2D interface. We also noticed that, soon after creating the walls, the participants at the 2D interface would proceed to place the doors and windows on them. This can also be noted from the break down shown in Fig. 6.

We noted all the different interactions (edits) that participants had with different objects that they created in the scene throughout the session. Since these edits constituted a major part of the total work done we used the logs to find the break up of these edits based on "location" within the design space such

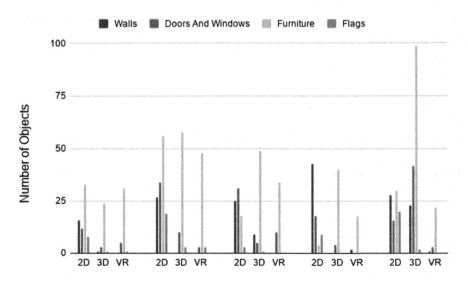

Fig. 6. Break down of all the design objects created by different participants

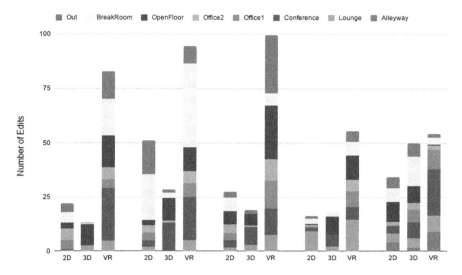

Fig. 7. Break down of edits done by different participants at different locations (Out refers to outside the boundaries of the floor plan)

as "Lounge Area" and so on. Figure 7 shows the results. We note two things from this break down. First, in some cases one of the participants worked almost entirely on particular parts of the space, for example, the VR user from trial 1 working solely on the conference room, whereas in certain cases such as in trial 5, all the participants shared the work in creating the conference room. Second, VR participants generally did more interactions. When seen together with the 2D participants mostly creating walls, the results suggest that task division is guided by the affordances of the different interfaces.

5   Discussion

From the video recordings of the sessions, we observed a few elements of the collaborative exchanges that happened between the participants. First, by the end of the practice session the participants concluded that certain aspects of the interfaces made it easy for performing particular roles within the collaboration and took this into account in working as a group. In every trial, the participants at the 3D interface took charge of leading the sessions. That is, they would assign different tasks to the other participants. For example, asking the 2D user to create walls, or asking the VR user to give feedback on a part of the designed space. We feel that this was largely due to the availability of an overview map, in addition to the in-person perspective for the 3D users, which gave them an advantage to choose between the two views that other two users had singly. This made them feel more in control to drive the session. Second, even though all interfaces are equipped with all the functionalities needed to complete the task at hand individually, the affordances of the different interfaces favored particular interfaces for specific functionalities. As we noted in the results, wall creation was perceived to be easiest for the 2D participant. The VR users were favored by the other participants for review of designed parts of the spaces, as they realized that realistic assessment of spaces was better done through VR. We noted several instances in these sessions where one of the other participants would ask the VR participant to give feedback on a part of the space that they had finished designing. They would place a flag at the location that they had worked on and notify the color of the flag to the VR user. The VR user would then quickly teleport to the flag, take a look around, comment on it, and go back to doing whatever they were doing. Third, having different representations did not impede the collaboration: Users were able to easily understand what a user was referring to whenever that user drew attention of others to some object within the scene. The recordings show that an average of 22 times (per session), users tried to draw the attention of each other to some object in the scene and this was immediately (less than 2 s) followed by an acknowledgement for their call and a response that confirmed to us that the other participants had correctly identified what was being referred to. Also, except for one trial, all the others made liberal use of flags (as seen in Fig. 6) to either draw each others' attention or to help others to navigate to a location. All these points help to reinforce our initial assumption that combining different interfaces with heterogeneous representations does not negatively affect the collaboration, but helps in making the group work more flexible.

6   Conclusion

We presented a hybrid collaborative virtual environment for designing architectural spaces that facilitates users to collaborate using different representations and asymmetric views. The results from a user study that we conducted demonstrated that our system enables users to freely collaborate despite having differences in interface affordances and scene representations. The results also bolstered our initial intuitive assumption that the limitations of one representation would be compensated by the affordances of another in carrying out the group task. As a future work, we plan to conduct a more detailed user study to gain insights into collaboration patterns that might emerge from our system.

Acknowledgement. We would like to thank the staff and students at the Electronic Visualization Laboratory, Department of Computer Science, University of Illinois at Chicago, for their support throughout the course of this work. This publication is based on work supported in part by National Science Foundation (NSF) awards # ACI-1441963 and # CNS-1625941.

References

1. Sugiura, Y., et al.: An asymmetric collaborative system for architectural-scale space design. In: Proceedings of the 16th ACM SIGGRAPH International Conference on Virtual-Reality Continuum and its Applications in Industry (VRCAI 2018), Article 21, pp. 1–6. Association for Computing Machinery, New York (2018). https://doi-org.proxy.cc.uic.edu/10.1145/3284398.3284416

2. Calabrese, C., Salvati, G., Tarini, M., Pellacini, F.: CSculpt: a system for collaborative sculpting. ACM Trans. Graph. **35**(4), 1–8 (2016). https://doi.org/10.1145/2897824.2925956

3. Marrinan, T., et al.: SAGE2: a new approach for data intensive collaboration using scalable resolution shared displays. In: 10th IEEE International Conference on Collaborative Computing: Networking, Applications and Worksharing, Miami, FL, pp. 177–186 (2014). https://doi.org/10.4108/icst.collaboratecom.2014.257337

4. Okuya, Y., Ladeveze, N., Gladin, O., Fleury, C., Bourdot, P.: Distributed architecture for remote collaborative modification of parametric CAD data. In: IEEE Fourth VR International Workshop on Collaborative Virtual Environments (3DCVE), Reutlingen, Germany, pp. 1–4 (2018). https://doi.org/10.1109/3DCVE.2018.8637112

5. Pick, S., Gebhardt, S., Weyers, B., Hentschel, B., Kuhlen, T.: A 3D collaborative virtual environment to integrate immersive virtual reality into factory planning processes. In: International Workshop on Collaborative Virtual Environments (3DCVE), Minneapolis, MN, pp. 1–6 (2014). https://doi.org/10.1109/3DCVE.2014.7160934

6. Leigh, J., Johnson, A.E.: CALVIN: an immersimedia design environment utilizing heterogeneous perspectives. In: Proceedings of the Third IEEE International Conference on Multimedia Computing and Systems, Hiroshima, Japan, pp. 20–23 (1996). https://doi.org/10.1109/MMCS.1996.534949

7. Fechner, T., Wilhelm, D., Kray, C.: Ethermap: real-time collaborative map editing. In: Proceedings of the 33rd Annual ACM Conference on Human Factors in Computing Systems (CHI 2015), pp. 3583–3592. Association for Computing Machinery, New York (2015). https://doi.org/10.1145/2702123.2702536

8. Salvati, G., Santoni, C., Tibaldo, V., Pellacini, F.: MeshHisto: collaborative modeling by sharing and retargeting editing histories. ACM Trans. Graph. **34**(6), 1–10 (2015). https://doi.org/10.1145/2816795.2818110

9. Smit, D., Grah, T., Murer, M., van Rheden, V., Tscheligi, M.: MacroScope: First-person perspective in physical scale models. In: Proceedings of the Twelfth International Conference on Tangible, Embedded, and Embodied Interaction (TEI 2018), pp. 253–259. Association for Computing Machinery, New York (2018). https://doi-org.proxy.cc.uic.edu/10.1145/3173225.3173276

10. Ibayashi, H., et al.: Dollhouse VR: a multi-view, multi-user collaborative design workspace with VR technology. In SIGGRAPH Asia: Posters (SA 2015), Article 24, p. 1. Association for Computing Machinery, New York (2015). https://doi.org/10.1145/2820926.2820948

11. Stafford, A., Piekarski, W., Thomas, B.: Implementation of god-like interaction techniques for supporting collaboration between outdoor AR and indoor tabletop users. In: Proceedings of the 5th IEEE and ACM International Symposium on Mixed and Augmented Reality (ISMAR 2006), pp. 165–172. IEEE Computer Society, USA (2006). https://doi-org.proxy.cc.uic.edu/10.1109/ISMAR.2006.297809

Smart Transportation

T2I-CycleGAN: A CycleGAN for Maritime Road Network Extraction from Crowdsourcing Spatio-Temporal AIS Trajectory Data

Xuankai Yang[1,2], Guiling Wang[1,2(✉)] , Jiahao Yan[1,2], and Jing Gao[1,2]

[1] School of Information Science and Technology,
North China University of Technology, Beijing 100144, China
[2] Beijing Key Laboratory on Integration and Analysis of Large-Scale Stream Data,
North China University of Technology, Beijing, China
wangguiling@ict.ac.cn

Abstract. Maritime road network is composed of detailed maritime routes and is vital in many applications such as threats detection, traffic control. However, the vessel trajectory data, or Automatic Identification System (AIS) data, are usually large in scale and collected with different sampling rates. And, what's more, it is difficult to obtain enough accurate road networks as paired training datasets. It is a huge challenge to extract a complete maritime road network from such data that matches the actual route of the ship. In order to solve these problems, this paper proposes an unsupervised learning-based maritime road network extraction model T2I-CycleGAN based on CycleGAN. The method translates trajectory data into unpaired input samples for model training, and adds dense layer to the CycleGAN model to handle trajectories with different sampling rates. We evaluate the approach on real-world AIS datasets in various areas and compare the extracted results with the real ship coordinate data in terms of connectivity and details, achieving effectiveness beyond the most related work.

Keywords: Unsupervised learning · Generative adversarial network · AIS data · Road network · Spatio-temporal data mining · Trajectory data mining

1 Introduction

With the increasing international trade, accurate and up-to-date maritime road network data becomes vital. Traditional road network generation and maintenance methods, such as field surveys, are not feasible. This is because the maritime road network is not man-made structure and cannot be obtained by observation. In recent years, with the rapid developing of mobile sensors and Cloud Computing, a large amount of crowdsourcing vessel trajectory data, or

© ICST Institute for Computer Sciences, Social Informatics and Telecommunications Engineering 2021
Published by Springer Nature Switzerland AG 2021. All Rights Reserved
H. Gao et al. (Eds.): CollaborateCom 2020, LNICST 350, pp. 203–218, 2021.
https://doi.org/10.1007/978-3-030-67540-0_12

Automatic Identification System (AIS) data, has been generated and collected, such historical crowdsourcing trajectory data provides a new opportunity to extract the underlying maritime road network.

Extraction of maritime road network is very challenging. In order to ensure the accuracy of the extraction results, road network extraction needs to use real map data, such as OpenStreetMap data, as the training set [26], or as a reference standard to distinguish the data in the road region from outside the road region [37]. However, maritime road networks are composed of customary routes for ship navigation rather than man-made roads, and the structure of the global maritime road network itself has not been fully documented [9]. These make it unrealistic to find enough real or reliable maritime road network data as experimental training sets for maritime road network extraction.

Some approaches have been proposed for extraction of urban and maritime road networks [4,6,7,12,15,17,19,23,26,29,33–37]. Detailed analysis of the related work is in Sect. 2. Among various kinds of the approaches, the deep learning based approaches are one of the most promising approaches that don't rely on many empirical parameters and have the potential to make good use of the prior knowledge in existing maps or regular routes [26]. The existing deep learning based approaches in this field adopt supervised learning which require enough high-quality paired samples for model training. While for maritime road network extraction, the supervised learning approaches cannot be borrowed directly, because it is unrealistic to obtain enough accurate maritime road network data corresponding to the original trajectory data as training set, as just analyzed. This is the main challenge that this paper focuses on.

Generative Adversarial Networks (GANs), which is introduced by Goodfellow et al. [13], have achieved outstanding success in the field of image-to-image translation, such as data augmentation, image inpainting, and style transfer. Their main goal is to generate fake images that is indistinguishable from the original images on the targeted domain. Recently, the advent of the cycle-consistency loss [39] has eliminated the need of paired sample data. By utilizing CycleGAN, the problem of not having access to paired training samples has been solved in many areas [10,21,38].

To overcome the challenge mentioned above, in this paper, we propose an unsupervised CycleGAN-based method named T2I-CycleGAN (Trajectory-to-Image CycleGAN) to generate maritime road network from trajectories.

The key contributions of this paper can be summarized as follows:

1) We propose an unsupervised learning-based maritime road network extraction model T2I-CycleGAN. The method doesn't rely on paired trajectory and road network samples for model training.
2) We use a trajectory data transition method to process trajectories with different sampling rates and add dense layers in CycleGAN to make it able to handle situations without recognized benchmark data as a reference standard.
3) Experimental results show that the proposed method can automatically effectively extract maritime road network from crowdsourcing spatio-

temporal AIS trajectory data without real or recorded maritime road network data as an experimental benchmark, by comparing with the related work.

The rest of this paper is organized as follows. Section 2 reviews related work. In Sect. 3, the basic concepts and definitions throughout the paper are introduced and the problems to be solved are described. Section 4 describes the trajectory data transition method. Section 5 describes the structure of T2I-CycleGAN model and details about its generator and discriminator. At last, Evaluation of our approach and comparison with the current methods are in Sect. 6 and Sect. 7 concludes the paper.

2   Related Work

The existing related research work on maritime and urban road network extraction from trajectory data can be classified into four categories:

1). Vector-based approaches: Transform trajectory points into vectors and then extract the road network information. Most of vector-based approaches are based on clustering algorithms. Point clustering methods cluster the trajectory points into way points and then find the information to connect them into roads or lanes [1,3,8,12,17,23,35]. Segment clustering methods cluster trajectory segments to extract road segments [15,18,28]. Others use statistical method of motion attributes to find way points and lines [11,22,24] or extract the road lines by incrementally inserting and merging tracks based on their geometric relations and/or shape similarity [30]. The above methods can be used to extract road lines or segments but is difficult to be used to extract lane boundaries or whole lane boundaries in a large area. Furthermore, the methods are applicable to densely sampling traces, but not robust to position points with noises and varying density.

2). Image-based approaches: Transform the location data of all the trajectory points into row and column coordinates of image, and then recognize and/or extract the lane information using digital image processing technology [5,20,27] to solve the problem in road network extraction.

3). Grid-based approaches: Transform trajectory points into grids, and then extract road information from grid data [19,34,36,37]. Some authors [34,36,37] apply Delaunay Triangulation to extract road boundaries and road centerlines. Others [19] use the number of trajectory points in grids (grid heat value) to determine the road direction, however the lane boundaries and junctions have not been considered in their paper. The grid-based approach can be used to extract both boundary information and centerlines, and is more robust to data with noises and varying density than the other two approaches. The existing research papers haven't discussed how to apply their algorithms on massive AIS data and how to extract a continuous and smooth lane in a large area, which is very challenging when the data volume is large, the data quality is low and the trajectory points density is quite uneven.

4). Grid-based Image Processing approaches: Construct features of trajectories grid by grid, then transform features of trajectory points into images. Deep learning methods such as CNN, GANs are then used to extract the lane information. Ruan et al. [26] combines trajectory image generation and deep learning, a road network road information extraction method based on convolutional neural network is proposed. This method uses road network road data and taxi trajectory data as the training set, and the model realizes the road region and Extraction of road centerline. Chuanwei et al. [7] proposes a road extraction model based on cGANs. The model includes two parts: generator and discriminator. Training is completed through confrontation between the two parts. This method realizes the extraction of two-way lanes on the road. The image-based methods are also applicable to densely sampling traces but not robust to noisy data and have efficient problem when applying to massive trajectory data.

Different from these related work, this paper proposed an unsupervised road network extraction method based on CycleGAN. The framework and approach proposed in this paper can deal with crowdsourcing Spatio-temporal trajectory data, which has different sampling rates and a lack of recognized benchmark data.

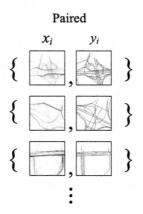

Fig. 1. Paired sample example

3   Definitions and Overview

In this section, we give several basic definitions, formalize the problem of maritime road network extraction, and outline our proposed approach.

3.1   Definitions

Definition 1 (Trajectory): A trajectory is the sailing route of a single ship, usually contains several attributes, such as MMSI (Maritime Mobile Service Identity), UTC (Coordinated Universal Time), longitude, latitude, velocity, and these attributes describe the status of the vessel. A trajectory is a sequence of spatiotemporal points, denoted as $t_r = (p_1, p_2, ..., p_n)$, $p_n = (M_n, x_n, y_n, t_n)$ indicates the MMSI of the vessel is M_n and vessel M_n locates at (x_n, y_n) at the time of t_n, points in a trajectory are organized chronologically.

Definition 2 (Grid): A grid is a square area in the 2D geographical space. 2D geographical space of the global is divided into many square areas through the horizontal and vertical directions. A grid has two attributes: *Code* and *Dsy*, where *Code* is the geohash code of the grid and *Dsy* is the density of the grid. A grid can be described as $Grid = (Code, Dsy)$.

Definition 3 (Paired Sample): Paired training data consists of training examples where the correspondence between them exists. For example, define $\{x_i, y_i\}_{i=1}^{N}$ as the training samples of road network. As shown in Fig. 1, x_i represents the trajectory points of the road network, and y_i represents the corresponding road network segments.

3.2   Problem and Approach Overview

Our goal in this paper is to extract maritime road network information from massive crowdsourcing spatio-temporal AIS trajectory data, specifically, given an AIS trajectory dataset $\{T_v\}$, we aim to use an unsupervised learning method to extract the maritime road network $\{G_v\}$ in a certain area.

As an unsupervised image-to-image translation method, CycleGAN can solve the problem of lack of paired training data in the target and real domains.

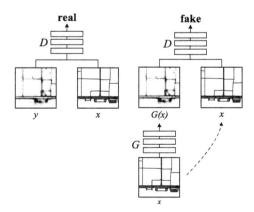

Fig. 2. Training procedure of cGANs

For image generation tasks, the GANs model usually consists of two parts: a generator and a discriminator. The generator is used to generate target objects, and the discriminator is used to determine the difference between the generated target objects and the real objects. Through the adversarial process between the generator and the discriminator, the dynamic balance between the models is realized, which means that the generator can generate images that can make it difficult for the discriminator to distinguish between real and fake.

The advantage of the GANs model is that it does not require prior knowledge and uses a noise distribution to directly sample, and gradually approximates the real data. However, due to the lack of prior information, uncontrollable training results will be produced, the researchers proposed the introduction of conditional GANs [16] with conditional variables model. The training procedure [16] of cGANs is diagrammed in Fig. 2, the generator G is trained to produce outputs that cannot be distinguished from "real" road network by a trained discriminator, D, which is trained to do as well as possible at detecting the generator's "fakes".

The CycleGAN developed based on the cGANs is an unsupervised model. For the problem of image transformation in two domains, CycleGAN can train the images in the two domains separately by using a consistent cycle structure, and does not require the two parts of the image to be pairs of samples corresponding to each other at the same location. Based on the unsupervised learning characteristics of CycleGAN, we propose a model for processing crowdsourcing spatio-temporal AIS trajectory data named T2I-CycleGAN.

The extraction procedure is presented in Fig. 3, which is comprised of two steps: 1) *Trajectory Data Transition*, which pre-process the raw trajectory data and extracts features from them; 2) *Maritime Road Network Extraction*, which trains the generator and discriminator of T2I-CycleGAN with the extracted features, and uses the model to extract the maritime road network.

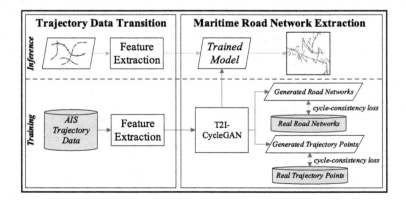

Fig. 3. Maritime road network extraction procedure with T2I-CycleGAN

4   Trajectory Data Transition

CycleGAN is a model that realizes image-to-image translation, and its input and output are grid data. The trajectory points of vessels and maritime road network data are generally vector data composed of vessel coordinates. Therefore, in order to achieve the extraction of the maritime road network, it is necessary to transform vector data to grid data. In addition, as crowdsourcing spatio-temporal trajectory data, the raw data has three problems that need to be resolved: 1) *Unsegmented trajectory points.* Trajectory points belonging to different trips but haven't been split into different segments; 2) *Noisy trajectory points.* Those points that cause the calculated speed deviating from normal reasonable ship speed and those too short trajectories can be seen as noisy data; 3) *Missing trajectory points.* If the time interval between two trajectory segments belongs to the same trip is larger than the average time interval between adjacent points, there must exist some missing points between these two trajectory segments. The trajectory data transition is to solve the above problems.

The first step is to pre-process the trajectory data for cleaning and interpolation. For the unsegmented trajectory points, we segment them by setting a maximum time interval threshold. If the time interval between two adjacent points is less than the threshold, they considered to belong to the same segment. For the noisy data, we simply delete them instead of correcting them. Since the volume of AIS data is quite large, deleting some points will not bring an apparent negative influence to the results. For missing points, we use linear interpolation to insert points into the trajectory segments to replace them, and this method also has no negative impact on the results.

After pre-processing, vector data needs to be converted into grid data. We use a grid data construction method based on the spatio-temporal feature extraction of trajectory data, it extracts and synthesizes the features of trajectory data in time and space to obtain the grid data corresponding to spatio-temporal trajectory data.

The first step of this method is to divide the research area into n grids of equal size, then map the trajectory data into n grids. In the light of [26] we then extract the following features that indicate the correspondence between grid and track points: 1) *Point*, represents the number of points mapped by the trajectory coordinates contained in each grid. This feature can most intuitively reflect the shape of the maritime road network. 2) *Line*, the Line feature of a grid represents the number of road segments formed by two consecutive trajectory points passing through the grid. This feature can establish a relationship between two trajectory points that are not in the same grid. 3) *Speed*, represents the average speed of the trajectory points in the grid. This feature can reflect the time and space characteristics of the trajectory data, and the coordinates on the same road basically have no difference in speed, so the feature can reflect the connection of the road. 4) *Direction*, which indicates the moving direction of the trajectory point in the grid. We consider eight directions such as north and northeast. First, count the occurrences of each direction for a movement between two consecutive points, and then normalize the counts into a histogram.

This feature can distinguish adjacent roads and depict the details of the road network data. 5) *Transition*, for a grid cell c, there exists a neighboring cells c'. We maintain a binary matrix to capture the transition of two consecutive points across two neighboring grid cells. The entry corresponding to c' is set to 1 if there exist two consecutive points which are from c' to c, and 0 otherwise. Besides, we consider those cells c'' such that there exist two consecutive points which are from c to c'' and capture the cells with another binary matrix similarly as we do for cells c.

5   T2I-CycleGAN Model

T2I-CycleGAN is an enhanced version of CycleGAN [39] architecture used for maritime road network extraction, with enhancements to the network structure. In practice, the trajectory features obtained through trajectory data transition could be so sparse, which is unfavorable for model training. To solve this problem, dense layers are added to the CycleGAN architecture. Therefore, after transforming the trajectory data into feature maps, T2I-CycleGAN first transforms the sparse feature matrix to obtain their dense representation, and then proceeds to the next CycleGAN-based model training. The structure of T2I-CycleGAN is shown in Fig. 4.

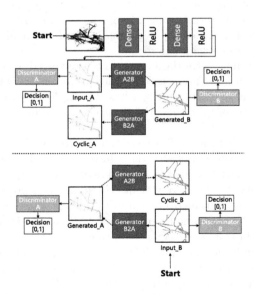

Fig. 4. Framework of T2I-CycleGAN

As shown in Fig. 4, generative adversarial networks we use in T2I-CycleGAN is a two-way cycle network structure composed of two mirrored cGANs [16] structures. The two cGANs share two generators, Generator A2B and Generator

B2A, and have two associated adversarial discriminators, Discriminator A and Discriminator B. In the forward process of the model training, Generator A2B realizes the conversion from Input A to Generated B, and generator B2A realizes the reverse conversion from Generated B to Cyclic A. The result, Generated B, is compared with the real images by Discriminator B, which can facilitate the generator to generate results that are indistinguishable from real images of maritime road network. In order to further standardize the image conversion function, two cycle consistency losses are used in the model [39]. The cycle consistency losses are obtained by comparing the result with the real image. And vice versa for real image Input B and Generated A.

The generator of T2I-CycleGAN generally uses a self-encoding model or its variants, such as the U-Net model [25]. The process is to first downsample to extract deep features, and then upsample to generate the structure of the target, where the depth of the model is limited by the convolution calculation in the downsampling process. According to the size of the input image, the maximum depth is limited. The generator network we use contains two stride-2 convolutions, several residual blocks [14], and two fractionally-strided convolutions. We use 6 blocks for 128×128 images and 9 blocks for 256×256 and higher-resolution training images and instance normalization [31] is used.

The discriminator that is used in the T2I-CycleGAN model is PatchGAN discriminator, which is to perform multiple convolution calculations on the input image, output a feature map of size N, and further judge whether each element on the feature map is real or fake, the final average is the probability that the input image is real or fake. The discriminator structure is a convolutional network. Pixel points on the output feature map correspond to the mapping area on the original input image, which is the receptive field.

We apply the adversarial loss [39] to both mapping functions. For the mapping function $G : X \to Y$, $F : Y \to X$, under the constraints of $x \in X$, $y \in Y$, and their discriminator D_Y, D_X, respectively. The objectives are expressed as:

$$\mathcal{L}_{\mathrm{GAN}}(G, D_Y, X, Y) = \mathbf{E}_{y \sim p_{\mathrm{data}}(y)} \left[\log D_Y(y) \right]$$
$$+ \mathbf{E}_{x \sim p_{\mathrm{data}}(x)} \left[\log \left(1 - D_Y(G(x)) \right) \right] \quad (1)$$

$$\mathcal{L}_{\mathrm{GAN}}(F, D_X, Y, X) = \mathbf{E}_{x \sim p_{\mathrm{data}}(x)} \left[\log D_X(x) \right]$$
$$+ \mathbf{E}_{y \sim p_{\mathrm{data}}(y)} \left[\log \left(1 - D_X(G(y)) \right) \right] \quad (2)$$

We use cycle consistency to guarantee that the learned function can map an individual input to a desired output:

$$\mathcal{L}_{\mathrm{cyc}}(G, F) = \mathbf{E}_{x \sim p_{\mathrm{data}}(x)} \left[\| F(G(x)) - x \|_1 \right]$$
$$+ \mathbf{E}_{y \sim p_{\mathrm{data}}(y)} \left[\| G(F(y)) - y \|_1 \right] \quad (3)$$

As shown in Fig. 5, for each image x from domain X, the image translation cycle should be able to bring x back to the original image, i.e., $x \to G(x) \to F(G(x)) \approx x$. Similarly, for each image y from domain Y, G and F should also satisfy the cycle consistency: $y \to F(y) \to G(F(y)) \approx y$.

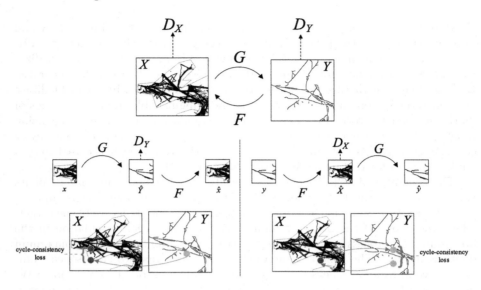

Fig. 5. Cycle consistency in T2I-CycleGAN

Full objective of T2I-CycleGAN is:

$$\mathcal{L}(G, F, D_X, D_Y) = \mathcal{L}_{\mathrm{GAN}}(G, D_Y, X, Y)$$
$$+ \mathcal{L}_{\mathrm{GAN}}(F, D_X, Y, X)$$
$$+ \lambda \mathcal{L}_{\mathrm{cyc}}(G, F) \tag{4}$$

where λ controls the relative importance of the two objectives. It aims to solve:

$$G^*, F^* = \arg\min_{G,F} \max_{D_x, D_Y} \mathcal{L}(G, F, D_X, D_Y) \tag{5}$$

6   Experiments

6.1   Dataset and Experimental Environment

The data we used in the experiment is the AIS data of all cargo ships in February 2016. Our target area is Bohai Sea. AIS data includes vessel's name, call sign, MMSI, IMO, ship type, destination, ship width and other static information such as UTC, latitude, longitude, direction, speed and status. The experiment uses four columns of MMSI, UTC, longitude and latitude from the dataset. A dataset sample is shown in Table 1.

The total amount of AIS data of Bohai sea cargo ships is 1.55G. The data contains over 109645 travel records of ships. The experiment runs in a CentOS release 7.0 environment. Model implementation and result analysis based on Python 3.6 and Pytorch 1.4.0 deep learning open source library. The model is trained with a single NVIDIA Tesla K40c GPU and the video memory is 12 GB.

Table 1. Sample AIS data

MMSI	UTC	Longitude	Latitude
445128000	1454633376	123.145330	37.031753
445128000	1454635569	123.138442	36.917635
. . .	. . .	. . .	. . .
445128000	1454637352	123.129790	36.824980

6.2   Results and Performance Comparison

First, the trajectory data is pre-processed, then the data features are extracted for trajectory data transition, and the results are divided proportionally to obtain the trajectory data sample set. At the same time, the road network style data is proportionally divided to obtain the maritime road network sample set, and finally the divided two parts of samples are input into the trained T2I-CycleGAN model to get the generated maritime road network. Figure 6 shows the extraction effect of maritime road network in the Bohai Sea area. In order to compare and evaluate the extraction results of the T2I-CycleGAN, this paper compares and evaluates CycleGAN [39], Simplified DeepMG [26] and the NiBlack-based extraction method [32].

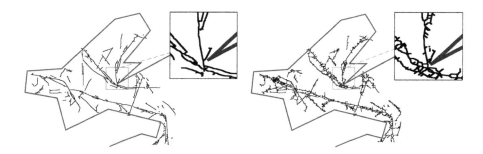

Fig. 6. Result of T2I-CycleGAN   **Fig. 7.** Result of CycleGAN

CycleGAN. After the trajectory data is mapped to the grid and then converted into image data, it is simply used as the trajectory data sample set of the CycleGAN model for training, and the final extraction result is shown in Fig. 7. It can be seen from the figure that although the shape of the extracted road network is basically the same as the result obtained by the T2I-CycleGAN model, there are more noisy road segments. The results obtained by the T2I-CycleGAN model are well optimized in noise processing.

Simplified DeepMG. The DeepMG framework uses a CNN-based supervised training method for road network extraction. Due to the small amount of ship trajectory data used in our experiment, overfitting occurs when these data are used for T2RNet training. In addition, we have not implemented the topology construction part of the DeepMG framework, which will be implemented in future work. The result is shown in Fig. 8. In the case of using the same test data, the results obtained by simplified DeepMG have poor integrity, but the details are richer.

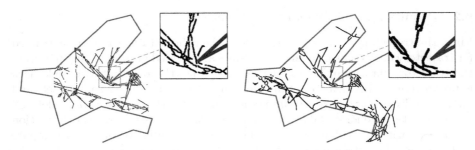

Fig. 8. Result of simplified DeepMG   **Fig. 9.** Result of the NiBlack-based method

NiBlack-Based Method. Figure 9 shows of the result of the NiBlack-based extraction method. The main idea of the method is to use a local sliding window filtering algorithm to filter those grids that are not on the marine lane. The processing result of this method generates many shortcuts that do not exist on a real map, and the T2I-CycleGAN Model has higher connectivity in processing the data.

Evaluation. Because of the lack of maritime road network benchmark data, we took the trajectory point data in the Bohai Sea area as the reference sample, and manually extracted the maritime road network data in this area as a reference, and the road length was about 880 nm. The quantitative evaluation method adopts the buffer method proposed in [2], and establishes a 0.26 nm radius buffer for the reference data. Then count the length of the road falling into the buffer zone and calculate the precision, recall and F1 score as evaluation indicators, we redefine these indicators based on the actual extraction meaning. 1) *Precision*, the ratio of the correctly extracted road length to the total length of the extracted road. 2) *Recall*, the ratio of the correctly extracted road length to the total road length. 3) F_1 *score*, weighted Harmonic mean of precision and recall, the calculation equation is

$$F_1 = \frac{2P \cdot R}{P + R} \tag{6}$$

In the target area, the maritime road network is extracted based on Cycle-GAN, Simplified DeepMG and the NiBlack-based extraction method. Table 2 respectively calculates the P, R and F_1 values of the comparison method. Table 2 shows that the results extracted using T2I-CycleGAN have better performance on each indicator. Using CycleGAN can improve the recall rate of the mar-itime road network, but the accuracy of the results obtained by simply using CycleGAN is lower. It shows that the dense layer added in T2I-CycleGAN can effectively reduce the error road in the extraction result. The extraction results of Simplified DeepMG and the NiBlack-based extraction method are relatively low.

Table 2. Model comparisons

Model	Precision	Recall	F_1 Score
T2I-CycleGAN	0.47	0.67	0.55
CycleGAN	0.32	0.67	0.43
Simplified DeepMG	0.38	0.37	0.37
NiBlack-based method	0.41	0.43	0.42

7   Conclusion

We proposed a novel maritime road network extraction approach T2I-CycleGAN model for large scale AIS data with different sampling rates and no reference benchmark. For the problem of lack of maritime road network benchmark, we introduce the unsupervised learning method CycleGAN, we use the cycle con-sistency of the model to achieve road network extraction. Not only that, we also added a dense layer to the model to solve the problem of sparse features of ship trajectory data. We evaluate the approach on real-world AIS datasets in various areas and compare the extracted results with the real ship coordinate data in terms of connectivity and details, achieving effectiveness beyond the most related work.

Acknowledgements. This work is supported by National Natural Science Founda-tion of China under Grant 61832004 and Grant 61672042. We thank the Ocean Informa-tion Technology Company, China Electronics Technology Group Corporation (CETC Ocean Corp.), for providing the underlying dataset for this research.

References

1. Agamennoni, G., Nieto, J.I., Nebot, E.M.: Robust inference of principal road paths for intelligent transportation systems. IEEE Trans. Intell. Transp. Syst. **12**(1), 298–308 (2011)

2. Ahmed, M., Karagiorgou, S., Pfoser, D., Wenk, C.: A comparison and evaluation of map construction algorithms using vehicle tracking data. GeoInformatica **19**(3), 601–632 (2015). https://doi.org/10.1007/s10707-014-0222-6

3. Arguedas, V.F., Pallotta, G., Vespe, M.: Automatic generation of geographical networks for maritime traffic surveillance. In: 17th International Conference on Information Fusion (FUSION), pp. 1–8 (2014)

4. Cao, L., Krumm, J.: From GPS traces to a routable road map. In: Proceedings of the 17th ACM SIGSPATIAL International Conference on Advances in Geographic Information Systems, pp. 3–12 (2009)

5. Chen, C., Cheng, Y.: Roads digital map generation with multi-track GPS data. In: 2008 International Workshop on Geoscience and Remote Sensing, vol. 1, pp. 508–511 (2008)

6. Chen, C., Lu, C., Huang, Q., Yang, Q., Gunopulos, D., Guibas, L.: City-scale map creation and updating using GPS collections. In: Proceedings of the 22nd ACM SIGKDD International Conference on Knowledge Discovery and Data Mining, KDD 2016, pp. 1465–1474. ACM, New York (2016)

7. Chuanwei, L., Qun, S., Bing, C., Bowei, W., Yunpeng, Z., Li, X.: Road learning extraction method based on vehicle trajectory data. Acta Geodaetica et Cartographica Sinica **49**(6), 692 (2020)

8. Dobrkovic, A., Iacob, M.E., van Hillegersberg, J.: Maritime pattern extraction and route reconstruction from incomplete AIS data. Int. J. Data Sci. Anal. **5**(2), 111–136 (2018). https://doi.org/10.1007/s41060-017-0092-8

9. Ducruet, C., Notteboom, T.: The worldwide maritime network of container shipping: spatial structure and regional dynamics. Glob. Netw. **12**(3), 395–423 (2012)

10. Engin, D., Genç, A., Kemal Ekenel, H.: Cycle-Dehaze: enhanced CycleGAN for single image dehazing. In: Proceedings of the IEEE Conference on Computer Vision and Pattern Recognition Workshops, pp. 825–833 (2018)

11. Etienne, L., Devogele, T., Bouju, A.: Spatio-temporal trajectory analysis of mobile objects following the same itinerary. Adv. Geo-Spatial Inf. Sci. **10**, 47–57 (2012)

12. Fernandez Arguedas, V., Pallotta, G., Vespe, M.: Maritime traffic networks: from historical positioning data to unsupervised maritime traffic monitoring. IEEE Trans. Intell. Transp. Syst. **19**(3), 722–732 (2018)

13. Goodfellow, I., et al.: Generative adversarial nets. In: Advances in Neural Information Processing Systems, pp. 2672–2680 (2014)

14. He, K., Zhang, X., Ren, S., Sun, J.: Deep residual learning for image recognition. In: Proceedings of the IEEE Conference on Computer Vision and Pattern Recognition, pp. 770–778 (2016)

15. Hung, C.C., Peng, W.C., Lee, W.C.: Clustering and aggregating clues of trajectories for mining trajectory patterns and routes. VLDB J. **24**(2), 169–192 (2015). https://doi.org/10.1007/s00778-011-0262-6

16. Isola, P., Zhu, J.Y., Zhou, T., Efros, A.A.: Image-to-image translation with conditional adversarial networks. In: Proceedings of the IEEE Conference on Computer Vision and Pattern Recognition, pp. 1125–1134 (2017)

17. Le Guillarme, N., Lerouvreur, X.: Unsupervised extraction of knowledge from S-AIS data for maritime situational awareness. In: Proceedings of the 16th International Conference on Information Fusion, pp. 2025–2032 (2013)

18. Lee, J.G., Han, J., Whang, K.Y.: Trajectory clustering: a partition-and-group framework. In: Proceedings of the 2007 ACM SIGMOD International Conference on Management of Data, SIGMOD 2007, pp. 593–604. ACM, New York (2007)

19. Li, J., Chen, W., Li, M., Zhang, K., Yajun, L.: The algorithm of ship rule path extraction based on the grid heat value. J. Comput. Res. Dev. **55**(5), 908–919 (2018)
20. Liu, X., Biagioni, J., Eriksson, J., Wang, Y., Forman, G., Zhu, Y.: Mining large-scale, sparse GPS traces for map inference: comparison of approaches. In: Proceedings of the 18th ACM SIGKDD International Conference on Knowledge Discovery and Data Mining, KDD 2012, pp. 669–677. ACM, New York (2012)
21. Lu, Y., Tai, Y.-W., Tang, C.-K.: Attribute-guided face generation using conditional CycleGAN. In: Ferrari, V., Hebert, M., Sminchisescu, C., Weiss, Y. (eds.) ECCV 2018. LNCS, vol. 11216, pp. 293–308. Springer, Cham (2018). https://doi.org/10.1007/978-3-030-01258-8_18
22. Naserian, E., Wang, X., Dahal, K., Wang, Z., Wang, Z.: Personalized location prediction for group travellers from spatial-temporal trajectories. Future Gener. Comput. Syst. **83**, 278–292 (2018)
23. Pallotta, G., Vespe, M., Bryan, K.: Vessel pattern knowledge discovery from AIS data: a framework for anomaly detection and route prediction. Entropy **15**(6), 2218–2245 (2013)
24. Wen, R., Yan, W., Zhang, A.N., Chinh, N.Q., Akcan, O.: Spatio-temporal route mining and visualization for busy waterways. In: 2016 IEEE International Conference on Systems, Man, and Cybernetics (SMC), pp. 849–854 (2016)
25. Ronneberger, O., Fischer, P., Brox, T.: U-Net: convolutional networks for biomedical image segmentation. In: Navab, N., Hornegger, J., Wells, W.M., Frangi, A.F. (eds.) MICCAI 2015. LNCS, vol. 9351, pp. 234–241. Springer, Cham (2015). https://doi.org/10.1007/978-3-319-24574-4_28
26. Ruan, S., et al.: Learning to generate maps from trajectories. In: Proceedings of the AAAI Conference on Artificial Intelligence, vol. 34, pp. 890–897 (2020)
27. Shi, W., Shen, S., Liu, Y.: Automatic generation of road network map from massive GPS, vehicle trajectories. In: 2009 12th International IEEE Conference on Intelligent Transportation Systems, pp. 1–6 (2009)
28. Spiliopoulos, G., Zissis, D., Chatzikokolakis, K.: A big data driven approach to extracting global trade patterns. In: Doulkeridis, C., Vouros, G.A., Qu, Q., Wang, S. (eds.) MATES 2017. LNCS, vol. 10731, pp. 109–121. Springer, Cham (2018). https://doi.org/10.1007/978-3-319-73521-4_7
29. Stanojevic, R., Abbar, S., Thirumuruganathan, S., Chawla, S., Filali, F., Aleimat, A.: Robust road map inference through network alignment of trajectories. In: Proceedings of the 2018 SIAM International Conference on Data Mining, pp. 135–143. SIAM (2018)
30. Tang, L., Ren, C., Liu, Z., Li, Q.: A road map refinement method using Delaunay triangulation for big trace data. ISPRS Int. J. Geo-Inf. **6**(2), 45 (2017)
31. Ulyanov, D., Vedaldi, A., Lempitsky, V.: Instance normalization: The missing ingredient for fast stylization. arXiv preprint arXiv:1607.08022 (2016)
32. Wang, G., et al.: Adaptive extraction and refinement of marine lanes from crowd-sourced trajectory data. Mobile Netw. Appl. **25**, 1392–1404 (2020). https://doi.org/10.1007/s11036-019-01454-w
33. Wang, S., Wang, Y., Li, Y.: Efficient map reconstruction and augmentation via topological methods. In: Proceedings of the 23rd SIGSPATIAL International Conference on Advances in Geographic Information Systems, pp. 1–10 (2015)
34. Wei, Y., Tinghua, A.: Road centerline extraction from crowdsourcing trajectory data. Geogr. Geo Inf. Sci. **32**(3), 1–7 (2016)

35. Yan, W., Wen, R., Zhang, A.N., Yang, D.: Vessel movement analysis and pattern discovery using density-based clustering approach. In: 2016 IEEE International Conference on Big Data (Big Data), pp. 3798–3806 (2016)
36. Yang, W., Ai, T.: The extraction of road boundary from crowdsourcing trajectory using constrained Delaunay triangulation. Acta Geodaetica Cartogr. Sin. **46**(2), 237–245 (2017)
37. Yang, W., Ai, T., Lu, W.: A method for extracting road boundary information from crowdsourcing vehicle GPS trajectories. Sensors **18**(4), 2660–2680 (2018)
38. Zhao, S., et al.: CycleEmotionGAN: emotional semantic consistency preserved cyclegan for adapting image emotions. In: Proceedings of the AAAI Conference on Artificial Intelligence, vol. 33, pp. 2620–2627 (2019)
39. Zhu, J.Y., Park, T., Isola, P., Efros, A.A.: Unpaired image-to-image translation using cycle-consistent adversarial networks. In: Proceedings of the IEEE International Conference on Computer Vision, pp. 2223–2232 (2017)

Where Is the Next Path? A Deep Learning Approach to Path Prediction Without Prior Road Networks

Guiling Wang[1,2](✉) (iD), Mengmeng Zhang[1,2], Jing Gao[1,2], and Yanbo Han[1,2]

[1] Beijing Key Laboratory on Integration and Analysis of Large-scale Stream Data,
North China University of Technology, No.5 Jinyuanzhuang Road,
Shijingshan District, Beijing 100144, China
wangguiling@ict.ac.cn
[2] School of Information Science and Technology,
North China University of Technology, No.5 Jinyuanzhuang Road,
Shijingshan District, Beijing 100144, China

Abstract. Trajectory prediction plays an important role in many urban and marine transportation applications, such as path planning, logistics and traffic management. The existing prediction methods of moving objects mainly focus on trajectory mining in Euclidean space. However, moving objects generally move under road network constraints in the real world. It provides an opportunity to take use of road network constraints or de-facto regular paths for trajectory prediction. As yet, there is little research work on trajectory prediction under road network constraints. And these existing work assumes prior road network information is given in advance. However in some application scenarios, it is very difficult to get road network information, for example the maritime traffic scenario on the wide open ocean. To this end, we propose an approach to trajectory prediction that can make good use of road network constrains without depending on prior road network information. More specifically, our approach extracts road segment polygons from large scale crowdsourcing trajectory data (e.g. AIS positions of ships, GPS positions of vehicles etc.) and translates trajectories into road segment sequences. Useful features such as movement direction and vehicle type are extracted. After that, a LSTM neural network is used to infer the next road segment of a moving object. Experiments on real-world AIS datasets confirm that our approach outperforms the state-of-the-art methods.

Keywords: Trajectory prediction · LSTM · Crowdsourcing trajectories · AIS data

1 Introduction

In recent years, with the development of Cloud Computing, IoT, Big Data, 5G, machine learning and location-based services (LBS) technologies, it is possible for

H. Gao et al. (Eds.): CollaborateCom 2020, LNICST 350, pp. 219–235, 2021.
https://doi.org/10.1007/978-3-030-67540-0_13

collecting massive spatio-temporal trajectory data of large scale moving objects and mining valuable hidden information from such crowdsourcing trajectory data. Among various trajectory mining research topics, trajectory prediction is one of the most interesting and significant topic. The trajectory prediction plays an important role to solve problems such as path planning and traffic management.

There is some research work on trajectory prediction for land transportation based on GPS trajectory data [4]. Most of the trajectory prediction approaches are based on Euclidean space. They simulate the moving trajectory through mathematical formulas or mine the moving pattern from historical trajectories. However, these approaches have obvious limitations. Most moving objects move on a restricted road network and are necessarily restricted by the network and cannot move freely in space. This is why some of the research work tries to take use of road network information for trajectory prediction. We all know that there exists relatively stable road networks for land transportation. For maritime traffic, however, it is very difficult to get stable and accurate road network information for maritime traffic. In fact, although there is no static road network for maritime traffic, the moving trajectories of a large number of ships show certain de-facto paths. This is due to several reasons such as people always select routes with more fuel-efficiency, ships would not pass through existing protected sea areas where maritime traffic is prohibited and ships always bypass areas with well known security threats.

Our previous research work [6,16] proposed a method for extracting channels as polygons from large scale crowdsourcing ship trajectories. And our method in [12] further extracts road centerline and construct road network from the crowdsourcing ship trajectories. Based on these previous work, we aim to solve the problem of trajectory prediction that can make good use of road network constrains without depending on prior road network information based on large scale crowdsourcing trajectory data. Throughout this paper, we take ship trajectory prediction on the open ocean as an example. And the ship trajectories we use are Automatic Identification System (AIS) data.

The contributions of this paper are:

1) We propose a deep learning approach to trajectory prediction based on LSTM (Long Short-Term Memory) model from large scale crowdsourcing trajectory data. The method can predict the next road segment that moving objects would reach or pass without any prior knowledge about the road network.

2) We label each trajectory with road segments and transform the trajectories into sequences of road segments. We also analyze the features of ships that affect the trajectory prediction and select type of ships and ship's direction of travel as features and joint with sequences of road segments as input of LSTM model.

3) We compare with the trajectory prediction method based on frequent sequence mining algorithm. Experiments show that our method is more effective.

The rest of this paper is organized as follows. Section 2 reviews related work. In Sect. 3, the basic concepts throughout the paper are introduced and the problems to be solved are described. Section 4 describes trajectory serialization feature extraction. Section 5 introduces the LSTM methods for prediction. Section 6 introduces the experiments and compares it with the trajectory prediction method based on frequent sequence mining. Section 7 concludes the paper and describes future work.

2   Related Work

At present, a series of research achievements have been made on trajectory prediction of moving objects. According to the prediction duration, the existing trajectory prediction methods can be divided into long-term prediction and short-term prediction. Long-term prediction generally focuses on the motion of the moving object after a few hours, while short-term prediction only focuses on the motion of the moving object within a few seconds.

According to different methods, long-term prediction can be divided into the following categories: (1) Deep learning based long-term trajectory prediction methods. Some methods obtain the destination set through clustering, and use the RNN model to predict the destination [3]. There are also some other methods that use LSTM prediction [20] model based on user similarity. (2) Long-term trajectory prediction methods based on frequent pattern mining. The idea is to mine frequent patterns of moving objects through analysis of historical data, and predict the future of moving targets by matching the sequence to be predicted and the frequent sequence movement track. This type of method mainly focuses on the selection of frequent pattern mining algorithms and the selection of matching strategies. Some researchers have proposed a frequent pattern mining algorithm based on prefix projection [5,8,9], and proposed three matching strategies: overall match, tail match, longest tail match. There are also some methods to sort the matching results according to confidence and output the top-k prediction results [17]. (3) Clustering based trajectory prediction methods. Some methods [19] use changes in simplified trajectory directions to identify inflection points, and use the DBSCAN algorithm to cluster inflection points to obtain turning nodes. Then the ant colony algorithm is used to find the optimal path from the starting turning node to the ending turning node to achieve the purpose of prediction. There are also some trajectory prediction methods of clustering trajectories based on trajectory similarity firstly, and then predicting through trajectory matching [15].

Short-term prediction methods can also be divided into multiple categories: (1) Deep learning based short-term trajectory prediction methods. Some researchers have proposed a method based on LSTM model to predict the trajectory of vehicles driving on the highway [1], this method takes into account the characteristics of the target vehicle and other vehicles in the vicinity, and can predict the trajectory in the next 10 seconds. Other methods adopt RNN [7] and ANN [21] models. (2) Short-term trajectory prediction methods based on

Gaussian regression model. These methods focus on the current motion state of moving objects, and achieve the purpose of prediction by establishing a moving model of moving objects. Some researchers have proposed a mixed model based probabilistic trajectory prediction method, which uses previously observed motion patterns to infer the probability distribution as a motion model, which can effectively predict the position of moving objects within 2 seconds [2,14]. Some researchers use Kalman filter [10] to estimate the state of the dynamic behavior of moving objects, update the estimation of state variables using the previous value and the observation value of the current time, and then predict the trajectory position at the next time. (3) Short-term trajectory prediction method based on Markov model [11,18]. The idea is to mine the hidden state and determine the parameters of Markov model for trajectory prediction from historical trajectories.

Only a few of the above prediction methods are based on road network constraints [5]. However, in addition to trajectory data, this approach relies on a priori knowledge of the road network. Different from these related work, this paper realizes the trajectory prediction only based on massive crowdsourcing trajectory data without relying on any prior knowledge of the road network. And different from [5], our approach is based on LSTM model instead of frequent pattern mining algorithm.

3   Definitions and Problem Description

3.1   Definitions

Several definitions involved in this article are as follows:

Definition 1 (*Trajectory*). A trajectory of a moving object (a ship or vehicle) is a set of trajectory points $T = \{t_1, t_2, t_3, ..., t_n\}$. Each ship or vehicle has its unique identification (i.e. Maritime Mobile Service Identify (MMSI) for ships), each trajectory is composed of multiple trajectory points, a trajectory point t_i is composed of a triple (utc_i, lon_i, lat_i), where utc_i is the time stamp of the occurrence of trajectory point t_i , lon_i is the longitude of track point t_i, lat_i is the latitude of track point t_i.

Definition 2 (*Road Boundary*). A road boundary is a plane polygon set $C = \{c_1, c_2, c_3, ..., c_n\}$. A road polygon represents the boundary of the ocean road, which is composed of multiple vertices, arranged in a clockwise sequence, which can be expressed as $c_i = \{p_0, p_1, ..., p_n\}$, $p_i = \{lon_i, lat_i\}$ represents the vertices of the road polygon. Sometimes there exists cavity polygons located inside the road polygon and represents an area (such as islands and reefs) that cannot be navigated.

Definition 3 (*Road Segment*). A road network is composed of multiple road segments. Each road segment is also a polygon. It is also composed of multiple triangles, which can be expressed as $S = \{tri_1, tri_2, tri_3, ...tri_n\}$. Each triangle consists of three points, which can be expressed as $tri_i = \{p_x, p_y, p_z\}$. We will introduce why to represent road segments as a set of triangles in Sect. 4.1.

Base on the above definitions, given a trajectory point, we are able to determine if it is inside a road segment or not. Generally speaking, most of the trajectory points in a trajectory locate in some road segments. Thus a trajectory can be represented as a sequence of road segments: $t_i = \{S_1, S_2, ..., S_m\}$.

3.2 Problem Description

Problem Statement. Given a set of trajectories $\mathcal{T} = \{t_1, t_2, ..., t_n\}$, for any trajectory of a moving object $t_i = \{S_1, S_2, ..., S_m\}$, infer the next road segment S_{m+1} this moving object will reach.

4 Geometry Translation

4.1 Trajectory Serialization Under Network Constrains

The aim of trajectory serialization under network constrains is annotating each trajectory with a sequence of road segments. We firstly extract road boundaries with road segments annotation from large scale crowdsourcing trajectory data.

As shown in Fig. 1(a), since the original trajectory data is collected from a large number of remote sensors, there will be a lot of quality problems such as errors, different sampling rates and missing positions. In order to improve the data quality, the MapReduce parallel computing framework is used in our previous work for trajectory sampling, interpolation, denoising, and segmentation [6, 16]. Figure 1(b) shows an intermediate result of the pre-processing process.

In order to extract the road boundaries, in our previous work [6, 13], a parallel grid merging and filtering algorithm was proposed. Firstly we split the region of interest into grids with a parallel GeoHash encoding algorithm, and get the density values of the grids. After that, we perform grid merging according to the density values of the grids, then filters the merged grid data based on a local sliding window mechanism to get the grids within road boundaries. Then we apply the Delaunay Triangulation on the center points of the grids which are within road boundaries (as shown in Fig. 1(c)), filter the triangles within road and generate the polygons of road boundaries (for more details please refer to the proposed CirShape algorithm in our previous work [13]).

The extracted road boundary polygons have a lot of jagged edges. We need to smooth the boundaries before further processing. For the details of the smoothing method please refer to our previous work [12] An example result after smoothing is shown as Fig. 1(d).

The detailed network constrained trajectory serialization method could be divided into the following two steps:

(1) Extract polygons of road segments

We apply the Delaunay Triangulation on the vertexes of the smoothed road boundary polygons as shown in Fig. 1(e). Then the triangles outside the road

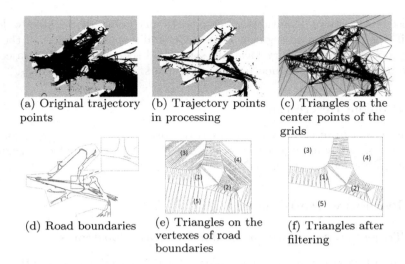

(a) Original trajectory points

(b) Trajectory points in processing

(c) Triangles on the center points of the grids

(d) Road boundaries

(e) Triangles on the vertexes of road boundaries

(f) Triangles after filtering

Fig. 1. Extraction of road boundaries and road segments

and inside the cavity (Area 2–5) are filtered. Among the left triangles (Area 1), those with three common edges are at the road intersection. The centers of gravity of these triangles are the nodes of road network and those triangles between two nodes are triangles on the road segments. The adjacent triangles are all traversed and all triangles on the road segments could be determined (For more details please refer to [12]). Then we can delete the inner edges of triangles on the road segments and only keep the edges forming the road segments. In another word, now we have polygons of road segments.

As shown in Fig. 2(a) below, there are a total of 5 road segments, and each road segment is composed of several triangles. Delete the inner edges of the road segment to get the road segment polygon, as shown in Fig. 2(b).

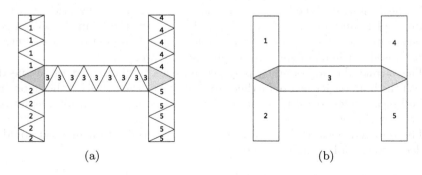

(a)   (b)

Fig. 2. Extract polygons of road segments: an example

(2) Trajectory annotation with road segments

Given a trajectory, for each point of this trajectory, we search the road segment polygon which the point locates inside respectively. Thus a trajectory is annotated with road segments. Then we merge the repeated segments and get the final serialization results.

As shown in Fig. 3, there are 5 road segments in the road network. The black dots represent the trajectory points as $T = <P_1, P_2, P_3, P_4, P_5, P_6, P_7, P_8, P_9, P_{10}, P_{11}, P_{12}, P_{13}, P_{14}>$. We firstly annotate it with road segments as $T_s = <1, 1, 1, 1, 3, 3, 3, 3, 3, 3, 3, 3, 5, 5>$, then get the final serialization result as $T_s = <1, 3, 5>$ after merging.

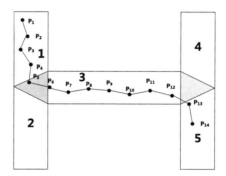

Fig. 3. An example trajectory serialization

4.2 Motion Feature Extraction

Data and features determine the upper limit of machine learning, and models and algorithms are only approaching that limit. In addition to the trajectory serialization, we extract some motion features from the trajectories to improve the accuracy of the prediction model.

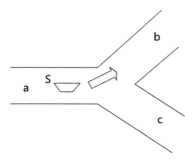

Fig. 4. Relationship between current movement direction and next road segment

Based on general observations of traffic, we believe that a vehicle's next segment destination is very related with its current moving direction when it travels across intersections. As shown in Fig. 4, if a ship S traveling on road segment a leaves the road segment in a direction shown by the arrow, the next road segment of S is likely to select the road segment b.

Here we use the Azimuth angle (Az) to measure the direction angel of movement. The calculation method of Az of two points is as follows: as shown in Fig. 5, each point is represented as longitude and latitude, and O is the earth's center. We calculate Az of point B relative to point A according formulas 1, 2, 3, 4 and 5.

$$\cos c = \cos(90 - B_{lat}) \cdot \cos(90 - A_{lat}) + \sin(90 - B_{lat}) \cdot \sin(90 - A_{lat}) \cdot \cos(B_{lon} - A_{lon}) \tag{1}$$

$$\sin c = \sqrt{1 - \cos^2(c)} \tag{2}$$

$$\sin A = \sin \frac{\sin(90 - B_{lat}) \times \sin(B_{lon} - A_{lon})}{\sin(c)} \tag{3}$$

$$A = \arcsin \frac{\sin(90 - B_{lat}) \times \sin(B_{lon} - A_{lon})}{\sin(c)} \tag{4}$$

$$A_z = \begin{cases} A, & Point\ B\ is\ in\ the\ first\ quadrant \\ 360 + A, & Point\ B\ is\ in\ the\ second\ quadrant \\ 180 - A, & Point\ B\ is\ in\ the\ third\ or\ fourth\ quadrant \end{cases} \tag{5}$$

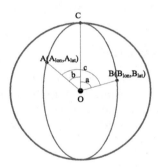

Fig. 5. Calculate the azimuth between two points A and B

The relationship between the direction angle and trajectories can be explained by comparing the direction of movement from the same segment to different segments. As shown in the Table 1, the distribution of direction always concentrates in a particular range.

Table 1. Statistics of direction distribution between different segments

Start-segment-ID	Arrival-segment-ID	Number-of-samples	Sailing-direction
30	31	157	5.4–6
30	59	356	2.5–3
60	61	421	5–6
30	105	477	2.5–3

For many reasons, such as climate, geographic environment and military, some road segments may only allow certain types of ships to travel. This article divides ships into six types, namely fishing vessels, cargo ships, oil tankers, passenger ships, container ships and other types. As shown in Table 2, we count the number of passes by different types of ships in different road segments. From the table, we can see that the number of different types of ships passing by the same segment is different, and different types of ships pass by the same segment. The proportion is different apparently. For some segments (e.g. segment 8), only certain types of ships will pass through. Therefore, road segments and types of ship are closely related. And the type of ship could be taken into considered as an important feature.

Table 2. Statistics of the number of different types of ships passing through road segments

Segment ID	Fishing boat	Cargo ship	Tanker	Passenger ship	Container Ship	Other types	Total number of passes
0	3	12	35	0	0	16	66
1	1	392	89	19	186	552	1239
2	0	38	4	0	9	6	57
3	1	2	2	0	0	4	9
4	3	22	6	0	0	15	46
5	1	19	3	0	1	5	29
6	0	37	12	0	9	55	113
7	1	34	1	0	0	5	41
8	0	3	0	0	0	3	6
9	2	101	13	0	9	52	177

5   Trajectory Prediction Based on LSTM Model

The single-layer LSTM architecture in this paper is shown in Fig. 6. First, the input layer receives tensors of fixed dimensions and length, then the LSTM layer consists of 256 neurons, and then one consists of 128 neurons. The fully connected layer uses Softmax as the activation function, and finally the output layer contains neurons with the same length as the output.

According to the characteristics of the input layer of the neural network, the input vectors needs to be divided into vectors of the same length. In this paper, the road segment sequences and features (direction and ship type) are selected as three-dimensional input vector units. We introduce the step size *step*, let *step* = *k* (such as *k* = 5), then get a normalized *k* * 3 two-dimensional input tensor.

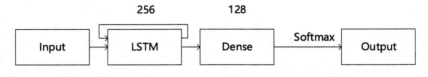

Fig. 6. LSTM model

6   Experiment and Analysis

6.1   Experimental Data

The dataset used in the experiment is the real AIS data from January 2017 to February 2017 in the Bohai Sea area. There are totally 15,725 ships in the dataset. After the sampling, denoising, interpolation and segmentation, a total of 10,668 trip records of 3574 ships were extracted. Table 3 gives the statistics for different types of ships:

Table 3. Dataset statistics for different types of ships

Ship type	Number of trips
Other	3644
Fishing boat	2197
Cargo ship	3295
Tanker	618
Passenger ship	376
Container Ship	538

Since there are some trajectory segments that cannot be matched with the road network, the trajectory segments should be filtered before serialization. We count the number of coordinate points in a trajectory segment located inside the road boundaries. If there are more than 80% of the coordinate points in a trajectory segment located inside the road boundaries, this segment can be serialized.

After preprocessing and serialization, there are a total of 3300 trips.

After serialization, the sequence of road segments is of varying lengths and need to be partitioned into vectors of the same length according to the requirement of the LSTM neural network input layer. As shown in Table 4, there is a sequence of road segments (a, b, c, d, e, f, g, h, i, j, k) that the ship passes through in a single trip. Each road segment passed is a three-dimensional vector, including the segment ID, ship type and travel direction. When we take $step = 5$, we get a $5 * 3$ tensor. So if we partition the complete sequence into several subsequences with step size of 5, we can get a three-dimensional normalized input tensor for any trajectory segment. The reason we take $step$ as 5 is analyzed in Sect. 6.3.

Table 4. Examples of road segment sequences normalization

Road segment sequence	Input data samples
(a, b, c, d, e, f, g, h)	(a, b, c, d, e)
	(b, c, d, e, f)
	(c, d, e, f, g)
	(d, e, f, g, h)

After the above processing, we get over 33,000 data samples. Then the dataset is partitioned into training set, validation set and test set according to the ratio of 6:2:2 using the hold-out method for model training, hyper parameters determination and model evaluation respectively. The division of the dataset generated according to the step size of 5 is shown in Table 5:

Table 5. Partitioning of the dataset

Total number of samples	Training set	Validation set	Test set
33037	19822	6607	6608

6.2 Experimental Settings

This paper uses three metrics: average precision ($Macro$-$Precision$), average recall ($Macro$-$Recall$) and average F_1 value ($Macro$-F_1) to evaluate the performance of the model. For each prediction result, find its Precision, Recall, and F_1 values, and then take the average as the average accuracy, average recall, and average F_1 value.

For each prediction result, it can be divided into true examples (TP, the prediction results and the true values are positive examples), false positive examples (FP, the prediction results are positive examples, the true values are negative examples), and false negative examples (FN, the prediction result is a negative example and the true value is a positive example) and the true negative example

(TN, the prediction result and the true value are both negative examples). The calculation methods of its precision rate, recall rate and F_1 value are as follows:

$$Precision = \frac{TP}{TP + FP} \tag{6}$$

$$Recall = \frac{TP}{TP + FN} \tag{7}$$

$$F_1 = \frac{2 \cdot Precision \cdot Recall}{Precision + Recall} \tag{8}$$

The calculation methods of average precision ($Macro\text{-}Precision$), average recall ($Macro\text{-}Recall$) and average F_1 value ($Macro\text{-}F_1$) are as follows:

$$Macro\text{-}Precision = \frac{1}{m} \cdot \sum_{i=1}^{m} Precision_i \tag{9}$$

$$Macro\text{-}Recall = \frac{1}{m} \cdot \sum_{i=1}^{m} Recall_i \tag{10}$$

$$Macro\text{-}F_1 = \frac{1}{m} \cdot \sum_{i=1}^{m} F_{1i} \tag{11}$$

The experimental environment of this article is shown in Table 6 below:

Table 6. Experimental environment

Operating system	Windows 7, 64-bit
CPU	Intel (R) Core (TM) i5-4200HQ CPU @ 1.60 GHz, 4 cores
RAM	4G
Tensorflow	Tensorflow-1.14.0
Keras	Keras-2.2.4
Java	1.8.0_241
Python	3.6.10

6.3 Results and Discussion

We perform the model tuning experiment firstly. Results of the experiment are shown in Table 7. The parameters of the model include loss function(Loss), activation function(Activation), optimizer(Optimizer), number of samples per batch (batch_size) and training rounds (Epochs). The last three columns in the table represent $Macro\text{-}Precision$, $Macro\text{-}Recall$, and $Macro\text{-}F_1$. We select *categorical crossentropy* as the Loss function.

Table 7. LSTM model tuning

Test number	Activation	Optimizer	Batch size	Epochs	Training time	M-P	M-R	M-F_1
1	softmax	adam	32	20	139.148	0.847	0.858	0.844
2	sigmoid	adam	32	20	137.008	0.847	0.849	0.836
3	relu	adam	32	20	142.427	0.342	0.420	0.303
4	softmax	sgd	32	20	139.906	0.430	0.622	0.462
5	softmax	rmsprop	32	20	139.964	0.837	0.845	0.830
6	softmax	adam	50	20	135.195	0.840	0.852	0.835
7	softmax	adam	100	20	118.606	0.830	0.836	0.820
8	softmax	adam	200	20	112.956	0.817	0.824	0.803
9	softmax	adam	400	20	117.774	0.796	0.823	0.797
10	softmax	adam	20	20	160.907	0.848	0.856	0.842
11	softmax	adam	10	20	211.774	0.847	0.851	0.840
12	softmax	adam	32	40	173.147	0.874	0.873	0.866
13	softmax	adam	32	60	350.734	0.876	0.878	0.870
14	softmax	adam	32	80	308.795	0.884	0.885	0.879
15	softmax	adam	32	100	351.054	0.876	0.879	0.871

For experiments 1–3, we choose the same Optimizer, Batch_size, and Epochs, and softmax, sigmoid, and relu as the Activation functions respectively. The experimental results show that under the same other parameters, the time used to train the model is not much different, but the model trained using softmax as the activation function has a better effect on the test set.

For experiment 2, 4 and 5, we choose different Optimizers, and other parameters are the same. The experimental results show that we can find the direction of gradient descent faster with adam as the Optimizer when other parameters are the same. The model using adam as the optimizer has a better effect on the test set, and the time for training the model is not much different.

Experiment 2, 6–11 choose different Batch_size, other parameters are the same. The experimental results show that the average F_1 value of the test set will increase with the increase of batch_size within a certain range when other parameters are the same, but if this range exceeds the average F1 value will decrease with the increase of batch_size. The model trained with batch_size 32 is better on the test set. The model training time will decrease as the batch_size increases.

Experiment 1, 12–15 choose different Epochs, other parameters are the same. The experimental results show that the average F_1 value of the test set will increase with the increase of training times within a certain range, but when the training times increase to a certain number, overfitting will occur. Although the accuracy of the training set and the verification set has increased, the average F_1 value of the test set has decreased. The duration of model training increases with the number of training sessions.

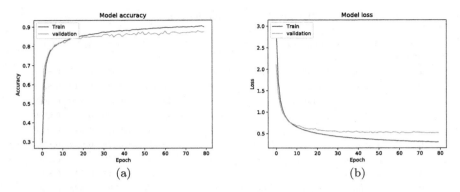

Fig. 7. Training curve of the 14th group

The results show that the 14th group of experiments is the best. The average accuracy rate is 0.884, the average recall rate is 0.885, and the average F_1 value is 0.879, with softmax as Activation function, adam as the Optimizer, 32 as the batch_size, and 80 as the Epochs times. The training curve is shown in Fig. 7. Figure 7-a is the model's Accuracy curve on the training set and validation set, and Fig. 7-b is the model on the training set and validation Loss curve on the set. We also use the controlled variable method to experiment with $step = \{3, 4, 5, 6, 7\}$. From the experimental results shown in Fig. 8 we can see that the effect is best when $step = 5$.

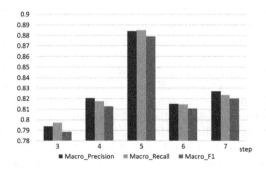

Fig. 8. Comparison of prediction effects for different step values

As analyzed in Sect. 2, only a few of the trajectory prediction methods are based on road network constraints. And the existing network constrained trajectory prediction approaches are based on frequent pattern mining algorithm, which is different with our work. So in the experiment, we compare our method with the trajectory prediction method based on frequent sequence pattern mining algorithm. We select 1,000 samples randomly as the test set, and other samples are used as the training set for frequent sequence pattern mining. Each sample contains only the road segment sequence, and does not include other features such as direction and ship type. In the experiment, by setting the support degree to 20, 2153 frequent sequences can be obtained. The number of frequent sequences with different lengths (n-th order) is shown in Table 8:

Table 8. Statistics of frequent sequences for different lengths.

Frequent sequence length	2	3	4	5	6	7	8	9	10
Quantity	216	211	207	193	187	177	168	154	137
Frequent sequence length	11	12	13	14	15	16	17	18	19
Quantity	123	107	88	67	51	35	20	9	3

The sequences to be predicted are matched against frequent sequences, and the frequent sequence with maximum support value is taken as the prediction result. The average accuracy rate of the prediction is 0.812, the average recall rate is 0.824, and the average F_1 value is 0.815. Compared with our work, the prediction results based on LSTM model are better than those based on frequent sequence (PF). We believe that this is because the inputs of LSTM model include more features, including road segment sequences, direction and ship type, while the PF method can only adopt road segment sequences as inputs.

7   Conclusion and Future Work

In this paper, we study the problem of long term trajectory prediction under road network constraints only based on massive crowdsourcing trajectory data and without any prior knowledge about the road network. We propose a deep learning-based trajectory prediction method based on LSTM model. We extract the road segment information from large scale crowdsouring trajectory data and annotate the position sequences of the original trajectories with road segment polygons. In this way we translate the original trajectory into a road segment sequences. We also analyze the useful features and select movement direction and ship type jointly union with the road segment sequences as the input of LSTM model. Experiments on the real-world datasets show our method outperforms the network constrained trajectory prediction approach based on frequent pattern mining algorithm.

In the future, we will try to use the deep learning method to solve other long-term trajectory problems without any prior knowledge about road network such as predicting the final destination except for next road segment in this paper. We expect the idea of extracting and annotating trajectories with road segment polygons can also perform well in these problems.

Acknowledgments. This work is supported by National Natural Science Foundation of China under Grant 61832004 and Grant 61672042. We thank the Ocean Information Technology Company, China Electronics Technology Group Corporation (CETC Ocean Corp.), for providing the underlying dataset for this research.

References

1. Altche, F., de La Fortelle, A.: An LSTM network for highway trajectory prediction. In: 2017 IEEE 20th International Conference on Intelligent Transportation Systems (ITSC), pp. 353–359 (2017). https://doi.org/10.1109/ITSC.2017.8317913
2. Dalsnes, B.R., Hexeberg, S., Flåten, A.L., Eriksen, B.O.H., Brekke, E.F.: The neighbor course distribution method with Gaussian mixture models for AIS-based vessel trajectory prediction, pp. 580–5877 (2018)
3. Endo, Y., Nishida, K., Toda, H., Sawada, H.: Predicting destinations from partial trajectories using recurrent neural network. In: Kim, J., Shim, K., Cao, L., Lee, J.-G., Lin, X., Moon, Y.-S. (eds.) PAKDD 2017. LNCS (LNAI), vol. 10234, pp. 160–172. Springer, Cham (2017). https://doi.org/10.1007/978-3-319-57454-7_13
4. Georgiou, H.V., et al.: Moving objects analytics: Survey on future location & trajectory prediction methods. CoRR abs/1807.04639 (2018). http://arxiv.org/abs/1807.04639
5. Guo, L., Ding, Z., Hu, Z., Chen, C.: Uncertain path prediction of moving objects on road networks. J. Comput. Res. Dev. **47**, 104–112 (2010). (in Chinese)
6. Li, Z., Wang, G., Meng, J., Xu, Y.: The parallel and precision adaptive method of marine lane extraction based on QuadTree. In: Gao, H., Wang, X., Yin, Y., Iqbal, M. (eds.) CollaborateCom 2018. LNICST, vol. 268, pp. 170–188. Springer, Cham (2019). https://doi.org/10.1007/978-3-030-12981-1_12
7. Liu, Q., Wu, S., Wang, L., Tan, T.: Predicting the next location: a recurrent model with spatial and temporal contexts. In: Proceedings of the Thirtieth AAAI Conference on Artificial Intelligence, AAAI 2016, pp. 194–200, AAAI Press (2016)
8. Qiao, S., et al.: Large-scale trajectory prediction model based on prefix projection technology. J. Softw. **28**, 3043–3057 (2017). (in Chinese)
9. Qiao, S., Han, N., Wang, J., Li, R.H., Gutierrez, L.A., Wu, X.: Predicting long-term trajectories of connected vehicles via the prefix-projection technique. IEEE Trans. Intell. Transp. Syst. **19**(7), 2305–2315 (2017)
10. Qiao, S., Han, N., Zhu, X.: Dynamic trajectory prediction algorithm based on Kalman filtering. Chin. J. Electron. **46**(2), 418–423 (2018). (in Chinese)
11. Qiao, S., Shen, D., Wang, X., Han, N., Zhu, W.: A self-adaptive parameter selection trajectory prediction approach via hidden Markov models, vol. 16, pp. 284–296. IEEE (2014)
12. Wang, G., Meng, J., Han, Y.: Extraction of maritime road networks from large-scale AIS data. IEEE Access **7**, 123035–123048 (2019)
13. Wang, G., Meng, J., Li, Z., Hesenius, M., Ding, W., Han, Y., Gruhn, V.: Adaptive extraction and refinement of marine lanes from crowdsourced trajectory data. Mob. Netw. Appl. 1392–1404 (2020). https://doi.org/10.1007/s11036-019-01454-w
14. Wiest, J., Höffken, M., Kreßel, U., Dietmayer, K.: Probabilistic trajectory prediction with Gaussian mixture models. In: 2012 IEEE Intelligent Vehicles Symposium, pp. 141–146. IEEE (2012)
15. Xie, B., Zhang, K., Zhang, Y., Cai, Y., Jiang, T.: Moving target trajectory prediction algorithm based on trajectory similarity. Comput. Eng. **44**(9), 177–183 (2018). (in Chinese)
16. Xu, Y., Li, Z., Meng, J., Zhao, L., Jianxin, W., Wang, G.: Extraction method of marine lane boundary from exploiting trajectory big data. Comput. Appl. **39**(1), 105–112 (2019). (in Chinese)
17. Yava, G., Katsaros, D., Ulusoy, Ö., Manolopoulos, Y.: A data mining approach for location prediction in mobile environments. Data Knowl. Eng. **54**, 121–146 (2005). https://doi.org/10.1016/j.datak.2004.09.004

18. Ye, N., Zhang, Y., Wang, R., Malekian, R.: Vehicle trajectory prediction based on hidden Markov model. Ph.D. thesis (2016)
19. Zhang, S., Shi, G., Liu, Z., Zhao, Z., Wu, Z.: Data-driven based automatic maritime routing from massive AIS trajectories in the face of disparity. Ocean Eng. **155**, 240–250 (2018)
20. Zhang, Z., Ni, G., Xu, Y.: Ship trajectory prediction based on LSTM neural network, pp. 1356–1364 (2020)
21. Zissis, D., Xidias, E.K., Lekkas, D.: Real-time vessel behavior prediction. Evolv. Syst. **7**(1), 29–40 (2015). https://doi.org/10.1007/s12530-015-9133-5

HMM-Based Traffic State Prediction
and Adaptive Routing Method
in VANETs

Kaihan Gao[1,2], Xu Ding[1,2], Juan Xu[1], Fan Yang[1], and Chong Zhao[1(✉)]

[1] School of Computer Science and Information Engineering,
HeFei University of Technology, Hefei, China
zhaochong@mail.hfut.edu.cn
[2] Institute of Industry and Equipment Technology,
HeFei University of Technology, Hefei, China

Abstract. As the number of vehicles increases, the traffic environment becomes more complicated. It is important to find a routing method for different scenarios in the vehicular ad hoc networks (VANETs). Although there are many routing methods, they rarely consider multiple road traffic states. In this paper, we propose a traffic state prediction method based on Hidden Markov Model (HMM), and then choose different routing methods according to different traffic states. Since we are aware that GPS may cause measurement errors, Kalman Filter is used to estimate the observation, which makes observation more accurate. For different road states, we can make appropriate methods to improve routing performance. When the road is in rush hour, we will use Extended Kalman Filter to predict vehicle information in a short time to reduce the number of broadcasts, which can alleviate channel load. The result show that our method is useful for reducing the number of packets and improving the delivery rate.

Keywords: VANETs · Kalman filter · HMM · Traffic state prediction

1 Introduction

With the rapid development of communication technologies, VANETs have become an important component of the Intelligent Transportation System (ITS) and played a very important role in many fields. The purpose of ITS is to provide reliable traffic management and make more coordinated, safer, and optimal use of transport networks. VANETs, an extension of MANETs, can be used to improve traffic management and reduce traffic accidents [1,2].

In recent years, with the increasing number of vehicular nodes, dealing with traffic congestion characterized by longer trip times, slower speeds, and complex

Supported by National Natural Science Foundation of China with grant number [61701162].

traffic environment have been an important issue. Thus, V2V communication plays an important role to reduce traffic congestion by exchanging road information among vehicles.

In VANETs, in order to better establish V2V communication, routing algorithm are divided into two types, position-based and topology-based [3]. Topology-based routing schemes generally require additional node topology information during the path selection process [4]. Geographic routing uses neighboring location information to perform the packet forwarding. The traditional position-based routing is Greedy Perimeter Stateless Routing (GPSR) [5]. In GPSR, the node used the information about the router's immediate neighbors to make greedy forwarding decision. However, GPSR may have unstable nodes and destroy network connectivity. Togou et al. [6] presented a distributed routing protocol, which computes end-to-end for the entire routing path and builds a stable route. In addition to information based on the location of vehicles on the road, some traffic information is also used in routing optimization methods. Vehicles can choose the best route to their destination based on real-time traffic information [7].

Because the vehicle moves at high speed on the road and changes direction frequently, which will cause communication link disconnection and information loss, some predictive routing methods and auxiliary methods have been proposed. Reza et al. [8] proposed a position prediction based multicast routing, which can alleviating the broadcast storm problem of multicast tree discovery and simultaneously minimizing the number of forwarding vehicles. The probability prediction-based reliable and efficient opportunistic routing algorithm is proposed [9]. Liu et al. [10] used Kalman Filter to predict the location of the vehicle to reduce the number of broadcasts, which can reduce the channel load. A centralized routing scheme with mobility prediction is proposed,which assisted by an artificial intelligence powered software-defined network controller [11]. Similarly, Bhatia et al. [12] also combined Software Defined Networking (SDN) with machine learning algorithms, proposed a datadriven approach for implementing an artificially intelligent model for vehicular traffic behavior prediction, which has the potential to predict real-time traffic trends accurately. Due to the regular movement of buses, a new routing scheme called Busbased Routing Technique (BRT) is proposed [13], which exploits the periodic and predictable movement of buses to learn the required time for each data transmission to Road-Side-Units (RSU) through a dedicated bus-based backbone to improve the delivery ratio and reduce end-to-end delay. The development of unmanned aerial vehicles (UAVs) also provides a new solution for VANETs. Oubbati et al. [14] designed a scheme that automatically reacts at each topology variation while overcoming the present obstacles while exchanging data in ad hoc mode with UAVs, the assistance of UAVs to vehicles can improve transmission performance. With the development of 5G networks, novel four-tier architecture for urban traffic management is proposed [15]. The novel architecture shows potential for improving the efficiency of traffic management.

In this paper, we analyze the possible road traffic status and get accurate observations based on Kalman Filter, and then predict the road status based on HMM. Different routing methods are used to improve routing performance according to different road traffic conditions.

Specifically, the main contributions of the paper are as follows: 1) Kalman Filter is used to get observations, which can help to get the road status more accurately; 2) HMM-based prediction is proposed for traffic conditions on the road; 3) Choosing the appropriate routing method according to different traffic status, which can reduce the number of packets and improve the delivery rate.

The remainder of this paper is organized as follows. In Sect. 2, the network model is given. In Sect. 3, observation based on Kalman Filter are obtained. In Sect. 4, we give a prediction model based on HMM and routing method. In Sect. 5, evaluates the performance of our scheme by simulations. In Sect. 6, we conclude this paper.

2   Network Model

In this research, multi-lane unidirectional traffic flow is considered. The road between two adjacent intersection is considered a road segment, which is assumed to be much longer than the transmission range. Consider a VANET with N vehicles, denoted by a set $N : \{c_1, c_2, ..., c_N\}$, where c is used to represent the vehicle. All vehicles will move along the prescribed road and will not exceed the speed V_{max}, where V_{max} is maximum speed limited on the road. Special public buses move on fixed routes, other types of vehicles move along the roads at its own will. Each vehicle is able to acquire its own instant state on the basis of the embedded GPS. The states of vehicle c at time t constitute a five-tuple, $s_c(t) : (id, v, a, p, t)$, representing the vehicle identification, velocity (v_{xt} and v_{yt}), acceleration, position (x and y coordinates), and timestamp. The states of road r at time t constitute a six-tuple, $s_r(t) : (id, \bar{v}, \bar{a}, \sigma^2, n, t)$, representing the road identification, average velocity, average acceleration, variance, number of vehicles and timestamp. Where $\bar{v}$, $\bar{a}$ and variance respectively calculated by Eq. (1), (2) and (3):

$$\bar{v} = \frac{\sum_{i=1}^{n} v_i}{n} \tag{1}$$

$$\bar{a} = \frac{\sum_{i=1}^{n} a_i}{n} \tag{2}$$

$$\sigma^2 = \frac{\sum_{i=1}^{n} (v_i - \bar{v})^2}{n} \tag{3}$$

The basic network model is shown in Fig. 1, and divided into four road sections according to traffic intersections and signal lights. Each vehicle is equipped with On Board Unit (OBU) for wireless communications. When the distance between two vehicles is less than their transmission range, they will communicate with each other. In the case where there is an obstacle, the communication is not possible. Therefore, vehicles on different roads cannot communication with each other as shown in Fig. 1. For the entertainment application or other data sharing applications, we assume that each node always has data to send. Emergency data is also transmitted once an accident occurs. The next model process is shown in Fig. 2, and a summary of main mathematical notations is provided in Table 1.

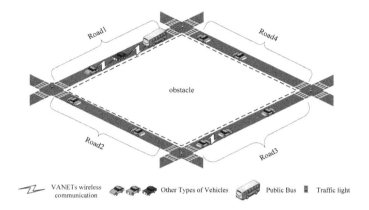

Fig. 1. Network model

Table 1. Summary of the main mathematical notations

Natation	Description
N	Road vehicle set
V_{max}	Maximum road speed limit
$\bar{v}$	Average speed of the vehicles
$\bar{a}$	Average acceleration of the vehicles
σ^2	Variance of vehicle speed
X_t	Vehicle state vector
Z_t	Measurement vector
F	State transition matrix
C	System input matrix
K_t	Optimal Kalman gain
H	Measurement matrix
π	Initial hidden state probabilities
A	Transition probabilities matrix
B	Emission probabilities matrix
O	Observation sequences

3   Observation Acquisition Based on Kalman Filter

It is necessary to correctly and promptly get the node state of VANETs for the road traffic prediction. In this section, Kalman Filter is proposed to obtain an accurate state, since the possible errors in the measured values will lead to the inaccurate prediction results.

3.1   Overview of Kalman Filter Model

Kalman Filter [16,17] is an efficient recursive filter that estimates the state of a linear dynamic system from a series of noisy measurement and can solve a set of mathematical equations for unknown state vectors in an optimal method that minimizes the estimated error covariance. Kalman Filter has two important vectors, state vector and measurement vector. We let X_t be the vehicle state vector. And the measurement vector Z_t is a measurement at time t. The Kalman Filter uses two equations: the process equation and the measurement equation. The process equation is defined as:

$$X_t = FX_{t-1} + Cu_t + \omega_t \tag{4}$$

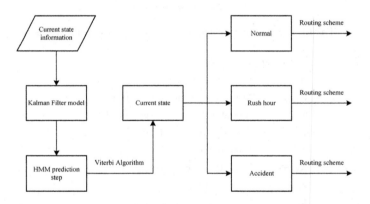

Fig. 2. Process of model

where F is a state transition matrix, C is the system input matrix and ω_t is the process noise which Gaussian distribution with zero mean and a covariance matrix Q.

The measurement equation is defined as:

$$Z_t = HX_t + \mu_t \tag{5}$$

where Z_t is the measurement vector at time t and H is the measurement matrix. μ_t is the measurement noise which is Gaussian distribution with zero mean and a covariance matrix R.

3.2   Kalman Estimation

When a node is added into the network, it will generate a state vector $SC_t = (x_t, v_{xt}, y_t, v_{yt}, a_{xt}, a_{yt})^\mathsf{T}$, where x_t and y_t represent the coordinates of the vehicle at time slot t, similarly, v_{xt} and v_{yt} represent the velocity of the vehicle on the x-axis and y-axis, a_{xt} and v_{yt} represent the acceleration of the vehicle on the x-axis and y-axis. According to the nodes state information on the road, the road state vector $X_t = (\bar{v}_{xt}, \bar{v}_{yt}, \bar{a}_{xt}, \bar{a}_{yt})^\mathsf{T}$ can be obtained. So the new speed in next step can be approximated by (6):

$$\begin{aligned}\bar{v}_{x(t+1)} &= \bar{v}_{xt} + \bar{a}_{xt}\Delta t \\ \bar{v}_{y(t+1)} &= \bar{v}_{yt} + \bar{a}_{yt}\Delta t\end{aligned} \tag{6}$$

Where acceleration is used as the system input, so we can get the 4×4 transitional matrix F and the 4×4 system input matrix C as follows in (7), (8):

$$F = \begin{bmatrix} 1 & \Delta t & 0 & 0 \\ 0 & 1 & \Delta t & 0 \\ 0 & 0 & 0 & 0 \\ 0 & 0 & 0 & 0 \end{bmatrix} \tag{7}$$

$$C = \begin{bmatrix} 0 & 0 & 0 & 0 \\ 0 & 0 & 0 & 0 \\ 0 & 0 & 1 & 0 \\ 0 & 0 & 0 & 1 \end{bmatrix} \tag{8}$$

Here t is the sampling interval and corresponds to the time interval.

The Kalman Filter uses the estimation of the previous state to make a prediction of the current state according to Eq. (9):

$$\widehat{X}'_t = F\widehat{X}_{t-1} + Cu_t \tag{9}$$

where $\widehat{X}'_t$ is the predicted value based on the previous step estimated state and $\widehat{X}_{t-1}$ is the estimated state obtained in the update step.

And it uses the measured values of current state to modify the prediction value which obtained in the prediction state to obtain a new estimate value that more closely matches the real value according to Eq. (10), (11) and (12):

$$\widehat{Z}_t = Z_t - H\widehat{X}'_t \tag{10}$$

$$K_t = P'_t H^\mathsf{T}(HP'_t H^\mathsf{T} + R)^{-1} \tag{11}$$

$$\widehat{X}_t = \widehat{X}'_t + K_t\widehat{Z}_t \tag{12}$$

where P'_t is the predicted covariance matrix based on the previous step estimated covariance matrix and K_t is optimal Kalman gain. The measurement only includes $\bar{v}_x t$ and $\bar{v}_y t$, so it is also a 2×1 vector. Because the state of the road X_t is a 4×1 vector, the measurement matrix H is defined as (13):

$$H = \begin{bmatrix} 1 & 0 & 0 & 0 \\ 0 & 1 & 0 & 0 \end{bmatrix} \tag{13}$$

The estimation problem begins with no prior measurements. Often, the value of the first state is chosen as the first measurement value, thus $X_0' = Z_0$. After the above steps, we can get more accurate observations for the next road traffic prediction.

4   HMM-Based Road State Prediction Routing Method

Due to the increase in the number of vehicles, road conditions are becoming more and more complex and time-varying. For better V2V communication, it is necessary to analyze and predict road traffic status. In this section, we present a road state prediction based on HMM and choose the appropriate routing scheme according to different road conditions to improve routing performance.

4.1   Prediction Based on Hidden Markov Model

Hidden Markov Model is a statistical Markov model. The system is modeled as a Markov process with unobserved states. In HMM, the state is not directly visible, while the observation that depends on the state is visible. Each state has a probability distribution over the possible observation states. Therefore, the HMM can obtain the state sequence based on a series of observation sequences.

In this paper, we consider the road state to be a hidden state, and state transitions are within these hidden states. For example, a road can change from a normal non-congested state to a jamming state, or it can recover from a jamming state to a normal state. Moreover, the road traffic state cannot be directly judged from the measured date. Each road traffic state corresponds to a variety of observation states, which is also in line with the characteristics of HMM, so we use HMM to predict the road traffic state.

HMM can be defined as

$$\lambda = \{\pi, A, B\} \tag{14}$$

where π is the initial hidden traffic state probabilities. A is the transition probabilities matrix between the hidden states. And B is matrix which the probabilities of the observable states in the hidden states.

We take the speed, acceleration, the number of vehicles and the variance of the speed as the observation states (See Fig. 3). In order to easily obtain the HMM model, the road traffic state is divided into three hidden states: normal, rush hour and accident. When the road is in a normal state, we consider that the vehicle can move at a higher speed, and also consider the possibility of a slower speed due to the driver's behavior when there are fewer vehicles. When the road traffic state is at the rush hour, the number of vehicles on the road increases, which will cause road congestion with a high probability. The accident state is a traffic accident on the road. In order to simplify the model, we believe that the accident will cause partial congestion on the road and will not result in the entire road paralyzed. Therefore, the variance of the vehicle speed will be larger. So as to distinguish it from the normal state when the number of vehicles is small, the number and speed of vehicles should also be taken into consideration. The HMM is based on the historical data on the road afterward.

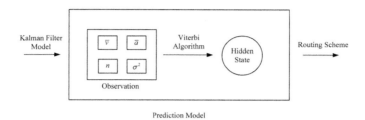

Prediction Model

Fig. 3. Prediction based on HMM

In this paper, Viterbi Algorithm is used to accurately predict the sequence of hidden states. We know model parameters λ and observation sequences $O = (o_1, o_2, ..., o_T)$. Where T is time. First, we initialize it according to Eqs. (15), (16):

$$\delta_1(i) = \pi_i b_i(o_1) \tag{15}$$

$$\psi_1(i) = 0 \tag{16}$$

where π_i is the initial hidden state i probabilities and $b_i(o_1)$ is the probabilities of the observable state o_1 in the hidden state j in the HMM. $\delta_t(i)$ is the most likely observation sequence when the hidden state is i at time t. $\psi_t(i)$ is the most likely hidden state at time $t - 1$.

Then recurse to $t = 2, 3, ..., T$:

$$\delta_t(i) = \max_{1 \leq j \leq N} [\delta_{t-1}(j)a_{ji}]b_i(o_t) \tag{17}$$

$$\psi_t(i) = \operatorname*{argmax}_{1 \leq j \leq N}[\delta_{t-1}(j)a_{ji}] \tag{18}$$

where N is the number of all hidden states and a_{ji} is the transition probabilities between the hidden state i and hidden state j in the HMM.

Finally, terminating algorithm and backtracking the optimal path, $t = T - 1, T - 2, ..., 1$. We can get the optimal path, which is the hidden state sequence in HMM. When the model parameters and the observed state sequence are known, we can get the most likely hidden state sequence according to the Viterbi Algorithm, so that the road state observed at multiple times can be used to predict the true road state. The pseudo-code for Viterbi is given in Algorithm 1.

Algorithm 1. Viterbi Algorithm.

Input: A sequence of observations $O = o_1, o_2, ..., o_T$;
Output: The most likely hidden state sequence $I = i_1, i_2, ..., i_t$;
1: **function** VITERBI(O):I
2:    **for** each state i **do**
3:        $\delta_1(i) \leftarrow \pi_i b_i(o_1)$
4:        $\psi_1(i) = 0$
5:    **end for**
6:    **for** $t \leftarrow 2, 3, ..., T$ **do**
7:        **for** each state s_j **do**
8:            $\delta_t(i) = \max_{1 \leq j \leq N}[\delta_{t-1}(j)a_{ji}]b_i(o_t)$
9:            $\psi_t(i) = \underset{1 \leq j \leq N}{argmax}[\delta_{t-1}(j)a_{ji}]$
10:        **end for**
11:    **end for**
12:    **for** $t \leftarrow T - 1, T - 2, ..., 1$ **do**
13:        $i_T = \underset{1 \leq j \leq N}{argmax}[\delta_T(i)]$
14:        $i_t = \psi_{t+1}(i_{t+1})$
15:    **end for**
16: **return** I
17: **end function**

4.2   Routing Method

If we know the hidden sequence, the appropriate routing algorithm will be chosen for different hidden states.

i) When the road is in a normal state, the vehicle is driving on the road at a high speed and the road is not congested. The target vehicle need to find other vehicles that can transmit within the communication range. If no vehicle is found within the transmission range, the vehicle will carry the information for a period of time slot s_1. During time s_1, if there is a suitable vehicle, the message will be transmitted, otherwise it will be discarded. If there is a bus in the transmission range, the data packet is sent to the bus first, because compared with other types of vehicles, the route of the bus is relatively stable, which is helpful for improving the transmission rate. As show in Fig. 4, $V1$ can communicate with $V2$. $V3$ will store the packet until it meets an appropriate vehicle or drop the packet upon reaching the maximum amount of time s_1 to hold the packet.

ii) When the road is in rush hour, there are more vehicles on the road, which will cause road congestion. Due to the excessive number of vehicles and slow speeds on the road, frequent broadcast messages will cause heavy channel load and severe packet loss. So Extended Kalman Filter (EKF) is used to predict the position of the vehicle and neighbor vehicles. KF is suitable for linear systems, while EKF can be used to solve nonlinear problems. Individual vehicle movement may be non-linear because of the complex road environment, So EKF is used to predict vehicle position. We also set a time slot s_2 and broadcast every time s_2 in order to prevent the error of EKF estimation from being amplified. As shown in Fig. 5, there are vehicles in the transmission range of $V1$. Communication is performed every time s_2 because of the large number of vehicles.

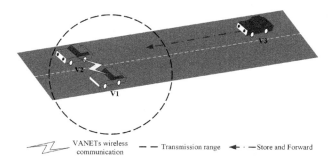

Fig. 4. Normal state

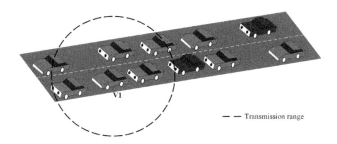

Fig. 5. Rush hour

iii) When an accident happens, it will introduce local congestion. Therefore, emergency data should be transmitted first. We will increase the priority of emergency information and extend the time to store packets in order to find a suitable next hop. As shown in Fig. 6, vehicles behind the accident site will suffer from congestion, and vehicles in other lanes will transmit emergency data first.

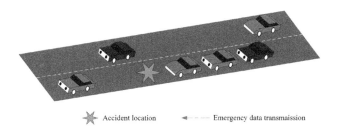

Fig. 6. Accident

Road traffic status are becoming more and more complicated due to the increase in vehicles. Encountered with different problems in different road environments, when the road is in a normal state, it is necessary to find a suitable next hop node to transmit packet. When the road is in the rush hour, there are always vehicles within the transmission range. However, due to the slow speed and the large number of vehicles, frequent broadcast messages will cause channel load and data loss. When a traffic accident occurs, emergency data should be transmitted first in order to restore traffic and reduce casualties. So we first predict the state of the road, then select the appropriate routing method to reduce the channel load and improve the packet delivery ratio. The pseudo-code for routing method selection is given in Algorithm 2.

Algorithm 2. Routing Method Selection.

Input: Current road traffic status I; The node set Φ;

```
 1: function ROUTING METHOD(I, Φ)
 2:     if I = NormalState then
 3:         for each node i ∈ Φ do
 4:             for any node j ∈ Φ ∥ i ≠ j do
 5:                 if Distance(i, j) < transmissionrange then
 6:                     Send Message
 7:                 else
 8:                     Store and Forward
 9:                 end if
10:             end for
11:         end for
12:     else if I = RushHour then
13:         for each node i ∈ Φ do
14:             if t = s₂ then
15:                 Broadcast message
16:             else
17:                 Extended Kalman Filter
18:             end if
19:         end for
20:     else
21:         if I = Accident then
22:             for node i ∈ Φ ∥ i ∉ Accident location do
23:                 Increase the priority of emergency information
24:                 Extend the time to store packet
25:             end for
26:         end if
27:     end if
28: end function
```

---

5   Simulation

In this section, some numerical results are given to show routing performance using routing scheme based on HMM prediction, which verifies the effectiveness of this method through the amount of packets and the packet delivery ratio.

5.1   Simulation Settings

We present a comparative performance study. Comparing our routing method with the hello protocol in AODV (HP-AODV) [18] and Kalman Prediction-based Neighbor Discover (KPND) [8]. The hello messaging protocol in AODV is a typical way that uses fixed time interval to broadcast. KPND uses Kalman Filter to predict the position of neighbor nodes. When the error is greater than the threshold, the neighbor table is refreshed through broadcasting. We assume that vehicles are gradually entering the road, so there are fewer vehicles on the road at the beginning of the simulation. Road traffic status will change with the increase of simulation time. When the vehicle reaches the end of the road, it randomly leaves the road or enters the next road. Public buses move on prescribed routes, and there is at most one bus on each road. All the vehicular nodes have the same transmission range. As in a typical entertainment and data sharing application of VANETs, every node always has data to transmit. All the mobile nodes have the same predefined maximum speed and the mobility generation parameters are given in Table 2. We conduct every simulation same times to achieve the average results in order to reduce the uncertainty from the random values of simulation parameters.

Table 2. Parameters for mobility generation

Parameter	Value
Simulation time (s)	200
X (m)	1000
Y (m)	1000
Max.speed (m/s)	10, 20, 30
Number of roads	4
Number of lanes	2
Max.acceleration (m/s^2)	2
Max.deceleration (m/s^2)	2
Transmission range (m)	200
Vehicle length (m)	4

5.2   Result

To measure how efficient the routing method performs, two metrics are added to analyze the routing method performance. One is the number of packets and the other is the packet delivery ratio. The results obtained by Kalman Filter are also shown. Figure 7 illustrates the error between the observed value obtained by Kalman Filter and the true value. The predicted value is calculated by the Eq. (4).

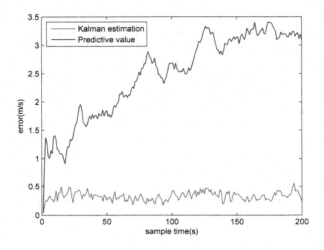

Fig. 7. Error compared to true value

We counted the total number of packets generated during the whole simulation. Figure 8 shows the number of packets generated during the simulation where the maximum speed is 30 m/s. Prediction based on HMM (PHMM) is the routing method we use. As shown in Fig. 9, the comparisons of the number of packets with different maximum speed is clearly depicted. Figure 10 shows the change in packet delivery rate over time.

5.3   Data Analysis

In Fig. 7, it can be clearly seen that the estimation based on Kalman Filter is more accurate than the predictive value. The error between the Kalman estimation and the true value is as low as 0.6 m/s, even if the increase in simulation time leads to an increase in the number of vehicles. The maximum error caused by the predictive value is 3.4 m/s, which is much larger than Kalman estimation. It is effective for obtaining accurate observation to use Kalman Filter.

At the beginning of the simulation, because there are fewer vehicles, fewer data packets are sent and the road is not congested. PHMM, AODV and KPND generate similar number of packets. When there are more vehicles, the state of

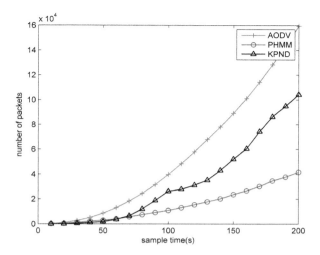

Fig. 8. Number of packets

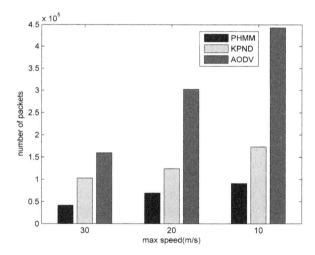

Fig. 9. The comparisons of the number of packets with different speed

the road traffic changes, and the data packets generated by PHMM are obviously less than AODV as shown in Fig. 8. As new vehicles continue entering the road, KPND needs to frequently send messages to refresh the neighbor table. So when the simulation time increases, the number of data packets sent is more than PHMM. From Fig. 9, it can be observed that PHMM can send fewer packets compared with AODV and KPND in different maximum speeds. The lower the maximum speed, the easier road becomes congested. And when the distance between vehicles become closer, it is easier to find relay nodes to transmit pack-

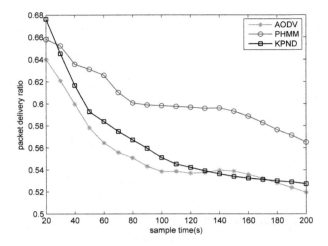

Fig. 10. Packet delivery ratio

ets, so when the maximum speed is lower, more date packets are sent. Therefore, when the maximum speed is 10 m/s, the most number of packets are sent, and when the maximum speed is 30 m/s, the least packets are sent. Among them, when the maximum speed is 20 m/s, compared with maximum speed of 30 m/s, KPND increases fewer packets, indicating that the Kalman Filter may perform well when the maximum speed is 20 m/s. As shown in Fig. 10, the value of packet delivery ratio decreased with the increase of sample time. When the number of vehicle nodes increases, the packet delivery ratio of AODV decreases faster than that of PHMM. With PHMM, we can obtain larger packet delivery ratio than using AODV. For example, under the simulation scenario of 60 sample times, the AODV results in an about 56% packet delivery ratio, but with PHMM, it yields delivery ratio of around 63%, which is 7% better than the AODV. During the whole simulation time, the packet delivery ratio of PHMM is about 5% higher than that of AODV on average. KPND shows a higher delivery ratio when there are fewer vehicles, but when there are more vehicles and the road traffic state changes, PHMM's delivery ratio is better than KPND. Because of the changing road traffic conditions, linear Kalman Filtering cannot accurately predict the position of the vehicles, and a large number of messages need to be sent to obtain the position of the neighboring vehicles. These comparisons show that PHMM is more adapted to the changing road environment.

6   Conclusion

In this paper, a routing method based on HMM is proposed to predict road status. The true state of the road is considered a hidden state in HMM. We take the speed of vehicle, acceleration, number of vehicles on the road and velocity variance as the observation states in the HMM. We can use these observation to predict whether the vehicles on the road are driving normal, congested due to rush-hours or traffic accidents. The state of road traffic is divided into three states to facilitate the establishment of HMM based on road historical data. In VANETs, to provide an accurate information transmission and a fast interaction, must find a suitable routing algorithm. But there is no single routing method which can adapt all road traffic conditions. Choosing the right routing method by predicting road conditions is a new idea. Simulation results show that compared with the AODV and KPND, our method increases the value of packet delivery ratio and decreases the number of packets in the changing road environment, and reduces the channel load.

References

1. Karagiannis, G., Altintas, O., Ekici, E., et al.: Vehicular networking: a survey and tutorial on requirements, architectures, challenges, standards and solutions. IEEE Commun. Surv. Tutor. **13**(4), 584–616 (2011)
2. Jerbi, M., Senouci, S.M., Rasheed, T., et al.: Towards efficient geographic routing in urban vehicular networks. IEEE Trans. Veh. Technol. **58**(9), 5048–5059 (2009)
3. Jayachandran, S., Jothi, J.D., Krishnan, S.R.: A case study on various routing strategies of VANETs. In: Krishna, P.V., Babu, M.R., Ariwa, E. (eds.) ObCom 2011. CCIS, vol. 269, pp. 353–362. Springer, Heidelberg (2012). https://doi.org/10.1007/978-3-642-29219-4_41
4. Bala, R., Krishna, C.R.: Performance analysis of topology based routing in a VANET. In: International Conference on Advances in Computing. IEEE (2014)
5. Karp, B., Kung, H.T.: GPSR: greedy perimeter stateless routing for wireless networks. In Proceedings of the 6th Annual International Conference on Mobile Computing and Networking, MobiCom 2000, pp. 243–254. ACM, New York (2000)
6. Togou, M.A., Hafid, A., Khoukhi, L.: SCRP: stable CDS-based routing protocol for urban vehicular ad hoc networks. IEEE Trans. Intell. Transp. Syst. 1–10 (2016)
7. Younes, M.B., Boukerche, A., Rom'An-Alonso, G.: An intelligent path recommendation protocol (ICOD) for VANETs. Comput. Netw. **64**(may 8), 225–242 (2014)
8. Reza, A.T., Kumar, T.A., Sivakumar, T.: Position Prediction based Multicast Routing (PPMR) using Kalman filter over VANET. In: IEEE International Conference on Engineering & Technology. IEEE (2016)
9. Ning, L., Jose-Fernan, M.O., Hernandez, D.V., et al.: probability prediction-based reliable and efficient opportunistic routing algorithm for VANETs. IEEE/ACM Trans. Netw. 1–15 (2018)
10. Liu, C., Zhang, G., Guo, W., et al.: Kalman prediction-based neighbor discovery and its effect on routing protocol in vehicular ad hoc networks. IEEE Trans. Intell. Transp. Syst. 1–11 (2019)
11. Tang, Y., Cheng, N., Wu, W., et al.: Delay-minimization routing for heterogeneous VANETs with machine learning based mobility prediction. IEEE Trans. Veh. Technol. **68**(4), 3967–3979 (2019)
12. Bhatia, J., Dave, R., Bhayani, H., et al.: SDN-based real-time urban traffic analysis in VANET environment. Comput. Commun. **149**, 162–175 (2019)
13. Chaib, N., Oubbati, O.S., Bensaad, M.L., et al.: BRT: bus-based routing technique in urban vehicular networks. IEEE Trans. Intell. Transp. Syst. **PP**(99), 1–13 (2019)
14. Oubbati, O.S., Chaib, N., Lakas, A., et al.: U2RV: UAV-assisted reactive routing protocol for VANETs. Int. J. Commun. Syst. **PP**(8), 1–13 (2019)

15. Liu, J., Wan, J., Jia, D., et al.: High-efficiency urban traffic management in context-aware computing and 5G communication. IEEE Commun. Mag. **55**, 34–40 (2017)
16. Kalman, R.E.: A new approach to linear filtering and prediction problems. J. Basic Eng. **82**(1), 35–45 (1960)
17. Mehra, R.: On the identification of variances and adaptive Kalman filtering. IEEE Trans. Autom. Control **15**, 175–184 (1970)
18. Perkins, C.: Ad hoc on-demand routing vector (AODV) routing. Rfc (2003)

A Hybrid Deep Learning Approach for Traffic Flow Prediction in Highway Domain

Zhe Wang[1,2]([✉]), Weilong Ding[1,2] [iD], and Hui Wang[3]

[1] School of Information Science and Technology, North China University of Technology,
Beijing 100144, China
wwwzzz7116@163.com
[2] Beijing Key Laboratory on Integration and Analysis of Large-Scale
Stream Data, Beijing 100144, China
[3] Beijing China-Power Information Technology Company Limited, Beijing 100192, China

Abstract. With the development of cities, intercity highway plays a vital role in people's daily travel. The traffic flow on the highway network is also increasingly concerned by road managers and participants. However, due to the influence of highway network topology and extra feature such as weather, accurate traffic flow prediction becomes hard to achieve. It is difficult to construct a multidimensional feature matrix and predict the traffic flow of the network at one time. A novel prediction method based on a hybrid deep learning model is proposed, which can learn multidimensional feature and predict network-wise traffic flow efficiently. The experiment shows that the prediction accuracy of this method is significantly better than existing methods, and it has a good performance during the prediction.

Keywords: Traffic flow prediction · Deep learning · Spatio-temporal data

1 Introduction

With the development of industrialization and infrastructure, the connection between cities have become more convenient. People's work and life are inseparable from the intercity highway. During the upgrades of road infrastructure and the increase in the number of vehicles, the traffic situation of highway has also been paid attention by city officer. In recent years, the rapid development of intelligent transportation systems has brought new solutions to traffic management in the domain of highway transportation [1]. By integrating and analyzing data generated by Internet of Things (IoT) devices, the intelligent transportation system can better provide suggestions and predictions for traffic management. One of the most important function in the intelligent transportation system is traffic flow evaluation. It can help managers better detect road conditions and make reasonable plans. So, many researchers are focusing on traffic flow prediction with different time and space scope.

With the development of IoT technology, types of data become diverse. These data can have more or less impact on the traffic flow [2]. For example, the impact of extreme

H. Gao et al. (Eds.): CollaborateCom 2020, LNICST 350, pp. 253–267, 2021.
https://doi.org/10.1007/978-3-030-67540-0_15

weather such as heavy rain and fog is obvious. As analogously for the weather data, the willing of travel on holiday and weekends is higher than that of weekdays. For weekdays, in morning and evening peak, traffic flow is significantly higher than the noon. Therefore, how to effectively employ the appropriate data type and fully explore the relationship between the data for traffic flow prediction has to be considered. At the same time, for the highway, it is a loop topology. The toll stations are closely related to distance. When a vehicle enters the highway from a certain station, it is necessary to pass the adjacent segments. Therefore, in highway domain, the benefits for the traffic flow prediction in the whole network are more valuable than the single one. So how to collect the data and predict the network traffic flow at the same time is the focus of the research.

In this paper, we propose a novel network traffic flow prediction method based on relevant data. We present a hybrid deep learning model to predict traffic flow. By collecting weather, date and other type data, we structure a multidimensional feature model. We predict the network traffic flow through fully convolution network (FCN) and the long short-term memory (LSTM) network. We solved the problems of multidimensional feature learning and network traffic flow prediction.

The rest of this paper is organized as follows: Sect. 2 explains the background of motivation and the related work; Sect. 3 introduces the feature modeling and the network traffic flow prediction method in detail; Sect. 4 evaluates the method with related experiments and results to show the performance. The fifth section summarizes the research and conclusion.

2 Background

2.1 Motivation

With the rapid growth of vehicle ownership, road congestion has become more serious. The intelligent transportation system allows road managers to manage traffic more intuitively and comprehensively. As an important function of the intelligent transportation system, traffic flow prediction has always been the focus of research and development. For example, Highway Big Data Analysis System in Chinese most populated province, Henan, can conduct business analysis by integrating historical toll station data [3]. Toll station data reveal the traffic flow around it. When the road becomes congested, the toll station data in the highway network will increase. Therefore, toll station data is the basis for us to predict traffic flow. Through the intelligent equipment and other information collection system installed in the toll station, we can get a record when vehicle passing toll station. As shown in Table 1 below.

As we can see, the toll record has three parts of dimension. In entity dimension, we use the vehicle details to identity the vehicle. In time dimension we can know the spatiotemporal attribute of vehicle. Like the time vehicle enter and exit the highway. In space dimension, we know the location information that the vehicle enters and exits the highway network. We can aggregate each record to determine the traffic flow at each toll station. It is very important for network traffic flow prediction. Through the data cleaning process, we can get the complete trajectory of the vehicles in highway network, including the static information of the vehicle and the dynamic information that vehicle passing through toll station.

Table 1. Highway toll station record

Attribute	Notation	Type
collector_id car_id vehicle_type etc_id	Toll collector identity Vehicle identity Vehicle type Vehicle ETC card identity	Entity
entry_time exit_time	Vehicle entry timestamp Vehicle exit timestamp	Time
entry_station entry_lane exit_station exit_lane	Identity of entry station Lane number of entry station Identity of exit station Lane number of exit station	Space

Using this data, Highway Big Data Analysis System can not only build individual profile for vehicles, but also makes overall planning and analysis for highway network. As shown in the Fig. 1 below, the system can calculate the traffic flow information of each station on the highway day by day.

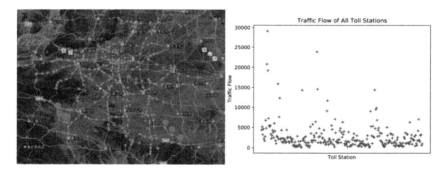

Fig. 1. Traffic flow of all toll stations in one day

As shown in the Fig. 1, the traffic flow is quite different at toll stations. Due to the agglomeration and planning of urbanization, the traffic flow around *zhengzhou*, the provincial capital of Henan Province, is significantly larger than that of its distant cities. Therefore, traffic flow prediction for key regional station can't represent the network situation. If we can predict traffic flow of all the toll stations in network at a time, it will provide global strategy support for highway managers. For network traffic flow prediction, due to the loop topology of the highway network, the traffic flow at the associated stations will also inevitably affect the traffic flow at a given station [3]. From entering the toll station to ending the journey, it will definitely pass adjacent segments which contain other toll stations. Those toll stations will also have impact on road traffic. Therefore, the traffic flow of associated station should be fully considered when predict. At the same time, there are other factors that will affect the traffic flow too.

One is the weather. When people are faced with heavy rain and fog, travel plans will be affected. It will lead to the fluctuation of traffic flow. Secondly, date type will also become one of influencing factors. From the daily analysis, the arrival of holidays will directly lead to a peak travel, and the traffic flow of that day will increase than usual. Therefore, how to fully consider weather, date, and other influencing factors, and find the correlation between data is the key issue of flow prediction. At the same time, choosing the appropriate prediction model will also let the traffic flow prediction of network achieved better results. In the era of rapid development of deep learning, shallow machine learning and deep machine learning have their own characteristics. How to choose a model that suitable for data and goals is also the focus of our research.

2.2 Related Work

As traffic flow prediction has gradually become a hot issue, more and more researchers have begun to use big data to predict traffic flow [4]. However, with the development of technology, lots of data in different types can be used. So, there are still challenges in traffic flow prediction domain. The first is the dimension of the features. Different data types will have different effects on traffic flow prediction. The second is prediction accuracy. With the development of deep learning, many researchers have begun to use deep learning methods to improve prediction accuracy. We divide the related work into two perspectives: feature dimension selection and model selection.

In feature dimension selection perspective, the traffic flow seems the traditional selection. Smith and Demetsky [5] use statistical and machine learning for prediction at first. It just used traffic flow data for analyze. However, with the development of science and technology, more and more dimensions of feature can be selected to improve the effect of traffic flow prediction. Now we often treat traffic flow prediction as a time series problem. Kumar and Vanajakshi [6] use seasonal Autoregressive Integrated Moving Average mode (ARIMA) model with limited input data for traffic flow prediction. Due to the structure of ARIMA model, only temporal dimension can be use. Now, more researchers begin to consider the impact of multiple features. Compared with a prediction model based on time series, these studies on the impact of multiple features are closer to the facts. For example, for prediction of air quality, Shengdong and Du [7] added meteorological data and population to compare and study the effect of these features on the experimental results. Meteorological data also can affect traffic flow. It gives us a inspire to ponder problem. Similar as highway domain, the prediction of pedestrian flow at railway stations is also based on loop topology. Niu [8] considers the impact of meteorological data and neighboring stations. In our research, associated stations also affect the traffic flow. In traffic flow study of Rong [9], he also referred to the driving speed and average speed of the vehicle. Cheng [10] regards the occupancy of roads as one of the features that affect the traffic flow. They all use multidimensional features for traffic flow analyzation and prediction, but their problem is that they can only predict one station one time. They lack the ability to learn multidimensional features on the entire network. Therefore, we chose meteorological data and date type for multidimensional feature. Moreover, in the domain of traffic flow prediction, the network traffic flow prediction has not been fully studied by researchers. Most researchers focus on the single station traffic flow, and then choose a better prediction method [8–10]. We focus on predicting the traffic flow of network at once. Because we want to learn the multidimensional feature of network precisely.

The second perspective is model selection. With the development of machine learning, model has become diverse. ARIMA is a typical time series learning model that can handle continuous data in the time dimension. Smith et al. introduced the construction and learning process of time series in detail [11–14]. But for the multidimensional feature matrix, the method of constructing time series is not suitable. Other machine learning methods such as Support Vector Regression (SVR), K-Nearest Neighbor (KNN) [15], etc. are also very popular. Luo [16–19] and others used KNN to learn the relationship between adjacent data points to predict traffic flow. Castro-Neto et al. [20, 21] used the kernel method in SVR to map the feature to high dimensions, and then made traffic flow prediction, which also achieved good results. However, with the development of deep learning, more researchers have begun to use deep learning methods to predict traffic flow, and achieved better results than shallow learning [22–25]. Zhang [26] et al. use deep belief network for feature learning and prediction. For traffic flow with time attributes, the time-step-oriented deep learning method LSTM is widely used, and for learning the feature matrix, the convolution method can also learn the relationship between features well. Therefore, a fully convolutional network and cooperates with the LSTM network is proposed as a hybrid deep learning model. It can fully learn the multidimensional feature, and predict the traffic flow of the whole network.

3   Method and Models

3.1   Feature Modeling

In highway traffic flow prediction domain, we find that there are many data types can affect it. Therefore, we propose a feature modeling method based on multiple types of data, which will uniformly model the various feature. We use variety of data to construct a multidimensional feature model suitable for network traffic flow prediction. Then we use a hybrid deep learning model to fit the feature model. Finally, we predict the traffic flow of network toll stations one time. For the prediction time range, we define daily traffic flow as the research background. The definition as follow:

Definition 1 (Daily traffic flow). For one toll station s in highway network, daily traffic flow is described as TF_s^t, represent the summation of vehicles exiting the toll station s in day t. Here, s is in the set of toll stations in highway network S, and t is the current day.

We choose the exiting record as the basis data for the summation of traffic flow. Because it is a completely record for the vehicle. In Chinese highway domain, only when the vehicle leaves toll station means tolls would be charged. After a complete record, we can still analyze the incoming information of the vehicle. Daily traffic flow is the main dimension for the prediction. But from elaborative observation, we found that other dimensions also affect the traffic flow of highway toll stations. Like weather, date type and related stations traffic flow, etc. those data types must be fully considered when modeling feature, so there a definition of feature modeling:

Definition 2 (multidimensional feature vector). In the day t, the multidimensional feature vector of daily traffic flow is $V_s^t = \left(W_s^t, D_t, Vol_s^t \right)$, toll station s $\in$ S in highway network. $W_s^t \in \{0, 1\}$ is weather category in date t, $D_t \in \{0, 1, 2\}$ is date type category in date t, $Vol_s^t = \left(TF_{s_1}^t, TF_{s_2}^t, TF_{s_3}^t, \ldots, TF_{s_n}^t \right)$, n $\in \mathbb{Z}^+$ represent *Daily traffic flow* on n nearest station's *Daily traffic flow* around s computed by the Euclidean distance in day t.

The following is further illustrations for the three features.

(1) In meteorological feature, the occurrence of extreme weather will affect the daily traffic flow of highway network. The raw weather data we have has the following properties: $Weather_s^t = (weather\ condition,\ temperature,\ wind\ speed\ and\ direction)$ in toll station s in highway network and day t. When weather condition is heavy rain, fog, or the degree of wind bigger than six-level, daily traffic flow of toll station s will drop. So, we define extreme weather $= (weather\ condition = rain \vee fog \vee snow) \vee (windspeed > 6)$. And in matrix of daily traffic flow, we define

$$W_s^t = \begin{cases} 1, & extreme\ weather \\ 0, & otherwise \end{cases}$$

(2) In date feature, due to the holiday period, the highway network will adjust the charging mode, which leads to a significant influence in traffic flow. Therefore, the date feature is also one of the important factors that affect the traffic flow. Based on existing statistics, we define D_t in date t as follow:

$$D_t = \begin{cases} 2, & if\ t\ is\ a\ holiday, \\ 1, & if\ t\ is\ a\ weekend, \\ 0, & otherwise. \end{cases}$$

(3) In the last feature, Vol_s^t represent the spatial related traffic flow with toll station s. As we all know, highway network is a loop topology. The network traffic flow is influenced by all toll stations. For the toll station s, its adjacent stations' daily traffic flow is considered. The adjacent station means the toll station nearby s in spatial dimension. The traffic flow of adjacent station will influence the traffic flow of s apparently. We calculate the Euclidean distance between each toll station, take n nearest toll stations as the spatial relationship for feature modeling in spatial dimension.

With multidimensional feature vector, we can build a feature matrix for network toll stations. It is construct by all toll station so that we can use it to reflect network traffic flow. The definition as follow:

Definition 3 (Network toll station feature matrix). In day t, the network toll station feature matrix $M^t = \left(V_1^t, V_2^t, V_3^t, \cdots, V_k^t \right)$. k is the number of all toll stations in highway network.

Network toll station feature matrix represent the highway network's multidimensional feature that can reflect the network traffic flow in day t. It also is the feature modeling for the traffic flow prediction method. After feature modeling, due to the difference in the value range between each dimension, features with small range will be ignored, so the normalization method is used to reduce the prediction error.

3.2 Traffic Flow Prediction

In the following, we describe a traffic flow prediction method, based on hybrid deep learning model. The structure is shown in Fig. 2.

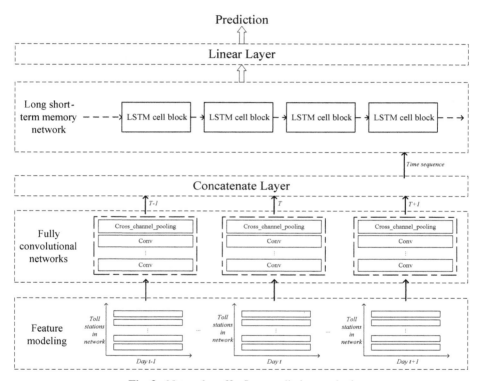

Fig. 2. Network traffic flow prediction method

The method is composed of fully convolution network (FCN) and the long short-term memory (LSTM). As we can see in Fig. 2, we propose to use FCN to fit multidimensional feature. Then we use concatenation layer to restrict the data into time series, then let LSTM to learn it. Finally, we use linear layer to restore network traffic flow. This combination considers both spatial and temporal of multidimensional feature. First of all, in feature modeling, the spatial relationship of toll stations in the highway network is fully considered. Based on one related work [27], three stations with min Euclidean distances from the toll station s have the most obvious impact, so n = 3 in *Definition 2*.

Three adjacent stations *daily traffic flow* was selected for feature modeling, plus weather and date type factors, which constituted the five-dimensional feature of each toll station. In order to learn the spatiality feature, we chose to use a special convolutional network. After that, we can see from Table 1 that the toll station data of the highway network has both spatial and temporal dimension, and it is not enough to only consider the spatiality. Therefore, after the convolution network, we add a concatenation layer to construct a time series and input to the LSTM network to learning temporal feature. Finally, a network traffic flow prediction method is formed through hybrid deep learning architecture to make prediction. The following is a detailed description of each component.

For feature modeling, we build *Network toll station feature matrix M^t* for all toll stations in network. This matrix is composed of all toll stations and *multidimensional feature vector* of each station. It contains spatial correlation feature and other dimension features that affect the traffic flow, like weather and date type. First, we use convolutional neural network for feature learning. Convolutional neural network (CNN) was first used in the field of image recognition. Through the calculation of pixels and convolution kernels, it learns the relationship of each pixel to achieve the role. The CNN involves many individual processing steps [28]. The general process includes convolution operation, pooling operation and the final fully connected layer with activation function. In our case, we use Cross-channel pooling instead of generate pooling operation. We complete CNN as follow functions.

$$o(M^t) = \text{Relu}(\text{Conv}(M^t, \text{kernel})) \tag{1}$$

$$c(M^t) = \text{Cross\_channel\_pooling}(o(M^t)) \tag{2}$$

where M^t is input feature matrix of day t. In Eq. 1, we choose hetero convolution kernel to calculate convolution. Usually, the pixels of the image are irregularly distributed. One feature in image often component by lots of pixels in different directions. The convolution kernel usually uses a square, so they can learn the feature form every direction. In recent years, with the development of convolutional neural networks, new technologies such as dilated convolution and deformable convolutional have emerged [29]. However, the design of the convolution kernel is essentially to learn the correlation feature in the matrix. We already considered the spatial dimension in *multidimensional feature vector*. There is no need to compute each station correlation on it. Therefore, in this paper we designed a hetero convolution kernel form, only compute the column direction of the feature matrix. In this way, the connection between each feature vector is cut, and only the respective multidimensional feature needs to be learned. First, we designed a set of 1×2 kernels to calculate with matrix. This is designed for the weather and date type dimension in the multidimensional feature vector. In feature modeling, the label encoding method is used for the feature of these two dimensions, which is different from the daily traffic flow dimension. After that, we use ReLU function as the activation function. Second, we also execute a convolution operation with a set of 1×2 convolution kernels, and continue to learn the features of each vector. Finally, we use a set of 1×3 convolution kernels to end the convolution operation, which reduces the dimensionality of the multidimensional feature vector of each station in the matrix in to a traffic flow value. After that, we uniformly perform pooling operations on the channel, as shown in

Eq. 2. Generally, pooling operations have maximum pooling or average pooling, which play a role in input to reduce the dimension. Cross-channel pooling is a method for reducing the size of state and number of parameters needed to have a given number of filters in the model [30]. After three times convolution operations, the multidimensional feature vector of each station in feature matrix is reduced to a one-dimensional daily traffic flow value. But on the channel, the number of convolution kernels causes channel increment. So, we use cross-channel pooling to reduce channel dimension. Our pooling method only averages channels, which we call channel subsampling, to achieve the purpose of channel dimensional reduction. So that, we can retain the feature matrix structure in one channel. At the same time, we discarded the fully connection layer of CNN, and keep the current feature matrix structure. After learning, feature matrix has turn to one-dimensional feature from multidimensional vector, as $c(M^t)$. So, the output $c(M^t)$ is the predicted value of traffic flow for each station in highway network in day t. We use real daily traffic flow as training label to training the FCN we designed. After training, we get daily traffic flow matrix on each day.

After getting the daily traffic flow matrix of all toll stations in the network, in order to use temporal of data, we will process the matrix in concatenation layer into time series. We set the learning time range as 15, so after concatenation layer we can get a time series data as input. Then we choose LSTM network for feature learning in time dimension. LSTM is an improved model of recurrent neural network (RNN). The classic LSTM cell block contains four components, namely input gate, forget gate, output gate and cell state. These four components together make LSTM cell block have memory and function of learning time dimension. The corresponding calculation steps are as follows:

$$i_t = \sigma\left(W_x^i c(M^t) + W_h^i h_{t-1} + b_i\right) \tag{3}$$

$$o_t = \sigma\left(W_x^o c(M^t) + W_h^o h_{t-1} + b_o\right) \tag{4}$$

$$f_t = \sigma\left(W_x^f c(M^t) + W_h^f h_{t-1} + b_f\right) \tag{5}$$

$$c_t = f_t \odot c_{t-1} + i_t \odot \tanh\left(W_h^c c(M^t) + W_h^c h_{t-1} + b_c\right) \tag{6}$$

$$h_t = o_t \odot \tanh(c_t) \tag{7}$$

where $c(M^t)$, c_t, and h_t are input, cell state, and cell output respectively at time t. W_* and b_* are weight and bias vectors connecting different gates. $\odot$ denotes an element-wise product. As shown in the above equation, the input gate can control how much new information enters the cell state, the forget gate can choose how much information is discarded, and the output gate can control how much information is passed into the next time step or output. Cell state records the state information of the cell with the change of time to complete the memory function of LSTM. By serializing the feature matrix of toll stations in whole network, LSTM can also learn the time dimension features of the data and improve the prediction accuracy. We put the time series data generated by the concatenation layer as input into LSTM. We use fifteen days feature matrix for training, and use the daily traffic flow of the next day as the label for learning. Finally, we get the predicted value as traffic flow of all toll stations in highway network. The combination of the above two deep learning model has fully studied the spatio-temporal and multidimensional feature of the data. At the same time, through the improvement of the convolution network, the method can predict the highway network traffic flow at one time.

4 Experiment

4.1 Settings

The experiment used real data from the toll stations of the highway network, which was derived from the Henan Highway Management System supported by the project. At the same time, it also collects the weather information of all toll stations. The data from May to September of Henan highway was used for model training and analysis. These large amounts of data are stored in a distributed cluster. The cluster has one master node and two slave nodes. The configuration is Intel (R) Xeon (R) CPU E5–4607 2.20 GHz, 32 GB RAM and 80 TB storage. Distributed computing frameworks such as Hadoop, Spark and HBASE are installed on it for data cleaning and feature modeling. The framework is built and trained on a computer with Intel (R) Core (R) CPU i7–7700 2.80 GHz, 32 GB RAM, 4 TB storage and two NVIDIA 1080 GPUs.

The open source deep learning framework TensorFlow 1.12.0 supports the construction of this method, and Scikit-learn 0.20.3 is used to build models of counterparts. We named our method as hybrid deep learning method (HDM) to compare with four machine learning models. They are: KNN is a classic supervised statistical machine learning model that classifies and regressions by analyzing the distance relationship of data; ARIMA is the most widely used prediction model for time series data; SVR is a machine learning model based on kernel method. it can map data to high dimensions through the kernel to make regression predictions; Gradient boost regression tree (GBRT) is an integrated machine learning method that learns by reducing the loss between different models. It has been widely used in recent years. Then we chose three predictive metrics to evaluate the method. First is root mean square error (RMSE) as Eq. 8; the second is mean absolute percentage error (MAPE) as Eq. 9 and R-square as Eq. 10 is the last metric.

$$\text{RMSE} = \sqrt{\frac{1}{n} \sum_{i=1}^{n} (\hat{y}_i - y_i)^2} \tag{8}$$

$$\text{MAPE} = \frac{100\%}{n} \sum_{i=1}^{n} \left| \frac{\hat{y}_i - y_i}{y_i} \right| \tag{9}$$

$$\text{R} - \text{square} = 1 - \frac{\sum_{i=1}^{n} (\hat{y}_i - y_i)^2}{\sum_{i=1}^{n} (y_i - \bar{y})^2} \tag{10}$$

where $\hat{y}_i$ represents the predicted value, y_i represents ground truth value and n is the number of test dataset.

4.2 Evaluation

We designed two experiments for method evaluation. The first is a comparative experiment to verify the effectiveness and accuracy of HDM by comparing with other methods. Secondly, we use HDM and predicted all toll stations traffic flow one time, and compare with traditional LSTM prediction method.

Experiment 1: Prediction Effects in a Station. The data from September 2017 was used as the test set. In the experiment, all 274 toll stations in Henan Province constituted the whole network. Among them, we randomly selected *zhengzhounan* toll station for analysis. The prediction results of HDM and the comparison model are shown in Fig. 3:

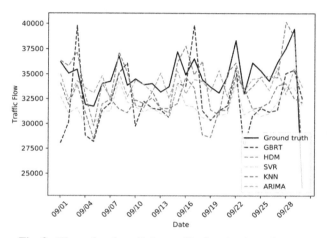

Fig. 3. The real and prediction traffic flow in *zhengzhounan*

Through Fig. 3, we can intuitively see that the HDM fits the real value best and has good performance in the time period of wave peak and wave trough. At the same time, the fluctuation of GBRT is worse than HDM. Although the overall trend is close to the true value, the prediction of the turning point is lacking. Compared with the SVR and ARIAM models, our method's prediction is more accurate. And like the KNN model, it can fit the real curve well. So, by analyzing the prediction results, we get Table 2:

Table 2. Prediction performances of different models

	HDM	GBRT	SVR	KNN	ARIMA
RMSE	1983.95	3933.77	3692.97	2583.35	3597.36
MAPE (%)	41.69	54.68	59.72	43.69	53.64
R-square	0.894	0.863	0.704	0.883	0.873

From Table 2, we can see that, comparing all models, the HDM has the best fitting effect on the traffic flow. The RMSE is kept within two thousand, indicating that the error between the predicted value and the true value is kept within an acceptable range. At the same time, the R-square value is closest to 1, better than the KNN model, which represents the best fitting effect on real traffic flow. Through this experiment, we can know that HDM has good prediction accuracy at the key station *zhengzhounan*, and better than other comparison models.

Experiment 2: Network Prediction. In this experiment, we use HDM to predict network traffic flow at once. And use normal LSTM model to predict traffic flow for comparison. We chose four key toll station for experiment, they are *puyangnan, zhengzhouxinqu, xinxiang and xuanyuanguli*. There daily traffic flow is different. We use LSTM four times. The single toll station prediction uses the same training set and uses LSTM neural network model in [16] for time series data learning. The experiment results are shown in Fig. 4:

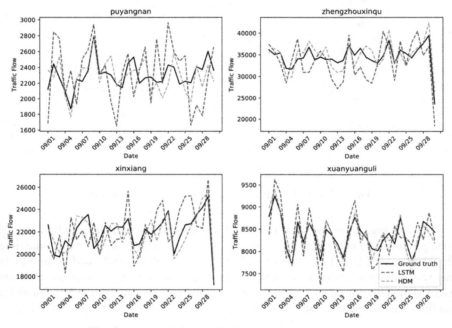

Fig. 4. The prediction traffic flow with HDM and LSTM

We can see from the Fig. 4 from the selected toll stations, the prediction performance of HDM is better than the single station traffic flow prediction. We calculate the average of the metrics of the four station to Table 3. According to the results in Table 3, the R-square is better than the LSTM model and is closer to the true value. It can be seen from the results that HDM method is not affected by the increase of the parameter of the feature matrix, but the learning ability is better than the single station prediction method. Therefore, the network traffic flow prediction method is fully verified, and the expected prediction effect can be achieved.

Table 3. Prediction performances

	HDM	LSTM
RMSE	1085.72	1489.56
MAPE (%)	32.71	38.96
R-square	0.843	0.794

5   Conclusion

In this paper, we proposed a new multidimensional feature-based network traffic flow prediction method. A hybrid deep learning model consists of a fully convolutional neural network and long-short term memory neural network is proposed. It can fully study the spatio-temporal of highway toll station data. We considered the spatial feature of data and some other feature can affect traffic flow like weather and date type. We use this for feature modeling. After learning this toll station feature matrix, the method can well predict the traffic flow of highway network. Experiments show that the method has a good performance compared with the shallow machine learning method. At the same time, it has a good effect on the learning of multidimensional feature. In the future research, we would further subdivide the dimensions of features, and consider more relationship between with traffic flow. We can also improve the performance of neural network in depth to predict more precise.

Acknowledgment. This work was supported by National Key R&D Program of China (No. 2018YFB1402500), National Natural Science Foundation of China (No. 61702014) and Beijing Municipal Natural Science Foundation (No. 4192020).

References

1. Avineri, E.: Soft computing applications in traffic and transport systems: a review. In: Hoffmann, F., Köppen, M., Klawonn, F., Roy, R. (eds.) Soft Computing: Methodologies and Applications, vol. 32, pp. 17–25. Springer, Heidelberg (2005). https://doi.org/10.1007/3-540-324 00-3_2
2. Ding, W., Zhang, S., Zhao, Z.: A collaborative calculation on real-time stream in smart cities. Simul. Model. Pract. Theory **73**(4), 72–82 (2017)
3. Wang, X., Ding, W.: A short-term traffic prediction method on big data in highway domain. J. Comput. Appl. **39**(01), 93–98 (2019)
4. Zheng, Y., Capra, L., Wolfson, O., et al.: Urban computing: concepts, methodologies, and applications. ACM Trans. Intell. Syst. Technol. (TIST) **5**(3), 1–55 (2014)
5. Smith, B.L., Demetsky, M.J.: Traffic flow forecasting: comparison of modeling approaches. J. Transp. Eng. **123**(4), 261–266 (1997)
6. Kumar, S.V., Vanajakshi, L.: Short-term traffic flow prediction using seasonal ARIMA model with limited input data. Eur. Transp. Res. Rev. **7**(3), 21 (2015)
7. Du, S., Li, T., Yang, Y., Horng, S.: Deep air quality forecasting using hybrid deep learning framework. IEEE Trans. Knowl. Data Eng. (2019). https://doi.org/10.1109/TKDE.2019.295 4510

8. Niu, Z., Sun, Q.: Study of railway passenger volume forecast based on grey forecasting model. In: 2016 International Conference on Logistics, Informatics and Service Sciences (LISS), pp. 1–4. IEEE (2016)

9. Rong, Y., Zhang, X., Feng, X., et al.: Comparative analysis for traffic flow forecasting models with real-life data in Beijing. Adv. Mech. Eng. **7**(12), 1687814015620324 (2015)

10. Cheng, A., Jiang, X., Li, Y., et al.: Multiple sources and multiple measures based traffic flow prediction using the chaos theory and support vector regression method. Phys. A: Stat. Mech. Appl. **466**, 422–434 (2017)

11. Smith, B.L., Williams, B.M., Oswald, R.K.: Comparison of parametric and nonparametric models for traffic flow forecasting. Transp. Res. Part C: Emerg. Technol. **10**(4), 303–321 (2002)

12. Sun, S., Zhang, C., Yu, G.: A Bayesian network approach to traffic flow forecasting. IEEE Trans. Intell. Transp. Syst. **7**(1), 124–132 (2006)

13. Li, L., He, S., Zhang, J., et al.: Short-term highway traffic flow prediction based on a hybrid strategy considering temporal–spatial information. J. Adv. Transp. **50**(8), 2029–2040 (2016)

14. Ghosh, B., Basu, B., O'Mahony, M.: multivariate short-term traffic flow forecasting using time-series analysis. IEEE Trans. Intell. Transp. Syst. **10**(2), 246–254 (2009)

15. Hong, W.C., Dong, Y., Zheng, F., et al.: Forecasting urban traffic flow by SVR with continuous ACO. Appl. Math. Model. **35**(3), 1282–1291 (2011)

16. Luo, X., Li, D., Yang, Y., Zhang, S.: Spatiotemporal traffic flow prediction with KNN and LSTM. J. Adv. Transp. **2019**, article ID 4145353, 10 p. (2019). https://doi.org/10.1155/2019/4145353

17. Cai, P., Wang, Y., Lu, G., Chen, P., Ding, C., Sun, J.: A spatiotemporal correlative k-nearest neighbor model for short term traffic multistep forecasting. Transp. Res. Part C: Emerg. Technol. **62**, 21–34 (2016)

18. Dell'acqua, P., Bellotti, F., Berta, R., De Gloria, A.: Time aware multivariate nearest neighbor regression methods for traffic flow prediction. IEEE Trans. Intell. Transp. Syst. **16**(6), 3393–3402 (2015)

19. Sun, B., Cheng, W., Goswami, P., Bai, G.: Short-term traffic forecasting using self-adjusting k-nearest neighbors. IET Intel. Transport Syst. **12**(1), 41–48 (2018)

20. Castro-Neto, M., Jeong, Y.-S., Jeong, M.-K., Han, L.D.: Online-SVR for short-term traffic flow prediction under typical and atypical traffic conditions. Expert Syst. Appl. **36**(3), 6164–6173 (2009)

21. Sun, Y., Leng, B., Guan, W.: A novel wavelet-SVM short time passenger flow prediction in Beijing subway system. Neurocomputing **166**, 109–121 (2015)

22. Ding, W., Xia, Y., Wang, Z., et al.: An ensemble-learning method for potential traffic hotspots detection on heterogeneous spatio-temporal data in highway domain. J. Cloud Comput. **9**(1), 1–11 (2020)

23. Moretti, F., Pizzuti, S., Panzieri, S., et al.: Urban traffic flow forecasting through statistical and neural network bagging ensemble hybrid modeling. Neurocomputing **167**, 3–7 (2015)

24. Priambodo, B.: Traffic flow prediction model based on neighboring roads using neural network and multiple regression. J. Inf. Commun. Technol. **17**(4), 513–535 (2020)

25. Jia, Y., Wu, J., Benakiva, M., et al.: Rainfall-integrated traffic speed prediction using deep learning method. IET Intel. Transp. Syst. **11**(9), 531–536 (2017)

26. Zhang, Y., Huang, G.: Traffic flow prediction model based on deep belief network and genetic algorithm. IET Intell. Transp. Syst. **12**(6), 533–541 (2018)

27. Ding, W., Wang, X., Zhao, Z.: CO-STAR: a collaborative prediction service or short-term trends on continuous spatio-temporal data. Future Gener. Comput. Syst. **102**, 481–493 (2020)

28. McLaughlin, N., del Rincon, J.M., Miller, P.: Recurrent convolutional network for video-based person re-identification. In: Proceedings of the IEEE Conference on Computer Vision and Pattern Recognition, pp. 1325–1334 (2016)

29. Yu, F., Koltun, V., Funkhouser, T.: Dilated residual networks. In: Proceedings of the IEEE Conference on Computer Vision and Pattern Recognition, pp. 472–480 (2017)
30. Goodfellow, I., Warde-Farley, D., Mirza, M., et al.: Maxout networks. In: International Conference on Machine Learning, pp. 1319–1327 (2013)

HomoNet: Unified License Plate Detection and Recognition in Complex Scenes

Yuxin Yang[1], Wei Xi[1], Chenkai Zhu[2], and Yihan Zhao[1(✉)]

[1] School of Computer Science and Technology, Xi'an Jiaotong University, Xi'an, China
yangdx6@gmail.com, weixi.cs@gmail.com, 18292047025@163.com
[2] SenseTime Group Ltd., Shenzhen, China
zhuchenkai@sensetime.com

Abstract. Although there are many commercial systems for license plate detection and recognition (LPDR), existing approaches based on object detection and Optical Character Recognition (OCR) are difficult to achieve good performance in both efficiency and accuracy in complex scenes (e.g., varying viewpoint, light, weather condition, etc). To tackle this problem, this work proposed a unified end-to-end trainable fast perspective LPDR network named HomoNet for simultaneous detection and recognition of twisted license plates. Specifically, we state the homography pooling (HomoPooling) operation based on perspective transformation to rectify tilted license plates. License plate detection was replaced with keypoints location to obtain richer information and improve the speed and accuracy. Experiments show that our network outperforms the state-of-the-art methods on public datasets, such as 95.58%@22.5 ms on RP and 97.5%@19 ms on CCPD.

Keywords: License plate · Keypoints location · HomoPooling

1 Introduction

With the development of deep learning, license plate detection and recognition (LPDR) has been widely applied in intelligent transportation and surveillance systems. In these scenarios, the high accuracy and fast speed must be met at the same time. Besides, cars run at high speeds in different directions in complex situations, causing the license plate (LP) may not face the camera, making it to be twisted, rotated, perspective and distorted, and posing a challenge to LPDR. Many researches have worked to improve performance in recent years.

Most systems [9, 11, 15, 25, 26] divided the LPDR into two independent parts: LP detection and LP recognition. These methods have three limitations: (1) The

This work was supported by National Key R&D Program of China 2018AAA0100500, NSFC Grant No. 61772413, 61802299, and 61672424.

detection result is a bounding box that can only use two points to locate the LP position. In complex scenes with different views, the boundary of LP in video surveillance is generally not a rectangle, which leads to misalignment between the actual LP region and the extracted region by a bounding box, and results in some important information missing. (2) The neural network is difficult to recognize tilted characters. The bounding box only can crop the region of LP ignoring its perspective change because of the direction to the camera. Thus methods lacking correction can not efficiently solve the problem of recognizing twisted LP in complex scenes. (3) There is a correlation between the detection and recognition which was ignore in these articles. The two subtasks can not train end-to-end jointly and spend more time on the inference process. Thus, Li et al. [15] proposed a unified deep neural network TE2E, as shown in Fig. 1a. The lack of correction, however, leads to poor performance on rotating and tilt license plates. Another problem is the detection module is anchor-based which limits the whole net speed.

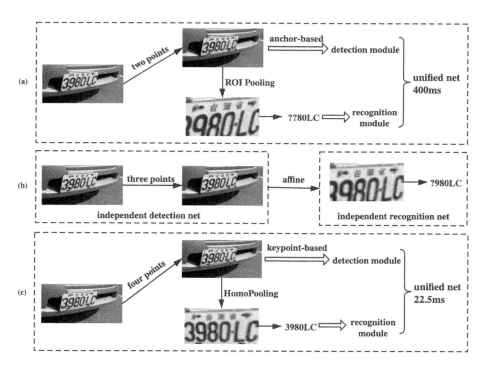

Fig. 1. Comparison between there methods: (a) TE2E, (b) warp-net, (c) HomoPooling.

Inspired from text spotting, affine transformation was introduced to solve the problem of LP distortion [10,15,18]. Affine transformation is equivalent to positioning three coordinates in the original picture, determining the transformation matrix, and then rectifying the license plate. This method can only improve the recognition problem of perspective character to some extent, but cannot fundamentally solve it. For example, as shown in Fig. 1b, character 9 is recognized benefiting from the transformation but character 3 is still difficult to identify due to the detection error of the upper left corner. What's more, the affine transformation is not differentiable in warp-net [26] leading to the issue that we have stated above.

In this paper, based on perspective transformation, a unified network with HomoPooling operation was proposed (see Fig. 1c). Even if there is an error points, four points can still extract all valid information and correct the twisted LP to improve the accuracy of the recognition. Moreover, HomoNet puts the detection, rectification and recognition into one network to ensure end-to-end training, improving the efficiency and accuracy.

The method of LP detection mainly learned from the object detection task. Anchor-based methods need the network to predict massive bounding boxes and non-maximum-suppression (NMS) operations and this method takes up computing resources and slows down the detection time. At the same time, it is hard to get high location accuracy. Different from the most object detection tasks that one image has multiple targets, a car only has one license plate. Thus, to improve the accuracy and speed of detection, we directly predict the heat map to get the key points of the LP, aspired by human skeleton key point detection. Particularly, experiments have been done to prove it. Regressing the four offsets of each pixel relative to the ground truth as in FOTS [18], the root mean square error in Road Patrol (RP) dataset was 13.6, while the proposed method was 3.4. As shown in Fig. 1a and Fig. 1c, benefiting from the keypoint-based method, the whole network speed improved from 400 ms to 22.5 ms with more accurate recognition results.

Till now the propose method is a novel network HomoNet, which can train end-to-end and correct images or features in the LPDR area. The source code of HomoPooling will be open in Github. The following are three main contributions of this paper:

- The proposed unified trainable network is used for fast perspective license plates detection and recognition. By sharing features and building relationships between the two subtasks, the HomoNet not only improves the inference speed but also promotes recognition accuracy in the whole.
- The HomoPooling is introduce which can rectify the distorted license plate or the convolutional features in more complex scenes. The quadrangular region allows it to get more LP information and handle LP from many different perspectives. The sample operation makes it to be differentiable and can be embedded in any network.
- The key point-based detection is apply to replace the bounding box regression. Through the knowledge of one car having one LP, labeling the images by 5 channels of heatmap improves the speed and accuracy of the LP location, as well as removing designed anchors and NMS.

2   Related Works

LPDR is aiming at detecting the position of the license plate and identifying each character in the license plate. Many traditional methods have been proposed respectively [7,9,19]. On account of deep learning (DL) achieves high precision in tasks such as image detection and recognition, researchers have begun to use convolutional networks in LPDR [11,14,15,23]. We will review the DL-based method on license plate detection, text rectification, and end-to-end systems.

2.1   License Plate Detection

The method of license plate detection mainly draws on the model in object detection. YOLO [20] detector is the most popular in this area [8,12,25,30]. Whereas YOLO appears to miss small objects, and YOLOv2 [21] needs artificial designed anchors. Safie et al. [23] used the retinanet algorithm to detect the license plate and solve the problem that the license plate is difficult to detect in scene environments. However, the same as the YOLO, retinanet also needs anchors. All these methods are suitable for multiple targets with different sizes in one image, so they need to design different anchors to match the corresponding target. Recently, researchers have proposed some anchor-free algorithms such as CornerNet [13] and FCOS [27]. CornerNet considered the bounding box as a pair of points and utilized the corner pooling to regress the heat map of each point, which has confirmed to be better than the other single step detector.

2.2   Text Rectification

In the text spotting task, some scholars have proposed different algorithms for correcting the deformed text. Current methods [10,18,26,32] can be classified into affine transformations and non-affine transformations. Spatial Transformer Networks (STN) [10] directly predicted 6 affine coefficients unsupervised trained and rectified input image or feature, adopting a sampler to make it differentiable to end-to-end trainable. Silva et al. [26] proposed the warp-net to regress the affine matrix with supervised information. Liu et al. [18] focused on the rotation of oriented text and proposed ROIRotate to align the text. Although these affine methods could solve some problems to some extent, they could not focus on more useful information. Yang et al. [32] proposed Symmetry-constrained Rectification Network (ScRN) based on Thin-Plate-Spline transformation to solve the curved text, but it was not suitable for rectangular objects.

2.3   End-to-End Systems

There were very few end-to-end networks in the LPDR so that we only found two frameworks. Based on faster-rcnn [22], after the several Convolutional Neural Network (CNN) layers, Li et al. [15] utilized region proposal network to extract the features of interest and applied to detect and identify the license plate at the same time. Li et al. also designed 6 scales of anchors to match each pixel which limited the inference speed. RPnet [31], based on SSD [17], selected three levels of features to predict the license plate position directly, then used ROI pooling [3] to concatenate the features to identify 7 LP characters which have constant quantities. The above two networks regressed the bounding box while they ignored the deformation of the license plate.

3   Method

HomoNet is an end-to-end trained network that supports license plate detection, rectification and recognition in different natural scenes. There are three submodules: key point detection module, HomoPooling module and recognition module. These three submodules will be

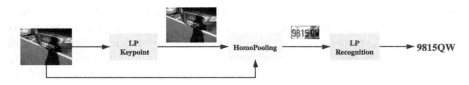

Fig. 2. Pipeline of the proposed method. Three submodules: key point detection module, HomoPooling module and recognition module.

3.1   Overview

The whole framework is shown in Fig. 2. The first is the keypoint detection module. When getting the license plate key points, it can apply them to the perspective transformation to correct the license plate to the horizontal direction, which is easier to recognize. HomoPooling can do the same thing, and it can be end-to-end trained. The last part is the recognition module, which using CNN and max pooling to encode text information, and using Connectionist temporal classification (CTC) [4] to decode the final result.

3.2   LP Keypoint Module

Same as the FOTS [18], LP keypoint module adopts the structure (see Fig. 3) whose backbone is ResNet [5] and Feature Pyramid Network (FPN) proposed in the paper [16]. Firstly, this module concatenates multi-level features from the input image. Only the level of FPN with the stride of 4 is used to extract the feature maps. Then, 4 same conv_gn_relu operations are followed. All convolutions are 3x3 with the stride of 2 and padding of 1. We use group normalization [29] with 32 groups and rule activate function refer to the paper. Finally, one convolution is applied to output 5 channels key points heat map. The first 4 channels

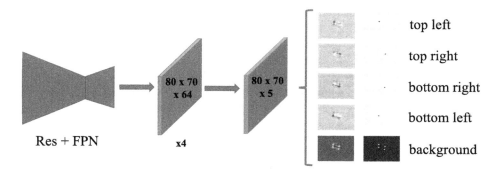

Fig. 3. The LP keypoint module.

predict the probability of every-pixel being the key point, whose order is left-top, right-top, right-bottom and left-bottom. The last channel is background heat map.

The heat map is set to 1 at the position of license corner and 0 at the other position in the first 4 channels, while the label is the opposite on the channel 5. In the experiments, it was found that the key points allow a little offset, which could improve the accuracy and robustness of the model. Thus, a radius of 1 is added at four different directions of ground truth (see Fig. 3). The loss function is binary cross entropy with logits loss, which can be formulated as follows:

$$L_{\text{kp}} = -\frac{1}{N} \sum_{i=1}^{HW} [y_i \log \sigma (x_i) + (1 - y_i) \log (1 - \sigma (x_i))] \tag{1}$$

3.3   HomoPooling

HomoPooling sample the quadrilateral areas determined by four key points to align the distorted license plate to the horizontal direction, as shown in Fig. 2. It consists of three steps. The first step is to get the homography matrix. It needs to normalize the source points and target points between -1 and 1, and then calculate the homography matrix through the four corresponding points, which can be described by the following formulas:

$$Q_S = \begin{bmatrix} -1 + \frac{2x_1}{w} & -1 + \frac{2x_2}{w} & -1 + \frac{2x_3}{w} & -1 + \frac{2x_4}{w} \\ -1 + \frac{2y_1}{h} & -1 + \frac{2y_2}{h} & -1 + \frac{2y_3}{h} & -1 + \frac{2y_4}{h} \\ 1 & 1 & 1 & 1 \end{bmatrix} \tag{2}$$

$$Q_T = \begin{bmatrix} -1 & -1 & 1 & 1 \\ -1 & 1 & 1 & -1 \\ 1 & 1 & 1 & 1 \end{bmatrix} \tag{3}$$

$$cQ_S = M_H Q_T \tag{4}$$

Table 1. The architecture of recognition module. The input height is 32 and the width is changeable. An example is given below.

Type	Configuration	Out size
conv_gn_relu	3×3, 1×1, 1×1, 64	32×96
max_pooling	2×2, 2×2, 0×0	16×48
conv_gn_relu	3×3, 1×1, 1×1, 128	16×48
max_pooling	2×1, 2×1, 0×0	8×48
conv_gn_relu	3×3, 1×1, 1×1, 256	8×48
conv_gn_relu	3×3, 1×1, 1×1, 256	8×48
max_pooling	2×1, 2×1, 0×0	4×48
conv_gn_relu	3×3, 1×1, 1×1, 512	4×48
conv_gn_relu	3×3, 1×1, 1×1, 512	4×48
max_pooling	2×1, 2×1, 0×0	2×48
conv_gn_relu	2×2, 0×0, 1×1, 512	1×47
fc	nc	nc

Where c is a non-zero constant, (x_i, y_i) represents the four ground truth keypoints of the license plate from top-left, top-right, right-bottom to left-bottom. (w, h) is the length of the input images or the feature maps, Q_T is the normalized points of the pooling target. With the relationship between Eq. 2, Eq. 3 and Eq. 5, the homography matrix (M_H) can be found through solving a linear system of equations [2].

The second step is to generate the sample grid. It needs to generate w_H points at equal intervals in the horizontal direction, and h_H points at equal intervals in the vertical direction. Then used the homography matric to calculate the sample grids in source features or images. (w_H, h_H) is the size of HomoPooling.

$$\begin{pmatrix} x_i^g \\ y_j^g \end{pmatrix} = M_H \begin{pmatrix} x_i^t \\ y_j^t \end{pmatrix} \tag{5}$$

Where i and j from -1 to 1, at intervals of w and h relatively.

The final step is to produce the rectified image through sampling from the source grids:

$$T_{ij}^c = \sum_n^h \sum_m^w S_{nm}^c k \left(x_{ij}^s - m; \Phi_x \right) k \left(y_{ij}^s - n; \Phi_y \right) \tag{6}$$

Where T_{ij}^c is the target rectified image value at the position of (i, j) in channel c, and S_{nm}^c is the source image value at the position (n, m) in channel c. h, and w is the height and width of source image. $k()$ is the sampling kernel, and $\Phi_x and \Phi_y$ are its parameters. We use the bilinear interpolation refers to STN [10]. The sample makes it differentiable, so the gradients can be backpropagated from the loss.

3.4   LP Recognition Module

LP recognition module aims at predicting a character sequence from the rectified images transformed by HomoPooling. Because the length of license plate numbers is different in some areas. For example, the new-energy plates have 8 characters while most multiple plates have 7 characters in China, we choose CTC decoder which is different from CCPD [31] and warp-net [26]. Furthermore, due to each character in the license plate is independent, CTC decoder has no connection with context in semantic, which makes us discard the long short term memory (LSTM) module [6] even though it was mostly used in license plate recognition area. What's more, the CTC decoder reduces the computation and speeds the inference.

Inspired by Convolutional Recurrent Neural Network (CRNN) [24], we design recognition module illustrated in Table 1. There are two main operators: conv_gn_relu and max pooling. The configuration means the parameters of convolutional layers and max pooling layers. The parameters are the size of kernel, stride, padding size in height and width as well as channels. nc represents the length of the encoder sequence after a fully-connect layer, which is equal to the number of character dictionary. The "out size" is the feature map size of convolutional layers or sequence length of the encoder layer.

The main differences between our method and CRNN are as follows: First, the group normalization replacing batch normalization which is sensitive to batch choice. Limited by end-to-end training and the memory of GPU devices, training the key points module needs large input size and small batch size. A study [29] indicated that when batch size is small, group normalization performs better than batch normalization. Second, two layers of the LSTM module is removed, which is not suitable for license plate recognition. Third, the input is colorful images or features while the CRNN is gray images.

To preserve the text information in the license plate, conv_gn_relu and max pooling operations are exploited to make the height of the feature maps reduce to 1 and keep the width axis reduce only once. Then, followed the fully-connected layer, we permute the features to time form as a sequence to encode the sequence in a probability distributions over the character dictionary. Finally, CTC is applied to decode the distributions to the recognition results. With the encode sequence $x \in \{x_1, x_2, ..., x_w\}$ and ground truth label $y^* = \{y_1, y_2, ..., y_L\}$, $L \leq w$, the recognition loss can be formulated as:

$$L_{recog} = -\frac{1}{N} \sum_{n=1}^{N} \log p\left(y_n^* | x\right) \tag{7}$$

$$p\left(y^* | x\right) = \sum_{\pi \in B^{-1}(y^*)} p(\pi | x) \tag{8}$$

where $p\left(y^* | x\right)$ is the conditional probability of a given labeling y^* as the sum of the probabilities of all the paths π. B is a many-to-one map from the set of possible labeling groups to the true label. N is the number of input images, and

y_n^* is the corresponding label. For given inputs, we'd like to train our model to maximize the log-likelihood of the summation of Eq. 7.

We trained the keypoint module and recognition module end-to-end, with the following objective function:

$$L = L_{kp} + \lambda L_{recog} \tag{9}$$

Where λ is a hyper-parameter controlling the balance of two losses, which was set to 1 in our experiments.

Fig. 4. Keypoints results of some examples. Images in the first line are from RP, the second line are from CCPD, the third line are from BGY. The images are cut from raw image for a better view.

3.5   Training Skills

HomoNet conducts a pre-trained model on the ImageNet [1] dataset. The Proposed training process involves two stages. In the first stage, it only train the key point module until it has a good performance on the test data. Then HomoPooling is applied to rectify the input image with the ground truth key points and recognition module is added to train jointly. On this basis, we froze the backbone layers and fine-tune it. In the whole training process, there is no any synthetic data or other license plate dataset.

6E2260	3989LC	A92746	1463ES
皖A3C159	皖AWP280	皖A055G5	皖A61660
川AF55111	鲁H7J102	桂BK7895	沪AD00806

Fig. 5. Rectification LPs from Fig. 4 and recognition results. First line LPs is scale in width for a better view.

4   Experiments

4.1   Datasets

The Application-Oriented License Plate (AOLP), CCPD and BGY datasets were used to verify the proposed model.

AOPL-RP: AOLP has 2046 Chinese Taiwan license plates which has three subsets: Access Control (AC), Traffic Law Enforcement (LE), and Road Patrol (RP). RP owns a wider range of variation in orientation. To compare with the Hsu et al. [9], AC and LE was consider as training set in this article.

CCPD: CCPD, collected from roadside parking in Anhui province of China with 250k images, which is the largest and the most diverse public dataset in LPDR. It has 8 subsets: Base, DB, FN, Rotate, Tilt, Blur, Weather and Challenge. Having great horizontal and vertical degrees, Rotate and Tilt subsets bring difficulty in recognition.

BGY: A new license plate dataset collected from China is introduced in this paper. The dataset contains three license plate colors: blue, yellow and green. There are a total of 2492 pictures. We divide the training set and test set according to 1247 and 1245 respectively. Compared with the public data set, the main features of BGY are: (1) license plates have different background colors; (2) the number of license plates is not fixed; (3) the view, scene, and environment are richer.

Table 2. The configuration of three datasets.

Dataset	Input size	LP input size	Backbone	IoU
RP	240×320	32×64	Resnet50	0.5
CCPD	480×320	32×96	Resnet18	0.7/0.6
BGY	512×416	32×96	Resnet50	0.5

4.2 Implementation Details

We adopt the Stochastic Gradient Descent (SGD) with hyper-parameters (momentum = 0.9, weight decay = 0.0001) to minimize the objective function. Warm up strategy is utilized at the first epoch to increase the learning rate (LR) from 0 to 0.05 in the first stage. Then LR is dropped by 10 times at the 16th epoch and 21st epoch until finishing the training at the 25th epoch. The batch size is set to 8 and the number groups of group normalization is 32. HomoPooling operation resizes the longer side of the input image to times of 32. The proposed model is implemented in PyTorch and trained on Nvidia GTX 1080TI graphics cards.

To improve the robustness of the HomoNet, the experiments involve the data augmentation during training. In the first stage we use the random rotate between $-5°$ to $5°$, and then utilize random disturbance at every key point in the second stage. To be emphasized, there are not any synthetic data.

For different datasets, the input size is changed, the corrected LP size and backbone, as given in Table 2. For a fair comparison with others, we exploit the same metrics for RP and BGY. When the intersection-over-union (IoU) is more than 0.5, the detection is right. When all characters of LP is right and detection is right, the end-to-end result is right. Differently for CCPD, while IoU is more than 0.7, the detection is right. Only when the IoU is more than 0.6 and all characters are right, the end-to-end result is right.

Table 3. Experiments results on the RP subset.

Model	IoU (%)	Speed (ms)	End-to-end (%)	Speed (ms)
Hsu et al. [9]	94	–	85.7	320
Li et al. [14]	95.6	–	88.38	1000
Li et al. [15]	98.2	–	83.63	400
Kessentini et al. [11]	99.2	36.7	90.42	48.7
Silva et al. [26]	–	–	93.29	200
Ours (no rectified)	**99.8**	**18.1**	80.85	**22.5**
Ours	**99.8**	**18.1**	**95.58**	**22.5**

4.3   Performance on RP Dataset

The first line in Fig. 4 shows some of the experimental results trained using our key point module. It can be seen that this module can effectively detect the key points of the license plate even in the dark, strong light, rotation, tilt, natural scene and other conditions. Although some key points have errors, the four key points can still contain the entire license plate information, such as the second in the first line of Fig. 4. At the same time, as illustrated in Table 3, our method achieved the highest performance (99.8%/18.1 ms), 0.6% higher and 18.6 ms faster than the state-of-the-art method (99.2% vs 99.8%, 36.7 ms vs 18.1 ms). The results indicate that Hsu et al. [9] used the expectation-maximization clustering based on the license plate boundary information, and Li et al. [14], regressed box based on the character information can not get high detection accuracy. Both Li et al. [15] who used region proposal network to extract the region of interests and then used anchors to return offset and Kessentini et al. [11] who adopted YOLOv2 detector can get better accuracy. Thanks to the prior information that only one car has only one license plate, we convert the license plate detection into license plate key point detection, which is without anchors and NMS. Our model achieves the best precision and speed.

As shown in Table 3. Extracted local binary patterns (LBP) features and used linear discriminant analysis (LDA) to classify characters [9] can not get a high accuracy. Method [14] which is based on LSTM [6] identifying the character sequence is too slow. The results of the bounding box detection [11,15], which are used to recognize the license plate number, can not get good performance too, while the correction network [26] can improve the recognition. However, affine transformation only can solve a part of the distortion problem. Benefited from rectifying the license plate by HomoPooling with the four points information, HomoNet achieves the highest end-to-end accuracy and the fastest speed. The pooling operation improves a large margin compared to baseline which is directly cropping the image with a bounding box. To further demonstrate the effectiveness of the HomoPooling, Fig. 5 shows the correction results after the HomoPooling operation detected in Fig. 4. It can be seen that the license plates that are perspective, rotated, tilted and distorted are corrected to the horizontal direction.

Table 4. The end-to-end results (percentage) of CCPD and speed is in milliseconds. HC is Holistic-CNN [33]. AP denotes average accuracy.

Model	Speed	AP	Base	DB	FN	Rotate	Tilt	Weather	Challenge
Wang et al. [28] +HC	34.5	58.9	69.7	67.2	69.7	0.1	3.1	52.3	30.9
Ren et al. [22] +HC	28.6	92.8	97.2	94.4	90.9	82.9	87.3	85.5	76.3
Liu et al. [3] +HC	27.8	95.2	98.3	96.6	**95.9**	88.4	91.5	87.3	83.8
Joseph et al. [21] +HC	76.9	93.7	98.1	96	88.2	84.5	88.5	87	80.5
Li et al. [15] +HC	333.3	94.4	97.8	94.8	94.5	87.9	92.1	86.8	81.2
Xu et al. [31] +HC	**16.4**	95.5	98.5	96.9	94.3	90.8	92.5	87.9	85.1
Zhang et al. [33]	–	95.4	98.4	**97**	90.6	92.7	93.5	86.9	84.8
Ours	19	**97.5**	**99.1**	96.9	**95.9**	**97.1**	**98**	**97.5**	**85.9**

4.4   Performance on CCPD Dataset

As shown in Table 4, HomoNet achieves the best results on all subsets except DB. Benefited from HomoPooling, our model achieves a large improvement in Rotate, Tilt and Weather, which is 4.4%, 4.5% and 9.6% respectively. Though the end-to-end speed is 2.6 ms slower than Xu et al. [31] due to the recognition module, our detection module speed is 4 ms faster. The visual results in the second line of Fig. 4 and Fig. 5 indicates the effectiveness of HomoPooling.

4.5   Performance on BGY Dataset

As illustrated in Table 5, we obtain a detection accuracy of 99.52% and an end-to-end recognition accuracy of 91.33% on the BGY dataset. The detection results and the corrected recognition results in the third line of Fig. 4 and Fig. 5 indicate that HomoNet can not only correct but also recognize LPs with multiple types and different character lengths.

Table 5. Experiments results on the BGY dataset.

Model	IoU (%)	Speed (ms)	End-to-end (%)	Speed (ms)
Ours	99.52	24.5	91.33	30.1

5   Conclusions

In this paper, a fast unfied network HomoNet that contains HomoPooling based on perspective transformation was proposed, which can correct oblique license plates. Using the a priori information of one plate for one car, the speed and accuracy of the detection were improved by predicting the heat map of the four points directly. HomoPooling can achieve more information and rectify the oriented LPs. Experiments on public datasets have demonstrated the advantage of the proposed method. In future, the proposed method will extend to the double line license plate to be jointly trained in this framework. Moreover, not all the characters in LP are able to recognize even for people in some situations. We plan to explore the representations of fuzzy LP for vehicle re-identification.

References

1. Deng, J., Dong, W., Socher, R., Li, L.J., Li, K., Fei-Fei, L.: ImageNet: a large-scale hierarchical image database. In: 2009 IEEE Conference on Computer Vision and Pattern Recognition, pp. 248–255. IEEE (2009)
2. Dubrofsky, E.: Homography estimation. Diplomová práce. Univerzita Britské Kolumbie, Vancouver (2009)
3. Girshick, R.: Fast R-CNN. In: Proceedings of the IEEE International Conference on Computer Vision, pp. 1440–1448 (2015)

4. Graves, A., Fernández, S., Gomez, F., Schmidhuber, J.: Connectionist temporal classification: labelling unsegmented sequence data with recurrent neural networks. In: Proceedings of the 23rd International Conference on Machine Learning, pp. 369–376. ACM (2006)
5. He, K., Zhang, X., Ren, S., Sun, J.: Deep residual learning for image recognition. In: Proceedings of the IEEE Conference on Computer Vision and Pattern Recognition, pp. 770–778 (2016)
6. Hochreiter, S., Schmidhuber, J.: Long short-term memory. Neural Comput. $9(8)$, 1735–1780 (1997)
7. Hou, Y., Qin, X., Zhou, X., Zhou, X., Zhang, T.: License plate character segmentation based on stroke width transform. In: 2015 8th International Congress on Image and Signal Processing (CISP), pp. 954–958. IEEE (2015)
8. Hsu, G.S., Ambikapathi, A., Chung, S.L., Su, C.P.: Robust license plate detection in the wild. In: 2017 14th IEEE International Conference on Advanced Video and Signal Based Surveillance (AVSS), pp. 1–6. IEEE (2017)
9. Hsu, G.S., Chen, J.C., Chung, Y.Z.: Application-oriented license plate recognition. IEEE Trans. Veh. Technol. $62(2)$, 552–561 (2012)
10. Jaderberg, M., Simonyan, K., Zisserman, A.: Spatial transformer networks. In: Advances in Neural Information Processing Systems, pp. 2017–2025 (2015)
11. Kessentini, Y., Besbes, M.D., Ammar, S., Chabbouh, A.: A two-stage deep neural network for multi-norm license plate detection and recognition. Expert Syst. Appl. $\mathbf{136}$, 159–170 (2019)
12. Laroca, R., et al.: A robust real-time automatic license plate recognition based on the YOLO detector. In: 2018 International Joint Conference on Neural Networks (IJCNN), pp. 1–10. IEEE (2018)
13. Law, H., Deng, J.: CornerNet: detecting objects as paired keypoints. In: Proceedings of the European Conference on Computer Vision (ECCV), pp. 734–750 (2018)
14. Li, H., Shen, C.: Reading car license plates using deep convolutional neural networks and LSTMs. arXiv preprint arXiv:1601.05610 (2016)
15. Li, H., Wang, P., Shen, C.: Toward end-to-end car license plate detection and recognition with deep neural networks. IEEE Trans. Intell. Transp. Syst. $20(3)$, 1126–1136 (2018)
16. Lin, T.Y., Dollár, P., Girshick, R., He, K., Hariharan, B., Belongie, S.: Feature pyramid networks for object detection. In: Proceedings of the IEEE Conference on Computer Vision and Pattern Recognition, pp. 2117–2125 (2017)
17. Liu, W., et al.: SSD: single shot multibox detector. In: Leibe, B., Matas, J., Sebe, N., Welling, M. (eds.) ECCV 2016. LNCS, vol. 9905, pp. 21–37. Springer, Cham (2016). https://doi.org/10.1007/978-3-319-46448-0_2
18. Liu, X., Liang, D., Yan, S., Chen, D., Qiao, Y., Yan, J.: FOTS: fast oriented text spotting with a unified network. In: Proceedings of the IEEE Conference on Computer Vision and Pattern Recognition, pp. 5676–5685 (2018)
19. Llorca, D.F., et al.: Two-camera based accurate vehicle speed measurement using average speed at a fixed point. In: 2016 IEEE 19th International Conference on Intelligent Transportation Systems (ITSC), pp. 2533–2538. IEEE (2016)
20. Redmon, J., Divvala, S., Girshick, R., Farhadi, A.: You only look once: unified, real-time object detection. In: Proceedings of the IEEE Conference on Computer Vision and Pattern Recognition, pp. 779–788 (2016)
21. Redmon, J., Farhadi, A.: YOLO9000: better, faster, stronger. In: Proceedings of the IEEE Conference on Computer Vision and Pattern Recognition, pp. 7263–7271 (2017)

22. Ren, S., He, K., Girshick, R., Sun, J.: Faster R-CNN: towards real-time object detection with region proposal networks. In: Advances in Neural Information Processing Systems, pp. 91–99 (2015)

23. Safie, S., Azmi, N.M.A.N., Yusof, R., Yunus, M.R.M., Sayuti, M.F.Z.C., Fai, K.K.: Object localization and detection for real-time automatic license plate detection (ALPR) system using RetinaNet algorithm. In: Bi, Y., Bhatia, R., Kapoor, S. (eds.) IntelliSys 2019. AISC, vol. 1037, pp. 760–768. Springer, Cham (2020). https://doi.org/10.1007/978-3-030-29516-5_57

24. Shi, B., Bai, X., Yao, C.: An end-to-end trainable neural network for image-based sequence recognition and its application to scene text recognition. IEEE Trans. Pattern Anal. Mach. Intell. **39**(11), 2298–2304 (2016)

25. Silva, S.M., Jung, C.R.: Real-time Brazilian license plate detection and recognition using deep convolutional neural networks. In: 2017 30th SIBGRAPI Conference on Graphics, Patterns and Images (SIBGRAPI), pp. 55–62. IEEE (2017)

26. Silva, S.M., Jung, C.R.: License plate detection and recognition in unconstrained scenarios. In: Ferrari, V., Hebert, M., Sminchisescu, C., Weiss, Y. (eds.) ECCV 2018. LNCS, vol. 11216, pp. 593–609. Springer, Cham (2018). https://doi.org/10.1007/978-3-030-01258-8_36

27. Tian, Z., Shen, C., Chen, H., He, T.: FCOS: fully convolutional one-stage object detection. arXiv preprint arXiv:1904.01355 (2019)

28. Wang, S.Z., Lee, H.J.: A cascade framework for a real-time statistical plate recognition system. IEEE Trans. Inf. Forensics Secur. **2**(2), 267–282 (2007)

29. Wu, Y., He, K.: Group normalization. In: Proceedings of the European Conference on Computer Vision (ECCV), pp. 3–19 (2018)

30. Xie, L., Ahmad, T., Jin, L., Liu, Y., Zhang, S.: A new CNN-based method for multi-directional car license plate detection. IEEE Trans. Intell. Transp. Syst. **19**(2), 507–517 (2018)

31. Xu, Z., et al.: Towards end-to-end license plate detection and recognition: a large dataset and baseline. In: Proceedings of the European Conference on Computer Vision (ECCV), pp. 255–271 (2018)

32. Yang, M., et al.: Symmetry-constrained rectification network for scene text recognition. In: Proceedings of the IEEE International Conference on Computer Vision, pp. 9147–9156 (2019)

33. Zhang, Y., Huang, C.: A robust Chinese license plate detection and recognition system in natural scenes. In: 2019 IEEE 4th International Conference on Signal and Image Processing (ICSIP), pp. 137–142. IEEE (2019)

Resource Management in Artificial Intelligence

Reactive Workflow Scheduling in Fluctuant Infrastructure-as-a-Service Clouds Using Deep Reinforcement Learning

Qinglan Peng[1], Wanbo Zheng[2(✉)], Yunni Xia[1(✉)], Chunrong Wu[1], Yin Li[3], Mei Long[4], and Xiaobo Li[5]

[1] Software Theory and Technology Chongqing Key Lab, Chongqing University, Chongqing 400044, China
xiayunni@hotmail.com
[2] Data Science Research Center, Kunming University of Science and Technology, Kunming 650031, China
zwanbo@163.com
[3] Institute of Software Application Technology, Guangzhou & Chinese Academy of Sciences, Guangzhou 511000, China
[4] ZBJ Network Co. Ltd., Chongqing 401123, China
[5] Chongqing Animal Husbandry Techniques Extension Center, Chongqing 401121, China

Abstract. As a promising and evolving computing paradigm, cloud computing benefits scientific computing-related computational-intensive applications, which usually orchestrated in terms of workflows, by providing unlimited, elastic, and heterogeneous resources in a pay-as-you-go way. Given a workflow template, identifying a set of appropriate cloud services that fulfill users' functional requirements under pre-given constraints is widely recognized to be a challenge. However, due to the situation that the supporting cloud infrastructures can be highly prone to performance variations and fluctuations, various challenges such as guaranteeing user-perceived performance and reducing the cost of the cloud-supported scientific workflow need to be properly tackled. Traditional approaches tend to ignore such fluctuations when scheduling workflow tasks and thus can lead to frequent violations to Service-Level-Agreement (SLA). On the contrary, we take such fluctuations into consideration and formulate the workflow scheduling problem as a continuous decision-making process and propose a reactive, deep-reinforcement-learning-based method, named DeepWS, to solve it. Extensive case studies based on real-world workflow templates show that our approach outperforms significantly than traditional ones in terms of SLA-violation rate and total cost.

Keywords: Workflow scheduling · IaaS cloud · Quality-of-Service · Pay-as-you-go · Reinforcement learning

ⓒ ICST Institute for Computer Sciences, Social Informatics and Telecommunications Engineering 2021
Published by Springer Nature Switzerland AG 2021. All Rights Reserved
H. Gao et al. (Eds.): CollaborateCom 2020, LNICST 350, pp. 285–304, 2021.
https://doi.org/10.1007/978-3-030-67540-0_17

1   Introduction

As a useful process modeling tool, workflow has been widely applied in scientific computing, big data processing, video rendering, business process modeling, and many other fields [2]. Workflow scheduling refers to an integration of workflow tasks, with varying non-functional properties in terms of Quality-of-Service (QoS), collectively orchestrated to automate a particular task flow or computational process. Scientific workflows, e.g., weather forecasting and high-energy-physics simulating workflows, are usually high time, resource, energy, and storage-requiring. Thus, effective scheduling algorithms and methods are in high demand for guaranteeing cost-effective and timely execution of complex scientific workflows.

However, the fast-growing data and computing requirements demand more powerful computing environments in order to execute large-scale workflows in a reasonable amount of time. Traditional grid and cluster systems are mostly dedicated and statically partitioned per administration policy. These traditional systems can be inefficient in adapting to today's computing demand. Fortunately, with the rapid development of the cloud ecosystem, large-scale workflow applications are able to leverage the dynamically provisioned resources from the cloud instead of using a dedicated server or node in grids or clusters [21]. As the cloud provides resources as maintenance-free and pay-as-you-go services, users are allowed to access resources on-demand at anytime and from anywhere. They only need to identify what type of resource to lease, how long to lease resources, and how much their applications will cost. Nevertheless, according to the study of Schad et al. [18], the performance of Virtual Machines (VMs) in Amazon EC2 cloud vary by 24% percentages under high workload. And such fluctuations of VMs' performance have a great possibility to impact the user-perceived quality of cloud systems, especially when the Service-Level-Agreement (SLA) constrains, e.g., workflow execution time, are violated. Therefore, how to schedule workflow tasks into the fluctuant Infrastructure-as-a-Service (IaaS) cloud resources to meet the SLA constraint while reducing the cost as much as possible has become the major problem.

As a practical problem, IaaS cloud-based workflow scheduling becomes a hot research topic since it is widely acknowledged that to schedule multi-task workflow on distributed platforms is a typical NP-hard problem [6], and various existing works [1,5,8,9,15,16] in this direction addressed the multi-objective-multi-constraint cloud workflow scheduling problem by using statistic, stochastic, heuristic, and meta-heuristic methods. However, these traditional solutions might be inefficient and incur high SLA-violation rate when encounter with massive requests and fluctuate cloud performance. Recent progresses in artificial intelligence [11,19] show that learning-based approaches, especially the reinforcement-learning ones, can be highly potential in dealing with such a cloud-based workflow scheduling problem as well. In this paper, we target at the online reactive workflow scheduling problem and propose a deep-reinforcement-learning-based method, shorts for DeepWS, to solve it.

The main contributions of this work are as follows: 1) we formulate the deadline-constrained-cost-minimization IaaS-cloud-workflow scheduling problem into a continuous decision-making process and use a Markov Decision Process (MDP) to model it; 2) we define the state space and action space of the proposed MDP problem, and design an SLA and cost-aware reward function; 3) we propose a deep-reinforcement-learning-based method to solve the reactive workflow scheduling problem on the fly. Extensive case studies based on various real-world workflow templates under different deadline constraint levels are conducted to verify the effectiveness of our approach.

The remainder of this paper is organized as follows. Section 2 introduces the DQN and how it incorporates with our problem. The system model and corresponding MDP problem formulation are described in Sect. 3. We propose a DQN-based method in Sect. 4. Our experimental results and discussions are given in Sect. 5. Section 6 reviews related works. Finally, Sect. 7 concludes this paper.

2 Preliminary

Reactive workflow scheduling can be regarded as a continuous decision-making process where the scheduling decisions of tasks are made on the fly, and we use deep-reinforcement-learning to train the scheduling model through a repeated trial-and-error learning strategy. In reinforcement learning (RL), an action-value function $Q(s, a)$ is often used to identify the discounted future reward of taking an action a at state s. The reward values are often stored in an action-value mapping table, shorts for Q-Table, and its update process can be expressed as follow:

$$Q(s, a) \leftarrow (1 - \alpha)Q(s, a) + \alpha[r + \gamma \max_{a'} Q(s', a')] \tag{1}$$

where α is the learning rate, r the direct reward of action a, s' the resulting state of tacking action a at state s. To overcome the limitation of state explosions,

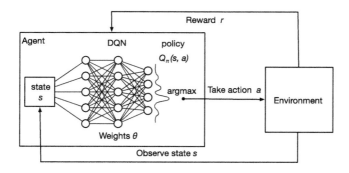

Fig. 1. Deep reinforcement learning

the action-value function is usually approximated by neural network for state-space saving:

$$Q(s, a; \theta) \approx Q(s, a) \tag{2}$$

Figure 1 shows the architecture of deep reinforcement learning. In this paradigm, an action-value function $Q(s, a; \theta)$ is often used to identify the discounted future reward of taking an action a at state s, and Deep Q-Network (DQN) [11] is used to fit the mapping between actions and reward. It can be trained through minimizing the loss function $L_i(\theta_i)$ at each iteration:

$$L_i(\theta_i) = E_{s,a \sim \rho(*)}[(y_i - Q(s, a; \theta_i))^2] \tag{3}$$

where

$$y_i = E_{s \sim \epsilon}[r + \gamma \max_{a'}(s', a'; \theta_{i-1} | s, a)] \tag{4}$$

where y_i is the target in the i-th iteration, $\rho(s, a)$ the probability distribution of state and action, and it can be through an ϵ-greedy method. The gradient can be derived based on the loss function as follows:

$$\nabla_{\theta_i} L_i(theta_i) = E_{(s,q \sim \rho(*); s' \sim \epsilon)}[(r + \gamma \max_{a'} Q(s', a'; \theta_{i-1})$$
$$- Q(s, a; \theta)) \nabla_{\theta_i} Q(s, a; \theta)] \tag{5}$$

Then, neural networks optimization methods, i.e., Adam, SGD, RMSprop, and so on, can be employed for updating the weights of neural network according to the gradients derived in Eq. (5). In this paper, we regard the status of workflow and VMs as the states, the schedules of tasks as actions, and propose a DQN-based method to train a scheduling model to yield reactive schedules in real-time.

3   System Model and MDP Problem Formulation

In this section, we first present our system model and the corresponding reactive deadline-constrained workflow scheduling problem, then we formulate it to a continuous decision-making process using MDP model.

3.1   System Model

A workflow can be described as a Directed Acyclic Graph (DAG) $W = (T, E, D)$, where $T = \{t_1, t_2, ...t_n\}$ is the set of tasks, $E = \{e_{ij} | i, j \in n, i \neq j\}$ the set of edges which represent the data dependencies between different tasks, edge e_{ij} indicates that the output of t_i is the input of t_j, D the user-defined deadline constraint. A task can only start when all its preceding ones are successfully executed and data transfers between them are accomplished. Figure 2 shows an workflow example with 8 tasks.

Generally, there are different types of VMs provided by IaaS providers available for custom selection, and these different types promise different prices and resource configurations in terms of, e.g., the number of cores of CPU, the size of

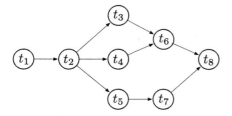

VMtype Task	Small	Medium	Large	XLarge	2XLarge
t_1	0.81	0.65	0.42	0.31	0.27
t_2	1.76	1.31	0.94	0.65	0.57
t_3	0.94	0.72	0.58	0.42	0.39
...	...	...	...	...	...

Fig. 2. An workflow example with 8 tasks and the execution of it tasks on different type of VMs.

memory and allocated storage. Therefore, for the same workflow task, its execution can vary in VMs with different types. The leased VMs can be regarded as a resource pool $P = \{VM_1, VM_2, ..., VM_m\}$. Due to the fact that the VM acquiring time is chargeable, therefore tasks are scheduled into idle VMs preferentially in the resource pool to reduce the cost. And the VMs which are in the status of idle for a certain period will be released to reduce the cost. We consider the violations to SLA are subject to penalties as well, and that penalty as a linear function similar to those proposed in [4,14,22]. Therefore, deadline-constrained-cost-minimized workflow scheduling problem over fluctuant IaaS cloud can be formulated as follow:

$$Min: \quad C = \sum_{i=1}^{n} h(g(i)) \times f_i^{g(i)} \tag{6}$$

$$s.t: \quad M \le D$$

where $g(i)$ is the function to identify which type of VM that task t_i is scheduled into, $h(j)$ the function to identify cost-per-unit-time of using a type j VM, $f_i^{g(i)}$ the actual execution time of task t_i on the VM with type $g(i)$, thus C denotes the cost of running the scientific workflow. D is the user defined deadline constraint, and M the actual makespan of the workflow. Note that, if a schedule violates the deadline constraint, additional penalty will be incurred to the total cost. Under this situation (i.e., $M > D$), cost C can be calculated as:

$$C = \sum_{i=1}^{n} h(g(i)) \times f_i^{g(i)} + \beta \sum_{t=D}^{M} h_t \tag{7}$$

where β the deadline constraint violation (i.e., SLA-violation) penalty rate, and h_t the cost incurred by all hired VMs at time slot t.

3.2 MDP Problem Formulation

A Markov decision process (MDP) is a discrete time stochastic control process [12], it can be seen as Markov Chain with additional actions and rewards. A MDP can be expressed as a 5-tuples (S, A, P_a, γ), where S is the set of finite states, A the set of finite actions, A_s the action set under s state,

$P_a(s, a, s') = \mathbf{Pr}(s_{t+1}|s_t = s, a_t = a)$ the transition probability from state s to s' if action a is taken at time t at state s, $R_a(s, a, s')$ the direct reward of the transition from state s to s' if an action a is performed, $\gamma \in [0, 1]$ the discount to decide the discount ratio of future reward to the previous steps. MDP provides a mathematical framework for modeling decision making in situations where outcomes are partly random and partly under the control of a decision-maker. It is a useful tool for studying a wide range of decision making problems solved via dynamic programming and reinforcement learning.

The workflow scheduling problem can also be formulated as an MDP problem. When a task in a workflow is ready to start, the scheduling agent will decide the action that next step to take based on evaluation of the current state in terms of, e.g., resource utilization and remaining execution time permitted, of the entire workflow. After the decided action is taken, the scheduling agent gets a reward calculated based on the resulting state of the workflow.

The training goal for the MDP-based workflow scheduling problem is thus to find a policy π which leads to an capable of maximizing the expected future rewards. It is usually assumed that the future reward is subject to a discount rate γ. To be specific, the total discounted reward at state s can be calculated as follow:

$$R_s = \sum_{s'=s}^{T} \gamma^{s'-s} R_{s'} \tag{8}$$

where T is the terminate state, which means there are no remaining tasks in the workflow to be scheduled (i.e., the workflow has finished), $R_{s'}$ the reward at state s'.

4   Reinforcement-Learning-Based Solution

To solve the MDP problem we formulated previously, in this section, we propose a deep-reinforcement-learning-based approach, short for DeepWS, to train the scheduling model to yield reactive schedules in real-time.

4.1   MDP Definitions

We first present the detailed MDP definitions, including its state space, observation, action space, and reward function, of the proposed reactive workflow scheduling problem.

1) State-space: In DeepWS, the state is defined to the status of the whole system when a task in the workflow is ready to be scheduled. Therefore, every task in a workflow has its corresponding state, and we use s_i to denote the state when task t_i is ready to be scheduled. For example, Fig. 3 shows an example of state transition in a whole scheduling process. Suppose that there is a workflow with 4 tasks to be scheduled, the first state of the system will be s_1 because the task t_1 is the entry task, and only t_1 is able to start at the beginning. Then the agent will schedule t_1 to a certain VM to be executed. After t_1 has accomplished, system will transfer into s_2 or s_3 to schedule t_2 and t_3. If data transfer $e_{(1,2)}$ is faster than $e_{(1,3)}$, then state will transfer from s_1 to s_2, otherwise, from s_1 to s_3. If data transfer $e_{(1,2)}$ and $e_{(1,3)}$ accomplish at the same time, we random choose an available state to step into, note that, the transition order $\{s_1 \rightarrow s_2 \rightarrow s_3\}$ is equal to $\{s_1 \rightarrow s_3 \rightarrow s_2\}$ under such a situation, because the schedule decisions will be made in millisecond level, and they can be regarded as simultaneous operations. Finally, if t_2 and t_3 have accomplished, system will step into s_4, which is obvious the terminal state, and the schedule process will end after t_4 is scheduled at s_4.

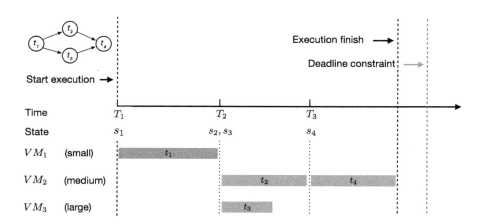

Fig. 3. An example of state transition and online scheduling.

2) Observation: In DeepWS, observation is the status of whole system at a certain state, it can be represented as a $(l + 5)$-tuple $o_i = \{t_i, \tau_1, \tau_2, ..., \tau_l, G, O, P, H\}$, where l is the number of VM types available for users to choose, t_i the task which need to be scheduled at current state, τ_j the estimated execution time of t_i in the VM with type j based on current performance of IaaS cloud, G the deadline left time, O the cumulative cost from the beginning to current state, P the time pasted from the beginning, H the longest remain time for task t_i to meet the deadline constraint, and it can be obtained as follow:

$$H = D - P - \sum_{t \in \{U - t_i\}} \tau_t^e \tag{9}$$

where U is the set of tasks in the crucial path of unfinished part of DAG, τ_t^e the estimated execution time of task t in most expensive type of VM.

When a task is ready to schedule, the agent first gets the observation of the system at the current state. Then it make a decision on next action, i.e., which type of VM this task is going to schedule to with the help of schedule policy π, i.e, trained scheduling model. Finally, the action will be performed and its corresponding reward will be evaluated to update current policy.

3) Action space: As we mentioned previously, we define the action a_i to the type of VM which task should be scheduled into at state s_i. For example, there are three types of VM available for custom selection: *small, medium,* and *large,* and we use number 1, 2, and 3 to identify them. Suppose that the action made by a agent at state s_i for task t_i is $a_i = 3$, and it indicates that t_i is going to be scheduled into a medium VM to be executed.

4) Reward function: The design of the reward function is the key step in reinforcement learning problem formulation. It tells agent that which kind of action under a certain state is proper and encouraging, and which kind of action is discourage and will be punished. The reward function in DeepWS is designed as follow:

$$R(s) = \begin{cases} \epsilon, & s \neq \mathbb{T} \\ \xi, & s = \mathbb{T} \land f > D \\ (\frac{f-F}{D-F})^\zeta, & s = \mathbb{T} \land f \leq D \end{cases} \tag{10}$$

where s is the current state, $\mathbb{T}$ the terminal state, f the final makespan the schedule, F the fastest estimated finish time of a workflow, it can be estimated by assuming that all tasks are scheduled into fastest VM instances, D the user defined deadline, and ζ the scale coefficient. In DeepWS, we define that only actions made at the terminal state can get nonzero rewards, and these rewards will effect previous actions' rewards in a discount way as shown in Eq. (8). Action series which result in missing deadlines will be punished, we use ξ, a negative real number which is infinitely close to 0, to present such a punishment to make a clear comparison with positive rewards got by those meeting deadline ones.

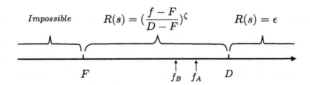

Fig. 4. Rewards distribution.

Figure 4 shows an illustrative example of the distribution of the reward function. The makespan of a workflow should be within $[F, +\infty)$ theoretically, once the deadline is missed, the reward will be ϵ, where $\epsilon \to 0^-$. The reward will be in the range of $(0, 1]$ when makespan within $[F, D]$, the more close makespan to D, the higher reward will get. Intuitively, this reward reflects how close does a makespan to the deadline constraint when the deadline is met. Figure 4 also shows an example of scheduling policy comparison, for the same workflow to be scheduled, the makespan of workflow scheduled by policy π_A/π_B are f_A/f_B, which both meeting the deadline constraint. We say that policy π_A is better than policy π_B, because the longer time left between actual makespan and deadline, the more likely to employ cheaper VM to reduce cost while meeting the deadline. The scheduling target in this paper is to reduce cost while meeting the deadline constraint, therefore the reward function is design to lead the agent to learn how to make full use of every possible time slot to reduce cost.

4.2   Environment Emulator Design

Agents in reinforcement learning learn policies over time through repeated interactions with the environment. For this purpose, we consider a simulated environment through which policies of scheduling workflow over IaaS clouds can gradually evolve. We use *env* to indicate a instance of the environment emulator, which has the following functions:

- *env*.Init(W): it initializes an environment instance with a workflow request W;
- *env*.GetTaskToSchedule(): it returns a set of tasks which are ready to execute/schedule;
- *env*.AssignTaskToVM(t_i, P, *VMType*): it assigns task t_i into a VM with type *VMType*. As shown in Procedure 1, it searches all idle VMs with the desired VM types in the resource pool. Otherwise, it simply releases a new VM with type *VMType* and allocates t_i to it;
- *env*.isDone(): It returns $True$ if the workflow is finished and otherwise $False$.

The design of emulator consider multiple factors, e.g., pay-as-you-go pricing, unlimited heterogeneous resources, performance fluctuation, VM acquiring and releasing delay, etc. Given the training set(i.e., historical workflow requests), the agent learns scheduling policies through repeatedly interact (e.g., make scheduling decisions) with the environment instance.

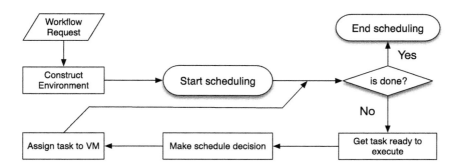

Fig. 5. The process of interacting with the emulator for an RL agent.

Figure 5 shows the process of interacting with the emulator for an RL agent, it can be seen that it begin with accepting a certain workflow request. Then, tasks ready to start are and assigned to VMs according to the evolving schedule policy, this process loops until all tasks in a workflow are scheduled.

4.3   Training Algorithm

In DeepWS, DQN is employed as an approximate function to map the state to the probability of different actions can be taken. The input of DQN is the

---

Procedure 1: AssignTaskToVM

Input: Task to schedule t; Resource pool P; VM type *VMType*
Output: VM instanace *vmi*;

1 *VMCancidates* $\leftarrow \varnothing$
2 **foreach** $vm \in P$ **do**
3 | **if** vm.isIdle() **and** vm.type $=$ *VMType* **then**
4 | | *VMCancidates*.append(vm)

5 **if** *VMCancidates* $= \varnothing$ **then**
6 | Find a new VM instance *vmi* with type of *VMType* and schedule task t into it to execute
7 **else**
8 | Select the VM $vmi \in VMCancidates$ with shortest idle time and schedule task t into it to execute
9 **return** *vmi*

Algorithm 2: Training algorithm for DQN

Input: Training workflow set W
Output: Scheduling model M

1 initialize ReplayMemory RM with random scheduling policy
2 initialize eval-DQN *eNN* with weights $\sim N(0, 0.1)$
3 initialize target-DQN *tNN* with weights $\sim N(0, 0.1)$
4 **foreach** $w \in W$ **do**
5 | $env \leftarrow$ Environment(w)
6 | **while** env.isDone() is **False do**
7 | | $tasks \leftarrow$ env.getTasksToSchedule()
8 | | **foreach** $task\ t \in tasks$ **do**
9 | | | $s_t \leftarrow$ env.getCurrentState()
10 | | | $a_t \leftarrow \epsilon$-greedy($\max\limits_{a} eNN(s_t, a),\ \epsilon$)
11 | | | $r_t \leftarrow$ env.ScheduleTaskToVM($t,\ a_t$)
12 | | | $s_{t+1} \leftarrow$ env.getCurrentState()
13 | | | RM.put(s_t, a_t, r_t, s_{t+1})
14 | | | $s_j, a_j, r_j, s_{j+1} \leftarrow$ mini batch randomly select from RM
15 | | | **if** env.isDone() **then**
16 | | | | $y = r$
17 | | | **else**
18 | | | | $y = r_j + \gamma \max\limits_{a'} tNN(s_{j+1}, a')$
19 | | | $loss \leftarrow (y - eNN(s_j, a_j))^2$
20 | | | Perform a gradient descent step on *loss* with respect to the *tNN*
21 | | | Every C steps reset $tNN \leftarrow eNN$

22 **return** *eNN*

---

observation of the current system, and the output is the probability of actions, i.e., which type of VM that the task should be scheduled into. A complete process (as shown in Fig. 5) of the scheduling of a workflow can be seen as an episode, and DQN is trained through such iterations of episodes.

To train DQN effectively, memory reply and delayed update [11] are employed to make it immune to the correlations between training set. Memory replay aims at making the agent learn from random experiences. All decision steps (s_i, a_i, r_i, s_{i+1}) are recorded into a memory set during training and a mini-batch of steps is randomly selected at each state to train the DQN. The delayed update aims at decreasing the fluctuation of the trained policy due to the frequent change of an action-value function. It maintains two DQN (target-DQN and eval-DQN) at the same time, eval-DQN is responsible for decision making and loss backpropagation, while target-DQN responsible for calculating the loss. The weights of eval-DQN are copied into target-DQN at stated intervals. The detail of the training procedure for workflow scheduling model is shown in Algorithm 2.

5   Experiment and Discussion

To verify the effectiveness of DeepWS, we conduct a series case studies to evaluate the performance of our method and its peers in terms of makespan, deadline hit rate and cost.

5.1   Experiment Setting

We consider SCS [9], PSO [16] and IC-PCP [1] these three state-of-art workflow scheduling algorithms as baseline algorithms since their scheduling targets are minimizing the cost while meeting the deadline constraint, which is similar with us. SCS is a dynamic heuristic scheduling method, scheduling decisions are pre-generated but task consolidations are performed at run time to reduce the total cost; while PSO is a static meta-heuristic scheduling method, which takes VM performance fluctuation into consideration and uses an overrated tasks execution time to cope with fluctuation; IC-PCP is a static heuristic scheduling method which does not consider VM performance fluctuation.

We consider there are five types of VM available (i.e., *small*, *medium*, *large*, *xlarge*, and *xxlarge*) in our experiments, and their configurations, pricing[1], and benchmark socres[2] (i.e., performance) are shown in Table 1. According to the study of Schad *et al.* [18], the performance of VM can vary up to 24%. In our experiments, the fluctuation of VM performance is set to follow a normal distribution with 15% mean and 20% standard deviation, which is same as the study of Rodriguez *et al.* [16]. To further evaluate the performance of DeepWS and its peers under various levels of the deadline constraint, we set four deadline constraint levels, i.e., *loose*, *moderate*, *medium*, and *tight*, for every cases. The actual deadline DL for a workflow is determined by deadline constraint factor $\alpha \in \{0.2, 0.4, 0.6, 0.8\}$ and it can be calculated as follow:

$$DL = F + (L - F) \times \alpha \tag{11}$$

---

[1] https://aws.amazon.com/cn/ec2/pricing/on-demand/.
[2] http://browser.geekbench.com/.

Table 1. VM configurations.

VM type	vCPU	Memory	Benchmark	Price (cent per hour)
t3.small	1	2G	4833	2.08
t3.medium	1	4G	4979	4.16
c5.large	2	4G	6090	8.5
c5.xlarge	4	8G	11060	17
c5.xxlarge	8	16G	17059	34

where F is the fastest estimated finish time of workflow request, L the longest estimated finish time, it can be estimated by assuming only the cheapest type of VM is available. Obviously, deadline constraint will never be met when $\alpha = 0$, thus we set four deadline constraint intervals to identify DeepWS and its peers' performance under different levels of deadline.

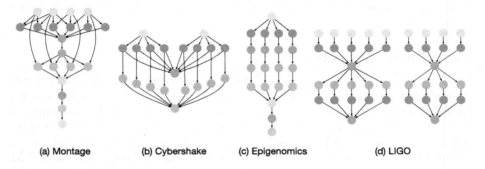

(a) Montage   (b) Cybershake   (c) Epigenomics   (d) LIGO

Fig. 6. Workflows used in experiments.

Pegasus project[3] has released a lot of real-world scientific workflow for academic research. As shown in Fig. 6, we use Montage, Cybershake, Epig, and LIGO these four cases to conduct experiments. For each case, 100 different workflows are tested by four algorithms and their performance in terms of makespan and cost are recorded.

We consider four layers of DQN with 8 neurons in the input layer, 256 neurons and 128 neurons in hidden layers, and 5 output neurons in out layer. We update DQN's weights by using the RMSProp algorithm with a learning rate of 0.001. The max number of training epochs is set to 200, the replay memory set length is set to 2000, the mini batch size is set to 32, and target network update frequency is set to 20. The ϵ value which is used in an ϵ-greedy is set to range from 0.3 to 0.05, and the scale coefficient ζ is set to 3. According to the latest Amazon

---

[3] https://confluence.pegasus.isi.edu/display/pegasus/WorkflowGenerator.

compute SLA[4], service credits will be compensated to users to cover their future bills if the included services do not meet the service commitment. And the service credit percentage for monthly uptime percentage ranged from 99% to 99.5% is 25%, which covers the most of SLA violation cases[3, 10], thus we set the ceiling of the penalty β is a quarter of VM real cost in this paper. The training environment emulator is implemented by Python and the training algorithm is implemented with the help of PyTorch. The training set for each case are from the Pegasus project but the task sizes are randomly generated.

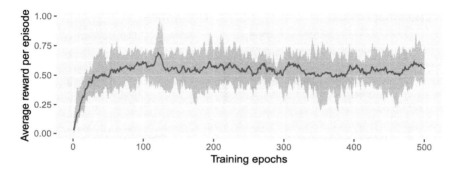

Fig. 7. Average reward got by agent during training.

Figure 7 shows the average reward got during the training in Montage workflow. We can see that the average reward shows its convergence after 100 epochs, which indicates the effectiveness of our designed reward function.

5.2 Makespan and Deadline Hit Rate Evaluation

We first evaluate the makespan and deadline hit rate. Figure 8 shows a makespan comparison between DeepWS and its peers for different workflows under different deadline constraints. The red dot line represents the average deadline of 100 different workflows for each case. It can be clearly seen that the makespan fluctuations of DeepWS is slight, and its Q3 values are always below but near to the average deadline constraint no matter what constraint levels are put, which guarantees good cost-effectiveness while meeting the deadline constraint.

Table 2 shows the deadline hit rates of different algorithms under different configurations. We can observe that our method, i.e., DeepWS, can achieve higher deadline hit rates in most cases. To be specific, our approach gets 82.5%, 43.75%, 72%, and 89.5% higher deadline hit rates than SCS on average; 61.25%, 33.5%, 50.25%, and 53.5% higher than PSO; and 11.25%, 11.75%, 25.25%, and 13.5% higher than IC-PCP. That is because as a reactive method, DeepWS is able to aware of the VM performance fluctuation at runtime and the scheduling decisions are made based on real-time system status. Besides, benefits from

---

[4] https://aws.amazon.com/compute/sla/.

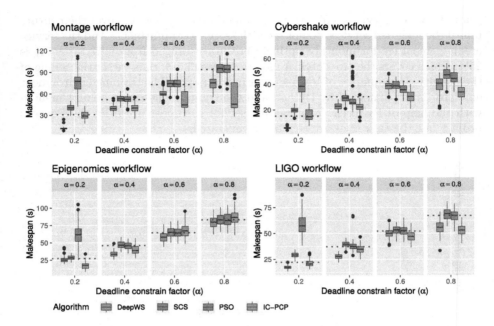

Fig. 8. Makespan comparison. (Color figure online)

Table 2. Deadline hit rates comparison

Algorithms	α	Montage	Cybershake	Epigenomics	LIGO
DeepWS	0.2	88%	100%	79%	100%
	0.4	98%	99%	77%	100%
	0.6	97%	91%	95%	80%
	0.8	100%	100%	100%	100%
SCS	0.2	0%	0%	0%	0%
	0.4	2%	61%	4%	0%
	0.6	10%	72%	12%	6%
	0.8	41%	82%	47%	16%
PSO	0.2	0%	0%	0%	0%
	0.4	47%	80%	45%	54%
	0.6	42%	85%	48%	57%
	0.8	49%	91%	57%	55%
IC-PCP	0.2	59%	51%	32%	72%
	0.4	98%	96%	41%	73%
	0.6	85%	96%	81%	83%
	0.8	96%	100%	96%	98%

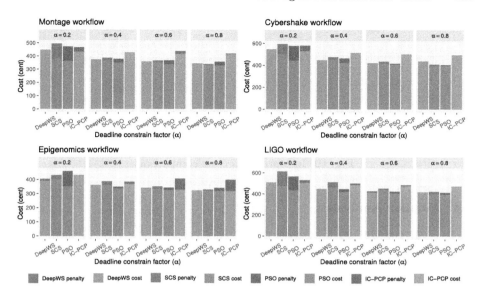

Fig. 9. Cost comparison.

deep-reinforcement-learning with strong decision making capability, it can make smarter scheduling decisions than its peers, especially when deadline constraint is tight.

5.3 Cost Evaluation

We also compare the total cost of DeepWS and its peers. As shown in Fig. 9, the costs incurred by deadline missing (i.e., SLA-violation) penalty are represented as the dark color, and the costs incurred by leasing VMs are represented as light color. If we ignore SLA-violation penalty, as shown in light-colored bars, DeepWS does not show its superiority in cost saving. But the the SLA-violation penalty is taken into consideration, we find that DeepWS can achieve lower cost in most of the cases. More specifically, our approach gets 3.32%, 4.68%, 2.45%, and 6.71% lower cost than SCS in for workflows on average; 3.21%, 4.54%, 2.06%, and 5.84% lower than PSO; and 6.79%, 8.32%, 4.87%, and 6.89% lower than IC-PCP. That is because DeepWS can always make scheduling decisions based on current VM performance, once the deadline is going to miss, it will employ faster but expensive VMs to avoid it. DeepWS beats IC-PCP because IC-PCP is too pessimistic, which represented in that there is a lot of spare time between workflow executions completed and final deadline as shown in Fig. 8. And that time has a great probability to be further transformed into more cost-saving while meeting the deadline constraint if we employ some proper cheaper VMs. The reward function of DeepWS is designed to make the full use of every possible time slot to reduce cost while meeting the deadline, and that makes our method outperforms than its peers.

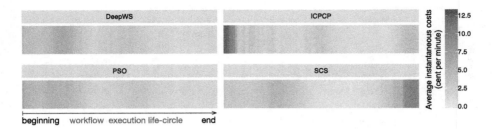

Fig. 10. Instantaneous costs comparison.

5.4 Further Analysis

To find out the difference in detail and further reveal where does the gain come from, we record the execution details of DeepWS and its peers including how many VMs have been leased to finish the workflows, what is the types of the leased VMs, and how much does their resource pools cost instantaneously.

Figure 10 shows the instantaneous costs of DeepWS and its peers when the deadline constraint level is set to medium, i.e., $\alpha = 0.6$. Instantaneous cost is the total cost incurred by VMs and EBS service in a pricing slot (one minute in this paper), which can be used to measure the pace of workflow execution, the integral of instantaneous cost over time come to the total cost. Intuitively, the more expensive VMs in resource pool are hired, the high instantaneous cost will be, and the faster makespan will be obtained. We can clearly see that the execution of schedules generated by PSO shows an excellent stability feature, where there are no sharp instantaneous cost rise or fall. That is because PSO is a kind of meta-heuristic method which generators schedules on the view of the whole workflow, thus its workflow execution pace presents a smooth process. DeepWS performs closely to PSO, there are no obvious fluctuations on instantaneous cost during the whole execution process. On the contrary, obvious changes are found in IC-PCP and SCS methods. IC-PCP performs too pessimistic at the beginning compared with DeepWS and PSO. VMs with high performance but expensive are leased at the beginning in order to meet the deadline, which leads to the high instantaneous cost at the beginning. Later, it realizes that there is still enough time to meet the deadline and thus decides to slow down the pace of execution to reduce cost. In the contrary, SCS performs too optimistic at the beginning which represented in the low instantaneous cost in the early part of its workflow execution process. Then it finds that it is hard to meet the deadline at the current pace of execution, and thus employs expensive VMs with higher performance to speed up the execution, which represented in the high instantaneous cost at the end of the execution process.

Figure 11 shows the proportions of different VM types hired by different methods. It can be clearly seen that the VMs leased by DeepWS and PSO are relatively cheap, mainly *small* type, *medium* type and *large* type, seldom relatively expensive VMs such as *xlarge* and *xxlarge* ones. On the contrary, these

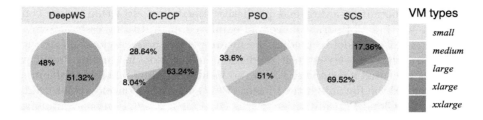

Fig. 11. VM type distribution comparison of rented VMs.

expensive VMs account for a considerable portion of VMs hired by IC-PCP and SCS approaches. Generally, expensive VMs with high performance have low cost-effective due to the fact that the price of VMs and the performance is nonlinear (as shown in Table 1). Therefore, the more expensive VMs are used the poor cost-effective a schedule will be. But employing expensive has the ability to speed up the execution to meet the deadline, which is essential especially when constraints are tight. Thus, the smart use of different types of VMs is the key to reduce total cost while meeting the deadline. Compared with baselines, expensive VMs (i.e., *large*, *xlarge*, and *xxlarge*) leased by DeepWS only takes up 0.68% of its resource pool, which is the lowest proportion among its peers, and that indicates that DeepWS is capable of well controlling the execution pace of workflows to achieve higher cost-effective while meeting the deadline constraints.

6   Related Work

According to the taxonomy by Rodriguez *et al.* [17], existing workflow scheduling algorithms fall into two major categories: static and dynamic (i.e., reactive). Static algorithms usually generate the schedule plans in advance and then schedule workflow step-by-step according to the plan, while the reactive ones see the decision-making for workflow scheduling as an evolvable and refinable process. We take recent static and reactive studies as related work and review them here.

6.1   Static Workflow Scheduling

Static scheduling methods are easy to implement and widely applied in traditional distributed computing systems, grids, cluster-based systems, and cloud-based workflows as well. E.g., Malawski *et al.* [8]. proposed workflow scheduling method for a muti-cloud environment with infinite heterogeneous VMs. They formalized the scheduling problem as an optimization problem aiming at minimization of the total cost under the deadline constraint and developed a Mixed-Integer-Program (MIP) algorithm to solve it. Rodriguez *et al.* [16] proposed Particle-Swarm-Optimization (PSO) based scheduling algorithm. They assumed infinite heterogeneous resources and elastic provisioning and Virtual Machine (VM) performance variation are considered in their system model. Li *et al.* [5]

proposed a fluctuation-aware and predictive scheduling algorithm. They consider time-varying VM performance and employed an Auto-Regressive-Moving-Average (ARMA) model to predict the future performance of cloud infrastructures. Then they employed such prediction information in guiding workflow scheduling and used a genetic algorithm to yield scheduling plans. Abrishami et al. [1] extended the Partial-Critical-Paths (PCP) algorithm, developed for utility grid, into the Cloud-Partial-Critical-Paths (IC-PCP). IC-PCP aims at minimizing execution costs while meeting deadline constraints. Zhu et al. [24] proposed a evolutionary multi-objective workflow scheduling method, named EMS-C. They modeled the scheduling problem aiming at make-span and cost reduction and developed a meta-heuristic algorithm based on NSGA-II to solve it.

To sum up, static methods can generate optimal or near-optimal schedules featured by a global optimization view. However, due to the fact that the schedules are generated in advance and they are unrefinable at run time, they methods suffer from frequent violations to SLA especially when the supporting cloud infrastructures are under high stress, prone to performance fluctuations.

6.2   Reactive Workflow Scheduling

Reactive methods see the scheduling process as a gradual and evolvable one. They consider the decision-making to be reactive to real-time events, e.g., performance fluctuations and losses of resource connectivity. For example, Wu et al. [20] employed a machine-learning-based method to predict the future moving trajectory of mobile users, then they develop a heuristic to achieve a mobility-aware online workflow scheduling. Rodriguez et al. [15] presented an Execution-Time-Minimization-Budget-Driven method, BAGS, for clouds with finer-grained pricing schemes. The proposed method first partition DAGs into bags of tasks (BoTs), then distributes budgets into these BoTs, provisions cloud resources according to the planed budgets, and finally executes the tasks maintained in a scheduling queue in a first-in-last-out manner. Mao et al. [9] proposed a Scaling-Consolidation-Scheduling (SCS) algorithm for workflow scheduling for IaaS clouds. They developed an auto-scaling mechanism that allows VMs to be allocated and deallocated dynamically based on system status. Zhou et al. [23] considered both on-spot and on-demand instances to support the running of workflows. They first developed an A*-based algorithm for assigning a combination of spot and on-demand instances for every task in a workflow, then they proposed a heuristic with the feature of instance consolidation and reusing to refine the schedule at the run time. Poola et al. [13] presented a heuristic, dynamic, and fault-tolerant scheduling approach. They considered both on-demand and on-spot instances in the resource pool, and develop a fault-tolerance-guaranteed heuristic to achieve fault-tolerant by switching on-spot instance to on-demand one when the deadline constraint is likely to be violated. Malawski et al. [7] proposed a Dynamic-Provisioning-Dynamic-Scheduling (DPDS) algorithm, that is capable of dynamically scaling the VM pool and scheduling a group of inter-related workflows under budget and deadline constraints. It creates an initial pool of homogeneous VMs with as many resources as allowed by the budget and updates it at runtime by monitoring resource utilization.

To sum up, reactive methods are able to make or refine scheduling decisions at the run time, they can thus achieve lower SLA-violation rate However, their limited task-level view of the problem restricts the possibility of designing an effective algorithm to get high-quality schedules as static ones. Fortunately, recent great progress made by deep-reinforcement-learning in complex sequential decision making brings us a new opportunity to this problem. In this paper, we propose a deep-reinforcement-learning-based method to get reactive schedules in real-time.

7   Conclusion and Future Study

This paper targets the reactive workflow scheduling problem over the IaaS cloud, where the performance of VMs fluctuates during the execution. We formulate the scheduling problem as an MDP problem and develop a deep-reinforcement-learning-based method, named DeepWS, to solve it. Experiments show that the proposed DeepWS can get lower running costs and SLA-violation rates.

The following issues should be well addressed as future work: 1) Some time-series prediction methods, e.g., ARIMA or LSTM neural networks can be used to predict the performance of VMs to achieve a further improvement; 2) this paper only consider on-demand VM instances as workers, resource pool constructed by both on-demand spot instances will be considered to further reduce cost.

Acknowledgement. This work is supported in part by the Graduate Scientific Research and Innovation Foundation of Chongqing, China (Grant No. CYB20062 and CYS20066), and the Fundamental Research Funds for the Central Universities (China) under Project 2019CDXYJSJ0022.

References

1. Abrishami, S., Naghibzadeh, M., Epema, D.H.: Deadline-constrained workflow scheduling algorithms for infrastructure as a service clouds. Future Gener. Comput. Syst. **29**(1), 158–169 (2013)
2. Belhajjame, K., Faci, N., Maamar, Z., Burégio, V., Soares, E., Barhamgi, M.: On privacy-aware eScience workflows. Computing 1–15 (2020)
3. Christophe, C., et al.: Downtime statistics of current cloud solutions. In: International Working Group on Cloud Computing Resiliency. Technical report (2014)
4. Irwin, D.E., Grit, L.E., Chase, J.S.: Balancing risk and reward in a market-based task service. In: Proceedings of the 13th IEEE International Symposium on High Performance Distributed Computing, pp. 160–169. IEEE (2004)
5. Li, W., Xia, Y., Zhou, M., Sun, X., Zhu, Q.: Fluctuation-aware and predictive workflow scheduling in cost-effective infrastructure-as-a-service clouds. IEEE Access (2018)
6. Li, X., Yu, W., Ruiz, R., Zhu, J.: Energy-aware cloud workflow applications scheduling with geo-distributed data. IEEE Trans. Serv. Comput. (2020)

7. Malawski, M., Juve, G., Deelman, E., Nabrzyski, J.: Algorithms for cost- and deadline-constrained provisioning for scientific workflow ensembles in IaaS clouds. In: International Conference for High Performance Computing, Networking, Storage and Analysis, pp. 1–11 (2012)
8. Malawski, M., Figiela, K., Bubak, M., Deelman, E., Nabrzyski, J.: Scheduling multilevel deadline-constrained scientific workflows on clouds based on cost optimization. Sci. Program. **2015**, 5 (2015)
9. Mao, M., Humphrey, M.: Auto-scaling to minimize cost and meet application deadlines in cloud workflows. In: 2011 International Conference for High Performance Computing, Networking, Storage and Analysis (SC), pp. 1–12. IEEE (2011)
10. Maurice, G., et al.: Downtime statistics of current cloud solutions. In: International Working Group on Cloud Computing Resiliency. Technical report (2012)
11. Mnih, V., et al.: Human-level control through deep reinforcement learning. Nature **518**(7540), 529 (2015)
12. Papadimitriou, C.H., Tsitsiklis, J.N.: The complexity of Markov decision processes. Math. Oper. Res. **12**(3), 441–450 (1987)
13. Poola, D., Ramamohanarao, K., Buyya, R.: Fault-tolerant workflow scheduling using spot instances on clouds. Procedia Comput. Sci. **29**, 523–533 (2014)
14. Rana, O.F., Warnier, M., Quillinan, T.B., Brazier, F., Cojocarasu, D.: Managing violations in service level agreements. In: Rana, O.F., Warnier, M., Quillinan, T.B., Brazier, F., Cojocarasu, D. (eds.) Grid Middleware and Services, pp. 349–358. Springer, Boston (2008). https://doi.org/10.1007/978-0-387-78446-5_23
15. Rodriguez, M.A., Buyya, R.: Budget-driven scheduling of scientific workflows in IaaS clouds with fine-grained billing periods. ACM Trans. Auton. Adapt. Syst. (TAAS) **12**(2), 5 (2017)
16. Rodriguez, M.A., Buyya, R.: Deadline based resource provisioning and scheduling algorithm for scientific workflows on clouds. IEEE Trans. Cloud Comput. **2**(2), 222–235 (2014)
17. Rodriguez, M.A., Buyya, R.: A taxonomy and survey on scheduling algorithms for scientific workflows in IaaS cloud computing environments. Concurr. Comput. Pract. Exp. **29**(8), e4041 (2017)
18. Schad, J., Dittrich, J., Quiané-Ruiz, J.A.: Runtime measurements in the cloud: observing, analyzing, and reducing variance. Proc. VLDB Endow. **3**(1–2), 460–471 (2010)
19. Vinyals, O., et al.: StarCraft II: a new challenge for reinforcement learning. arXiv preprint arXiv:1708.04782 (2017)
20. Wu, C., Peng, Q., Xia, Y., Lee, J.: Mobility-aware tasks offloading in mobile edge computing environment. In: 2019 Seventh International Symposium on Computing and Networking (CANDAR), pp. 204–210. IEEE (2019)
21. Wu, Q., Zhou, M., Zhu, Q., Xia, Y., Wen, J.: MOELS: multiobjective evolutionary list scheduling for cloud workflows. IEEE Trans. Autom. Sci. Eng. **17**(1), 166–176 (2019)
22. Yeo, C.S., Buyya, R.: Service level agreement based allocation of cluster resources: handling penalty to enhance utility. In: IEEE International Cluster Computing, pp. 1–10. IEEE (2005)
23. Zhou, A.C., He, B., Liu, C.: Monetary cost optimizations for hosting workflow-as-a-service in IaaS clouds. IEEE Trans. Cloud Comput. **4**(1), 34–48 (2016)
24. Zhu, Z., Zhang, G., Li, M., Liu, X.: Evolutionary multi-objective workflow scheduling in cloud. IEEE Trans. Parallel Distrib. Syst. **27**(5), 1344–1357 (2016)

BPA: The Optimal Placement of Interdependent VNFs in Many-Core System

Youbing Zhong[1,2,3], Zhou Zhou[1,2,3]([✉]), Xuan Liu[4], Da Li[5], Meijun Guo[6], Shuai Zhang[1,2,3], Qingyun Liu[1,2,3], and Li Guo[1,2,3]

[1] Institute of Information Engineering, Chinese Academy of Sciences, Beijing, China
zhouzhou@iie.ac.cn
[2] School of Cyber Security, University of Chinese Academy of Sciences, Beijing, China
[3] National Engineering Laboratory of Information Security Technologies, Beijing, China
[4] School of Information Engineering, Yangzhou University, Yangzhou, China
[5] Department of Electrical and Computer Engineering, University of Missouri, Columbia, USA
[6] School of Mathematical Sciences, University of Chinese Academy of Sciences, Beijing, China

Abstract. Network function virtualization (NFV) brings the potential to provide the flexible implementation of network functions and reduce overall hardware cost by running service function chains (SFCs) on commercial off-the-shelf servers with many-core processors. Towards this direction, both academia and industry have spent vast amounts of effort to address the optimal placement challenges of NFV middleboxes. Most of the servers usually are equipped with Intel X86 processors, which adopt Non-Uniform Memory Access (NUMA) architecture. However, existing solutions for placing SFCs in one server either ignore the impact of hardware architecture or overlook the dependency between middleboxes. Our empirical analysis shows that the placement of virtual network functions (VNFs) with interdependency in a server needs more particular consideration. In this paper, we first manage the optimal placement of VNFs by jointly considering the discrepancy of cores in different NUMA nodes and interdependency between network functions (NFs), and formulate the optimization problem as a Non-Linear Integer Programming (NLIP) model. Then we find a reasonable metric to describe the dependency relation formally. Finally, we propose a heuristic-based backtracking placement algorithm (BPA) to find the near-optimal placement solution. The evaluation shows that, compared with two state-of-art placement strategies, our algorithm can improve the aggregate performance by an average of 20% or 45% within an acceptable time range.

Keywords: NFV · SFCs plaecment · Interdependent NFs · NUMA

© ICST Institute for Computer Sciences, Social Informatics and Telecommunications Engineering 2021
Published by Springer Nature Switzerland AG 2021. All Rights Reserved
H. Gao et al. (Eds.): CollaborateCom 2020, LNICST 350, pp. 305–319, 2021.
https://doi.org/10.1007/978-3-030-67540-0_18

1   Introduction

The emerging of Network Function Virtualization (NFV) has been an innovative technology to network architecture in recent years [1]. It is devoted to addressing the limitations of traditional middleboxes [2]. Unlike the tradition network architecture hosted on dedicated physical equipments or customized hardware, NFV runs network functions (NFs) using the software on the commodity servers with general processors, such as Intel X86, by leveraging underlying virtualization technology. What's more, NFV brings substantial flexibility for the distribution of network service and cost reduction. Telecom operators can deploy NFs according to requirements dynamically. A network service is usually composed of a series of NFs, which are often referred to as Service Function Chain (SFC). The traffic generated by request traverses the NFs one by one.

To achieve desired performance for NFV, the commodity servers to host Virtualized Network Functions (VNFs) are equipped with high-performance processors with many cores, which we refer to as many-core systems [3]. This kind of powerful server usually has enough capacity to accommodate an entire SFC or even multiple SFCs completely [4].

In the current stage, most of the studies focus on the placement of SFCs across multiple physical servers [5,6]. Almost all of the researches treat each server as an entire node. That's to say, all cores in the server are equal in the allocation of various NFs. However, CPU with many cores usually adopts Non-Uniform Memory Access (NUMA) architecture. For instance, most of Intel x86 CPU cores are partitioned into several nodes. Each node contains multiple cores and its own local memory. The performance of assigning CPU cores to host SFC under cross-node and Intra-node cases is significantly different. One of the main factors is that local memory access is much faster than remote access through Intel QuickPath Interconnect (QPI).

Furthermore, the placement of SFCs is also constrained by the dependency relation that may or may not exist between NFs [2]. For instance, there may or may not exist dependency between middleboxs. For instance, the IPSec decryptor is usually deployed before a NAT gateway [7], while the VPN proxycan be deployed either before or after a firewall [8]. Besides, NFs are usually heterogeneous and diverse. Different NFs have different computation resource requirements. The dependency relation of different types of middleboxes further complicates the placement of SFCs. We refer the NFs with interdependency as NF couple (two NFs).

We use the following example to illustrate the dependency relation effects for SFCs placement problem. In this paper, we treat all CPU cores in one NUMA node as equal. Consider a SFC including six NFs co-locating on two nodes as shown in Fig. 1(a). Here, we first consider this chain with computation cost: NF1 (100 units), NF2 (300 units), NF3 (300 units), NF4 (200 units), NF5 (100 units), NF6 (200 units). In this paper, we assume that the flow path is predetermined(NF1 $\rightarrow$ NF2 $\rightarrow$ NF3 $\rightarrow$ NF4 $\rightarrow$ NF5 $\rightarrow$ NF6). Namely, there exists dependency between each other, and all of NFs are totally-ordered. Each node has 600 computation units. For separated NFs, this placement makes use of the most available resources on the nodes. NF1, NF2, and NF4 share LLC and local memory in the local node and NF3, NF5, NF6 can be considered as sharing QPI with the remote node (Fig. 1(b)). However, if there exists a sequential order between NF2 and NF3, namely flow traversing through NF2 before NF3, cross-node chaining between these two NFs can change the directly available resource subsequent NFs can get. We can see that there almost no LLC benefit and CPU cores have to pass through QPI to get data from remote memory for three times, which could incur performance drop dramatically (Fig. 1(c) red line). Nevertheless, if we place these two interdependent NFs in the same remote node, it can make subsequent NFs (i.e., NF5) get more high chance of cache hits and benefit the performance improvement from local packet memory access (Fig. 1 (d)). Moreover, the flow with the same sequence only needs to traverse the QPI for only twice.

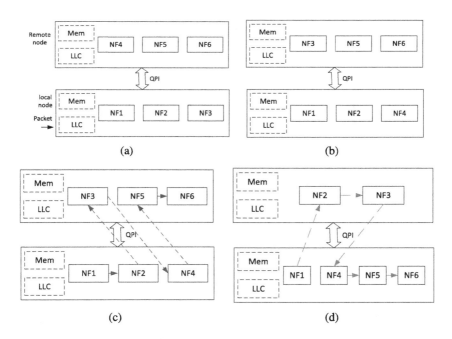

Fig. 1. Different placement for the same SFC. (Color figure online)

Therefore, we can see that more specific considerations are required in the placement of SFCs in the many-core system due to the dependency relation between NFs. The above observations motivate us to design an optimal strategy for the problem to achieve maximized aggregate throughput of all SFCs.

In this paper, we study the optimal placement of SFCs within a server. Compared with existing work in Sect. 2, we propose comprehensive solutions that consider the inequality of CPU cores in different nodes as well as different types of NFs dependency relation. To understand the key factors in the optimization problem, we first formulate the problem with a Non-Linear Integer Programming (NLIP) model. Due to the NP-hardness of our problem, we then propose a heuristic-based backtracking placement algorithm to find an optimal or near-optimal solution in the range of reasonable time. Our design philosophy is to prevent existing NF couples from being placed on separated nodes as possible to avoid remote memory access and cache miss, but instead, let them be an entirety to be placed in the same node.

The remaining of this paper is organized as follows. Section 2 provides a brief overview of related work. We formulate the Interdependent VNFs Placement in Many-core System (IVPMS) problem in Sect. 3. Section 4 shows our proposed algorithm. The experiments and result are presented in Sect. 5. Section 6 gives a conclusion for our work.

2 Related Work

This section summarizes the state-of-art research on the placement of SFC in NFV.

As the promising domain for the next generation network infrastructure, NFV has attracted vast significant research attention. The institutions and enterprises are spending a huge amount of effort in this field. Most of the studies devoted to NFs and SFC placement in NFV. To the best of our knowledge, we are the first to formally study the SFC placement problem by combining the dependency relation between NFs among SFC with the characteristic NUMA architecture. We study and model performance affection by interdependency between NFs in many-core systems thoroughly and proposed an efficient algorithm to solve this problem.

There have been amounts of researches on NFs or SFCs placement in NFV. Rami et al. presented bi-criteria solutions for the actual placement of the virtual functions within the physical network [9]. It can reach constant approximation with respect to the overall performance and adhere to the capacity constraints of the network infrastructure. Defang et al. formulated the problem as an Integer Linear Programming (ILP) model and designed a two-stage heuristic solution to minimize the number of used physical machines [5]. [10] jointly considered the VNF placement and path section to utilize the network better. However, all the existing works above optimized SFC placement by mapping VNFs to the right servers. The server in this models is formulated as an entire node. They treat all CPU cores as equal without considering NUMA architecture in many-core systems. Thus, consideration without this factor will not be suitable and performance degradation will be incurred in actual situations.

Meanwhile, to make up for the insufficiency, many research efforts have been devoted to addressing the placement of SFC in a many-core system to achieve optimal execution performance. Zhilong et al. considered the inequality of CPU cores in respective NUMA node and proposed an NFV orchestrator to achieve maximum aggregate throughput of all SFCs in many-core systems [11]. Nevertheless, it overlooked the relationship of NFs among SFC. [12–15] relied on frequent migration of threads to optimize thread-to-core placement to maximize task execution performance. However, comparing with threads, migrating NF progresses across cores or nodes brings significant performance overhead [16]. Besides, the heterogeneity of NFs could worsen the problem in NFV context and each running status of SFC contains a amout of flow information, resulting to the impossibility of migration. Moreover, to achieve high performance, advanced NFV systems bond NFs on dedicated CPU cores that cannot be scheduled by the operating system. This makes it difficult to migrate NFs among cores.

At the same time, it has been recognized as a challenge to place SFCs with certain dependency constraints between NFs. Wenrui et al. formulated the interdependent middleboxes placement as a graph optimization and proposed an efficient heuristic algorithm for the general scenario of a partially-ordered middlebox set [17]. [18] considered the dependency between NFs and presented a heuristic method to coordinate the composition of VNF chains and their embedding into the substrate network. Chaima et al. organized the VNFs placement in smaller interdependent subproblems and designed an efficient dynamic programming algorithm that runs in polynomial time [19]. Although considering the dependency within NFs, they did not in view of the fact that the deployment of same SFC on different core sets could result in significantly different throughput.

Different from the above ones, the formal model defined in this paper considers not only the NUMA architecture in modern CPU but also different types of NF dependency relations. Compared with the extensive studies on general SFC placement, we propose comprehensive solutions and conduct evaluations on hardware and software architectures.

3 Problem Formulation and Analysis

In this section, we present the system model and mathematical formulation of the IVPMS problem. We aim to find an optimized placement scheme for the SFCs, including NF couples in a many-core system.

The placement of SFCs is usually described as the assignment of a sequence of VNF instances to substrate nodes. Each node has limited available resources to host NF instances. To analyze the problem quantitatively and identify the key factors for problem-solving, we formulate the problem with the NLIP model and find a metric to measure the NF couple's effect formally.

3.1 Formulation of the Optimal Placement Problem

Formally, for many-core systems on the market, there are multiple nodes with an incremental number. Each node contains an identical number of CPU cores. We assume that there is a set of nodes V with C cores for each in a many-core system. A set of SFCs is denoted as S. For each SFC $s \in S$, let f_i^s be the ith VNF, $1 \le |i| \le |s|$. Note that different SFCs may share the same VNF in practice. However, the path that traffic traverses can be considered separately. Moreover, flows tend not to arrive at the same time exactly, and multiple flows are processed one by one [5]. Thus, we can overlook the impact and specify that chains in S are disjoint in our model.

We use $\leftarrow$ to describe the dependency relation between NFs. It is defined as a strict partial order. For $f_i, f_j \in s$, if $f_i \leftarrow f_j$, we say that the VNF f_j depends on VNF f_i. That's to say f_j should be chained after f_i. If $f \in s$ depends on no others, i.e., $\forall f' \in s, f' \nleftarrow f$, we say that f has no interdependent VNF. Besides, the dependency relation has transitivity. If $f_i \leftarrow f_j$ and $f_j \leftarrow f_l$, then $f_i \leftarrow f_l$. We can see that a NF may has multiple interdependent NFs. For easy representation, we define $r(i, j) = 1$ if $f_i \leftarrow f_j$, and 0 otherwise.

In addition, we present a placement scheme, denoted as $P(s, i, k)$, to indicate whether NF i of chain s is located at the node k. It is a decisive binary variable defined as:

$$p(s, i, k) = \begin{cases} 1 \; if \; the \; i^{th} \; NF \; of \; chain \; s \; placed \; on \; node \; k \\ 0 \; otherwise \end{cases} \tag{1}$$

When all the NFs belonging to the same SFC s co-locate on the same node, the performance they achieve by sharing LLC and local memory access is the most optimal performance, which we set it as ϕ_s^{best}. We assume that the effect of contention can be ignored if the node has enough capacity to host all the NFs since NFs are stationed on specified cores. As involved in some work [4,10], the optimal performance can be measured by placing an entire SFC on the same node. Furthermore, when NFs included in the same SFC are co-located across nodes, the performance is referred to as attenuation performance and is defined as ϕ_s, of which $s \in S$. It's obvious that the attenuation performance is a referred value of optimal performance. We can see that the difference between the above two cases is cross-node placement.

The cross-node interdependent NFs flow traverses through are the vital key factor that affects the performance. Therefore, we introduce correlation index (denoted by $\varphi(i, j)$), which represents the effect of cross-node NF couple(i.e., (i, j)) to relate the optimal performances and attenuation performance. It should be noted that the objective is to maximize the aggregate performance across all required SFCs when deploying. Hence, we can formulate it as:

$$Max \sum_{s \in S} \phi_s \tag{2}$$

where:

$$\phi_s = \prod_{i,j \in s} \varphi(i, j) \phi_s^{best} \quad \forall i, j \in s, s \in S, \; r(i, j) = 1 \tag{3}$$

subject to the following constraints:

$$\left(\sum_{k \in V} P(s, j, k) * k - \sum_{k \in V} P(s, i, k) * k \right) * r(i, j) \geq 0 \quad if \; i \leftarrow j \tag{4}$$

$$\sum_{s \in S, i \in s} P(s, i, k) \leq C \quad \forall k \in V \tag{5}$$

$$\sum_{k \in V} P(s, i, k) = 1 \tag{6}$$

A brief explanation of this model is as follows. Equation (2) defines the optimization objective. Equation (3) describes the relationship between optimal performance, attenuation performance, and the effect of cross-node interdependent NFs. Constraints (4) enforces the dependency relation between NFs. In other words, flow must traverses i no later than j if the former is depended on by the latter. Equation (5) states that the resource consumed should not exceed the capacity of a many-core system during deployment. Equation (6) specifies that an NF should be deployed once and only once.

3.2   Formulation Analysis

The formulation of this problem gives us an intuitive description for finding optimal placement for S. However, we can see that the critical parameter, namely $\varphi(i, j)$ need to be figured out for problem-solving. Nevertheless, computing the correlation index is not-trivial due to some reasons.

As mentioned in Sect. 1, an SFC may contain multiple pairs of NF couples. If the NF couple co-locates in the same node, then there does not exist cross-node remote memory access through QPI, and the effect can be ignored (i.e., $\varphi(i, j) = 1$). Howeveer, for the estimation of placement, the prior knowledge we have are only the types and the chained sequence before acutal deployment. It's troublesome to adopt this information directly for the computation of the correlation index since they are not quantifiable. Of course, we can refer to some work [20] and manually analyze the system performance metric value to quantify the NF couple. However, metric value calculation could be platform depedent and burdensome.

Furthermore, this problem could be worse in the NFV context because NFs are usually heterogeneous and diverse, making the computation of $\varphi(i, j)$ are onerous. Moreover, there exists the contention of QPI resource if multiple pairs of interdependent NFs in the same SFC or among different SFCs adopt the scheme of cross-node placement. Besides, the chaining order for same SFC can change the count of the dependent NFs pair.

Consequently, we can see that more specific considerations are required for placing SFCs containing NF couples in many-core system due to above analysis.

Obviously, to address above problems, we should find some performance metrics to describe an NF and then describe the NF couple indirectly. Fortunately, there are many potential system-level metrics can be adopted [14, 21]. Therefore, we should determine which one and automatically detect the appropriate metrics by existing tools. To achieve this goal, we first need to qualitatively analyze the characteristic these metrics should present.

Since the metrics should be able to describe an NF properly, its value should not vary with a large margin when located in different node. Besides, the metric value should be sensitive when NF change. That's to say, different NFs or same NF with different intrinsic properties, for instance, implementation or program languages, should have different metric values. Therefore, the metric values shoud be different when the performance are different for a set of NFs. In another word, the performance and metric value for an NF should exist strong correlation when NF run solely.

To figuratively express this correlation relationship, we assume P_i and m_i denote the ideal performance and metric value of NF i. Following this work [15,21], there exist linear correlation coefficient between the performance and metric value. Here we adopt mathematical Pearson Correlation Coefficient [22] to compute the correlation.

$$\rho\left(P_i, m_i\right) = \frac{Cov\left(P_i, m_i\right)}{\sigma_{P_i}\sigma_{m_i}} \tag{7}$$

$Cov\left(P_i, m_i\right)$ is the covariance for NF i between performance P_i and metric $m_[i$. σ_{P_i} and σ_{m_i} correspond to the varianes of performance and metric.

Then, we can deduce that the function of $\varphi\left(i, j\right)$ should be consistent with following function in form.

$$\varphi\left(i, j\right) = \frac{\alpha_i m_i \alpha_j m_j}{\alpha_i' m_i' \alpha_j' m_j'} \tag{8}$$

where $\alpha_i m_i$ and $\alpha_j m_j$ are the performance of NF couple i and j when they are located in the same node. α_i and α_j are the coefficient. Similarly, the denominator are the performance of NF couple i and j when they are placed across node. Besides, if assigning the processing order of the flow by NFs in advance, we can determine the sequence and the amount of NF couple. In simpler terms, we can get knowledge of the dependency relation inside a SFC.

As mentioned above, we cannot use limited existing knowledge as the input directly to expound the effect of interdependent NFs to IVPMS problem and the parameter is not known. Instead, we need to find some performance metrics to represent the effect of NF couple's cross-node placement formally. Undoubtedly, compared with local nodes, remote access through QPI in many-core system is the one of the main factors. Hence, we focus on the difference between local access and remote access across nodes. Remote access could cause LLC misses, QPI contention and so on. Single metric cannot profile all of factors clearly. Nevertheless, we find that the metric $resource_{stalls}$ profiled by OProfile [23] provides aggregate embodiment with the help of other research findings [11]. This metric including multiple sub indicators is appropriate to profile the aggregate performance. Consequently, we measure the metric of downstream NF among NF couples under above two cases. Then the function (i.e., $\varphi\left(i, j\right)$) maps the metric value of NFs to attention performance (i.e., ϕ_s).

To explore this function, we first check the relationship between the metric and performance. With more interdependent NF couples placed across nodes, $resource_{stalls}$ increases gradually and performance decreases, which shows negative correlation to remote access. Moreover, the function (i.e., $\varphi(i,j)$) shows positive correlation to local access if they are co-located in the same node. From this observation, we can approximately formulate this functions as:

$$\varphi(i,j) \approx \frac{resource\_stalls_{local}}{resource\_stalls_{remote}} \qquad (9)$$

$$f(\varphi(1,2),\ldots\varphi(i,j)\ldots) = \alpha \prod_{i,j \in s} \varphi(i,j) + \beta \qquad (10)$$

$$s.t.\ \alpha + \beta = 1 \qquad (11)$$

Where α and β are tuning parameters to the aggregate effect of cross-node placement of interdependent NF couples to the SFC (the metric where NF couples placed in the local node is denoted as $resource\_stalls_{local}$ and other are relatively $resource\_stalls_{remote}$). Equation (10) states that if all of NFs (including NF couples) are co-located in the same node, the result of continued multiplication is equal to one, namely no effect to performance. Since multiple SFCs could be placed in all nodes, there should be distinct parameter groups due to the heterogeneity of chains. Thus, we define (α_s, β_s) for the case that chain s is deployed in the nodes.

We can see that if measuring samples are given, including the predetermined path of SFCs, accompanying the number of NF couple and the correlation index of each NF couple, it is easy to approximate the parameters (α, β) for each SFC. However, because of following reasons, we do not need to figure out them or find some mathematical method to approximate them. Instead, we propose a more efficient algorithm based on above analysis. One time-consuming but definite approach is to exhaust all possible schemes.

4   Backtracking Placement Algorithm

So far, we have constructed the concrete function formulation for correlation index between optimal performance and attenuation performance, which make IVPMS problem resolvable. However, it is not suitable for practical deployment. As can be seen, its time complexity is $O(K^C)$ (K is the number of available nodes and C is the number of total NFs), which is not satisfied the floor threshold of response time for operator. Furthermore, the relaxed model of IVPMS problem is NP-hard [17]. Therefore, there is no polynomial-time approximation algorithm to find an optimal placement. A alternative approach to finding optimal solution is brute-force search. Although it must can find the optimal placement, the time complexity is same with abvoce, which cannot tolerable for practical application. Thus, we propose a heuristic-based backtracking placement algorithm (BPA), to find the optimal or near-optimal solution.

The basic idea behind our heuristic algorithm is to place all of NF couples in the same node as possible while minimizing the number of remote access across nodes. Instead of finding placement for each NF, we first treat all of NFs in the same SFC as entirety and place an entire chain on the same node. It is based on an important empirical observation that the performance will drop dramatically as the number of cross-node placement event in the chain increase. (This is because cross-node access bring the augment of $resouce\_stalls_{remote}$, minishing the value of $\varphi(i,j)$). Then, we move NFs selectively to remote node due to resource constraints. The selection of NFs following two priority principle: (1) search NFs that have nothing to do with position in chain and remove them to remote nodes. If this NFs have no interdependent relation with other NFs inside chain, they can be placed in an arbitrary order; (2) remove NF couples at the end of chain to remote nodes. After removing NFs without dependency to remote node, the remains are completely-dependent and totally-ordered. If the local node cannot hold the rest of NFs yet, we still need to remove subchain in the tail of SFC from local node to remote node. The split point in the rest chain to form a subchain is the NF among NF couples with least performance degradation, denoted as LPD-NF (i.e., $\varphi(i,j)$ with the highest value).

The pseudo-code of placement algorithm is shown in Algorithm 1. It first places all of SFCs in the same local node without considering the resource limitation. In the following steps, the main work is to remove the out-of-limit NFs to remote node. In such a case, we choose NFs based on above two principles. We search the ones that no dependency exists between NFs and remove them to remote node in priority. Furthermore, we iterate each SFC to move subchain from local node to remote node. We adopt deep-first search (DFS) to find the LPD-NF for each chain and update the potential optimal placement, meanwhile examine whether the capacity of local node could cover the requirement. Once reaching the lower bound, we will terminate this iteration. Finally, we will find the potential optimal solution with high performance in all possible placements.

5   Performance Evaluation

In this paper, we use a high performance NFV platform, OpenNetVM [4] to run NFs and SFCs. We select several type of NFs for evaluation, including IPv4 Router (Rout), Firewall (FW), deep packet inspection (DPI) and Switch (SW). We run our algorithm and OpenNetVM on the same server. The computer is a two-NUMA-node server, which is equipped with Intel(R) Xeon(R) Silver 4114 CPU, 256G RAM and two dual-port DPDK compatible NICs located on node 0. The operation system server runs is Centos 7.2 with kernel version 3.10.0-693.el7.x86 64. The input traffic is generated by the DPDK Pktgen tool [24].

Before proceeding with our evaluation, we need to measure the metric value of each NF. To facilitate the computation of $\varphi(i,j)$ and speed up the search of split point, we assume that there is dependency relation between each other. To demonstrate the correlation index, we measure the metric value under two different deployment situation, namely co-locating placement or cross-node placement. We measure the metric for many times and finally get the average result.

Algorithm 1: Backtracking placement algorithm

Input: S:SFC sets, C: the number of cores in a node
Output: O: Placement of each NF in the SFCs

1 //init placement
2 $O_{init} \leftarrow place\ all\ of\ SFCs\ on\ the\ local\ node$
3 $Max_{perf} \leftarrow 0$ //Optimal performance
4 Sort all of SFCs based on ϕ_s^{best} in ascending order
5 **while** $|s_{local}| \geq C$ **do**
6      $(s_{local}, s_{remote}) \leftarrow initialization$ // $|s_{remote}| = 0$
7      $r_c = C$
8   **foreach** $s \in S$ **do**
9         $NF_s \leftarrow find\ NFs\ that\ without\ dependency$
10   **if** $|NF_s| \leq r_c$ **then**
11           $(s_{local}, s_{remote}) \leftarrow (s_{local} - NF_s, s_{remote} + NF_s)$
12           $r_c \leftarrow r_c - |NF_s|$
13   **if** $|s_{local}| \leq C$ **then**
14   Set flag
15   UpdatePlacement(Max_{perf}, s, NF_s)
16   **if** *flag* **then**
17   break
18   DFS($s_{local}, s_{remote}, S, r_c, Max_{perf}$)
19 **Function** DFS($s_{local}, s_{remote}, S, r_c, perf$):
20   **foreach** $s \in S$ **do**
21        $LPD\_NF \leftarrow find\ split\ point\ with\ max\varphi\,(i,j)$
22        $s_{subchain} \leftarrow (LPD\_NF : NF_{tail})$
23   **if** $r_c > |s_{subchain}|$ **then**
24           $(s_{local}, s_{remote}) \leftarrow (s_{local} - s_{subchain}, s_{remote} + s_{subchain})$
25           $r_c \leftarrow r_c - |s_{sunchain}|$
26   UpdatePlacement($perf, s, s_{subchain}$)
27   **if** $|s_{local}| \leq C$ **or** $|s| = 0$ **then**
28   **continue**
29   DFS($s_{local}, s_{remote}, S, r_c, perf$)
30 **Function** UpdatePlacement($perf, s, NF_s$):
31   **if** $perf < Perf_{measured}$ **then**
32        $perf \leftarrow Perf_{measured}$
33        $s \leftarrow s - NF_s$

Figure 2(a) shows the average metric value for each type of NF. We can see that when multiple NFs are co-located in the same node, the metric value varies with small margin. Figure 2(b) shows the average correlation index for different NF couple. By this way, we can reduce the search time enormously.

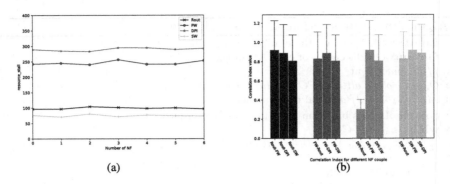

Fig. 2. Metric value for NFs and NF couples.

To demonstrate that our heuristic algorithm can improve the aggregate performance by finding optimal or near-optimal placement solution, we also implement two placement strategies: (1) greedy placement(GP): it co-locates the NFs on the local node in priority until exhaust the cores, then places the remains on the other nodes; and (2) fairness placement(FP); it attempts to place SFCs on the all nodes in sequence. According to [25,26], the length of currently deployed service chains do not exceed seven. Thus, the number of VNFs in a chain can range from 2 to 7. In this paper, we use 5 sets of SFC. Each set includes 3 SFCs, which consist of above NFs.

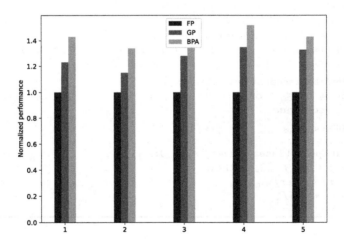

Fig. 3. Aggregate performance for three placement schemes under different number of SFC sets.

Figure 3 shows the normalized performance of above three placement algorithms. We can see that in all SFC sets, our algorithm can achieve the best aggregate performance among three placement strategies. Comparing with greedy

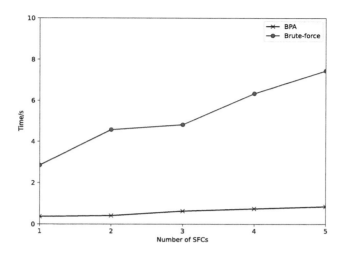

Fig. 4. Time cost under different numbers SFC sets.

placement, BPA improves the aggregate performance by an average of about 46%. Also, BPA improves the aggregate performance by an average of about 20% compared with fairness placement.

We also evaluation the efficiency of our algorithm. We compare the backtracking placement algorithm with the brute-force method. We randomly select two sets from above SFC sets as input and increase the number of placed SFC gradually. We repeat the experiment for multiple times and record the cost time. Figure 4 shows that our algorithm can find a solution less than two second, even in the worst case with five SFCs. Comparing with brute-force search strategy, our algorithm can reduce the time to find the near-optimal placement tremendously.

6 Conclusion

In this paper, we study the SFCs placement in a many-core system to improve the aggregate performance. In order to address the placement problem, we formulate it with a NILP model. Besides, to reveal the relation between optimal performance and attention performance, we find a reasonable metric to weigh the effectiveness of NF couple systematically and describe the relation with formulation. Furthermore, we present a heuristic-based backtracking placement algorithm to search optimal or near-optimal placement for SFC in a many-core system. Our evaluation built on OpenNetVM shows that BPA can improve the aggregate performance significantly and has high efficiency to find optimal or near-optimal solutions.

References

1. Han, B., Gopalakrishnan, V., Ji, L., Lee, S.: Network function virtualization: challenges and opportunities for innovations. IEEE Commun. Mag. **53**(2), 90–97 (2015)
2. Mehraghdam, S., Keller, M., Karl, H.: Specifying and placing chains of virtual network functions. In: 2014 IEEE 3rd International Conference on Cloud Networking (CloudNet), pp. 7–13. IEEE (2014)
3. Palkar, S., et al.: E2: a framework for NFV applications. In: Proceedings of the 25th Symposium on Operating Systems Principles, pp. 121–136. ACM (2015)
4. Zhang, W., et al.: OpenNetVM: a platform for high performance network service chains. In: Proceedings of the 2016 Workshop on Hot Topics in Middleboxes and Network Function Virtualization, pp. 26–31. ACM (2016)
5. Li, D., Hong, P., Xue, K., et al.: Virtual network function placement considering resource optimization and SFC requests in cloud datacenter. IEEE Trans. Parallel Distrib. Syst. **29**(7), 1664–1677 (2018)
6. Sefraoui, O., Aissaoui, M., Eleuldj, M.: OpenStack: toward an open-source solution for cloud computing. Int. J. Comput. Appl. **55**(3), 38–42 (2012)
7. Cisco: Cisco: Nat order of operation (2014). http://www.cisco.com/c/en/us/support/docs/ip/network-address-translation-nat/6209-5.html
8. TechNet, M.: Microsoft technet: VPNs and firewalls (2015). https://technet.microsoft.com/en-us/library/cc958037.aspx
9. Cohen, R., Lewin-Eytan, L., Naor, J.S., Raz, D.: Near optimal placement of virtual network functions. In: 2015 IEEE Conference on Computer Communications (INFOCOM), pp. 1346–1354. IEEE (2015)
10. Kuo, T.W., Liou, B.H., Lin, K.C.J., Tsai, M.J.: Deploying chains of virtual network functions: on the relation between link and server usage. IEEE/ACM Trans. Netw. (TON) **26**(4), 1562–1576 (2018)
11. Zheng, Z., et al.: Octans: optimal placement of service function chains in many-core systems. In: IEEE Conference on Computer Communications, IEEE INFOCOM 2019, pp. 307–315. IEEE (2019)
12. Zhuravlev, S., Blagodurov, S., Fedorova, A.: Addressing shared resource contention in multicore processors via scheduling. ACM Sigplan Not. **45**, 129–142 (2010)
13. Lepers, B., Quema, V., Fedorova, A.: Thread and memory placement on NUMA systems: asymmetry matters, pp. 277–289 (2015)
14. Rao, J., Wang, K., Zhou, X., Xu, C.: Optimizing virtual machine scheduling in NUMA multicore systems, pp. 306–317 (2013)
15. Liu, M., Li, T.: Optimizing virtual machine consolidation performance on NUMA server architecture for cloud workloads. In: International Symposium on Computer Architecture, vol. 42, no. 3, pp. 325–336 (2014)
16. Gemberjacobson, A., et al.: OpenNF: enabling innovation in network function control. ACM Spec. Interest Group Data Commun. **44**(4), 163–174 (2015)
17. Ma, W., Sandoval, O., Beltran, J., Pan, D., Pissinou, N.: Traffic aware placement of interdependent NFV middleboxes. In: IEEE Conference on Computer Communications, IEEE INFOCOM 2017, pp. 1–9. IEEE (2017)
18. Beck, M.T., Botero, J.F.: Coordinated allocation of service function chains, pp. 1–6 (2014)
19. Ghribi, C., Mechtri, M., Zeghlache, D.: A dynamic programming algorithm for joint VNF placement and chaining. In: Proceedings of the 2016 ACM Workshop on Cloud-Assisted Networking, pp. 19–24. ACM (2016)

20. Dobrescu, M., Argyraki, K., Ratnasamy, S.: Toward predictable performance in software packet-processing platforms, p. 11 (2012)
21. Nathuji, R., Kansal, A., Ghaffarkhah, A.: Q-clouds: managing performance interference effects for QoS-aware clouds, pp. 237–250 (2010)
22. Statistics solutions: Pearson's correlation coefficient (2015). https://www.statisticssolutions.com/pearsons-correlation-coefficient/
23. Sourceforge: Oprofile (2018). https://oprofile.sourceforge.io/news/
24. Online: DPDK pktgen (2019). http://pktgen-dpdk.readthedocs.io/en/latest/
25. Haeffner, W., Napper, J., Stiemerling, M., Lopez, D., Uttaro, J.: Service function chaining use cases in mobile networks. Internet Engineering Task Force (2015)
26. Kumar, S., Tufail, M., Majee, S., Captari, C., Homma, S.: Service function chaining use cases in data centers. IETF SFC WG 10 (2015)

Cooperative Source Seeking in Scalar Field: A Virtual Structure-Based Spatial-Temporal Method

Cheng Xu[1,2,3], Yulin Chen[1,2,3], Shihong Duan[1,2,3(✉)], Hang Wu[1,2,3], and Yue Qi[1,2,3]

[1] School of Computer and Communication Engineering,
University of Science and Technology Beijing, Beijing, China
duansh@ustb.edu.cn
[2] Shunde Graduate School,
University of Science and Technology Beijing, Beijing, China
[3] Beijing Key Laboratory of Knowledge Engineering for Materials Science,
Beijing, China

Abstract. Source seeking problem has been faced in many fields, especially in search and rescue applications such as first-response rescue, gas leak search, etc. We proposed a virtual structure based spatial-temporal method to realize cooperative source seeking using multi-agents. Spatially, a circular formation is considered to gather collaborative information and estimate the gradient direction of the formation center. In terms of temporal information, we make use of the formation positions in time sequence to construct a virtual structure sequence. Then, we fuse the sequential gradient as a whole. A control strategy with minimum movement cost is proposed. This strategy rotates the target formation by a certain angle to make the robot team achieve the minimum moving distance value when the circular team moves to the next position. Experimental results show that, compared with state-of-the-art, the proposed method can quickly find the source in as few distances as possible, so that the formation can minimize the movement distance during the moving process, and increase the efficiency of source seeking. Numerical simulations confirm the efficiency of the scheme put forth. Compared with state-of-the-art source seeking methods, the iterative steps of our proposed method is reduced by 20%, indicating that the method can find the signal source with higher efficiency and lower energy consumption, as well as better robustness.

Keywords: Cooperative computing · Gradient estimation · Source seeking · Circular formation · Spatial-temporal information

1 Introduction

In the last decade, *cooperative source seeking* based on multi-agents (or multi-robots) has been drawing more and more attention and widely used in many

© ICST Institute for Computer Sciences, Social Informatics and Telecommunications Engineering 2021
Published by Springer Nature Switzerland AG 2021. All Rights Reserved
H. Gao et al. (Eds.): CollaborateCom 2020, LNICST 350, pp. 320–336, 2021.
https://doi.org/10.1007/978-3-030-67540-0_19

fields, such as oil exploration [1–3], odor source search [4], environmental monitoring [5,6], pollution detection [7], and first-response search and rescue (SAR) tasks [8–10], etc. The target signal source can generally be an electromagnetic signal, an acoustic signal, or a chemical or biological signal. For example, in the event of a gas leak, rescuers need to be dispatched to find the source of the leak. At the same time, the lives of the participating rescuers must be protected as much as possible to avoid causing serious safety accidents. For the sake of safety and efficiency, robots (also known as agents) replace search and rescue personnel to enter dangerous areas and perform SAR operations. For example, when looking for a missing tour pal with a positioning signal device in the wild, in the face of complex terrain conditions, drones or unmanned vehicles can be used to perform search and rescue missions. To sum up, a source search algorithm with higher accuracy and consistency is of practical significance in many aspects, such as industrial production, military security, and civil security.

Many existing signal source seeking algorithms have evolved from the inspiration of biological behavior. Based on observations of spiny lobster behaviors, Consi et al. [11] proposed an agent that simulates the behavior of lobsters. In [12], inspired by silkworm moths, Kuwana et al. proposed a sourcing method that imitated the behaviors of silkworm moths seeking odor sources. Russell et al. [13] also verified the applicability of the above-mentioned algorithm to chemical source tracing in airflow environments.

However, the aforementioned biometric methods are based on the inspiration of individual biological behavior and still face obvious limitations. They rely on the information generally collected by only one single agent, such as in [11] and [12]. The agent uses its own sensor to perform the sourcing task, which lacks information interaction with other agents, making it difficult for researchers to apply it to scenarios where multiple agents work together. Although the efficiency of source seeking methods has been improved, the limitations of single agents will lead to the insufficient information collection and low robustness [11,12].

Many organisms in the natural world forage and reproduce through group behaviors [14–16]. Clustered organisms can efficiently find food and avoid natural enemies through an individual division of labor and information exchange. It has the characteristics of high efficiency and strong adaptability. Through the observation of biological populations, researchers proposed multiple swarm optimization algorithms, such as ant colony [14], bee colony [15], and wolf colony [16]. In 1992, Marco Dorigo [14] proposed an ant colony optimization algorithm by simulating the principles of ant social division of labor and cooperative foraging. Based on the inspiration of bee colonies to find nectar sources, Karaboga proposed the Artificial Bee Colony (ABC) [15] in 2005, which has the advantages of high accuracy and fewer control parameters. The algorithm was successfully applied to many fields such as artificial neural network training and combination optimization [17]. In [16], Wu et al. were inspired by the cooperative hunting behavior of wolves and proposed the Wolf Pack Algorithm (WPA). The swarming behavior of fish schools has also attracted researchers' attention. In [22,23], Wu et al. and Said Al-Abri et al. were inspired by the behavior of fish swarm clusters to study a cooperative collaborative mobile strategy called *acceleration-deceleration*. It can simulate the swarm behavior of fish schools avoiding light, and move the agent team to the position of the signal source. Kennedy and Eberhart et al. [18] firstly proposed Particle Swarm Optimization (PSO), which was originally proposed to simulate the motion of bird swarms. Jatmiko et al. [19,20] applied the particle swarm optimization algorithm to the field of odor source localization, which uses multiple agents to find stationary odor sources. Li et al. [21] proposed an improved probabilistic particle swarm optimization algorithm for source seeking in a ventilated environment.

As intelligent robots and sensors work very differently from real creatures, biological behavior-inspired methods also have limitations. Besides, uncoordinated agents can cause resource competition and conflicts, which affects the overall performance of multiple agents.

Gradient-based methods are also widely considered in source seeking applications, which can be mainly divided into single-agent methods and multi-agents cooperative ones. As for a single agent, in [24–26], researchers used random gradient estimation to make the agent randomly move in the signal field, measure the spatial information of the signal field, and calculate the gradient direction of the signal field. Krstic et al. [24,25] applied extreme value search control to make a single agent move to the local signal maximum in a noise-free signal field. Anatasov et al. [26] made a single agent calculate the gradient by random movement, and drove the agent to signal source with gradient information. However, the above-mentioned single-target-based gradient source seeking method still faces the following main problems:

1) Random gradient estimation can avoid local extremes to a certain extent, but the agent needs to constantly move back and forth to measure the signal strength, so as to calculate the gradient. This may lead to the inefficiencies of searching process.
2) The sudden failure of the single robot may lead to failure of the whole source seeking task, which means a low fault tolerance and robustness.

Due to above-mentioned drawbacks, more and more researchers are focusing on multi-agent cooperative source seeking. Petter Ögren et al. [27] used a coordinated movement strategy to make the agent team form a sensor network, and drive the team to find a signal source by the least square method. Zhu et al. [28] utilized the *leader − follower* strategy combined with least squares to calculate the gradient direction. In [29], Li et al. used the method of least squares estimator to enable the agent team to collaboratively calculate the gradient. The method proposed by Ruggero Fabbiano et al. in [30] enables the agents to sustain a circular formation by maintaining the same relative angle, and drive the agent team to the signal source with gradient descent algorithm. In [31,32], Lara-Brinon et al. proposed a circular formation method, which enables the team to maintain a uniformly distributed circular formation. With use of this means, agents calculate the gradient and drive themselves to the signal source.

However, the above-mentioned cooperative source seeking methods take the agent-formation into consideration, but only make use of the spatial information of the scalar signal field. They do not effectively use the information in the time sequence alongside the gradient direction. Signal strength measurement and intelligent formation control will inevitably have errors. The source seeking results obtained by gradient estimation using error signals and position information are often of poor accuracy.

In this paper, we aim to propose a source seeking method that utilize both spatial and temporal information in the scalar signal field, which can improve the source seeking efficiency. The rest of this paper is organized as follows. Section 2 puts forward the specific definition of a cooperative source seeking problem. Section 3 focuses on the details of a proposed virtual structure-based method. Section 4 demonstrates the experimental verification and analysis. Conclusions are drawn in Sect. 5.

2 Problem Formulation

The problem of source seeking mainly refers to the searching of signal source the unknown scalar fields. In this section, we proposed a virtual structure-based method with the fusion of spatial and temporal information collected by agents in formation. The agents advance along the gradient descent direction of the signal field and finally reach the position of the signal source.

We assume that the agent team moves in a two-dimensional space. For each agent i in the team, its dynamic equations could be denoted as follows:

$$\dot{p}_i = v_i \tag{1}$$

$$\dot{v}_i = u_i \tag{2}$$

where p_i is the position vector in the two-dimensional plane, v_i is the velocity vector and u_i is the acceleration input.

The nodes are collected into a vertex set $\mathcal{V}$, while links between nodes are collected in an edge set $\mathcal{E}$. Communication and measurements between nodes are bidirectional, so that $(\mathcal{V}, \mathcal{E})$ forms an undirected graph $\mathcal{G}$. The agents can communicate with each other within the communication topology and generally include the following information: the coordinates, the speed, and the measured signal strength information, etc. For simplicity, we assume that the communication between robot pairs are bidirectional and fully connected, and any pair of agents can exchange information with each other. The performance of the proposed algorithm under limited communication conditions is not within the scope of this paper and left for further studies.

The scalar signal field distribution function is represented as $z(p) : \mathbb{R}^2 \to \mathbb{R}$, which does not change over time. Its independent variable is coordinates of the sampling points denoted as p, and the signal strength reaches a maximum value at the source location p_s.

3   Virtual Structure Based Method

In this section, we detailed the proposed virtual structure based source seeking method, which combines the spatial and temporal information of the signal strength in the signal field. We firstly give a brief introduction to circular formation based collective gradient estimation criterion, so as to utilize the spatial distribution of the signal field. Then, we fuse the temporal information of the signal field by sequential sampling. Finally, we recursively perform the above steps, calculating the direction of the gradient until we find the source.

3.1   Gradient Estimation with Spatial Information

In order to better exploit spatial information in the signal field for gradient estimation, we considered a circular formation for multi-agents. A vector is formed from each agent to the center of the formation. The uniform and symmetric distribution of the circular formation makes the sum of these vectors a zero, which means we could make use of the individual measurement of each agent to accomplish spatial information fusion to estimate the gradient as a whole.

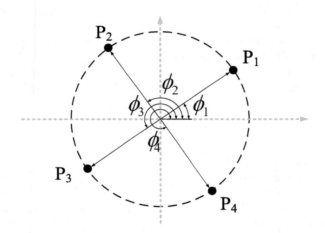

Fig. 1. The robots are evenly distributed on circular formation

We consider a robot team consisting of N agents, and they are organized in a circular formation. These N robots are evenly distributed on a circle with radius R, each of whose coordinates is represented as $p_i = (x_i, y_i)$, where x and y are respectively the abscissa and ordinate of robot i. A typical example is shown as Fig. 1, in which condition N is set as 4. The agents in the team are uniformly distributed on a circle with a radius R and a center position of $c(x_c, y_c)$. Therefore, the ith robot among the evenly distributed agents could be observed as a position p_i in the circular formation, which could be formulated as follows:

$$p_i = c + RD(\phi_i) \tag{3}$$

where R is the radius of the circular formation, $D(\phi_i)$ is the direction function of the ith agent in the circular formation, and $\phi_i = \phi_0 + \frac{2\pi i}{N}$ represents the azimuth angle of the ith agent, which could be referred to Fig. 1. It could be easily obtained that $D(\phi_i) = (\cos\phi_i, \sin\phi_i)$.

All agents in this circle formation are full-connected. The swarm agent team can approximate the signal field gradient through the weighted average of the signal strength collected by each individual agent [31], so as to achieve the purpose of spatial information fusion. Consider each agent measures the signal strength $z(p_i)$ at its current position $p_i(x_i, y_i)$ in the working space W. The gradient direction of the circular formation center c is denoted as $\widehat{\nabla}z(c)$, which could be calculated by combining the measured values of multiple agents around the circular formation center c with a radius R, namely

$$\frac{2}{NR^2} \sum_{i=1}^{N} z(p_i)(p_i - c) = \widehat{\nabla}z(c) + o(R) \tag{4}$$

where N is the number of agents in formation, and the approximation error term $o(R)$ is bounded by

$$\|o(R)\| \le \lambda_{max}(H_z)R \tag{5}$$

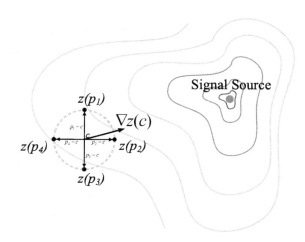

Fig. 2. Gradient estimation of circular formations

Proof: We assume that the agent team is evenly distributed in the circular formation, then it could be easily concluded that $\sum_{i=1}^{N}(p_i - c) = 0$. Using the first-order Taylor expansion of each measurement $o_i(R)$ about the point

c and recalling that $\|p_i - c\| = R$, then the following equation holds for all $i = 1, 2, ..., N$:

$$z(p_i) - z(c) = \nabla z(c)^{\mathrm{T}}(p_i - c) + o_i(R) \tag{6}$$

where $o_i(R)$ denotes the remainder of the Taylor expansion. Multiplying Eq. 6 by $\frac{2}{D^2 N}(p_i - c)$ and averaging the sum, we get

$$\frac{2}{ND^2}\sum_{i=1}^{N} z(p_i)(p_i - c) + \frac{2}{ND^2}\sum_{i=1}^{N} z(c)(p_i - c) =$$

$$\frac{2}{ND^2}\sum_{i=1}^{N}\nabla z(c)^{\mathrm{T}}(p_i - c)(p_i - c) + \frac{2}{ND^2}\sum_{i=1}^{N} o_i(R)(p_i - c).$$

Since the agents are uniformly distributed along a fixed circle, then we have $\sum_{i=1}^{N}(p_i - c) = 0$ and thus

$$\frac{2}{ND^2}\sum_{i=1}^{N} z(p_i)(p_i - c) =$$

$$\frac{2}{ND^2}\sum_{i=1}^{N}[(p_i - c)(p_i - c)]\nabla z(c)^{\mathrm{T}} + o(R),$$

where $o(R) = \frac{2}{ND^2}\sum_{i=1}^{N} o_i(R)(p_i - c)$. We analyze the second term of the previous equation using Eq. 4 to express the position of the agents p_i to obtain

$$\sum_{i=1}^{N}(p_i - c)(p_i - c)^{\mathrm{T}} = \sum_{i=1}^{N} D(\phi_i)D(\phi_i)^{\mathrm{T}}$$

$$= R^2 D(\phi_0)(\sum_{i=1}^{N} D(2\pi i/N)D(2\pi i/N)^{\mathrm{T}})D(\phi_0)^{\mathrm{T}}$$

$$= R^2 D(\phi_i)\sum_{i=1}^{N}\begin{bmatrix} \cos^2(2\pi i) & 0.5\sin(4\pi i) \\ 0.5\sin(4\pi i) & \sin^2(2\pi i) \end{bmatrix} D(\phi_i)^{\mathrm{T}}$$

$$= R^2 D(\phi_0)(\frac{N}{2}I_2)D(\phi_0)^{\mathrm{T}} = \frac{NR^2}{2}I_2.$$

Since $\cos^2\phi = 0.5(1 + \cos(2\phi))$, $\sin^2\phi = 0.5(1 - \cos(2\phi))$, and $\sum_{i=1}^{N}\cos(2\frac{2\pi i}{N}) = \sum_{i=1}^{N}\sin(2\frac{2\pi i}{N}) = 0$ for $N > 2$, where $I_2 \in \mathbb{R}^{2\times2}$ represents the identity matrix. Thus, the equality of (4) is satisfied. Thanks to the Taylor's Theorem cite each remainder $o_i(R)$ satisfies the inequality

$$|o_i(R)| \le \frac{1}{2}\lambda_{max}(H_z)\|p_i - c\|^2, \forall i.$$

Therefore, the function $o_i(R)$ can be bounded as

$$\|o_i(R)\| \leq \frac{2}{R^2 N} |o_i(R)| \|p_i - c\| \leq \lambda_{max}(H_z)R.$$

∎

As shown in Fig. 2, the gradient direction calculated by Eq. 4, incorporates the signal strength information of N agents at different spatial positions, so that the agent team can effectively estimate the gradient direction.

3.2   Gradient Estimation with Temporal Information

The above-mentioned gradient estimation method with circular formation, makes full use of the spatial information in the cooperative network. To some extent, it can avoid the noise of individual source detection, and improve the source seeking accuracy and robustness. However, existing studies, including the spatial fusion method in the above section, do not take the time-series information into consideration. It is with this in mind that in this section, we proposed a virtual structure-based method to reduce the influence of noise on the gradient estimation by introducing time-sequential information to the circular formation. Numerical simulations confirm the efficiency of the scheme put forth.

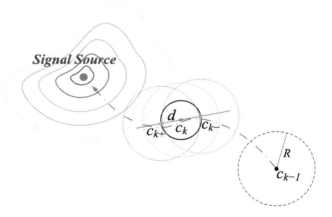

Fig. 3. Illustration of two-point gradient estimation

To integrate the time-sequential, we simulate two virtual positions (denoted as c_{k+} and c_{k-} respectively) where the agents in formation may appear. Then, we fuse these two estimations to obtain more accurate outputs. The detailed calculation process is listed as follows.

Firstly, when the agent team calculates the next formation center in time slot k (i.e., the target position c_k), we randomly select a point c_{k-} on the formation circle, centering at the target position c_k as the center, with a diameter of d, as shown in Fig. 3.

Then, as shown in Fig. 4 alongside the diameter we obtain another intersect c_{k+}. The circular formation passed these two points in succession, and the gradient directions of the circular team at these two points are respectively calculated. The weighted average of these two is used as the next gradient estimation, namely

$$\widehat{g}_k = \frac{1}{2}\left(\nabla z(c_{k-}) + \nabla z(c_{k+})\right) \tag{7}$$

Finally, the center position c_{k+1} of the circular agent team in $(k+1)\,th$ round is given by

$$c_{k+1} = c_k - a_k\widehat{g}_k \tag{8}$$

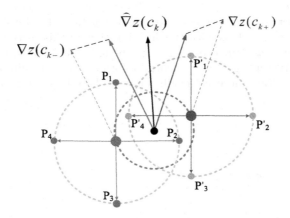

Fig. 4. Illustration of virtual structure based method.

where a_k is the step coefficient. The agent team continuously generates three series of points $\{c_{k-}, c_k, c_{k+}\}$ through above steps, and gradually moves to the vicinity of the signal source.

As for the step coefficient, it is generally denoted as

$$a_k = \frac{a}{(k+1+s)^\alpha}, \quad k = 0, 1, \dots \tag{9}$$

where a is a positive constant. s is a stability factor, which makes the algorithm have a large step size in the early iteration without causing instability, and should be set to 5% or 10% of the expected iterations of the algorithm. α controls the attenuation rate of the gain, and should be set to $\alpha = 0.602$, as suggested in [33].

However, we need to modify a_k. Because if the numerator a is constant, the gain coefficient a_k decreases monotonically, which is not desirable. As the step length of the agent will gradually decrease, it may be trapped in where the gradient estimation value is small. In our proposed method, we use a variable instead, which is inversely proportional to the magnitude of the gradient estimate. When the gradient estimation value is large, the agent travels in smaller steps; if the magnitude of the gradient estimation decreases, the gain coefficient increases by increasing the step size. Even if the signal field is very flat, the agent can perceive the gradient by increasing the step size. According to the above principles, the expression of a_k is formulated as

$$a_k = \frac{\text{r} \times (1+s)^\alpha}{\frac{1}{\omega} \sum_{j=k-\omega+1}^{k} \frac{1}{n} \|\widehat{g}_j\|} \tag{10}$$

where r is the greedy coefficient, and the larger the value of r, the larger the step size coefficient of each step.

3.3   Control Strategy with Minimal Moving Cost

An algorithm based on gradient descent can drive the agent team to the target signal source position. However, without a good control strategy, the gradient information cannot be well used by the agents. In order to more efficiently complete the source seeking task, we propose a minimal moving distance control strategy.

During the movement of the agent team, the control strategy in this paper does not require the agent to maintain a fixed circular formation all the time. It aims to minimize the overall movement distance from the start point to the source. To achieve the control goal, after the next formation center coordinates are calculated, the circular agent team can rotate an arbitrary angle around the formation center, and then we use optimization methods to find the angle that minimizes the moving distance.

We denote the distance that the ith agent need move from current position p_i to the next as $L(i) = \|p_i - p'_i\|$, where p'_i is the next step position of agent i. Our purpose is to minimize the overall moving cost of the agent team, so for the robot movement strategy we display as

$$L(\mathcal{P}) = \underset{p'_i \in \mathcal{P}}{arg\,min} \sum_{i}^{N} \|p_i - p'_i\| \tag{11}$$

where $\mathcal{P} = \{p'_1, ..., p'_N\}$.

To sum up, based on our proposed spatial-temporal source seeking method, together with the control strategy, the general seeking process could be formulate as displayed in Algorithm 1.

4   Experimental Verification and Analysis

In order to verify the effectiveness, we compared our method proposed with two state-of-the-art methods [15,21] in the numerical experiments. The method in

---

Algorithm 1. Virtual Structure based Method

Input: Initialize the coordinates of formation center c_0 and signal source p_s, initialize iteration indicator $k = 1$.

Output: Estimated position c

 1: Calculate the next position c_1 at initial position c_0 using (9), (11) and (10), then sample c_{1-} and c_{1+} on a circle of radius R along an arbitrary diameter;
 2: **Repeat**
 3: Move the agents to the position c_{k-} and c_{k+} successively considering control strategy with minimal moving cost by optimizing (11).
 4: Calculate the gradients $\nabla z(c_{k-})$ and $\nabla z(c_{k+})$;
 5: Calculate calculate the next target position c_{k+1} using (9), (11) and (10);
 6: Sample $c_{(k+1)-}$ on the circle centered at c_k;
 7: The other point on the diameter of the circle is taken as $c_{(k+1)+}$, which is the opposite of $c_{(k+1)-}$ on the circle centered at c_k.
 8: **if** $\|c_{(k+1)-} - c_{k+}\| > \|c_{(k+1)+} - c_{k+}\|$ **then**
 9:     $c_{(k+1)+} \leftarrow c'_{(k+1)}, \ c_{(k+1)-} \leftarrow c''_{(k+1)}$;
10: **end if**
11: **Untill** $\|c_k - p_s\| < 1$
12: Let $c \leftarrow c_k$;
    return c

---

[26] has only one robot for the source seeking task, while the method in [31] presented a circular team of four agents for source seeking.

A scalar signal source simulates a working space W with the coordinates of signal source $p_s = (5m, 25m)$. The collective information, as well as signal strength in this experiment, received by each agent in the working space is given by Eq. 12. The overall signal source model used in the experiment is denoted as follows [26]:

$$z(p) = -20.05 - 20\log_{10}\|p - p_s\| - N \qquad (12)$$

where p is the current position of the robot, p_s is the position of the signal source, and N is the signal noise [15]. Thereinto, $N = \sqrt{\alpha^2 + \beta^2}$, and α and β both have normal distributions respectively, $\alpha \sim \mathcal{N}(\nu\cos\theta, \sigma^2)$, $\beta \sim \mathcal{N}(\nu\sin\theta, \sigma^2)$, σ is the standard deviation, and ν and θ are related-factors with the mean of the distribution [26]. The initial positions of the center of the robot and the robot team are $p_{init} = (25m, 5m)$. This article conducts experimental verification and analysis under the above experimental conditions. In the experimental verification and analysis part, we compare the method proposed in this article with the methods in [26] and [31]. Numerical simulations confirm the efficiency of the scheme put forth.

4.1 Typical Experimental Results

As it could be seen from Fig. 5, the two comparative methods are more easily affected by signal noises, as their trajectories (green and magenta ones) fluctuated more heavily and are more tortuous. In contrast, the trajectory curve

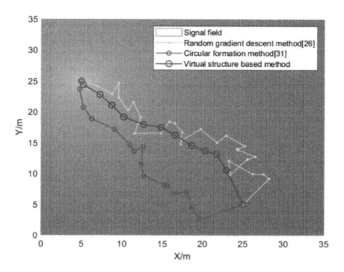

Fig. 5. Comparison of trajectories of three methods. The red trajectory is the trajectory of the robot formation center of the virtual structure based method proposed in this paper. The magenta trajectory is the trajectory of the center of the robot formation of the circular formation, and the green trajectory is the trajectory of the single robot center in the random method.

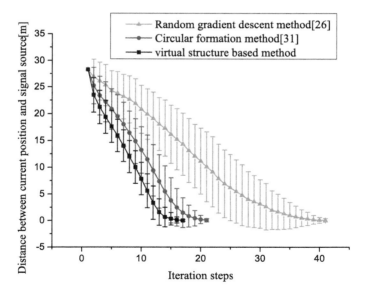

Fig. 6. The distance from the center of the robot to the source varies with the number of iteration steps.

obtained by the method described in this paper (red one) is smoother and simpler, which implicit that our approach is more efficient and robust to various changing signals. The method proposed in this paper reduces the interference of noise on the gradient calculation, and makes the trajectory more consistent with the direction of the gradient increase in the signal field. This shows that our proposed virtual structure based method has better stability and anti-noise characteristics.

Furthermore, the two comparative methods, as well as our own proposed one, are performed 100 times in the same experimental scenario. The signal field used is given by Eq. 12. For the generated noise term R, we make its parameters to be $\nu = 2$, $\theta = \frac{\pi}{3}$, $\sigma = 1$. These parameters are used to generate the experimental signal field. The averages of 100 experimental results are taken as the final experimental output. The variance of calculated total movement distances are drawn as error bars in Fig. 6, which demonstrates the detailed the trend of distance variance along with the iteration steps.

When the distance between the source and the agents center approaches to 0, we take it as the agents in formation successfully find the source. The speed of finding signal sources is one of the important indicators to measure the efficiency of the algorithm. The fewer iteration steps that the agents take to reach the signal source, the faster the distance converges to zero, as well as the higher the algorithm's efficiency.

Figure 6 shows the distance between the current agent formation center and the signal source, along with the number of iteration steps. It can be seen that our proposed virtual structure-based method convergences faster than comparative methods. The circular formation method that is superior to the [31] is also far superior to the random approximation method of [26], indicating that the cooperative estimation among agents outperforms independent work, and could solve the source seeking problem in a more efficient way.

4.2 Cumulative Distances with Noise Variance

The cumulative distances represent the total movement distances made by the agents, traveling from the starting position to the signal source. The lower the cumulative distances, the shorter the distances the robots move to find the signal source, which indicates that the source seeking method is more efficient.

The control variable of this experiment is the noise variance of the signal field, that is, the noise term N, in Eq. 12. We vary the noise variance of the signal field, and then compare the differences of the cumulative distances of the agents calculated by the three different source seeking methods. Let $\sigma = \{1, 1.5, 2, 2.5, 3, 3.5, 4, 4.5, 5, 5.5\}$. In different signal fields generated by different noise parameters σ, we performed three algorithms 100 times in various signal fields, and then recorded the averages and variances of each experiment.

Figure 7 shows the robot's cumulative distances of all three methods as the noise variance of σ changes. The error bars in the figure are given by the variance of 100 experimental results. It can be seen from Fig. 7 that our proposed virtual structure-based method has significantly lower cumulative distances than those in [26] and [31] under the same noise condition. The virtual structure-based method enables the agents to find the signal source with a shorter movement distance, and is of much higher search efficiency.

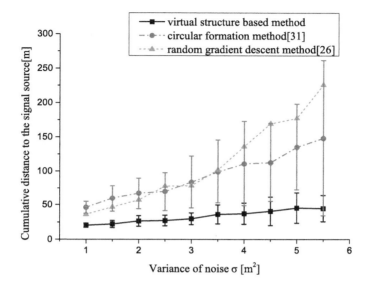

Fig. 7. Cumulative distance changes with noise variance σ.

On the other hand, as the noise variance σ increases, the curves of the three methods all have an upward trend. The growth rate of the proposed virtual structure-based method is significantly lower than that of the others, and the gap with the other two methods is getting wider and wider. It indicates that the proposed method is significantly more robust to noise interference than the methods in [26] and [31]. In addition, the amplitude of the error bars of virtual structure-based method is also smaller, which means more stable and robust to noise characteristics.

5   Conclusions

In this paper, we proposed a virtual structure-based method for multi-robot cooperative source seeking, in order to fuse both spatial and temporal information in the scalar field. Compared with state-of-the-art multi-robot collaborative methods, our proposed method enables to find signal sources with smaller iteration steps and cumulative distances, which can effectively reduce task overhead and improve efficiency. Our future work will focus on reducing the formation error of the robot team, which may provide a more accurate gradient estimation with circular or arbitrary formation.

Acknowledgments. This work is supported in part by National Postdoctoral Program for Innovative Talents under Grant BX20190033, in part by Guangdong Basic and Applied Basic Research Foundation under Grant 2019A1515110325, in part by Project funded by China Postdoctoral Science Foundation under Grant 2020M670135, in part by Postdoctor Research Foundation of Shunde Graduate School of University of Science and Technology Beijing under Grant 2020BH001, and in part by the Fundamental Research Funds for the Central Universities under Grant 06500127 and FRF-GF-19-018B.

References

1. Waleed, D., et al.: An in-pipe leak detection robot with a neural-network-based leak verification system. IEEE Sens. J. **19**(3), 1153–1165 (2019)
2. Berjaoui, S., Alkhatib, R., Elshiekh, A., Morad, M., Diab, M.O.: Free flowing robot for automatic pipeline leak detection using piezoelectric film sensors. In: International Mediterranean Gas and Oil Conference (MedGO). Mechref 2015, pp. 1–3 (2015)
3. Watiasih, R., Rivai, M., Penangsang, O., Budiman, F., Tukadi, Izza, Y.: Online gas mapping in outdoor environment using solar-powered mobile robot. In: 2018 International Conference on Computer Engineering, Network and Intelligent Multimedia (CENIM), Surabaya, Indonesia, pp. 245–250 (2018)
4. Che, H., Shi, C., Xu, X., Li, J., Wu, B.: Research on improved ACO algorithm-based multi-robot odor source localization. In: 2018 2nd International Conference on Robotics and Automation Sciences (ICRAS), Wuhan, pp. 1–5 (2018)
5. Cao, X., Jin, Z., Wang, C., Dong, M.: Kinematics simulation of environmental parameter monitor robot used in coalmine underground. In: 2016 13th International Conference on Ubiquitous Robots and Ambient Intelligence (URAI), Xi'an, pp. 576–581 (2016)
6. Shin, H., Kim, C., Seo, Y., Eom, H., Choi, Y., Kim, M.: Aerial working environment monitoring robot in high radiation area. In: 2014 14th International Conference on Control, Automation and Systems (ICCAS 2014), Seoul, pp. 474–478 (2014)
7. Shin, D., Na, S.Y., Kim, J.Y., Baek, S.: Fish robots for water pollution monitoring using ubiquitous sensor networks with sonar localization. In: 2007 International Conference on Convergence Information Technology (ICCIT 2007), Gyeongju, pp. 1298–1303 (2007)
8. Kanwar, M., Agilandeeswari, L.: IOT based fire fighting robot. In: 2018 7th International Conference on Reliability, Infocom Technologies and Optimization (Trends and Future Directions) (ICRITO), Noida, India, pp. 718–723 (2018)
9. Chen, X., Zhang, H., Lu, H., Xiao, J., Qiu, Q., Li, Y.: Robust SLAM system based on monocular vision and LiDAR for robotic urban search and rescue. In: IEEE International Symposium on Safety, Security and Rescue Robotics (SSRR), Shanghai 2017, pp. 41–47 (2017)
10. Denker, A., İşeri, M.C.: Design and implementation of a semi-autonomous mobile search and rescue robot: SALVOR. In: International Artificial Intelligence and Data Processing Symposium (IDAP), Malatya, pp. 1–6 (2017)
11. Consi, T., Atema, J., Goudey, C., Cho, J., Chryssostomidis, C.: AUV guidance with chemical signals. In: Proceedings of the 1994 Symposium on Autonomous Underwater Vehicle Technology, AUV 1994, pp. 450–455. IEEE (1994)

12. Kuwana, Y., Nagasawa, S., Shimoyama, I., Kanzaki, R.: Synthesis of the pheromone-oriented behaviour of silkworm moths by a mobile robot with moth antennae as pheromone sensors1. Biosens. Bioelectron. **14**(2), 195–202 (1999)
13. Russell, R.A., Bab-Hadiashar, A., Shepherd, R.L., Wallace, G.G.: A comparison of reactive robot chemotaxis algorithms. Robot. Auton. Syst. **45**(2), 83–97 (2003)
14. Dorigo, M., Maniezzo, V., Colorni, A.: Ant system: optimization by a colony of cooperating agents. IEEE Trans. Syst. Man Cybern. Part B (Cybern.) **26**(1), 29–41 (1996)
15. Karaboga, D.: An idea based on honeybee swarm for numerical optimization. Technical Report TR06. Erciyes University, Engineering Faculty, Computer Engineering Department (2005)
16. Wu, H.S., Zhang, F., Wu, L.: New swarm intelligence algorithm-wolf pack algorithm. Syst. Eng. Electron. **35**(11), 2430–2438 (2013)
17. Neto, M.T.R.S., Mollinetti, M.A.F., Pereira, R.L.: Evolutionary artificial bee colony for neural networks training. In: 2017 13th International Conference on Natural Computation, Fuzzy Systems and Knowledge Discovery (ICNC-FSKD), Guilin, pp. 44–49 (2017)
18. Kennedy, J., Eberhart, R.: Particle swarm optimization. In: Proceedings of ICNN 1995 - International Conference on Neural Networks, Perth, WA, Australia, vol. 4, pp. 1942–1948 (1995)
19. Jatmiko, W., Sekiyama, K., Fukuda, T.: Apso-based mobile robot for odor source localization in dynamic advection-diffusion with obstacles environment: theory, simulation and measurement. IEEE Comput. Intell. Mag. **2**(2), 37–51 (2007)
20. Jatmiko, W., et al.: Robots implementation for odor source localization using PSO algorithm. WSEAS Trans. Circuits Syst. **10**(4), 115–125 (2011)
21. Li, F., Meng, Q.-H., Bai, S., Li, J.-G., Popescu, D.: Probability-PSO algorithm for multi-robot based odor source localization in ventilated indoor environments. In: Xiong, C., Huang, Y., Xiong, Y., Liu, H. (eds.) ICIRA 2008. LNCS (LNAI), vol. 5314, pp. 1206–1215. Springer, Heidelberg (2008). https://doi.org/10.1007/978-3-540-88513-9_128
22. Wu, W., Zhang, F.: A speeding-up and slowing-down strategy for distributed source seeking with robustness analysis. IEEE Trans. Control Netw. Syst. **3**(3), 231–240 (2016)
23. Al-Abri, S., Wu, W., Zhang, F.: A gradient-free three-dimensional source seeking strategy with robustness analysis. IEEE Trans. Autom. Control **64**(8), 3439–3446 (2019)
24. Liu, S.-J., Krstic, M.: Stochastic source seeking for nonholonomic unicycle. Automatica **46**(9), 1443–1453 (2010)
25. Cochran, J., Krstic, M.: Source seeking with a nonholonomic unicycle without position measurements and with tuning of angular velocity part I: stability analysis. In: 2007 46th IEEE Conference on Decision and Control, New Orleans, LA, pp. 6009–6016 (2007)
26. Atanasov, N., Le Ny, J., Michael, N., Pappas, G.J.: Stochastic source seeking in complex environments. In: 2012 IEEE International Conference on Robotics and Automation, Saint Paul, MN, pp. 3013–3018 (2012)
27. Ogren, P., Fiorelli, E., Leonard, N.E.: Cooperative control of mobile sensor networks: adaptive gradient climbing in a distributed environment. IEEE Trans. Autom. Control **49**(8), 1292–1302 (2004)
28. Zhu, S., Wang, D., Low, C.B.: Cooperative control of multiple UAVs for moving source seeking. In: 2013 International Conference on Unmanned Aircraft Systems (ICUAS), Atlanta, GA, pp. 193–202 (2013)

29. Li, S., Kong, R., Guo, Y.: Cooperative distributed source seeking by multiple robots: algorithms and experiments. IEEE/ASME Trans. Mechatron. **19**(6), 1810–1820 (2014)
30. Fabbiano, R., Garin, F., Canudas-de-Wit, C.: Distributed source seeking without global position information. IEEE Trans. Control Netw. Syst. **5**(1), 228–238 (2018)
31. Briñón-Arranz, L., Renzaglia, A., Schenato, L.: Multi-robot symmetric formations for gradient and hessian estimation with application to source seeking. IEEE Trans. Robot. **35**, 782–789 (2019)
32. Briñón-Arranz, L., Seuret, A., Pascoal, A.: Circular formation control for cooperative target tracking with limited information. J. Franklin Inst. **356**, 1771–1788 (2019). https://doi.org/10.1016/j.jfranklin.2018.12.011
33. Spall, J.: Intro to Stochastic Search and Optimization. Wiley, Hoboken (2003)

BiC-DDPG: Bidirectionally-Coordinated Nets for Deep Multi-agent Reinforcement Learning

Gongju Wang[1], Dianxi Shi[1,2($\boxtimes$)], Chao Xue[1,2($\boxtimes$)], Hao Jiang[3],
and Yajie Wang[3]

[1] Artificial Intelligence Research Center (AIRC),
National Innovation Institute of Defense Technology (NIIDT),
Beijing 100166, China
dxshi@nudt.edu.cn, xuec11@tsinghua.org.cn
[2] Tianjin Artificial Intelligence Innovation Center (TAIIC),
Tianjin 300457, China
[3] College of Computer, National University of Defense Technology, Changsha
410073, China

Abstract. Multi-agent reinforcement learning (MARL) often faces the problem of policy learning under large action space. There are two reasons for the complex action space: first, the decision space of a single agent in a multi-agent system is huge. Second, the complexity of the joint action space caused by the combination of the action spaces of different agents increases exponentially from the increase in the number of agents. How to learn a robust policy in multi-agent cooperative scenarios is a challenge. To address this challenge we propose an algorithm called bidirectionally-coordinated Deep Deterministic Policy Gradient (BiC-DDPG). In BiC-DDPG three mechanisms were designed based on our insights against the challenge: we used a centralized training and decentralized execution architecture to ensure Markov property and thus ensure the convergence of the algorithm, then we used bi-directional rnn structures to achieve information communication when agents cooperate, finally we used a mapping method to map the continuous joint action space output to the discrete joint action space to solve the problem of agents' decision-making on large joint action space. A series of fine grained experiments in which include scenarios with cooperative and adversarial relationships between homogeneous agents were designed to evaluate our algorithm. The experiment results show that our algorithm out performing the baseline. -

Keywords: Multi-agent deep reinforcement learning · Large discrete joint action space · Cooperative · Mapping method

This work was supported in part by the Key Program of Tianjin Science and Technology Development Plan under Grant No. 18ZXZNGX00120 and in part by the China Postdoctoral Science Foundation under Grant No. 2018M643900.

H. Gao et al. (Eds.): CollaborateCom 2020, LNICST 350, pp. 337–354, 2021.
https://doi.org/10.1007/978-3-030-67540-0_20

1   Introduction

Multi-agent problems appear common both in nature and human society, especially in computer science and robotics. The study of multi-agent problems is a hot issue both in academia and industry. A series of algorithms based on the mathematical basis of Markov decision process game theory and so on has been formed, such as communication neural network (CommNet) [21], bidirectionally-coordinated network (BiC-Net) [15], QMIX [16], Value-Decomposition Networks (VDN) [22], Counterfactual Multi-Agent Policy Gradients (COMA) [6], Multi-Agent Deep Deterministic Policy Gradient (MADPPG) [13].

In recent years, with the extensive application of deep neural networks (DNN) supported by the powerful computing capabilities, the combination of DNN and multi-agent reinforcement learning has provided new ideas for solving multi-agent cooperation and confrontation problems, among them AlphaStar [25] defeated professional StarCraft players and has exceeded 99.8% of active players online. Alphastar uses multi-agent reinforcement learning based methods to train its model leveling both the human game data and the agent game data. In particular, several league pools are designed to continuously learn better policies and countermeasures. The multi-agent reinforcement learning algorithms also performs well in autonomous cars [7] and the coordination of robot swarms [8].

The purpose of multi-agent reinforcement learning algorithms is to learn an optimal policy π^* to get the maximum reward expectation as (1) based on the current state s_t.

$$V^*(s) = E_{\pi^*}[R_t | s_t = s] \tag{1}$$

The scenarios of multi-agent reinforcement learning (MARL) can be divided into three types: fully cooperative scenarios, fully competitive scenarios [12], mixed competition/cooperative scenarios. For fully cooperative scenarios, agents need to work together to maximize global reward expectation. The optimal policy can be expressed as (2) which means that every single agent selects its own optimal action under the assumption that other agents have already selected their optimal actions.

$$\pi^{i^*}(s) = \underset{u^i}{argmax} \underset{u^1,..,u^{i-1},u^{i+1},...,u^n}{max} Q^{\pi^*}(s, u) \tag{2}$$

On the contrary, for fully competitive scenarios [10], agent need to maximize their own reward expectation and minimize other agents' reward expectation. The optimal state-value function can be expressed as (3), which $-i$ means other agents except agent i.

$$V^{i^*}(s) = \underset{\pi^i(s,\cdot)}{max} \underset{u^{-i} \in u^{-i}}{min} \sum_{u^i \in U^i} Q^{i^*}\left(s, u^i, u^{-i}\right) \pi^i\left(s, u^i\right), \tag{3}$$

$$i = 1, 2...$$

For the mixed scenarios [11], the learning of agent's policy must consider not only maximizing its own reward expectations, but also cooperating with other agents to obtain the maximum global reward expectation, which will eventually form a game of Nash equilibrium. It can be expressed as (4).

$$
V^{i^*}(s) = \max_{\pi^1(s,\cdot),\cdots,\pi^{n_1}(s,\cdot)} \min_{o^1,\cdots,o^{n_2}\in O^1\times\cdots\times O^{n_2}} \sum_{u^1,\cdots,u^{n_1}\in U\times\cdots\times U^{n_1}}
$$
$$
Q^i\left(s,u^1,\cdots,u^{n_1},o^1,\cdots,o^{n_2}\right)\pi^1\left(s,u^1\right),\cdots,\pi^{n_1}\left(s,u^{n_1}\right) \tag{4}
$$

Traditional single-agent reinforcement learning algorithms usually performs poorly in the scenarios mentioned above. This is due to the lack of Markov properties caused by partial observation of the environment, the neglect of other agent's action, and the large joint action space. The MADDPG algorithm gives each agent an actor-critic network, where the actor network relies on agent's own observation to make decisions, and the critic network will consider the actions of other agents when calculating the Q value, which can be seen as an information sharing mechanism between agents, the algorithm can make the agents show a clear tendency to cooperate. As for COMA, each agent still has its own actor network, but it introduces a centralised critic network to uniformly calculate the Q value of the state. The algorithm uses the global state provided by the environment in training, which can guarantee the Markov property and promote the convergence of the algorithm. However, in multi-agent cooperation scenarios, with the increase in the number of agents, the action space and overall state space of the cooperative agent will increase exponentially. How to learn an effective policy in fully competitive scenarios is a challenge.

In order to solve the problem of policy learning in fully cooperative scenarios, we propose a new multi-agent algorithm called Bidirectionally-Coordinated Deep Deterministic Policy Gradient (BiC-DDPG), we shed light on this problem by modifying it into decentralized partially observable Markov decision progress (Dec-POMDP) and used a policy-based model to deal with the value-based problem, in our algorithm we propose 3 novel methods to meet the challenge of information sharing between agents, lack of Markov property, and action space explosion respectively. First, we use the bidirectional rnn [18] to solve the problem of agents information sharing with each other in training, and it decouples the number of agents from the complexity of training. Then we use the global state training critic. Finally, we map the joint actions of the agents output by the actor network from continuous space to discrete space, so that the problem can be solved by using policy-based reinforcement learning algorithm deep deterministic policy gradient which is called DDPG for short [9].

The rest of the paper was arranged as follows: Sect. 2 first introduces the classification of reinforcement learning, then introduces the existing multi-agent reinforcement learning algorithm based on Dec-POMDP problem modeling. Section 3 undertakes the problem modeling of the Sect. 2 and introduces our algorithm from three aspects: neural network design, action generation process and training algorithm method. Section 4 introduces our experimental setup, including the introduction to the simulation environment, the setting of observations and rewards, and the experimental design of the evaluation algorithm.

Section 5 shows the experimental results of our algorithm compared with other baselines. Section 6 makes our conclusion.

2   Related Work and Preliminary

Significant progress has been made in research on reinforcement learning based on single agents, including the Go agent AlphaGo [20], the poker robot Libratus [1] that defeated the human master. Reinforcement learning algorithms can be divided into value-based reinforcement learning algorithms, policy-based reinforcement learning algorithms and search-supervision based reinforcement learning algorithms.

One of the representative algorithms of the value-based algorithm is the Deep Q-Network (DQN) [14]. Its performance in the experimental environment of various video games such as Atari video games even exceeds the level of human masters. The contribution of DQN lies in the combination of Convolutional Neural Network (CNN) and Q-learning algorithm [27] in traditional reinforcement learning, which is more focused on solving decision problems of small-scale discrete action space.

Policy-based reinforcement learning algorithm uses deep neural networks for parametric representation, and uses policy gradient algorithms to optimize its own behavior policy. The representative algorithms include actor-critic (AC) and its variant, DDPG is to deal with the problem of agent learning in continuous action space. It draws on the experience replay mechanism of DQN and draws on AC, The algorithm thought to establish a deterministic policy network and value function, and uses the gradient information on the value function of the action to update the policy network. The goal of optimization is to add up the cumulative rewards for discounts, it performs well in continuous control scenarios, and the convergence speed is much higher than the DQN algorithm.

The typical representative work of supervision based reinforcement learning algorithms is AlphaGo [20], It adopts the idea of combining human experience, Monte Carlo tree search (MCTS) [3] and the trial and error process of reinforcement learning to prove reinforcement learning has the potential to solve single agent control problems in complex state spaces.

Traditional single-agent reinforcement learning algorithm does not perform well in multi-agent scenarios, and most of the typical multi-agent algorithms are based on actor-critic to improve, so we first introduce the modeling of multi-agent problems, then introduce The characteristics of the AC algorithm, finally introduce the multi-agent algorithm in turn.

2.1   Decentralized Partially Observable Markov Decision Progress

Decentralised unit micromanagement scenarios refer to the low-level, short-term control of the combat unit during the battle against the enemy. For each battle, the agents on one side are fully cooperative, so each battle can be regarded as a confrontation between N agents and M enemies, in addition, the enemies usually are built-in AI written by rules. This can be modeled as Dec-POMDP, $s \in S$ can describe the state of the current environment, every agent $a \in A \equiv \{1, \ldots, n\}$ in every game step choose an action $u^a \in U$ form joint action of agents $u \in U \equiv U^n$, In addition, joint action $u \in U$ corresponds to an n-dimensional vector, such that $u \in R^n$, This vector provides information related to the joint action. It will transform the environment according to the state transition function $P(s'|s, \mathbf{u})$: $S \times \mathbf{U} \times S \to [0, 1]$, all agents share global rewards $r(s, \mathbf{u}) : S \times \mathbf{U} \to R$, and $\gamma \in [0, 1)$ represents the discount factor.

We also need to describe the partially observable scenario of the agent $z \in Z$ according to observation function $O(s, a) : S \times A \to Z$, every agent has an action-observation history $\tau^a \in T \equiv (Z \times U)^*$ It represents the policy of the agent. $\pi^a(u^a|\tau^a) : T \times U \to [0, 1]$. The joint policy also has a joint action-value function $Q^\pi(s_t, \mathbf{u}_t) = {}_{s_{t+1:\infty}, \mathbf{u}_{t+1:\infty}}[R_t|s_t, \mathbf{u}_t]$, the cumulative rewards can be expressed as $R_t = \sum_{i=0}^{\infty} \gamma^i r_{t+i}$.

2.2 Actor-Critic and Deep Deterministic Policy Gradient

Both the Actor-Critic algorithm [23] and the Deep Deterministic Policy Gradient algorithm can be classified as policy-based algorithms. The main idea is to adjust the parameters ξ the direction $\nabla_\xi J(\xi)$ to maximize the expected reward $J(\xi) = E_{s \sim p^\pi, u^a \sim \pi_\xi}[R]$. The gradient direction of the AC algorithm can be expressed as (5).

$$\nabla_\xi J(\xi) = \pi_\xi(u^a|s)_{s \sim p^\pi, u^a \sim \pi_\xi} [\nabla_\xi \log \pi_\xi(u^a|s) Q^\pi(s, u^a)] \tag{5}$$

The AC algorithm can learn an approximate distribution of the value function $Q^\pi(s, \mathbf{u})$ through time difference error. This distribution is the critic net in AC algorithm. The DDPG algorithm combines the advantages of both the DQN and AC algorithms. It uses experience replay technology and dual neural network structure and AC algorithm to fit the idea of distribution for training. It has the advantages of both algorithms, but if the DDPG is directly deployed under the problem of multi-agent cooperative scenarios, it will also fail to ensure convergence due to the loss of Markovianness in the environment. It is also because the participation environment of multiple agents is unstable.

Inspired by the single agent reinforcement learning algorithm, the multi-agent reinforcement learning algorithm has also made good progress.

2.3 CommNet

The CommNet acquiesces agents use the full connection mechanism within a certain range to share information, using $N(j)$ to represent the set of agents j. Hidden state h_j^i and exchange information c_j^i are used to get the exchange information of next moment, the calculation of h_j^i and c_j^i are as (6) and (7).

$$c_j^{i+1} = \frac{1}{|N(j)|} \sum_{j' \in N(j)} h_{j'}^{i+1} \tag{6}$$

$$h_j^{i+1} = f^i\left(h_j^i, c_j^i, h_j^0\right) \tag{7}$$

We can find that the shared information is the average value of the hidden state of all agents at the last time, and the hidden state is also derived from the communication information about the last time and the hidden states of the last time. The hidden state output action of the last layer, but with the effect of the increasing number of agents, algorithm has not been demonstrated and is the goal of its next step. This learning algorithm with implicit communication is a good baseline to evaluate our algorithm.

2.4 Bidirectionally-Coordinated Network

The work of BiC-Net can be seen as an extension of CommnNet. It consists of a multi-agent actor network and a multi-agent critic network. Both the policy network (actor) and the Q network (critic) is based on a bidirectional rnn structures [15] policy network to receive shared observations, return actions for each individual agent. Since the bidirectional recursive structure can not only be used as a communication channel, but also as a local memory protection program, each agent can maintain its own local state and share information about collaborators, and the structure of bidirectional rnn helps alleviate the increase in the number of agents and the training problems caused by.

2.5 QMIX and VDN

The QMIX and VDN algorithm is belonged to value function approximation algorithms, and the agents get a global reward. The algorithms adopt a framework of centralized learning and distributed execution of applications. VDN integrates the value function of each agent to obtain a joint action value function. Let $\tau = (\tau^1, ..., \tau^n)$ be the joint action-observation history, where $\tau^i = \left(u_0^i, o_1^i, \cdots, u_{t-1}^i, o_t^i\right)$ is the action-observation history, which $u = (u^1, ..., u^n)$ be the joint action. Q^{tot} is a joint action value function and $Q^i\left(\tau^i, u^i\right)$ is a local action-value function of agent i. The local value function only depends on the local observation of each agent. The joint action value function adopted by VDN is direct addition and summation as (8).

$$Q^{tot} = \sum_{i=1}^n Q^i\left(\tau^i, u^i\right) \tag{8}$$

QMIX and VDN are also committed to obtaining the joint action value function of all agents, and the expression of QMIX is more abundant. In order to get a decentralized policy that has the same effect as the centralized policy training, there is no need to add the Q value like VDN, and the policy adopted is to make the global value equal to the argmax of each agent's decentralized Q as (9).

$$\text{argmax}Q^{tot}(\tau, \mathbf{u}) = \begin{pmatrix} \text{argmax}_{u^1} Q^1\left(\tau^1, u^1\right) \\ \vdots \\ \text{argmax}_{u^n} Q^n\left(\tau^n, u^n\right) \end{pmatrix} \tag{9}$$

This allows each agent a to select the Q value associated with it to select the action, and it is very easy to obtain the argmax of each local Q. QMIX proposed a mixed neural networks.The processing is through inputting each agent's action-observation history to achieve the agent's communication with an rnn network composed of GRU cells and finally selecting the local Q of each agent. Then input the above Q combination into a mixing network structure to get the Q^{tot} the QMIX algorithm will be deployed in a larger number of agent confrontation environments as its next goal.

2.6   COMA

The COMA algorithm is also based on the AC algorithm. A centralized critic network is used to estimate the Q value of the joint action u in state s, and the Q function is compared to each agent. The counterfactual baseline keep the actions of other agents u^{-a} and marginalizes the action u^a of agent a. The advantage function is defined as (10).

$$A^a(s, \mathbf{u}) = Q(s, \mathbf{u}) - \sum \pi^a \left(u'^a | \tau^a\right) Q \left(s, \left(\mathbf{u}^{-a}, u'^a\right)\right) \tag{10}$$

In order to solve the problem of multi-agent credit allocation, a separate baseline $A^a(s, \mathbf{u})$ is calculated for each agent. This process uses the counterfactuals inferred by the critic network to learn directly from the individual experience of the agent without relying on additional simulations, but it is not performed under the condition of a larger number of agents experiment verification record.

3   Bidirectionally-Coordinated Deep Deterministic Policy Gradient

We make a comprehensive elaboration of the algorithm BiC-DDPG we proposed, BiC-DDPG was purposed to solve the problem of policy learning in multi-agent scenarios. We will introduce the solutions proposed by the algorithm in order to overcome the three problems: environmental stability, information sharing between agents, and large joint action space.

3.1   Environmental Stability

The modeling of this work is based on Dec-POMDP, for each agent, due to the limitation of its own observation range, its observation of the environment is incomplete, resulting in the environment is unstable for a single agent. In order to solve this problem, the current mainstream is to adopt the idea of centralized training distribution execution to design the algorithm. In other words, the decentralized actor network makes actions based on part of the agent's observation, while the centralized critic network is based on the state output Q value to update the actor network.

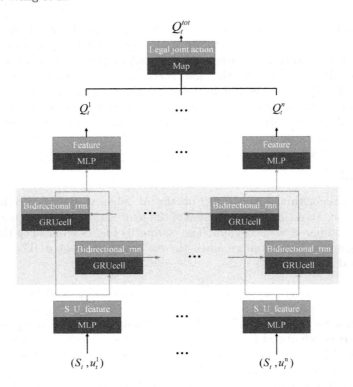

Fig. 1. Critic net of Bic-DDPG.

In the simulation environment of this paper, at each time step of training we can know the actions taken by all agents. So based on this fact we can assume that the simulation environment satisfies the Markov property as (11)

$$P_{ss'} = \mathbb{E}\left(s_{t+1} = s' \mid s_t = s, u_t = u\right) \tag{11}$$

Under assumption of (11), we designed the critic network as shown in the Fig. 1. The input data is global state and every agent's legal action in order to make the centralized critic function more quickly and efficiently fit the state transition function P. We use bidirectional rnn network structure to accept the last layer's data features. According to the features of bidirectional rnn, the output of each time step is not only related to the input of other locations but also related to the input of its corresponding position. So each output of the time step implies the evaluation of the joint action, we finally merge these outputs into Q^{tot}. In order to better approximate the state transition function P, we minimize the loss as follows:

$$L\left(\theta\right) = \frac{1}{N}\sum_t \left[y_t - Q_\theta\left(\mathbf{s}_t, \mathbf{u}_t\right)\right]^2 \tag{12}$$

$$y_t = r_t + \gamma Q_{\theta'}\left(s_{t+1}, \pi_{\xi'}\left(o_{t+1}\right)\right) \tag{13}$$

In addition, the calculation of the value function is designed into two steps. First, the local value function of each agent is calculated as the value Q_t^n, and then the values are used to calculate the Q^{tot}. In a scenario where only a global reward is given, Q^{tot} can be used for calculation, and if the scene gives each agent's own local reward, the Q value of each agent can be used to participate in the calculation and fitting.

3.2   Information Sharing

We inspired by the work of CommNet to add the bidirectional rnn structure to the actor network as shown in Fig. 2. Since the bidirectional rnn structure is an extended version of the rnn structure, its output combines the results of forward propagation and backward propagation in the structure, which enables the agent to implicitly consider the situation of other agents when making decisions. Equation (14) shows that the hidden state of the agent i at step t in the forward propagation structure depends on its own observation o_i^t, hidden state fo_{i-1}^t calculated by forward propagation, and hidden state bk_{i+1}^t calculated from backward propagation. And each agent also continues to pass forward fo_i^t and backward hidden state bk_i^t at the same time, providing a basis for other agents to make decisions. So we can think that the decision of each agent is based on the local observations of all agents. We use variable U and W to represent the parameters of the corresponding position in the process of forward propagation

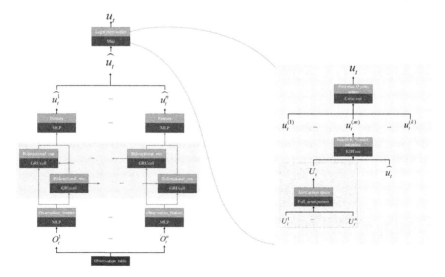

Fig. 2. Actor net of BiC-DDPG. The right part shows the mapping process from proto joint action to legal joint action.

and backward propagation, we use variable b represents the bias and ϕ represents activation function. The calculation of the forward hidden state can be expressed as Eq. (14) and the calculation of the backward hidden state can be expressed as Eq. (15), both of them have the same structure but the propagation direction is opposite.

$$fo_i^t = \phi(U_i o_i^t + W_i(fo_{i-1}^t) + b_i) \tag{14}$$

$$bk_i^t = \phi(U_i' o_i^t + W_i'(bk_{i+1}^t) + b_i') \tag{15}$$

The final output is the combination of forward propagation output in (16) and the backward propagation output in (17), as shown in (18).

$$fo\_out_i^t = V_i * fo_i^t + c_i \tag{16}$$

$$bk\_out_i^t = V_i' * bk_i^t + c_i' \tag{17}$$

$$out_i^t = [fo\_out_i^t, bk\_out_i^t] \tag{18}$$

3.3 Large Joint Action Space

A method of using policy-based reinforcement learning algorithm DDPG to handle large-scale discrete actions when training single agents was proposed by G. Dulac-Arnold et al. [5]. It provides new ideas and algorithms for solving the problem of modeling based on value-based. In order to solve the problem of large joint action space, inspired by their work, we propose a method of action mapping. We adopt a deterministic policy and the agent's decision is based on their own observations, our policy for generating actions should be:

$$\pi_\xi(o) = \hat{u} \tag{19}$$

However, the proto joint action $\hat{u}$ of the output agent may not be an executable joint action at this time, so $\hat{u} \notin U$. Therefore, we need to map the proto joint action to a legal discrete joint action space U.

$$f_k(\hat{u}) = \arg\min_{u \in U} |u - \hat{u}|_2 \tag{20}$$

So we used an action mapping method as (20), Where f_k is the k nearest neighbors to $\hat{u}$ in the discrete joint action space, the distance measure is the Euclidean distance, in order to improve the efficiency of searching, the paper [5] uses the K-Nearest Neighbor [4] algorithm to search. But the time complexity of knn querying neighbors is generally estimated to be $O(D * N * N)$, where D represents the data dimension and N represents the number of samples, In order to query of near legal joint actions more quickly, this paper uses the method of

KDTree to find it, and it can also find the result in logarithmic time which is $O(log(D * N * N))$. The proto joint action maps to the k number of legal joint actions we specify. To ensure that the selected joint action always evaluates the highest profit, the candidate joint action set needs to be evaluated. And we choose the joint action with the highest possible return which is shown in Fig. 2.

$$u = argmaxQ(s, u)|_{u=[u^{(1)}, u^{(2)}, ..., u^{(k)}]} \tag{21}$$

The goal of the training algorithm is to find an optimal policy π_{ξ^*}, and the most important thing of the work is to optimize the parameters ξ so that the cumulative return obtained by it reaches the maximum value. The optimization goals for actor parameters are defined as (22).

$$\nabla_\xi J \approx \frac{1}{N} \sum_i \nabla_u Q_\theta(s, \hat{u})|_{s=s_i, \hat{u}=\pi(o_i)} \nabla_\xi \pi(o)|_{o=o_i} \tag{22}$$

We summarize the algorithm flow as the pseudo code shown in as follows.

Algorithm 1. BiC-DDPG algorithm

1: Initialise actor network and critic network with ξ and θ
2: Initialise target network and critic network with $\xi' \longleftarrow \xi$ and $\theta' \longleftarrow \theta$
3: Initialise replay buffer with buffer R
4: **for** $epsiod = 1, 2, ..., E$ **do**
5:    initialise a random process μ for action exploration
6:    receive initial local observation o_1 and global state s_1
7:    **for** $t = 1, 2, ..., T$ **do**
8:       for each agent a, get select action $\hat{u}_t^a = \pi_\xi(o_t^a) + \mu_t^a$
9:       all agent actions form a joint action $\hat{u}_t$
10:      get legal joint action set $[u_t^{(1)}, u_t^{(2)}, ..., u_t^{(k)}]$
11:      get the max Q legal joint action
         $u_t = argmaxQ(s_t, u)|_{u=[u_t^{(1)}, u_t^{(2)}, ..., u_t^{(k)}]}$ and excute
12:      receive reward r_t and new local oberservation o_t and global state s_t
13:      Store transition $([o_t, s_t], u_t, r_t, [o_{t+1}, s_{t+1}])$ in R
14:      Sample a random minibatch of N transitions
         $([o_i, s_i], u_i, r_i, [o_{i+1}, s_{i+1}])$ from R
15:      Set $y_i = r_i + \gamma Q_\theta(s_{t+1}, \pi_{\xi'}(o_{t+1}))$
16:      Update the critic by minimizing the loss:
17:      $L(\theta) = \frac{1}{N}[\sum_i(y_i - Q_\theta(s_i, u_i))]^2$
18:      Update the actor using the sampled gradient:
19:      $\nabla_\xi J \approx \frac{1}{N} \sum_i \nabla_u Q_\theta(s, \hat{u})|_{s=s_i, \hat{u}=\pi(o_i)} \nabla_\xi \pi(o)|_{o=o_i}$
20:      Update the target networks:
21:      $\xi' \longleftarrow \tau\xi + (1 - \tau)\xi'$
22:      $\theta' \longleftarrow \tau\theta + (1 - \tau)\theta'$
23:   **end for**
24: **end for**

---

4 Experiments

In this section, we will describe the simulation environment for evaluating our algorithm and the arrangement of experiments.

4.1   StarCraft II Cooperative Scenarios

Real-time strategy (RTS) is the main research platform and tool for multi-agent reinforcement learning [2,19,24], StarCraft II is a typical RTS game, which contains the confrontation with a variety of heterogeneous agents, providing sufficient materials for simulation experiments, and existing researchers had developed API wrappers for researchers to carry out on this platform Reinforcement learning research, which is representative of pysc2 [26] and SMAC [17]. These efforts alleviate the pressure of researchers on the environment interface call work, allowing them to focus more on the optimization of algorithms.

We use SMAC as the interface for the algorithm to interact with the environment, It specifies the agent's discrete action space: move[direction], attack[enemy id], stop, and no-op. An agent can only move in four directions: east, west, south, and north. An agent is only allowed to attack when the enemy is within its range.

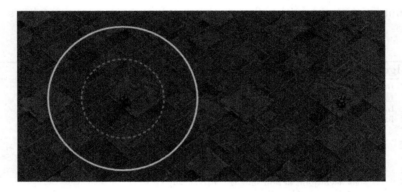

Fig. 3. Partial observation of the agent, where the red dotted circle represents the attack range of the agent, and the orange solid circle represents the observation range of the agent. (Color figure online)

Part of the observability is achieved by restricting the unit of view of the field of view, which limits the agent to observe the information about the enemy or friendly agent outside the field of view as shown in Fig. 3. In addition, the agent can only observe living enemy units, and cannot distinguish between units that are dead or units that are out of range.

In order to understand the properties and performance of our proposed BiC-DDPG algorithm, we conducted experiments with different settings in the cooperative scenarios of StarCraft II, Similar to Tabish Rashid's work [16].

BiC-DDPG controls a group of agents to try to defeat enemy units controlled by built-in AI. We use the winning rate and reward in training as the sole criterion for evaluating the algorithm.

4.2 State, Local Observation, Reward

The local observation value of each agent is a circular observation range centered on the unit it controls, and 10 is a radius. The observation contains information about all agents in the range: distance from the target, attribution of target, type of target, The relative x coordinate of the target, the relative y coordinate of the target, the health, the shield, and the cooling of the weapon. All attribute items are normalized and preprocessed according to their maximum values, which is convenient to maintain a stable amplitude when the neural network is learning.

The global state is used to train the critic network. It contains information about all units on the map. Specifically, the state vector includes the coordinates of all agents relative to the center of the map, and the unit features that appear in the observation.

As for reward calculation: at each time step, the agents receiving a global reward equal to the damage to the target unit plus minus half of the damage taken. Killing an enemy will get a 10 point reward, and winning the game will get a reward equal to the team's total remaining health plus 200. Compensation calculation followed Table 1. This damage-based reward signal uses the same mechanism as the QMIX algorithm. The benefit is that the results of joint action are evaluated using a shared reward. Since the absolute value of the maximum reward obtained in different scenarios is different, we finally limit the range of rewards to between 0 and 20.

Table 1. SMAC's rewards calculation

Name	Description	Reward
Kill	Kill an enemy	10
Win	All enemies are eliminated	200+remaining health
Attack	Attack an enemy once in a time step	Damage to target
Be attacked	Be attacked by an enemy once in a time step	Half of damage taken

4.3 Homogeneous Agents Cooperative Scenarios

We have designed experiments with different features in order to evaluate the performance of various aspects of the algorithm. Our experiments can be divided into three categories, We will explain the purpose of the experiment settings.

Cooperation with a Smaller Number of Homogeneous Agents. In this type of experiment, we avoid the impact on the heterogeneity of the agent, and test the learning ability of the algorithm for fine-grained combat in a small number of cases. Compare its performance with the baseline algorithms. The test environment is selected as 3 marines (3m) (Fig. 4).

Fig. 4. n marines scenarios(n = 3, 5, 8) and red is our agents. In these scenarios, the number and types of our agents are equal to those of the enemy. The purpose of this is to evaluate the willingness to cooperate between agents. (Color figure online)

Cooperation with a Large Number of Homogeneous Agents. In this type of experiment, we also avoid the influence of the heterogeneity of the agents, but we increase the number of agents to make the confrontation complicated. This mainly evaluates the learning degree of the algorithm for the general fire policy, The test environment is selected as 5 marines (5m), 8 marines (8m).

Self-contrast Experiment. In the self-comparison experiment, we will compare the performance curve of the algorithm with the change of the number of homogeneous agents. The test environment is Selected as 3m, 5m, 8m.

All scenarios are run 20,000 epochs, and the reward and win rate are recorded every 100 epochs. The above process loops 10 times to evaluate the overall performance of the algorithm.

5   Results and Discussion

The 3m scenario's experiment results are as shown in Fig. 5, we can find that all algorithms can update their policies on the direction on increasing win rate and rewards. The BiC-DDPG algorithm can achieve win rate of about 60% in the first

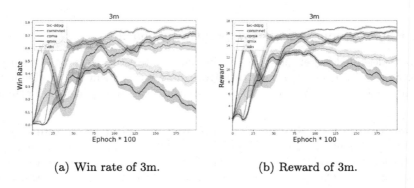

(a) Win rate of 3m.   (b) Reward of 3m.

Fig. 5. Results of cooperation with a smaller number of homogeneous agents

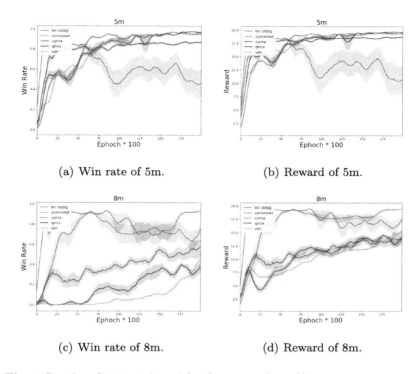

(a) Win rate of 5m.   (b) Reward of 5m.

(c) Win rate of 8m.   (d) Reward of 8m.

Fig. 6. Results of cooperation with a larger number of homogeneous agents.

800 epochs, baselines except COMA algorithm seems to find the right update direction on the gradient, other algorithms are still looking for a better gradient update direction. From the global observation of the convergence of the policies, BiC-DDPG can update the policy on a good direction, it is slightly better than COMA and CommNet. Although the COMA algorithm has experienced the wrong gradient update direction, it can correct the direction in time and its final win rate is concentrated on 70%. CommNet's convergence curve trend is similar to BiC-DDPG, its overall performance is weaker than BiC-DDPG algorithm and its final win rate is around 60%. The QMIX and VDN algorithms update parameters in the wrong direction and their final win rate are around 11% and 40%. The final results of CommNet, COMA and BiC-DDPG algorithms show that they have the ability to converge policy learning to a relatively high win rate in a short training time. QMIX and VDN are not good at handling the problem of fine-grain agents control in a short period of training time. The trend of the reward curve is like the trend of the win rate (Fig. 6).

In the 5m experiment, we can find that the performance of the COMA algorithm is still good, it can upgrade the policy on a higher level in the early stage of training, and its parameter optimization speed is second only to CommNet, its final win rate is stable at about 81%. The final results of both QMIX, VDN, and BiC-DDPG are close and their win rates are stable at 90%, The optimization speeds of the three algorithm are similar. In the range of 5,000 to 9,000 epochs, the BiC-DDPG obviously surpasses other algorithms. CommNet's policy improvement speed in the early stage is the fastest, but the parameter update in the later training seems to be in the wrong direction, resulting in its worst performance, but still have a 50% win rate. The trend of the reward is similar to the winning percentage (Fig. 7).

With the increase in the number of agents again in 8m experiment, the final performance of baselines have decreased significantly relative to the 5m experiment. The VDN algorithm and QMIX algorithm have once again fallen to 30% and 43% win rate. Although the COMA algorithm has a relatively large drop rate, it eventually stabilized at 58% win rate, and in this scenario our algorithm's win rate suddenly rose to 90% win rate near the end of the experiment, and the second best performing CommNet algorithm was only about 78%.

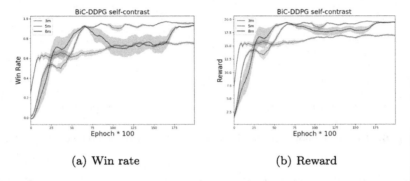

(a) Win rate   (b) Reward

Fig. 7. Self-Contrast, in order to observe the changes in the performance of the algorithm with the number of agents, it can be found that the performance is the best in the 5m scenario.

In the scenarios of self-contrast result, we found that the algorithm performs best in the 5m scenario, and the final performance in the 8m scenario is close to it. The performance in the 3m scenario is obviously not as good as the above two, which proves that the algorithm is not very good at policy learning in small-scale scenarios.

6   Conclusion

This paper proposes a deep multi-agent reinforcement learning algorithm BiC-DDPG, which allows for end-to-end learning of decentralized policy on a centralized environment and the effective use of additional state information.

BiC-DDPG allows the use of policy-based reinforcement learning algorithms to solve value-based decision-making problems. This is achieved by mapping the actions of the agent in the continuous space to the discrete space. The results of our decentralized unit micromanagement scenarios of StarCraft II show that, compared with other multi-agent algorithms, BiC-DDPG improves the problem of the explosion of the decision space dimension caused by the increase in the number of agents.

This paper evaluates the effectiveness of the algorithm in a simulation environment where the number of agents is less than 8, because the simulation environment provides a variety of multi-agent cooperation scenarios. Therefore, In the following work, we also intend to verify the effect of the algorithm in the cooperation of heterogeneous agents in a similar way, and also verify the impact of changes in the number of agent types on the algorithm. Then we will verify the effectiveness of our algorithm in the scenarios of larger number of homogeneous agents and heterogeneous agents.

In addition, BiC-DDPG will be deployed in the experimental environment of the full-length confrontation with StarCraft II, as a micro-management module added to the algorithm of the full-length game.

References

1. Brown, N., Sandholm, T.: Safe and nested endgame solving for imperfect-information games. In: Workshops at the Thirty-First AAAI Conference on Artificial Intelligence (2017)
2. Bubeck, S., Cesa-Bianchi, N.: Regret analysis of stochastic and nonstochastic multi-armed bandit problems. Found. Trends Mach. Learn. $5(1)$, QT06, 1–7, 9–21, 23–43, 45–65, 67–105, 107–115, 117–127 (2012)
3. Coulom, R.: Efficient selectivity and backup operators in Monte-Carlo tree search. In: van den Herik, H.J., Ciancarini, P., Donkers, H.H.L.M.J. (eds.) CG 2006. LNCS, vol. 4630, pp. 72–83. Springer, Heidelberg (2007). https://doi.org/10.1007/978-3-540-75538-8_7
4. Dudani, S.A.: The distance-weighted k-nearest-neighbor rule. IEEE Trans. Syst. Man Cybern. **SMC-6**, 325–327 (1976)
5. Dulac-Arnold, G., et al.: Deep reinforcement learning in large discrete action spaces. http://arxiv.org/abs/ArtificialIntelligence (2015)
6. Foerster, J., Farquhar, G., Afouras, T., Nardelli, N., Whiteson, S.: Counterfactual multi-agent policy gradients (2017)
7. Liu, J., Li, P., Chen, W., Qin, K., Qi, L.: Distributed formation control of fractional-order multi-agent systems with relative damping and nonuniform time-delays. ISA Trans. **93**, 189–198 (2019)
8. Kolokoltsov, V.N., Malafeyev, O.A.: Multi-agent interaction and nonlinear Markov games (2019)
9. Lillicrap, T.P., et al.: Continuous control with deep reinforcement learning. Comput. Sci. $8(6)$, A187 (2015)
10. Littman, M.L.: Markov games as a framework for multi-agent reinforcement learning. In: Machine Learning Proceedings 1994, pp. 157–163. Elsevier (1994)
11. Littman, M.L.: Friend-or-foe Q-learning in general-sum games. ICML **1**, 322–328 (2001)
12. Littman, M.L.: Value-function reinforcement learning in Markov games. Cogn. Syst. Res. **2**, 55–66 (2001)

13. Lowe, R., Wu, Y., Tamar, A., Harb, J., Abbeel, P., Mordatch, I.: Multi-agent actor-critic for mixed cooperative-competitive environments. ArXiv abs/1706.02275 (2017)

14. Mnih, V., et al.: Playing Atari with deep reinforcement learning. ArXiv abs/1312.5602 (2013)

15. Peng, P., et al.: Multiagent bidirectionally-coordinated nets: emergence of human-level coordination in learning to play StarCraft combat games (2017)

16. Rashid, T., Samvelyan, M., de Witt, C.S., Farquhar, G., Foerster, J.N., Whiteson, S.: QMIX: monotonic value function factorisation for deep multi-agent reinforcement learning. In: ICML (2018)

17. Samvelyan, M., et al.: The StarCraft multi-agent challenge. In: AAMAS (2019)

18. Schuster, M., Paliwal, K.: Bidirectional recurrent neural networks. IEEE Trans. Sig. Process. **45**(11), 2673–2681 (1997)

19. Seabold, S., Perktold, J.: Statsmodels: econometric and statistical modeling with Python (2010)

20. Silver, D., et al.: Mastering the game of Go with deep neural networks and tree search. Nature **529**(7587), 484–489 (2016)

21. Sukhbaatar, S., Szlam, A., Fergus, R.: Learning multiagent communication with backpropagation. ArXiv abs/1605.07736 (2016)

22. Sunehag, P., et al.: Value-decomposition networks for cooperative multi-agent learning. In: AAMAS (2018)

23. Sutton, R.S., Barto, A.G.: Reinforcement learning: an introduction. IEEE Trans. Neural Netw. **16**, 285–286 (1998)

24. Tavares, A.R., Azpurua, H., Santos, A., Chaimowicz, L.: Rock, paper, StarCraft: strategy selection in real-time strategy games. In: AIIDE (2016)

25. Vinyals, O., et al.: Grandmaster level in StarCraft II using multi-agent reinforcement learning. Nature **575**(7782), 350–354 (2019)

26. Vinyals, O., et al.: StarCraft II: a new challenge for reinforcement learning (2017)

27. Watkins, C.J.C.H.: Learning from delayed reward. Ph.D. thesis, Kings College University of Cambridge (1989)

FocAnnot: Patch-Wise Active Learning for Intensive Cell Image Segmentation

Bo Lin[1], Shuiguang Deng[1(✉)], Jianwei Yin[1(✉)], Jindi Zhang[1], Ying Li[1], and Honghao Gao[2,3(✉)]

[1] College of Computer Science and Technology, Zhejiang University, Hangzhou 310027, China
{rainbowlin,dengsg,zjuyjw,zjindiss,cnliying}@zju.edu.cn
[2] School of Computer Engineering and Science, Shanghai University, Shanghai 200444, China
gaohonghao@shu.edu.cn
[3] Gachon University, Seongnam, Gyeonggi-Do 461-701, South Korea

Abstract. In the era of deep learning, data annotation becomes an essential but costly work, especially for the biomedical image segmentation task. To tackle this problem, active learning (AL) aims to select and annotate a part of available images for modeling while retaining accurate segmentation. Existing AL methods usually treat an image as a whole during the selection. However, for an intensive cell image that includes similar cell objects, annotating all similar objects would bring duplication of efforts and have little benefit to the segmentation model. In this study, we present a patch-wise active learning method, namely FocAnnot (focal annotation), to avoid such worthless annotation. The main idea is to group different regions of images to discriminate duplicate content, then evaluate novel image patches by a proposed cluster-instance double ranking algorithm. Instead of the whole image, experts only need to annotate specific regions within an image. This reduces the annotation workload. Experiments on the real-world dataset demonstrate that FocAnnot can save about 15% annotation cost to obtain an accurate segmentation model or provide a 2% performance improvement at the same cost.

Keywords: Active learning · Intensive cell image · Duplicate annotation · Semantic segmentation

1 Introduction

Semantic segmentation is a fundamental and challenging task in computer vision. Given a single image, it aims to distinguish and localize each predetermined object at the pixel level. Owing to the rapid development of deep learning in recent years, advanced data-driven models such as fully convolutional network (FCN) [17] and Deeplab [4] can automatically discriminate multiple objects in an intricate image with promising results, which are faster and more accurate than old approaches. Many applications have introduced semantic segmentation techniques to enhance

© ICST Institute for Computer Sciences, Social Informatics and Telecommunications Engineering 2021
Published by Springer Nature Switzerland AG 2021. All Rights Reserved
H. Gao et al. (Eds.): CollaborateCom 2020, LNICST 350, pp. 355–371, 2021.
https://doi.org/10.1007/978-3-030-67540-0_21

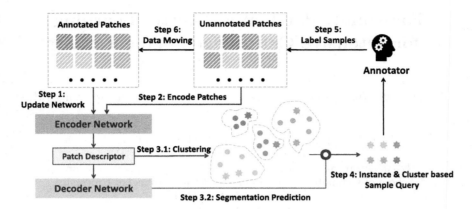

Fig. 1. Overview of our approach. All annotated and unannotated images are partitioned into patches before sending to the loop. Six steps are executed in order and repeated until reaching stop criteria.

automation level, such as remote sensing monitoring, autonomous vehicles, and auxiliary diagnosis [1,3,7,27]. Besides the superiority of the model structure and learning algorithm, the success of deep learning also relies on high-quality labeled data. Unfortunately, this data requirement cannot be satisfied in many practical problems, for instance, the biomedical image segmentation task. Annotating this kind of images is very costly and time-consuming because only pathologists are able to identify tissues or lesions, and mark their contours.

To this end, active learning (AL), which intends to maximize the model performance with minimum cost in labeling data, becomes an emerging research hotspot. In other words, an AL method iteratively runs a query strategy to select the most valuable samples for the annotation and helps classifiers achieve high accuracy from limited data points [2,14,22]. Many studies have introduced how to combine AL to the biomedical image segmentation task to reduce annotation cost [6,19,20,29]. In this work, we focus on the intensive cell image, within which cell objects are close together. We observe that cells in this kind of image have relatively fixed and similar contours. Thus, existing AL methods that run an image-level query strategy during the data selection would bring duplicate annotation because similar objects provide limited information in model training. This inconsistency raises an interesting problem: *Is the image-level query strategy efficient enough for the intensive cell image segmentation?* Based on the observation, we assume that further cost reduction can be achieved by measuring different regions of images separately and only selecting the most critical regions within images for the annotation.

Consequently, we present a patch-wise active learning method named FocAnnot (focal annotation) for the intensive cell image segmentation, as illustrated in Fig. 1. There are six steps in a selection iteration. In Step 1, annotated images are partitioned into patches with fixed size to initialize an encoder-decoder convolutional network (ED-ConvNet) for image segmentation. The unannotated candidates with the same patch size are then fed to the trained ED-ConvNet to get segmentation results (Step 3.2), as well as their latent representations mapped by the encoder part of ED-ConvNet (Step 2). After that, a clustering method is adopted to cluster candidates into distinct groups (Step 3.1). In Step 4, we propose a novel query strategy that integrates candidate information at both instance and cluster level generated in Step 3. Based on query results, we select valuable regions within images to experts for the annotation (Step 5). Finally, these new samples are added to enlarge the annotated dataset in Step 6. The selection procedure is repeated until reaching predefined conditions.

The main contributions of this work can be summarized as follows:

- We propose FocAnnot, a patch-wise active learning method, to reduce duplicate annotation and save cost in the intensive cell image segmentation task. FocAnnot measures distinct regions within an image and only asks experts to annotate a part of valuable regions.
- We propose a cluster-instance double ranking query strategy consisting of two cluster-based criteria that estimate the importance of different image patch groups, and an instance criterion incorporated with traditional uncertainty for the patch selection.
- FocAnnot is evaluated on a real-world cell-intensive dataset. Up to 15% cost saving or 2% performance improvement can be achieved in the segmentation task.

The remainder of this paper is organized as follows. In Sect. 2, we review the related work on biomedical image segmentation and active learning. Section 3 describes the details of the proposed FocAnnot. Experimental results on a real-world dataset are reported in Sect. 4, and conclusions are given in Sect. 5.

2 Related Work

Active Learning. The main task of AL is to design a query strategy that measures the value of unlabeled data for different task objectives [6, 10, 14, 16]. Recent studies can be concluded in three categories, i.e., single model, multi-model, and density-based methods. In the class of single model, the most informative samples are picked according to the probabilistic outputs of a trained classifier. This strategy is also known as uncertainty sampling [8, 30]. Similarly, the multi-model approach, or query-by-committee, also leverages predicted labels but in an ensemble way. A sample with the most disagreements among multiple classifiers is considered as an informative instance [12]. Kullback–Leibler (KL) divergence, entropy, and top-k best are some criteria in common use that measure amount of additional information brought by selected data. The density-based method aims to find data points that are uncertain as well as representative. The idea is re-ranking queried samples based on the similarity to their neighbors or directly querying from the pre-clustered sets [23]. Moreover, Zhou et al. [34] use a pre-trained CNN to predict augmented data based on the assumption that images generated from the same seed are expected to have similar predictions. Entropy and KL divergence are employed to evaluate uncertainty and prediction consistency among augmented images, respectively. The study in [18] incorporates the generative adversarial network into the AL framework to generate informative data, while [15] introduces a learnable query strategy that estimates expected error reduction by a regressor. Yang *et al.* [29] applies uncertainty sampling to choose several candidates and discards duplicate selections with high similarity. Their proposed suggestive annotation method achieves state-of-the-art segmentation performance in an intensive cell image dataset.

Biomedical Image Segmentation. Automatic segmentation brings benefits to the medical field that enhances lesion identification, surgery planning, and evaluation of treatment effects. Recent advances in biomedical image segmentation have covered many organs, such as liver [11], brain [21], and prostate [33]. There are also many efforts on other human tissues, including cells [26], nucleus [28], and melanoma [32]. Technically, most of the state-of-the-art segmentation models are based on the convolutional neural network (CNN) with an encoder-decoder architecture. The encoder network applies convolution and down-sampling operation to images, which compresses raw inputs to learn latent features. The decoder network then deconvolves and up-samples latent features to predict each pixel in images. Many studies have designed new components to improve the robustness and generalization of CNN models. Ronneberger *et al.* [24] build skip connections between the down-sampling and up-sampling path to enhance the sharing of local information. Dilated convolution [31] is adopted to increase the receptive field, which works as the alternative to pooling operation but reduces the model size. Jointly learning knowledge from multi-scale data is an effective strategy as well, for instance, multiresolution inputs [5], multi-scale latent representations [4], and sequential structure of multiple networks [9].

3   Our Approach

Well-annotated data empowers segmentation models to achieve promising results but is also costly, especially for the biomedical image. With the constraint of annotation costs, active learning aims to retrieve the most valuable images from the unannotated dataset for the specific tasks. We suppose annotators are experts who provide high-quality labeling. Hence our task becomes to get similar segmentation performance of full supervision by selecting limited data with minimum cost. To this end, we propose FocAnnot, a patch-wise active learning method that only assigns parts of the region within an image to human annotators, to further reduce annotation costs of intensive cell images compared to existing image-level AL methods.

3.1   Overview of FocAnnot

The overview of our approach is shown in Fig. 1, including six steps. Images are first partitioned into small patches before sending them to the loop. The details

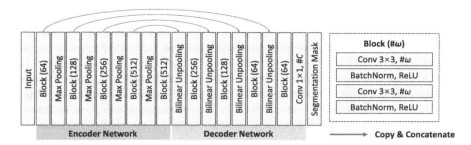

Fig. 2. Architecture of the ED-ConvNet implemented in this work. The symbols C and ω denote the number of classes and the number of kernels, respectively.

will be described in the next section. Initially, annotated image patches are fed to ED-ConvNet for network training. The encoder of ED-ConvNet learns latent representations of patches in low dimensional space and the decoding part makes predictions to each pixel. After that, we put all unannotated patches into well-trained ED-ConvNet. Besides decoding results, outputs of the encoder network are taken as well (*patch descriptors* for short). As mentioned earlier, cells within an image are similar. Instead of the overall profile, we focus on the local difference and group high-level features of patches, i.e., patch descriptors, into clusters in the third step. Each cluster can be regarded as regions with similar types of contents. Based on this, we propose a double ranking query strategy from perspectives of the image patch itself (instance-level) and its kind (cluster-level). Patches with the highest-ranking are selected for labeling and moved to the annotated dataset in the last two steps. Again, parameters of ED-ConvNet are updated by the enlarged training set. FocAnnot runs all these steps sequentially until predefined conditions are achieved, for instance, budget or accuracy of segmentation.

Furthermore, we illustrate an implementation of the ED-ConvNet constructed in FocAnnot. As shown in Fig. 2, ED-ConvNet contains a simple block design that executes a 3×3 convolution followed by a batch normalization [13] and ReLU activation, in two iterations. In the encoding part, a max-pooling layer is set between each of two blocks. An image patch is processed in such stacked layers and its patch descriptor is obtained, that is, the outputs of the second "Block (512)" in this example. The internal outputs of blocks are copied to the decoder network as one of the inputs in corresponding blocks. The decoding part has a symmetrical structure to the encoder network that uses bilinear interpolation for up-sampling to reconstruct image patches. At last, a 1×1 convolution layer is applied to predict the class of each pixel and gives final segmentation results.

3.2 Image Partitioning and Clustering

This preprocessing step aims to alleviate the problem of duplicate annotation in existing image-level AL methods. For cell-intensive images, the same types of biomedical objects are similar, with partial differences. Hence we can partition images into patches to focus on details at the region level. The advantage of

patch-wise learning is that each patch is considered to be of different importance. Compared to the image-wise annotation, experts are only required to
annotate the most valuable regions in an image and ignore other parts, which
save expenses.

We assume patch size and step size are two critical factors affecting outcomes,
i.e., trade-offs between cost and accuracy. Patch size controls the integrity of partitioned objects. It should be neither too big nor too small, which goes against the
idea of patch-wise learning or loses object information, respectively. Step size aims
to increase the richness and novelty of partitioned objects. Similarly, small step
size is not desirable because it increases the computational efforts of the query.

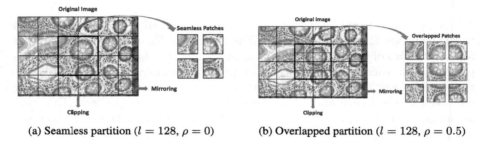

(a) Seamless partition ($l = 128$, $\rho = 0$)   (b) Overlapped partition ($l = 128$, $\rho = 0.5$)

Fig. 3. Two strategies for image patches partition. Each patch is square with the side
length of l, and ρ is ratio of overlapping between neighbors.

In this study, we set two partition strategies for cell-intensive images, as shown
in Fig. 3. Patch size l and overlapping rate ρ are two parameters to control the partitioning. The patch size denotes the side length of each rectangular region while
ρ indicates how many overlaps are between two adjacent patches. Figure 3a illustrates a seamless strategy that outputs 24 patches of 128×128 pixels. We zoom in
a subregion involving four patches (bold black box) and show them on the right
side of the image. Complete objects are scattered in different patches, but they
are almost distinguishable from the background. In other words, the appropriate
patch size has a limited impact on task difficulty. Figure 3b shows the overlapped
partition strategy with $\rho = 0.5$. We add red lines to the same subregion to demonstrate the boundaries of additional patches. On the right side, extra five patches
are obtained by half overlapping, which enriches the novelty of the unannotated
dataset. It also helps alleviate the potential object integrity problem brought by
the seamless strategy. Nevertheless, the side effect is the time of the sample query
in Step 4 will increase because of the doubled data. In both two strategies, remaining areas on the bottom and rightmost of an image (gray shadows) will be dropped
or expanded. An image is padded to fill the missing by mirroring regions if the
required size is smaller than $l/2$, otherwise just clipping the area for alignment.

After partitioning, we are able to measure different regions separately to
avoid duplicate annotation by the image-level selection. Various types of patches

are categorized into groups based on their similarity by calculating Euclidean distance between their patch descriptors and group centers:

$$\|\overline{\mathbf{F}} - \mathbf{M}\|^2 \tag{1}$$

where $\overline{\mathbf{F}}$ and $\mathbf{M}$ are channel-wise average pooling of a patch descriptor and the centroid of a group, respectively. In the remainder of this article, we use $\overline{\mathbf{F}}$ to denote the patch descriptor.

To improve query efficiency, we adopt the mini-batch k-means algorithm for the clustering. The number of groups depends on the complexity of cell objects. We have mentioned that the contours of some cells are quite simple. So a small k is enough to distinguish different types of patches for most cases. Detailed performance comparisons of the group number will be discussed in the experiment.

3.3   Cluster-Instance Double Ranking

Many criteria have been studied to determine which images are valuable for the annotation. Among them, uncertainty and diversity are two concepts to evaluate so-called "worthiness". The term uncertainty indicates how confident predictions are given by a model, while diversity considers the degree of dissimilarity among samples. Most of the current works focus on query strategy at the instance level but ignore the cluster structure. In this study, we propose two cluster-wise criteria that measure patch types based on uncertainty and diversity. At the instance level, we also propose a Wasserstein distance-based diversity criterion incorporated with an instance uncertainty criterion.

For an image patch $\mathbf{x}_n$, $p_j(\mathbf{x}_n)$ denotes the predicted probability of each pixel belonging to the j-th object by the trained ED-ConvNet. The segmentation results are close to 0.5 if the model does not have enough knowledge, for example, representative training data, to identify a patch. This kind of patch is informative and can be regarded as a candidate for the annotation. It is a good choice to use entropy to capture the degree of information involved in samples. The higher value suggests more uncertainty of a patch to the model. Thus, we define *instance uncertainty* (IU) as:

$$\mathcal{H}_n = -\sum_{j=1}^{C} p_j(\mathbf{x}_n) \log(p_j(\mathbf{x}_n)) \tag{2}$$

where C is the number of predefined objects.

Besides, we also want to select a couple of samples, among which are dissimilar. Two patches with high uncertainty could also cause duplicate annotation if they provide similar information. In our case, a patch descriptor can be seen as a probability distribution of high-level features. Thus, we introduce Wasserstein distance as a diversity measurement that estimates differences between two probability distribution P and Q. Compared to KL divergence, Wasserstein distance is a symmetric metric and satisfies the triangle inequality, which is suitable for the similarity calculation. Another advantage of Wasserstein distance is the ability to measure two distributions with little overlap. For example, two patch descriptors would show quite different distribution even in the same group with similar IU. Jensen–Shannon divergence [34] is hard to measure the diversity of patches in this situation, while Wasserstein distance provides a better estimation. Let Ω be a metric space with distance function D and collection $Z(P,Q)$ denotes all possible joint distributions on $\Omega \times \Omega$. The τ-Wasserstein distance is formalized:

$$W_\tau(P,Q) = \left(\inf_{\zeta \in Z(P,Q)} \int_{\Omega \times \Omega} D(u,v)^\tau \, d\zeta(u,v) \right)^{1/\tau} \tag{3}$$

Here ζ is a joint distribution of P and Q, and (u,v) is a data point sampled from ζ. Specially, we calculate Wasserstein distance between patch descriptors to measure the diversity. Supposing $\overline{\mathbf{F}}_n$ and $\overline{\mathbf{F}}_{n'}$ are two d-dimensional descriptors of image patches $\mathbf{x}_n$ and $\mathbf{x}_{n'}$, the *instance diversity* (ID) is defined as their 2-Wasserstein distance:

$$W_{n,n'} = \min_{\mathcal{V} \in \mathfrak{S}(d)} \left(\sum_{m=1}^d \|\overline{\mathbf{F}}_{n,m} - \overline{\mathbf{F}}_{n',\mathcal{V}_m}\|^2 \right) \tag{4}$$

where $\mathfrak{S}(d)$ is all permutations of indices $\{1, \ldots, d\}$, and $\mathcal{V}$ is one of the permutations. The Euclidean norm is adopted as the distance function.

With IU and ID, each image patch can be scored and ranked. Only the first few candidates with the highest scores are selected. To get better results, the reweighting technique is applied for rank adjustment during the selection. For an image patch $\mathbf{x}_n$, its score is defined as the uncertainty of model predictions weighted by average diversity to other Q candidates:

$$S_n = \mathcal{H}_n \times \frac{1}{Q} \sum_{q=1}^Q W_{n,q} \tag{5}$$

Besides criteria at the instance level, we further describe two measurements that consider the characteristics of clusters. After applying the mini-batch k-means algorithm, patches with similar contents are grouped to the same cluster. We suppose that the degree of aggregation, or density, is relevant to the amount of information involved in a cluster. A dense cluster is less-informative because patch descriptors in corresponding metric space tend to be compact. On the contrary, a large distance between patch descriptors implies more novelty and uncertainty in a sparse cluster. To estimate informativeness of a cluster, we denote *cluster uncertainty* (CU) by the average distance of a single group:

$$\mathcal{I}_a^{(c)} = \frac{1}{|\mathcal{K}^{(c)}|} \sum_{\overline{\mathbf{F}}_n \in \mathcal{K}^{(c)}} \|\overline{\mathbf{F}}_n - \mathbf{M}_c\|^2 \tag{6}$$

where $\overline{\mathbf{F}}_n$ is a patch descriptor in the cluster $\mathcal{K}^{(c)}$, and $\mathbf{M}_c$ is the cluster centroid. The $|\cdot|$ indicates the size of a set.

Similar to the instance diversity, clusters should be dissimilar as well. Directly excluding limited clusters is infeasible. Alternatively, we determine the number of candidates to be provided in each group. A cluster that is far from others is asked for more patches because it is quite different from other groups and has a higher probability of providing valuable data. For a cluster $\mathcal{K}^{(c)}$, the mean distance to other centroids are defined as the *cluster diversity* (CD):

$$\mathcal{I}_r^{(c)} = \frac{1}{k-1} \sum_{v=1}^{k} \|\mathbf{M}_c - \mathbf{M}_v\|^2 \tag{7}$$

Instead of uniform sampling, we select the most valuable image patches from clusters proportionately based on the importance. Two cluster-wise criteria defined on Eq. (6) and Eq. (7) are parameterized by λ to give an estimation of cluster importance:

$$\mathcal{I}^{(c)} = \lambda \mathcal{I}_a^{(c)} + (1-\lambda)\mathcal{I}_r^{(c)} \tag{8}$$

At last, the proposed cluster-instance double ranking query strategy is described in Algorithm 1. The importance $\mathcal{I}^{(c)}$ of each cluster is calculated and normalized. Recall that $\sum_{c=1}^{k} \mathcal{I}^{(c)} = 1$, which means $\mathcal{I}^{(c)}$ can be treated as the probability of a cluster to provide informative data. For the double ranking strategy, we use $\mathcal{I}^{(c)}$ to confirm the number of required patches in each ranking step. In the first round, we filter out less informative data and retain $Q^{(c)}$ patches from cluster $\mathcal{K}^{(c)}$ based on IU. Corresponding reweighted scores $\mathcal{S}^{(c')}$ of only $Q^{(c)}$ patches are then calculated in the second step. Finally, we pick top $T^{(c)}$ rankings from refined cluster $\mathcal{K}^{(c')}$ for the annotation.

Algorithm 1
Cluster-instance double ranking query strategy

Input:
    Patch descriptors in k groups $\mathcal{K}^{(1)}, \ldots, \mathcal{K}^{(k)}$;
    The number of uncertain patches Q;
    The number of required patches T in a query
Output:
    Set of selected patches $\mathcal{A}$ for the annotation
1: $\mathcal{A} = \varnothing$
2: **for** $c = 1, \ldots, k$ **do**
3:     Compute $\mathcal{I}^{(c)}$ of $\mathcal{K}^{(c)}$ by Eq. (6), (7), (8)
4:     $Normalize(\mathcal{I}^{(c)})$
5:     $T^{(c)} = \mathcal{I}^{(c)} \times T$
6:     $Q^{(c)} = \mathcal{I}^{(c)} \times Q$
7:     Let $\mathbf{X}^{(c)} = \{\mathbf{x}_n | \overline{\mathbf{F}}_n \in \mathcal{K}^{(c)}\}$ where $\mathbf{x}_n$ is the image patch of $\overline{\mathbf{F}}_n$
8:     $\mathcal{H}^{(c)} = \{\mathcal{H}_n | \mathbf{x}_n \in \mathbf{X}^{(c)}\}$ by Eq. (2)
9:     $\mathcal{K}^{(c')} = TopRank(\mathcal{K}^{(c)}, \mathcal{H}^{(c)}, Q^{(c)})$
10:    $\mathcal{S}^{(c')} = \{\mathcal{S}_n | \mathcal{H}^{(c')}, \overline{\mathbf{F}}_n \in \mathcal{K}^{(c')}\}$ by Eq. (5)
11:    $\mathcal{A} = \mathcal{A} \cup TopRank(\mathcal{K}^{(c')}, \mathcal{S}^{(c')}, T^{(c)})$
12: **end for**
13: **return** $\mathcal{A}$

---

4   Experiment

We implemented our FocAnnot in Python using deep learning framework PyTorch and machine learning framework Scikit-learn. All experiments run on an Ubuntu server with eight cores of 2.20 GHz Intel Xeon E5-2630 and two NVIDIA GTX 1080 Ti GPU.

4.1   Dataset

The proposed active learning method is evaluated on a real-world cell-intensive dataset provided by the 2015 MICCAI Gland Segmentation Challenge (GlaS) [25], which contains 165 images of colon histology. As an example, Fig. 3 shows several glands involved in an image. According to the rules of GlaS, images are divided into a training set with 85 images and two test sets with 80 images in total (60 in Part A and 20 in Part B, P-A and P-B for short). In order to eliminate influence by such man-made split, we also generate a random train-test split from 165 images (Mixed for short) using 80% images as the training set and the remaining 20% for testing. All experiments will run on the three different train-test pairs.

Table 1. List of seven query strategies for comparison.

#Strategy	Partitioning	Clustering	Criteria
1	✗	✗	IU
2	✗	✗	SA
3	✗	✗	ED-ConvNet-SA
4	✓	✗	IU
5	✓	✗	IU+ID
6	✓	✓	IU+ID
7	✓	✓	IU+ID+CU+CD

4.2   Experimental Settings

In Algorithm 1, we set $Q = 5\%$ and $T = 2.5\%$ for a single selection, that is, FocAnnot queries 5% patches as candidates once from the training set and half of them are finally selected for the annotation. In each round, ED-ConvNet is retrained on the newly updated dataset, and then the segmentation performance is evaluated on the test set without image partitioning. In the experiment, the annotation cost is calculated as the number of pixels in the selected images. The parameter λ of Eq. 8 is fixed to 0.5. We repeat the train-test step until half of the training set is selected. Furthermore, we intend to explore the influence of parameter settings on the model performance, such as patch size, overlapping rate, and the number of clusters. Totally 12 combinations of $l = \{64, 128, 256\}$, $\rho = \{0, 0.5\}$, $c = \{3, 5\}$ are investigated on GlaS dataset. Two metrics, i.e., Dice coefficient and volumetric overlap error (VOE), are used for performance evaluation. Among them, Dice is preferred by the biomedical image segmentation task, so we choose it as the primary metric in comparisons. Formally, Dice and VOE are defined as:

$$Dice = \frac{2 \times |\mathbf{y} \cap \mathbf{y'}|}{|\mathbf{y}| + |\mathbf{y'}|} \tag{9}$$

$$VOE = 1 - \frac{|\mathbf{y} \cap \mathbf{y'}|}{|\mathbf{y} \cup \mathbf{y'}|} \times 100\% \tag{10}$$

Here $\mathbf{y}$ and $\mathbf{y'}$ are the ground truth and predicted segmentation result, respectively. The $|\cdot|$ is the number of pixels involved in this image. The larger Dice value or smaller VOE value indicates better performance of the model.

In Table 1, we list seven query strategies, including the state-of-the-art method in this scenario (suggestive annotation [29], SA for short), its variant ED-ConvNet-SA (replacing FCN in SA by our implemented ED-ConvNet) and our proposed criteria. Strategies 1–3 are baselines using the whole image while strategies 4–7 run a patch-wise selection. Specially, the comparison of strategies 1 and 4 investigate the effectiveness of image partitioning based on the traditional IU criterion. For strategies 2 and 3, we don not set contrast tests because SA and ED-ConvNet-SA both require a similarity computation with $\frac{1}{2}\mathcal{O}(N^2)$ time complexity, which is unacceptable on partitioned patches. In strategy 5, the proposed instance diversity is incorporated with instance uncertainty as a new selection criterion. In addition to the instance-level strategy, patches are selected from clusters equally in strategy 6, and proportionately in strategy 7. Note that query strategy 7 is a full implementation of cluster-instance double ranking strategy described in Algorithm 1.

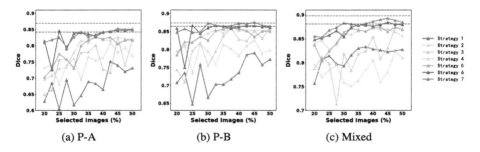

(a) P-A   (b) P-B   (c) Mixed

Fig. 4. Comparisons between seven query strategies evaluated on three train-test pairs. Red and blue dash lines are segmentation performance of ED-ConvNet and FCN module in SA using full training data, respectively.

4.3   Results

We run seven query strategies listed in Table 1 using the same experimental settings. In strategies 4–7, the side length l, overlapping rate ρ, and group number k are set to 128, 0.5, and 3, respectively. For results in Fig. 4, the annotation cost of full-size images and image patches is aligned to the same scale based on the number of revealed pixels. It needs to be emphasized that the number of training images in Mixed is different from P-A and P-B, so the percentage of selected data cannot be compared directly. In the following, we first summarize our observations.

Strategy 1 vs. 2 vs. 3: Performance of Baselines. As we can see in Fig. 4a and Fig. 4b, SA indeed improves the segmentation performance compared to traditional IU. Surprisingly, in Fig. 4c, SA fails on mixed data and even perform worse than IU criterion. We believe the man-made interference in the provided train-test split leads to over-fitting and reduces the generalization of SA. In strategy 3, we replace all FCNs in SA by our ED-ConvNet to get its variant, i.e., ED-ConvNet-SA. Again, ED-ConvNet-SA cannot surpass IU on Mixed with less than 40% data. This also explains the limited generalization of the SA-based strategy. From a model point of view, the results show that ED-ConvNet-SA outperforms original SA in all three testing sets, which validates the robustness and effectiveness of ED-ConvNet.

Strategy 1 vs. 4: the Benefit of Image Partitioning. Different from traditional strategies querying at the image level, we introduce the image partitioning approach to generate dozens of patches that separates informative regions from the whole image. Strategy 4 achieves considerable improvement (~6% better in P-A & P-B and ~4% better in Mixed) and even slightly better than ED-ConvNet-SA. This result answers our question in Sect. 1: the image-level query strategy is not good enough for the intensive cell image segmentation task. We can get better performance by selecting parts of valuable regions within an image rather than the whole.

Strategy 4 vs. 5: the Effect of Instance Diversity. We compare the proposed instance diversity criterion that reweights IU, to IU-only on image patches. Results show that our reweighting approach improves the Dice score to a certain extent. The reason is that the ID criterion avoids selecting too many uncertain image patches with high similarity, aka duplicate annotation. This helps refine the selection procedure to obtain more diverse patches.

Strategy 5 vs. 6 vs. 7: the Effect of Cluster-Wise Selection. In addition to image partitioning, we also explore the effect of clustering in the selection. Strategy 6 applies the mini-batch k-means algorithm and select image patches equally from clusters. It improves almost 1% Dice score compared to non-clustering strategy 5. As for the proposed cluster-instance double ranking algorithm in strategy 7, it has a further improvement for about 0.8%. This verifies the advantages of cluster importance estimation described in Eq. 8.

In general, FocAnnot is able to select less than half of the data to surpass FCN of full supervision. It also approximates (in Mixed) or even outperforms (in P-B)

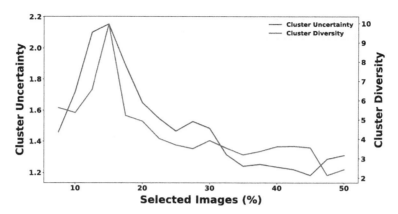

Fig. 5. Changes of averaged cluster uncertainty and cluster diversity during the selection.

ED-ConvNet using full training data. Specially, strategies 6 and 7 can achieve similar performance as ED-ConvNet-SA by selecting only 25% training data and obtain a good enough model with 30% data. All these experiments indicate that the combination of instance-level and cluster-level criteria is more potent than the image-level query strategy.

4.4   Visualization and Discussion

In Fig. 5, we visualize the changes in cluster uncertainty and cluster diversity during the selection on average. In the first few rounds, CU and CD are increasing and reach a peak when obtaining 15% of training data. This indicates that cluster-level criteria may not help much in the beginning. FocAnnot mainly depends on instance-level criteria to select valuable image patches at that time. After that, the segmentation model has learned enough knowledge to distinguish distinct patch descriptors. We can see that CU and CD decrease, which means cluster-level criteria begin to work at this stage. Representative image patches are continuously selected from groups. Hence clusters become denser and closer to each other. More concretely, CU and CD have a rapid drop between 15% to 25% and reach a relatively low position at 30%. This supports the previous conclusion, that is, the proposed query strategy can quickly improve segmentation accuracy and tends to be stable with around 30% training data.

FocAnnot has been proved effective and our query strategy is superior to the competitor. Moreover, we would like to investigate the model generalization in different parameter settings. As shown in Table 2, 12 parameter combinations are grouped into three categories based on patch sizes. The overlapped partition approach is always better than the seamless partition. We believe overlapped patches enrich local details of images and provide more valuable choices during the selection. Based on the overlapped setting, $k = 3$ is preferred except for two comparisons in P-A. Three clusters are enough to describe structures of cell

Table 2. Analysis of parameter combinations based on query strategy 7 with 50% selected training data. The best performances in each group are in bold. The underline indicates the setting outperforms all other combinations.

Parameters			Dice			VOE (%)		
l	ρ	k	Mixed	P-A	P-B	Mixed	P-A	P-B
64	0	3	0.857	0.805	0.854	24.40	31.51	24.72
64	0	5	0.852	0.806	0.843	24.92	31.47	26.51
64	0.5	3	**0.876**	0.836	**0.861**	**21.24**	27.36	**23.82**
64	0.5	5	0.866	**0.839**	0.855	22.76	**26.81**	24.55
128	0	3	0.856	0.802	0.848	24.17	32.00	25.63
128	0	5	0.867	0.815	0.869	22.69	30.00	22.60
128	0.5	3	<u>0.893</u>	0.848	**0.876**	<u>18.85</u>	25.40	**21.30**
128	0.5	5	0.887	**0.852**	0.872	19.80	**24.68**	22.02
256	0	3	0.875	0.792	0.855	21.49	32.55	24.66
256	0	5	0.870	0.820	0.860	22.22	29.24	23.67
256	0.5	3	**0.891**	<u>0.857</u>	<u>0.887</u>	**18.87**	<u>23.87</u>	<u>19.63</u>
256	0.5	5	0.886	0.843	0.864	19.81	26.30	23.31

objects in this dataset because their contours are less complicated. Too many categories would lead to ambiguous groups instead. The selection of patch size is critical. As we discussed in Sect. 3.2, a small patch size will lose content details and degrade model performance, for example, $l = 64$ vs. $l = 128$. On the other hand, a larger size brings limited benefits to the segmentation as well. Thus, a moderate size, i.e., $l = 128$, could be a good trade-off to reduce annotation cost with an acceptable accuracy at the same time.

In Fig. 6, we show the ten most valuable images selected by SA. Then, we mark ten patches per image in red, which are considered important based on FocAnnot. All these images are from the training set of Mixed, and the following names are their origin ID of GlaS. We observe that our query strategy prefers some interesting patterns:

- Most of the regions are located on the object boundaries, especially crossing two glands (#4, #50, #58, #84 in the training set). This pattern benefits the segmentation task on biomedical images with multiple and intensive objects. The model pays more attention to detailed differences among close glands.
- Contours with rough shape are identified and recommended for the annotation (#10, #19, #29 in the training set and #42 in test Part A). It is natural to select rare types of contours to improve model generalization and so our method does.
- Poorly differentiated epithelial cell nuclei and lumen in malignant glands are also useful to the model (#10, #29, #58 in the training set). This type of patch shows one of the most significant differences between benign and malignant glands in clinical practice.

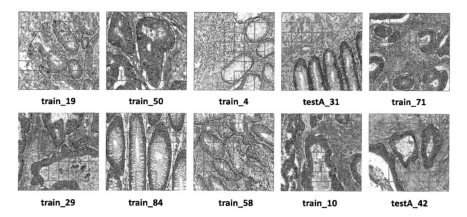

Fig. 6. Visualization of top 10 valuable images selected by SA. For each image, the first 10 queried patches by FocAnnot are marked as red rectangles. The blue curves show contours of glands.

According to these particular regions, more effective active learning methods can be developed by applying well-designed constraints.

5 Conclusion and Future Work

We have proposed a patch-wise active learning method, namely FocAnnot, for the intensive cell image segmentation problem based on an encoder-decoder convolutional network. The key idea is to partition images into patches and cluster them based on their high-level features. Hence, we can evaluate each patch separately to avoid duplicate annotation of similar cells within an image. Two criteria, i.e., cluster uncertainty and cluster diversity, are proposed to estimate the importance of each group. We also present an instance diversity criterion incorporated with the instance uncertainty to seek valuable data within clusters. The experimental results on a real-world cell-intensive dataset demonstrate that our method is able to reduce an additional 15% annotation cost or improve 2% segmentation performance compared to the competitor.

In the future, we would like to extend our query strategy in a multi-scale manner that combines local content with global knowledge. We are also interested in introducing object detection techniques to replace the hard partitioning approach with arbitrary size.

Acknowledgment. This work was supported in part by the National Key Research and Development Program of China under Grant 2017YFC1001703, in part by the National Natural Science Foundation of China under Grant 61825205, Grant 61772459, and in part by the National Science and Technology Major Project of China under Grant 50-D36B02-9002-16/19.

References

1. Badrinarayanan, V., Kendall, A., Cipolla, R.: SegNet: a deep convolutional encoder-decoder architecture for image segmentation. IEEE Trans. Pattern Anal. Mach. Intell. **39**(12), 2481–2495 (2017)
2. Bonnin, A., Borràs, R., Vitrià, J.: A cluster-based strategy for active learning of RGB-D object detectors. In: ICCV Workshops, pp. 1215–1220 (2011)
3. Chen, B.k., Gong, C., Yang, J.: Importance-aware semantic segmentation for autonomous driving system. In: IJCAI, pp. 1504–1510 (2017)
4. Chen, L.C., Papandreou, G., Kokkinos, I., Murphy, K., Yuille, A.L.: DeepLab: semantic image segmentation with deep convolutional nets, atrous convolution, and fully connected crfs. IEEE Trans. Pattern Anal. Mach. Intell. **40**(4), 834–848 (2018)
5. Chen, L.C., Yang, Y., Wang, J., Xu, W., Yuille, A.L.: Attention to scale: Scale-aware semantic image segmentation. In: CVPR, pp. 3640–3649 (2016)
6. Chyzhyk, D., Dacosta-Aguayo, R., Mataró, M., Graña, M.: An active learning approach for stroke lesion segmentation on multimodal MRI data. Neurocomputing **150**, 26–36 (2015)
7. Dou, Q., Chen, H., Jin, Y., Yu, L., Qin, J., Heng, P.-A.: 3D deeply supervised network for automatic liver segmentation from CT volumes. In: Ourselin, S., Joskowicz, L., Sabuncu, M.R., Unal, G., Wells, W. (eds.) MICCAI 2016. LNCS, vol. 9901, pp. 149–157. Springer, Cham (2016). https://doi.org/10.1007/978-3-319-46723-8_18
8. Dutt Jain, S., Grauman, K.: Active image segmentation propagation. In: CVPR, pp. 2864–2873 (2016)
9. Eigen, D., Fergus, R.: Predicting depth, surface normals and semantic labels with a common multi-scale convolutional architecture. In: ICCV, pp. 2650–2658 (2015)
10. Gal, Y., Islam, R., Ghahramani, Z.: Deep Bayesian active learning with image data. In: ICML, pp. 1183–1192 (2017)
11. Hoogi, A., Subramaniam, A., Veerapaneni, R., Rubin, D.L.: Adaptive estimation of active contour parameters using convolutional neural networks and texture analysis. IEEE Trans. Med. Imaging **36**(3), 781–791 (2017)
12. Iglesias, J.E., Konukoglu, E., Montillo, A., Tu, Z., Criminisi, A.: Combining generative and discriminative models for semantic segmentation of CT scans via active learning. In: Székely, G., Hahn, H.K. (eds.) IPMI 2011. LNCS, vol. 6801, pp. 25–36. Springer, Heidelberg (2011). https://doi.org/10.1007/978-3-642-22092-0_3
13. Ioffe, S., Szegedy, C.: Batch normalization: Accelerating deep network training by reducing internal covariate shift. In: International Conference on Machine Learning, pp. 448–456 (2015)
14. Konyushkova, K., Sznitman, R., Fua, P.: Introducing geometry in active learning for image segmentation. In: ICCV, pp. 2974–2982 (2015)
15. Konyushkova, K., Sznitman, R., Fua, P.: Learning active learning from data. In: NeurIPS, pp. 4228–4238 (2017)
16. Lin, C.H., Mausam, M., Weld, D.S.: Re-active learning: active learning with relabeling. In: AAAI (2016)
17. Long, J., Shelhamer, E., Darrell, T.: Fully convolutional networks for semantic segmentation. In: CVPR, pp. 3431–3440 (2015)

18. Mahapatra, D., Bozorgtabar, B., Thiran, J.-P., Reyes, M.: Efficient active learning for image classification and segmentation using a sample selection and conditional generative adversarial network. In: Frangi, A.F., Schnabel, J.A., Davatzikos, C., Alberola-López, C., Fichtinger, G. (eds.) MICCAI 2018. LNCS, vol. 11071, pp. 580–588. Springer, Cham (2018). https://doi.org/10.1007/978-3-030-00934-2_65

19. Mahapatra, D., Buhmann, J.M.: Visual saliency based active learning for prostate MRI segmentation. In: Zhou, L., Wang, L., Wang, Q., Shi, Y. (eds.) MLMI 2015. LNCS, vol. 9352, pp. 9–16. Springer, Cham (2015). https://doi.org/10.1007/978-3-319-24888-2_2

20. Mahapatra, D., et al.: Active learning based segmentation of Crohn's disease using principles of visual saliency. In: ISBI, pp. 226–229 (2014)

21. Mansoor, A., et al.: Deep learning guided partitioned shape model for anterior visual pathway segmentation. IEEE Trans. Med. Imaging 35(8), 1856–1865 (2016)

22. Möller, T., Nillsen, I., Nattkemper, T.W.: Active learning for the classification of species in underwater images from a fixed observatory. In: ICCV (2017)

23. Nguyen, H.T., Smeulders, A.: Active learning using pre-clustering. In: ICML, p. 79 (2004)

24. Ronneberger, O., Fischer, P., Brox, T.: U-net: convolutional networks for biomedical image segmentation. In: Navab, N., Hornegger, J., Wells, W.M., Frangi, A.F. (eds.) MICCAI 2015. LNCS, vol. 9351, pp. 234–241. Springer, Cham (2015). https://doi.org/10.1007/978-3-319-24574-4_28

25. Sirinukunwattana, K., et al.: Gland segmentation in colon histology images: the glas challenge contest. Med. Image Anal. 35, 489–502 (2017)

26. Song, Y., et al.: Accurate cervical cell segmentation from overlapping clumps in pap smear images. IEEE Trans. Med. Imaging 36(1), 288–300 (2017)

27. Volpi, M., Tuia, D.: Dense semantic labeling of subdecimeter resolution images with convolutional neural networks. IEEE Trans. Geosci. Remote Sens. 55(2), 881–893 (2017)

28. Xing, F., Xie, Y., Yang, L.: An automatic learning-based framework for robust nucleus segmentation. IEEE Trans. Med. Imaging 35(2), 550–566 (2016)

29. Yang, L., Zhang, Y., Chen, J., Zhang, S., Chen, D.Z.: Suggestive annotation: a deep active learning framework for biomedical image segmentation. In: Descoteaux, M., Maier-Hein, L., Franz, A., Jannin, P., Collins, D.L., Duchesne, S. (eds.) MICCAI 2017. LNCS, vol. 10435, pp. 399–407. Springer, Cham (2017). https://doi.org/10.1007/978-3-319-66179-7_46

30. Yang, Y., Ma, Z., Nie, F., Chang, X., Hauptmann, A.G.: Multi-class active learning by uncertainty sampling with diversity maximization. Int. J. Comput. Vis. 113(2), 113–127 (2015)

31. Yu, F., Koltun, V.: Multi-scale context aggregation by dilated convolutions. arXiv preprint arXiv:1511.07122 (2015)

32. Yu, L., Chen, H., Dou, Q., Qin, J., Heng, P.A.: Automated melanoma recognition in dermoscopy images via very deep residual networks. IEEE Trans. Med. Imaging 36(4), 994–1004 (2017)

33. Yu, L., Yang, X., Chen, H., Qin, J., Heng, P.A.: Volumetric ConvNets with mixed residual connections for automated prostate segmentation from 3D MR images. In: AAAI, pp. 66–72 (2017)

34. Zhou, Z., Shin, J., Zhang, L., Gurudu, S., Gotway, M., Liang, J.: Fine-tuning convolutional neural networks for biomedical image analysis: actively and incrementally. In: CVPR, pp. 7340–7349 (2017)

Short Paper Track

EFMLP: A Novel Model for Web Service QoS Prediction

Kailing Ye[1,2], Huiqun Yu[1,3(✉)], Guisheng Fan[1], and Liqiong Chen[4]

[1] Department of Computer Science and Engineering, East China University of Science
and Technology, Shanghai 200237, China
yhq@ecust.edu.cn
[2] Shanghai Key Laboratory of Computer Software Evaluating
and Testing, Shanghai 201112, China
[3] Shanghai Engineering Research Center of Smart Energy, Shanghai, China
[4] Department of Computer Science and Information Engineering,
Shanghai Institute of Technology, Shanghai 201418, China

Abstract. With the emergence of service-oriented architecture, quality of service (QoS) has become a crucial factor in describing the non-functional characteristics of Web services. In the real world, the user only requests limited Web services, the QoS record of Web services is sparsity. In this paper, we propose an approach named factorization machine and multi-layer perceptron model based on embedding technology (EFMLP) to solve the problem of sparsity and high dimension. First, the input data will be sent to embedding layer to reduce the data dimension. Then, the embedded feature vector will send to the factorization machine. After that, the first-order and second-order weights of the factorization machine are used as the initial weights of the first layer of the multi-layer perceptron. And the multi-layer perceptron is trained to adjust the weights. Finally, 1,974,675 pieces of data from an open dataset is used as experiment data to validate the model, and the result shows that our EFMLP model can predict QoS value accurately on the client side.

Keywords: Embedding model · Factorization machine · Neural network · QoS prediction

1 Introduction

A mount of Web services has been provided in the world. How to choose a web service which satisfies the user has been an urgent problem. Usually, when users access the network, a number of services with the same function can be provided by multiple service providers. Choosing a best Web service to provide users with a Web service recommendation system can greatly improve the user's Internet experience.

In real world, it's also hard to obtain the QoS value by calling the corresponding service from the client side. First, not all Web services are free, some Web services require a fee. Second, the number of services is huge. It will be a huge overhead in

H. Gao et al. (Eds.): CollaborateCom 2020, LNICST 350, pp. 375–385, 2021.
https://doi.org/10.1007/978-3-030-67540-0_22

time when each user requests each web service. Finally, most users are not experts, they do not have the experience and skills to measure Web services, which results in the measurement results not being credible.

In brief, the main contributions of this paper are as follows:

First, we propose the EFMLP model. The factorization machine and the multi-layer perceptron based on embedding model are combined to predict the QoS value.
Second, a real-world dataset is used to verify the effectiveness of the EFMLP model. Experimental results show that the method is more accurate than traditional collaborative filtering and matrix decomposition methods.

The rest part of this paper is organized as follows. The second section describes the missing QoS prediction problem and gives the framework of the EFMLP model. Section 3 introduces the EFMLP model to predict personalized QoS values. Section 4 discusses and analyzes the experimental results. Section 5 contains related work, and Sect. 6 summarizes the full paper.

2   The Overview of EFMLP Model

In this section, the prediction problem of missing QoS value is first described in Sect. 2.1. Then we introduce the framework of EFMLP model in Sect. 2.2.

2.1   Problem Description

Our aim is using the historical QoS information to accurately predict the missing QoS value. The existing data usually contains a user-item matrix, where $U = \{u_1, u_2, u_3, \ldots . u_i,\}$ is the user set, containing i users, $S = \{s_1, s_2, s_3, \ldots . s_j,\}$ is a collection of j Web services. v_{ij} represents the QoS value of the service j observed by the service from user i. Therefore, we construct an $i \times j$ user-service sparse matrix $M \in R^{i \times j}$. In this user-service matrix, each item v_{ij} represents a QoS attribute value. Figure 1 shows a tiny example. Figure 1(a) shows the existed QoS attribute values in the user-service matrix, which represented by set H. When the number of users and services is large, the sparsity of the matrix will be huge. Figure 1(b) shows the complete QoS value after prediction, which is represented by set M.

	S_1	S_2	S_3	S_4	S_5
U_1	5.982		0.228		
U_2		0.262		0.652	
U_3	0.854		0.649		0.115

	S_1	S_2	S_3	S_4	S_5
U_1	5.982	5.394	0.228	0.453	0.527
U_2	2.134	0.262	0.328	0.652	0.554
U_3	0.854	0.352	0.649	0.238	0.115

(a) User-Service Matrix   (b) Predicted User-Service Matrix

Fig. 1.   A tiny example for missing QoS prediction

The missing QoS value can be represented by the set $P = M - H$. The predicted QoS value is to obtain the set P by using the existing set H. The EFMLP model is based on the factorization machine model and the neural network, which is trained through the existing data, finally predict the missing QoS. Then a complete and accurate QoS value set M is obtained.

2.2 The Framework of EFMLP Model

In this section, we show the framework of EFMLP model. As shown in Fig. 2, the EFMLP model training is divided into two stages. In the first stage, the input data is converted into feature vectors by one-hot encoding, and then the embedding layer is used for dimensionality reduction and obtain information existing between input vectors. In the second stage, the output of the embedding layer is sent to the factorization machine for training, and the first-order weights and second-order weights of the factorization machine are used as the initial weights of the first layer of the multi-layer perceptron. Then train the multi-layer perceptron.

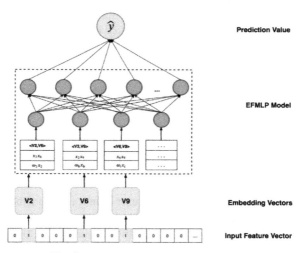

Fig. 2. The framework of EFMLP model

The following we will introduce the components of the model in detail.

(1) Obtain the feature vector. The input data includes user id and service id, both of them are integer types. One-hot encoding is applied to user id and service id respectively.

(2) Embedding vector. The Embedding layer compresses the input feature vectors into low-dimension, dense real-valued vectors, as to keep the feature information and ease the training pressure of the model in the next stage.

(3) EFMLP model. The factorization machine is used to model the embedding vector of the user id and the embedding vector of the service id, thus obtaining the interactive relationship between the user id and the service id. The first and second order coefficients of the factorization machine are used to initialize the first layer parameters of the multilayer perceptron to train the multilayer perceptron.

(4) Obtain the predicted value. After using all historical data to train EFMLP model, the model parameters will be determined. The model will output the predicted QoS value using the new data.

3   Details of EFMLP Model

The prediction process of EFMLP model can be divided into extracting embedding features phase, sending the embedding feature vectors to factorization machine phase. Finally, train the multilayer perceptron. The model is implemented with Pytorch1.3.1. We train the model on a workstation with Ubuntu18.04 LTS, Intel Xeon 8 cores processor and 32 GB memory.

3.1   Extracting Embedding Features

The input data is the user id and service id, and the data type is integer. First, the user id and service id will encode with one-hot layer, one-hot uses a bit status register to encode a state. The total number of item types is represented by a bit value of 1 and all other bit values of 0. For example, there are three users, u_1, u_2, and u_3. Figure 3 is a toy example of one-hot encoding in three users.

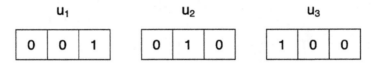

Fig. 3. The toy example of user id one-hot encoding

After obtaining the one-hot encoding of the user id and service id, the one-hot encoded vector is used as the input of the fully connected layer to output the embedding vector of the user id and service id. The role of the embedding layer is to map the original data from the source space to the target space and retain its structure. The data dimension sent to the model is kept at a low dimension, to speed up the model training process.

3.2   The Core of EFMLP Model

First, the factorization machine part of the EFMLP model is used for training, after that the factorization machine model trained weight is used as the initial weight of the first layer of the multi-layer perceptron to train the multi-layer perceptron. Finally, the model is combined with the factorization machine model to predict the missing QoS value.

The one-hot encoded user id and service id are sent to the embedding layer, and the embedding layer outputs low-dimension embedding features, then the embedding user id and service id are sent to the factorization machine. The expression of the factorization machine here is:

$$y_{FM}(x) = \omega_0 + \sum_{i=1}^{N} w_i x_i + \sum_{i=1}^{N} \sum_{j=i+1}^{N} \langle V_i, V_j \rangle x_i x_j \tag{1}$$

Here, we use W_0^i to represent the weight value of the i-th layer of the MLP in the EFMLP model. W_0^i is initialized by the deviation term w_i and the vector v_i (for example, $W_0^i[0]$ is initialized by w_i, $W_0^i[1]$ initialized by v_i^1, $W_0^i[2]$ initialized by v_i^2). $z_i \in R^{K+1}$ is the parameter vector of the i-th field in the factorization machine.

$$z_i = W_0^i x[start_i : end_i] = \left(w_i, v_i^1, v_i^2, v_i^3, \ldots, v_i^K\right) \tag{2}$$

Where $w_0 \in R$ is a global scalar parameter, and n is the total number in all fields.

$$z = (w_0, z_1, z_2, \ldots z_i, \ldots, z_n) \tag{3}$$

Here $W_1 \in R^{M \times j}$, $b_1 \in R^M$ and $z \in R^J$.

We choose the sigmoid function as the activation function of the multilayer perceptron, where $sigmoid(x) = \frac{1}{1+e^{-x}}$, $W_1 \in R^{M \times J}$, $b_1 \in R^M$ and $z \in R^J$. Here are:

$$l_1 = sigmoid(W_1 z + b_1) \tag{4}$$

The predicted QoS value is:

$$\hat{y} = W_2 l_1 + b_2 \tag{5}$$

Where $W_2 \in R^{L \times M}$, $b_2 \in \mathbb{R}^L$ and $l_1 \in R^M$.

In the EFMLP model, the factorization machine part can establish a model for data expression in the latent space, which is conducive for further learning. In order to measure the effect of the EFMLP model, a loss function must be used to evaluate the error between the predicted value and the original value. Therefore, the loss function of the EFMLP model is defined as follows:

$$\min_{\theta} L(y, \hat{y}) = \frac{1}{2} \sum_{i=1}^{m} \sum_{j=1}^{n} I_{ij}\left(y_{ij} - \hat{y}_{ij}\right)^2 \tag{6}$$

I_{ij} is an indicator function. When I_{ij} is 1, it indicates that user u_i invoked the web service v_j, otherwise not invoked. The optimization of the loss function refers to minimizing the sum of squared errors. The initial weight of the factorization machine is learned through stochastic gradient descent (SGD). Using the chain rule in back propagation, the weights in the factorization machine are effectively updated. The weight update formula in the factorization machine is:

$$\frac{\partial L(y,\hat{y})}{\partial W_0^i} = \frac{\partial L(y,\hat{y})}{\partial z_i} \frac{\partial z_i}{\partial W_0^i} = \frac{\partial L(y,\hat{y})}{\partial z_i} x[start_i : end_i] \tag{7}$$

$$W_0^i \leftarrow W_0^i - \eta \cdot \frac{\partial L(y,\hat{y})}{\partial z_i} x[start_i : end_i]. \tag{8}$$

4 Experiment

In this chapter, we conduct a series of experiments on the EFMLP model using a real dataset, validate the effectiveness of the EFMLP model, compare with 7 methods, and analyze the results.

4.1 Description of Dataset

WS-Dream is a client side QoS dataset publicly released on the Internet. It contains 339 users and 5825 Web services. Table 1 is the statistical data of WS-Dream dataset.

Table 1. Statistics of WS-dream dataset

Statistics	Value
Number of service users	339
Number of web services	5825
Number of web services invocations	1,974,675
Range of response time	1–20 s
Range of throughput	1–1000 kbps

As mentioned above, there are many attributes of QoS. Although we focus on response time and throughput in this experiment, the EFMLP model can be easily apply to other attributes.

4.2 Metrics

In this experiment, we use Mean Absolute Error (MAE) and Root Mean Square Error (RMSE) to measure the error between the predicted QoS value and the real QoS value.
     MAE is given by:

$$MAE = \frac{\sum_{ij}|\hat{p}_{ij} - p_{ij}|}{N},$$ (9)

and RMSE is defined as follows:

$$RMSE = \sqrt{\frac{\sum_{ij}(\hat{r}_{ij} - r_{ij})^2}{N}}.$$ (10)

4.3 Performance Comparison

In order to verify the performance of EFMLP model on predicting the missed client's QoS value, the following methods are used as baselines.

(1) UMEAN (User Average). Calculate the average QoS value of a service that a user has called to predict the QoS value of a service has not been called.
(2) IMEAN (item average). This method calculates the average QoS of a Web service that has been called by other users as the QoS value of a user who has not called this service.
(3) UPCC (user-based collaborative filtering method using Pearson correlation coefficient) [9]. This method predicts the QoS value by looking for users with high similarity.

(4) IPCC (Project-based collaborative filtering method using Pearson correlation coefficient) [17]. This method is widely used by companies in industry, such as Amazon. In this paper, we look for similar Web services (items) to predict QoS values.

(5) UIPCC [5]. Combine user-based collaborative filtering with item-based collaborative filtering to predict QoS values based on similar users and similar Web services.

(6) PMF (Probability Matrix Factorization) [18]. This method was proposed by Salakhutdinov and Minh. QoS prediction is performed by using a user-item matrix through probability matrix decomposition.

(7) NIMF (Neighbor Synthesis Matrix Factorization). This method first finds a series of similar users for the current user, aggregates the information of these similar users and the available QoS values, and applies the concept of user-collaboration to make QoS predictions.

In this experiment, the sparsity of the matrix is controlled by removing part of the QoS entities from the dataset and filling with 0 to simulate the real situation of client service invocation. For example, a 20% matrix sparsity means that 20% of the user-service matrix will be removed. For the EFMLP model, 80% of the QoS records are used as the training set, and 20% of the QoS records are used as the test set. The greater the sparsity of the matrix, the smaller the amount of data can be used as the training set. The results of QoS missing value prediction are shown in Table 2 and Table 3. The EFMLP model is more accurate than the seven existing prediction methods in both response time and throughput QoS attributes.

Table 2. QoS prediction accuracy comparison (Response Time)

QoS properties	Methods	Matrix sparsity = 90%		Matrix sparsity = 10%	
		MAE	RMSE	MAE	RMSE
Response time	UMEAN	0.8776	1.8542	0.8728	1.8522
	IMEAN	0.6892	1.6416	0.6770	1.5256
	UPCC	0.5561	1.3092	0.4010	1.0851
	IPCC	0.5962	1.3423	0.3476	1.0280
	UIPCC	0.5836	1.3298	0.3447	1.0178
	PMF	0.4865	1.3130	0.3738	1.0514
	NIMF	0.4782	1.2914	0.3674	1.0397
	EFMLP	**0.4274**	**1.1987**	**0.2786**	**0.9730**

4.4 Impact of Matrix Sparsity

For QoS prediction, matrix sparsity is a factor that affects the accuracy of prediction. The lower sparsity means that more user-service entity information is available. In order

Table 3. QoS prediction accuracy comparison (Throughput)

QoS properties	Methods	Matrix sparsity = 90%		Matrix sparsity = 10%	
		MAE	RMSE	MAE	RMSE
Throughput	UMEAN	53.8810	110.3452	53.6560	109.7987
	IMEAN	26.8735	64.8038	26.5214	63.2873
	UPCC	22.6036	54.5220	13.5389	38.8792
	IPCC	26.1821	60.2451	16.7591	42.9217
	UIPCC	22.3639	54.3285	13.1787	38.1428
	PMF	15.8972	48.1782	11.9224	35.8713
	NIMF	15.1392	47.0476	11.7735	35.5175
	EFMLP	**14.7826**	**42.0129**	**8.2694**	**28.1795**

to study the influence of matrix sparsity, the matrix sparsity was adjusted from 10% to 90%. At the same time, we set the embedding size with 80 for response time and 400 for throughout, the number of hidden layer units of multilayer perceptron was 128. The following are the experimental results:

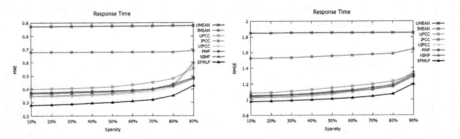

Fig. 4. Impact of matrix sparsity (QoS property is response time)

The experiment results in Fig. 4 and Fig. 5 show that when the matrix sparsity is set from 90% to 50%, the prediction error shows a trend of decline. When the sparsity of

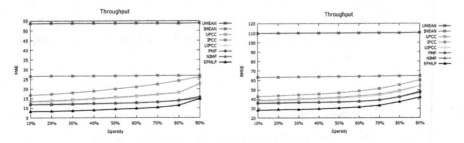

Fig. 5. Impact of matrix sparsity (QoS property is throughput)

the matrix is between 50% and 10%, the prediction error continues to decrease, but the rate of decline becomes slow. The lower the sparsity of the matrix, the more amount of information known as QoS. Therefore, the amount of QoS information will affect the prediction accuracy. When the QoS information reaches a certain level (such as 50%), the effect of QoS information on the prediction accuracy will not be obvious.

4.5   Impact of Dimensionality

The dimension of the embedding layer is a hyper parameter. The dimension determines the number of feature vectors extracted from the user id and service id. In the experiment, the dimension of the embedding layer for response time is set from 50 to 100 with an interval of 10, and the dimension of throughput is set from 200 to 800 with an interval of 100. The matrix sparsity is 90% in the experiment. Figure 6 and 7 are the experiment results of response time and throughput respectively.

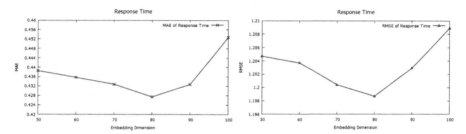

Fig. 6. Impact of embedding vector dimensionality (QoS property is response time)

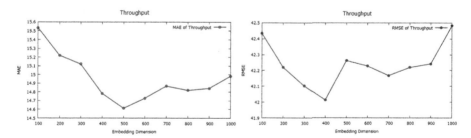

Fig. 7. Impact of embedding vector dimensionality (QoS property is throughput)

The experimental results show that different embedding dimension value satisfy the different QoS attributes. Figure 6 shows that when the dimension changes from 50 to 80, the error of the response time shows a downward trend, which means that the prediction accuracy is improving; when the dimension changes from 80 to 100, the prediction accuracy shows an upward trend. For the throughput, when we choose the MAE as evaluation standard, the dimension changes from 100 to 1000, the trend of both the MAE and RMSE are roughly the same. It is obviously that when the dimension of the RMSE changes from 400 to 500, the accuracy has decreased, indicating that the model is overfitting when the dimension is set to 500. The reason why the embedding dimension of the response time is smaller than the throughput due to the range of the response time is relatively small.

5 Related Work

In this section, we introduce some existing methods about the QoS value prediction research. Most of these methods have been applied to service-oriented architecture scenarios, here are a lot of investigations. Ardagna and Pernici [15] use five QoS attributes (for example, execution time, availability, price, data quality, and reputation) to facilitate flexible configuration during the adaptive service composition process. By studying common QoS attributes and QoS attributes of designated domains, Alrifai and Risse [16] proposed an effective service composition method. This article focuses on the prediction method of the QoS value that the client can observe. In the study of client QoS missing value prediction, the most extensive and in-depth research is collaborative filtering.

The memory-based, model-based and hybrid collaborative filtering methods have two disadvantages:

(1). Missing QoS value, the relevant research fields have been lacking real datasets as experimental research materials.
(2). The dataset has the characteristics of high dimension and huge sparsity.

6 Conclusion

According to the existing QoS attribute data has the features of huge sparsity and high dimensions, we propose a model called EFMLP. Embedding is used to reduce the dimensionality of sparse data, and a factorization machine and multi-layer perceptron are used to build a model for QoS missing value prediction and verify it on a public dataset with lots of records. The results show that the EFMLP model has better accuracy.

In the future, the model can be used to predict missing values of other QoS attributes on the client side (such as fault tolerance, compatibility, reliability.). In addition, other client side factors can be incorporated into the model, to improve model performance.

Acknowledgments. This work was supported by the National Natural Science Foundation of China (No. 61702334, 61772200), the Project Supported by Shanghai Natural Science Foundation (No. 17ZR1406900, 17ZR1429700) and the Planning Project of Shanghai Institute of Higher Education (No. GJEL18135).

References

1. Ding, S., Li, Y., Wu, D., Zhang, Y., Yang, S.: Time-aware cloud service recommendation using similarity-enhanced collaborative filtering and RIMA model. Decis. Support Syst. **107**, 103–115 (2018)

2. Zeng, L., Benatallah, B., Ngu, A.H., Dumas, M., Kalagnanam, J., Chang, H.: QoS-aware middleware for Web services composition. IEEE Trans. Softw. Eng. **30**(5), 311–327 (2004)
3. Wu, Y., Xie, F., Chen, L., Chen, C., Zheng, Z.: An embedding based factorization machine approach for web service QoS prediction. In: Maximilien, M., Vallecillo, A., Wang, J., Oriol, M. (eds.) ICSOC 2017. LNCS, vol. 10601, pp. 272–286. Springer, Cham (2017). https://doi.org/10.1007/978-3-319-69035-3_19
4. Su, K., Ma, L., Xiao, B., Zhang, H.: Web service QoS prediction by neighbor information combined non-negative matrix factorization. J. Intell. Fuzzy Syst. **30**(6), 3593–3604 (2016)
5. Zheng, Z., Ma, H., Lyu, M.R., King, I.: QoS-aware Web service recommendation by collaborative filtering. IEEE Trans. Serv. Comput. **4**(2), 140–152 (2011)
6. Zheng, Z., Ma, H., Lyu, M.R., King, I.: Collaborative web service QoS prediction via neighborhood integrated matrix factorization. IEEE Trans. Serv. Comput. **6**(3), 289–299 (2013)
7. Rendle, S.: Factorization machines. In: 2010 IEEE 10th International Conference on Data Mining (ICDM), pp. 995–1000. IEEE (2010)
8. Harris, D., Harris, S.: Digital design and computer architecture, 2nd edn. Morgan Kaufmann, San Francisco (2012). p. 129. ISBN 978-0-12-394424-5, (2012-08-07)
9. Shao, L., Zhang, J., Wei, Y., Zhao, J., Xie, B., Mei, H.: Personalized QoS prediction for web services via collaborative filtering. In: IEEE International Conference on Web Services (ICWS 2007), Salt Lake City, UT, pp. 439–446 (2007)
10. Athman, B., et al.: A service computing manifesto: the next 10 years. Commun. ACM **60**, 4 (2017)
11. Kim, M., Oh, B., Jung, J., Lee, K.H.: Outlierrobust web service selection based on a probabilistic QoS model. Int. J. Web Grid Serv. **12**(2), 162–181 (2016)
12. Linden, G., Smith, B., York, J.: Amazon.com recommendations: item-to-item collaborative filtering. IEEE Internet Comput. **7**(1), 76–80 (2003)
13. Zhu, J., He, P., Zheng, Z., Lyu, M.R.: Online QoS prediction for runtime service adaptation via adaptive matrix factorization. IEEE Trans. Parallel Distrib. Syst. **28**(10), 2911–2924 (2017)
14. Liu, A., et al.: Differential private collaborative Web services QoS prediction. World Wide Web **22**(6), 2697–2720 (2018). https://doi.org/10.1007/s11280-018-0544-7
15. Ardagna, D., Pernici, B.: Adaptive service composition in flexible processes. IEEE Trans. Softw. Eng. **33**(6), 369–384 (2007)
16. Alrifai, M., Risse, T.: Combining global optimization with local selection for efficient QoS-aware service composition. In: Proceedings of 18th International Conference on World Wide Web (WWW 2009), pp. 881–890 (2009)
17. Resnick, P., Iacovou, N., Suchak, M., Bergstrom, P., Riedl, J.: GroupLens: an open architecture for collaborative filtering of Netnews. In: Proceedings of ACM Conference Computer Supported Cooperative Work, pp. 175–186 (1994)
18. Salakhutdinov, R., Mnih, A.: Probabilistic matrix factorization. In: NIPS, vol. 1, pp. 2–1 (2007)
19. Zhang, W., Du, T., Wang, J.: Deep learning over multi-field categorical data – a case study on user response prediction. In: ECIR (2016)

Automated Detection of Standard Image Planes in 3D Echocardiographic Images

Wei Peng[1], XiaoPing Liu[2(✉)], and Lanping Wu[3]

[1] Information Technology Support, East China Normal University, Shanghai 200062, China
wpeng@admin.ecnu.edu.cn
[2] Computer Center, East China Normal University, Shanghai 200062, China
xpliu@cc.ecnu.edu.cn
[3] Shanghai Children's Medical Center, Shanghai Jiaotong University, Shanghai 200240, China

Abstract. During the diagnosis and analysis of complex congenital heart malformation, it is time-consuming and tedious for doctors to search for standard image planes by hand from among the huge amounts of patients' three-dimensional (3D) ultrasound heart images. To relieve the laborious manual searching task for echocardiographers, especially for non-physicians, this paper focuses on the auto-detection of five standard image planes suggested by experts in the 3D echocardiographic images. Firstly, the four-chamber (4C) image plane is auto-detected by template matching, and then the other standard image planes are obtained according to their spatial relation with the 4C image plane. We have tested our methods on 28 normal and 22 abnormal datasets, and the error rates are 7.1% and 13.6%, respectively. With low computational complexity and simple operation, the method of auto-detection of standard planes in 3D echocardiographic images shows encouraging prospects of application.

Keywords: 3D ultrasound image · 3D echocardiographic image · Template matching · Image retrieval

1 Introduction

The congenital heart disease (CHD) is a group of cardiac anomalies in a three-dimensional (3D) space. Its diagnosis is always a difficult problem in cardiology. Today, real-time 3D echocardiography (RT3DE) [1–3] has become an important tool to diagnose CHD. With RT3DE, you can see a beating heart and choose any cross section to observe the internal structure of the heart. To facilitate the diagnosis of complex congenital heart malformation, five standard image planes (SIP) [4] are used in both research and clinical practice [5–7]. They are the long-axis planes of left and right ventricular (L1, L2), the short axis planes of atria and ventricular (S1, S2) and the four-chamber plane (H), as shown in Fig. 1. However, it is time-consuming and boring for doctors to manually search standard image planes with RT3DE systems or off-line software. Automatic detection of SIP in 3D echocardiographic (3DE) images will relieve the laborious

H. Gao et al. (Eds.): CollaborateCom 2020, LNICST 350, pp. 386–395, 2021.
https://doi.org/10.1007/978-3-030-67540-0_23

manual searching for echocardiographers, especially for non-physicians. In addition, it can be a prior process for 3DE image registration, fusion, analysis, etc., and can help non-physicians with rapid processing for research purposes.

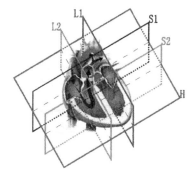

Fig. 1. Diagram of the five standard image planes (SIP) to visualize the heart.

3DE images are usually acquired from multiple different locations, and the image content thus obtained in different location is distinct. The image acquired at the left apical, which is called the apical dataset, is most commonly used in assessing the mitral or tricuspid valve disease and atrioventricular septal defects, and in measuring left ventricular parameters. As an initial work, this paper focuses on the detection of the five SIPs in the apical datasets.

The main innovations of this paper are as follows:

- Since few investigations on the problem have been reported at home and abroad, the automatic detection of the best cross-sections in 3-D echocardiac image is essential and valuable.
- As a preliminary work, we retrieve the 4C plane from the cross-sections extracted from the 3-D volume data with a 4C plane template, referring to the doctor's knowledge. Subsequently, the other best cross-sections could be detected, according to their spatial relation with the 4C plane.
- A coarse-to-fine approach is applied to retrieve the 4C plane with inspiring experimental results.

2 Related Work

Previous work closely related to our study includes techniques to distinguish the 4C image. Zhou et al. presented a method to differentiate between the apical 2C and 4C views, and it requires a pre-processing phase in which the left ventricle is identified by human [8]. To recognize the 4C image automatically, the number of horizontal and vertical borders in the image was used in [9]. With an intensity-based rigid registration method, Leung et al. registered the 3DE data set acquired in the stress stage, the data in the rest stage had already manually selected the two-dimensional (2D) planes as reference, and they detected the two-chamber (2C), 4C and short-axis planes [10]. However, the computational complexity for registration of ultrasound volume is high and it is troublesome to perform 3DE image registration for different subjects.

All the above papers have discussed the detection of the 4C image from enormous 2D echocardiographic (2DE) images. Inspired by these methods, we attempt to detect the 4C image plane in the apical dataset with a template matching based method. Then we try to obtain the other planes according to their spatial relation with the 4C plane.

3   Materials and Methods

3.1   Apical 3DE Datasets

To detect the SIPs in the apical dataset, we should first study the structure of these datasets. An apical dataset is composed of a sequence of volume data. An end-diastolic volume data is shown in Fig. 2. As shown in Fig. 2, each volume data includes the original ultrasound acquisition data shaped approximately like a pyramid and the background data. The pyramid's azimuth and elevation angular spans is 60°. And the values of the black background voxels are zero. Since the four-chamber view in the end-diastolic data is the easiest to identify, we have extracted the end-diastolic volume data from the apical dataset. According to the characteristic of the end-diastolic volume data shown in Fig. 2, we assume that the pyramid apex is the left apex of the heart, and the central axis passing through the cube data is the long axis of the heart. Based on these assumptions, we have established a coordinate system as shown in Fig. 3. We set the pyramid apex to the origin O of the coordinate system, and the central axis OZ passes through the origin O, as shown in Fig. 3. In this way, a 4C cross-section can be found in a series of cross-sections that rotate about the axis OZ.

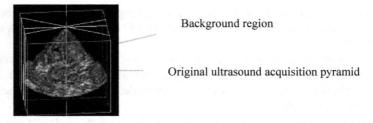

Background region

Original ultrasound acquisition pyramid

Fig. 2. The 3DE image of the end-diastolic volume data

3.2   Four-Chamber Image Detection

To find the 4C image plane in the 3D volume by template matching, we take the following steps. Firstly, a 4C image plane in the end-diastolic volume data of a normal person is selected as the template image. Secondly, a series of cross sections around the central axis are extracted from the end-diastolic volume data of an apical dataset. Thus an image library composed of these cross-sections is established. In the third step, by a coarse-to-fine strategy, the image most similar to the template image is retrieved from the image library, which is considered to be the 4C image plane.

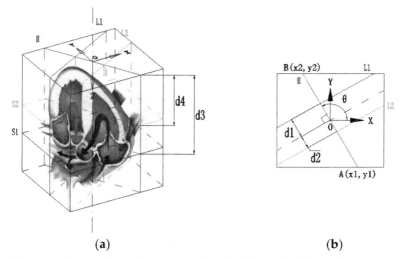

(a)   (b)

Fig. 3. Diagram of the location of the heart and its five SIPs in the 3D coordinate system of the apical dataset. (a) Diagram of the heart and its five SIPs in the 3D coordinate system. (b) Diagram of the projection of the long axis of L1 and L2 in the XOY plane.

3.2.1 Template Selection

The commonly used simple template matching method is difficult to apply to echocardiographic images. Due to changes of the size and shape of the heart, deviation of the positions to collect data, differences of the ultrasound image intensities, and the multiple deformities in the congenital heart disease, it is necessary to build a big template library, which will lead to a high computational cost. Schlomo et al. match the unknown sample image onto known templates by a multi-scale elastic registration using only one or very few templates to identify the 4C image [12].

In our research, we choose a rectangular area from the center of the 4C cross-section of a normal end-diastolic data as the template image, which is described in Fig. 4. The characteristics of the septum and the chamber in the rectangular region of the end-diastolic four-chamber view are obvious, which is significantly different from the distribution of other image planes, and the feature is almost independent of the size, shape and image intensity of the heart, etc. Consequently, it is feasible to detect the most similar 4C image without a large template library. Furthermore, it saves time compared with matching the 4C image by elastic registration since a 3D dataset contains hundreds of 2D images.

3.2.2 Image Library Establishment

An image library is created by extracting a series of cross-sections from the 3D end-diastolic volume data. According to the coordinate system established in Fig. 3, this series of cross-sections share a common central axis OZ, and the projection of any

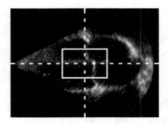

Fig. 4. Diagram of the central rectangular area in the apical 4C image

cross-section on the XOY plane is a line such as AB shown in Fig. 3(b). The coordinates of the point P on a certain cross-section satisfy Eq. 1:

$$\begin{cases} x = x1 + l \cos \theta \\ y = y1 + l \sin \theta \end{cases} (0° \leq \theta \leq 180°) \tag{1}$$

In Eq. 1, θ represents the included angle between line AB and the positive axis OX, ranging from 0 to 179, as shown in Fig. 3(b). And l denotes the Euler distance from the projection of point P on the XOY plane to point A, along the direction of line AB. l is smaller than or equal to L. L is the length of line AB, and is also the width of the cross section, which is defined as:

$$L = \overline{AB} = \sqrt{(x2 - x1)^2 + (y2 - y1)^2} \tag{2}$$

In Eq. 2, the coordinates of point A and point B are $(x1, y1)$ and $(x2, y2)$, respectively. In Fig. 3, the ordinate value of the boundary point A, $y1$, can be calculated by the distance from the origin O to the boundary line of the volumetric data passing through point A. And the abscissa of point A, $x1$ can be calculated by the product of the value of $y1$ and the tangent function value of θ.

In this way, according to angle θ (ranging from 0 to 179), cross-sections are extracted per degree, and a total of 180 slices are obtained to form an image library. The size of each slice is $L \times n$, and n is the size of the volume data in the OZ direction. In the paper, the units of the sizes are pixels.

3.2.3 4C Image Retrieval

A coarse-to-fine strategy is utilized in the 4C image retrieval. Due to the high speckle-noise, non-constant intensities and discontinuous structural edges, it is infeasible to segment the septum of the heart from the echocardiographic images. The coarse-to-fine strategy improves the speed and accuracy of the 4C image retrieval.

Firstly, to reduce the index size and retrieval time, a coarse retrieval is implemented. Due to the movement of the probe and patient respiration, the heart septum tissue will inevitably offset from its ideal position. As a transformation, rotation and scale invariant feature, image histogram is relatively robust to the misalignment. We adopt the cumulative histogram of the rectangular region as the feature. As shown in Fig. 3, the geometric center of this region is the same as that of the cross section. Given two images G and S, their dissimilarity measure is defined as

$$dissim(G, S) = \sum_{i=0}^{255} |g(i) - s(i)| \tag{3}$$

Here, i is the image intensity ranging from 0 to 255. $g(i)$ and $s(i)$ are cumulative histograms of G and S, respectively. $g(i)$ is defined as:

$$g(i) = \sum_{k=0}^{i} \frac{n_k}{N} \tag{4}$$

in which, N is the total number of the pixels of the image. n_k is the number of the pixels whose intensity values are equal to k. $dissim\ (G, S)$ is set to zero if two images are identical.

Next, in order to eliminate the misalignment error, we apply a rigid registration method to match the optimal rectangular area in each cross-section image of the decreased image set preprocessed by the coarse retrieval. Before the fine search, we take the cross-correlation of the two images as the similarity measure, and use the simplex algorithm [13] as the optimization strategy to achieve the optimal match. As a result, the sensitive factors of scale, transformation and rotation changes of the wavelet feature in the fine retrieval are deleted to have good results.

Finally, the fine retrieval is implemented by catching the difference between the template image and the best matches of the images of the reduced image set. Since the wavelet transform has a multi-resolution property, the image transformed by wavelet can be indexed in a hierarchical form [14]. We can extract direction information from sub-bands in various directions to improve index performance [15]. The Daubechies wavelets [16], based on the work of Ingrid Daubechies, are a family of orthogonal wavelets defining a discrete wavelet transformation and characterized by a maximal number of vanishing moments for some given support. In order to detect the local detailed information of the rectangular area image, we perform wavelet transformation on the image by the wavelet function db2 and decompose the image into four sub-bands, LL1, HL1, HV1 and HD1. We calculate the mean E and variance V of the wavelet coefficients of the three high frequency sub-bands, HL1, HV1 and HD1. E and V are respectively defined by Eq. 5 and Eq. 6.

$$E = \frac{1}{MN} \sum_{m=1}^{M} \sum_{n=1}^{N} |x(m, n)| \tag{5}$$

$$V = \frac{1}{MN} \sum_{m=1}^{M} \sum_{n=1}^{N} \left| x(m, n)^2 - E^2 \right| \tag{6}$$

In Eq. 5 and Eq. 6, the sizes of the sub-bands are M by N. Here, x is the wavelet coefficient matrix of the high-frequency sub-bands. We design $\vec{f}$ (EHL1, VHL1, EHV1, VHV1, EHD1, VHD1) as the feature vector of the image. The dissimilarity of two images is measured by the Euclidean distance of $\vec{f}$. Finally, the image with the minimum Euclidean distance is detected as a 4C image.

3.3 Other Planes' Detection

According to Fig. 3, the other SIPs including L1, L2, S1 and S2 could be assumed perpendicular to the 4C image plane, H. The distance from L1, L2, S1 and S2 to the origin, point O, is respectively d1, d2, d3 and d4 as shown in Fig. 3. If the coordinate (x, y, z) of point P can satisfy any of the following Eq. 7, 8, 9 or 10, point P is located on the corresponding planes L1, L2, L3 or L4.

$$L1: x \cos\theta + y \sin\theta = d1 \tag{7}$$

$$L2: x \cos\theta + y \sin\theta = -d2 \tag{8}$$

$$S1: z = d3 \tag{9}$$

$$S2: z = d4 \tag{10}$$

The positions of the standard image planes should not be thought of as single and precisely fixed. They are approximate and flexible. Though the heart size of a 3DE image usually ranges approximately from 0.7 to 1.3 times the size in the template 3DE image, d1, d2, d3 and d4 could be equal to those in the template 3D volume which are pre-defined.

4 Experimental Results

The experimental datasets used in our study were the full-volume data collected by Philips Sonos 7500. The experimental data included 28 normal and 22 abnormal full-volume data. At different time, the normal datasets were collected from 16 normal children aged from 1 month and 12 years old, and the abnormal datasets were collected from 14 patients aged from 3 months to 13 years old. Figure 5 lists the detected 4C images of the 28 normal and 22 abnormal datasets.

To further quantitatively analyze our experimental results, we use the angle θ in Fig. 3 to indicate the position of the four-chamber section. Since the 4C image plane is not precisely fixed at a certain position, image planes located nearby may also be regarded as a 4C image. Therefore, image planes with a value of angle θ within a reasonable range can also be considered as valid 4C image plane.

Firstly, we asked three experienced sonographers to manually search the valid 4C image planes in each individual data. Each 4C image plane is quantified by θ. The image plane with a value of θ equal to the average values of θ of all the 4C image planes is considered to be the reference 4C image plane. To determine a reasonable range of the value of θ, a 4C image planes' library is established, which includes image planes from all datasets with d ranging from 0 to 20. d is defined as the absolute angular deviation from the reference 4C image plane. After checking all the images in the 4C image planes' library, $d \leq 10°$ was chosen to be a reliable range of the valid 4C cross section since the echocardiographers were 99% confident that the 4C image planes in this range should be accepted.

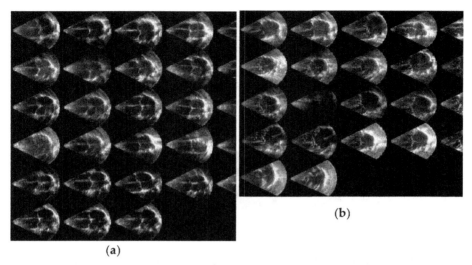

(b)

(a)

Fig. 5. Experimental result of the detected 4c image in 3-DE images; (a) Experimental results of 28 normal data; (b) Experimental results of 22 abnormal data.

Table 1 lists the error rates with different ranges of d. As shown in Table 1, with d ranged from 8 to 10°, the correct rates of our method are 93% for 28 normal datasets and 91% for 22 CAVC datasets, which are acceptable in clinical applications. However, the results are still affected by factors such as image noise, etc. For example, we find in the experiment that the template image with lower speckle noise can obtain a higher accuracy rate than a blurred template image.

Table 1. Error rates list with different scope of d

d	Error rates of normal datasets	Error rates of CAVC datasets
0–3	32%	36%
3–6	25%	18%
6–8	7%	14%
8–10	7%	9%
>10	4%	9%

We compared this coarse-to-fine algorithm with the 3d rigid registration method [16] on the same computer and get the correct rates of the 4C image plane as listed in Table 2.

With the detected 4C image plane, the other standard image planes can be found by Eq. 7–10. Figure 6 lists examples of the detected standard image planes of a normal and an abnormal datasets, respectively.

Table 2. List of the correct rate of the two methods

Method	Correct rate	
	Normal datasets	CAVC datasets
Coarse-to-fine method	93%	91%
3d rigid registration method	85%	52%

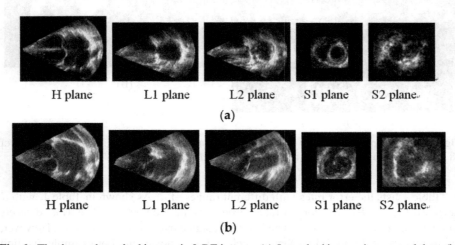

H plane   L1 plane   L2 plane   S1 plane   S2 plane

(a)

H plane   L1 plane   L2 plane   S1 plane   S2 plane

(b)

Fig. 6. The detected standard images in 3-DE images; (a) 5 standard images in a normal data; (b) 5 standard images in an abnormal data.

5 Conclusions

We detect the 4C image plane with a template matching method, and the other four SIPs L1, L2, S1 and S2 according to their positional relationship with each other. The accuracy of the detection of the 4C image plane is critically important. In order to improve the accuracy, we adopt a coarse to fine retrieval strategy. During our experiments, we find that a clear and high-quality template image improves the accuracy of our algorithm. Currently, our method only applies to the dataset collected by the left apex. How to extract the SIPs in the 3DE images collected in other positions is an important issue to be solved in our future research.

References

1. Adriaanse, B.M.E., van Vugt, J.M.G., Haak, M.C.: Three- and four-dimensional ultrasound in fetal echocardiography: an up-to-date overview. J. Perinatol. **36**(9), 685–693 (2016)
2. Dave, J.K., Mc Donald, M.E., Mehrotra, P., Kohut, A.R., Eisenbrey, J.R., Forsberg, F.: Recent technological advancements in cardiac ultrasound imaging. Ultrasonics **84**, 329–340 (2018)
3. Joao, P., Daniel, B., Nuno, A., Olivier, B., Johan, B., Jan, D.: Cardiac chamber volumetric assessment using 3D ultrasound - a review. Curr. Pharm. Des. **22**(1), 105–121 (2016)

4. Sun, K., Chen, S., Jiang, H.: A methodological study on three-dimensional echocardiographic sectional diagnosis for complex congenital heart malformation. Zhongguo Chaosheng Yixue Zazhi **15**(2), 84–88 (1999)

5. Zhou, J., et al.: Clinical value of fetal intelligent navigation echocardiography (5D Heart) in the display of key diagnostic elements in basic fetal echocardiographic views. Chin. J. Ultrason. **26**(7), 592–598 (2017)

6. Li, J.: 3D Visualization and Computer Aided Diagnosis Based on Heart Images. Shanghai Jiaotong University, Shanghai (2015)

7. Chen, G.: Methodological Study and Application of Real-Time Three-Dimensional Echocardiography in Children with Complex Congenital Heart Disease. Fudan University, Shanghai (2006)

8. Zhou, S.K., Park, J., Georgescu, B., Comaniciu, D., Simopoulos, C., Otsuki, J.: Image-based multiclass boosting and echocardiographic view classification. In: Proceedings of the 2006 IEEE Computer Society Conference on Computer Vision and Pattern Recognition, vol. 2, pp. 1559–1565 (2006)

9. Otey, M.E., et al.: Automatic view recognition for cardiac ultrasound images. In: Proceedings of the 1st International Workshop on Computer Vision for Intravascular and Intracardiac Imaging at Annual Conference on Medical Image Computing and Computer-Assisted Intervention (2008)

10. Leung, K.Y.E., et al.: Proceedings of the SPIE (2006)

11. Xiaoping, L., Xin, Y., Lanping, W., Xiao, T.: 2010 International Conference on Bioinformatics and Biomedical Technology (2010)

12. Aschkenasy, S.V., et al.: Unsupervised image classification of medical ultrasound data by multiresolution elastic registration. Ultrasound Med. Biol. **32**(7), 1047–1054 (2006)

13. Lagarias, J.C., Reeds, J.A., Wright, M.H., Wright, P.E.: Convergence properties of the Nelder-Mead simplex method in low dimensions. SIAM J. Optim. **9**(1), 112–147 (1998)

14. Mandal, M.K., Aboulnasr, T., Panchanathan, S.: Fast wavelet histogram techniques for image indexing. Comput. Vis. Image Underst. **75**(1–2), 99–110 (1999)

15. Laine, A., Fan, J.: Texture classification by wavelet packet signatures. IEEE Trans. Pattern Anal. Mach. Intell. **15**(11), 1186–1191 (1993)

16. Daubechies, I.: The wavelet transform, time-frequency localization and signal analysis. IEEE Trans. Inf. Theory **36**(5), 961–1005 (1990)

Distributed Color-Based Particle Filter for Target Tracking in Camera Network

Yueqing Jing and Yanming Chen[✉]

Anhui University, Hefei 230000, China
jingyueqing@stu.ahu.edu.cn, cym@ahu.edu.cn

Abstract. Color-based particle filters have appeared in some literatures. However, there are still some important drawback in tracking targets, such as illumination changes, occlusion and low tracking accuracy. To solve these problems, in this paper, we propose a distributed color-based particle filter (DCPF) for target tracking, which can track targets accurately in a large-scale camera network with less data transmission and less computation. Compared with the previous algorithms, the algorithm proposed in this paper has two obvious advantages. First, the DCPF framework merges color features into the target's state to obtain better robustness. Second, it considers the situation where the target is disappear in some cameras because of limited field of view (FoV). Convincing results are confirmed that the performance analysis of the proposed algorithm in this paper is very close to the centralized particle filter method.

Keywords: Consensus algorithm · Particle filter · Camera network

1 Introduction

With the decrease of cost for surveillance equipments, cameras have been deployed on a large scale in our city. They have been widely used in the wide-area surveillance, disaster response, environmental monitoring, etc. [1]. Multiple cameras can observe a wider area and provide different angles of view, which make information fusion possible. In the target tracking application, it may be got a more robust tracking method by information fusion among different cameras. There are two types information fusion methods: the centralized solution and the distributed solution. In the wide-area surveillance, the distributed solution is usually selected because of its scalability to a large number of sensors, easy replacement, and strong robustness. In view of the above reasons, this paper mainly proposes a distributed color-based particle filter (DCPF) for target tracking in camera networks, which can track targets accurately in a large-scale camera network with less data transmission and less computation. our method can deal with target occlusion and target disappearing in some cameras. The term 'distributed' means that each camera processes its own data and interaction information among neighboring nodes. There is no centralized unit.

© ICST Institute for Computer Sciences, Social Informatics and Telecommunications Engineering 2021
Published by Springer Nature Switzerland AG 2021. All Rights Reserved
H. Gao et al. (Eds.): CollaborateCom 2020, LNICST 350, pp. 396–406, 2021.
https://doi.org/10.1007/978-3-030-67540-0_24

There are already some distributed methods, which may encounter some difficulties for applying to camera networks. For example, the data obtained by the camera are video streaming, which are relatively large and not easy to transmit. If a centralized method is adopted, it will be difficult to process, transmit, and store data in a large scale camera network. Therefore the centralized method is not suitable for a large scale camera network. In addition, there is another problem. The camera has a limited field of view. When the data from a single camera, the target may be lost because of the limited field of view. Therefore, we need fuse data from different cameras. However, due to limited processing power, bandwidth, and real-time processing requests, etc., we need adopt a distributed method with less data transmission and less computation. The method proposed in this paper adopts the idea of combining local filtering and fusion filtering, and uses local filtering to process the video to get smaller data. Therefore, it can effectively alleviate the data congestion phenomenon in terms of data transmission, storage, and operation.

Figure 1 depicts a scene application for tracking people in the camera networks. Although some cameras do not observe the target, which can also obtain the estimated state of the target by the distributed algorithm.

Fig. 1. Cameras' FoVs and network connectivity

The **main contributions** of this paper are proposing a distributed color-based particle filter for target tracking in camera networks, taking special care of the issues of huge data transmission and less computation, and using the proposed algorithm to track target in camera networks.

1.1 Description and Solution of the Problem

Many distributed methods can obtain good tracking accuracy with the observations of each node in sensor networks [2–4], but they cannot handle the scenario

shown in Fig. 1, where each node can only observe a small part of the target trajectory. The method proposed by Kamal et al. [5] does not process video data.
The method proposed by Mohammadi et al. [6] use Kalman to track multiple targets in video data. The experimental results are quite good. However, this
method may not be suitable when the state model is nonlinear. The method proposed by Kamal et al. [7–9] use a single video to track objects, the experimental
results are particularly good, and the robustness is extreme. But it is necessary
to have observation data for the target, which is very challenging. Video target tracking method needs to solve following issues in camera networks, such as
limited field of view, transmission delay, data fusion, and computational overhead for processing video data. The method proposed in this paper directly uses
the color feature as the measurements. After local calculation, the processed
data is transmitted, which is much smaller than the video data. The method
significantly eases the pressure of transmission, storage, bandwidth, and delay.
The experimental results show that the accuracy is very close to the centralized
particle filter without the communication delay [4].

1.2   Related Work

Distributed particle filter has attracted the attention of many scholars in recent
years. It has better performance than the centralized particle filter. The centralized particle filter requires a fusion centre [10], which receive the measurements from all sensor nodes and use them to update their weights. But it has
a severe problem that once the fusion centre is broken down, the entire sensor
network will be paralyzed. Distributed particle filtering is different. Each sensor node is a fusion centre. It communicates with its neighbors and transmits
the processed data. After multiple iterations [5], consensus results are obtained.
The consensus method turns each sensor node into a fusion center. Even if a
sensor node is broken down, the entire sensor network still maintains regular
operation. Many tracking methods based on distributed particle filter have been
proposed [3,11,12], but they cannot solve the particular situation related to the
camera network. Such as some cameras have no target. Due to delays and weak
processing requests, the previous expectations and variances are merged [6]. In
cluttered environment, features need to be added to improve tracking accuracy.
Such as color, texture, angle, gradient, motion [8,13–16], the color feature is a
very convenient feature. The method proposed in this paper solves the problem
of some cameras missing targets by adding color features.

The rest of the paper is organized as follows: the Sect. 2 introduces centralized
particle filter, distributed particle filter consensus method, and color characteristics. The DCPF algorithm proposed in this paper is described in the Sect. 3.
The Sect. 4 is the experimental results. Finally, we summarize the paper in the
Sect. 5.

2   Background

We consider a camera network which composed of N nodes. The state variable is $x(k) = [X(k), Y(k), \dot{X}(k), \dot{Y}(k)]^T$. Where $(1 \leq l \leq N)$, T represents the transpose, $(X(k), Y(k))$ is the position, $\dot{X}(k), \dot{Y}(k)$ are the velocity, node l generates a measurement $z^{(l)}(k)$ at time k. The overall space is shown as follows:

$$\text{State model}: x(k) = f(x(k-1)) + \xi(k) \tag{1}$$

$$\text{Observation model}: z^{(l)}(k) = g(x(k)) + \zeta^{(l)}(k) \tag{2}$$

The $\xi(k), \zeta^{(l)}(k)$ represent the noise error of the state model and observation model, respectively.

2.1   Centralized Particle Filter (CPF)

In CPF [4], observations from all sensors are transmitted to a centralized unit. The centralized unit contains n weighted particles $(\chi^{(j)}(k-1), \omega^{(j)}(k-1))(j = 1, ... n)$, where n is the number of particles. It merging the measurements of multiple sensors by increasing the dimensions of the measurements when the centralized unit receives the measurements of the sensor $z^{(l)}(k)$ at time k. It uses CPF to predict particles $\chi^{(j)}(k)$ and update weights $\omega^{(j)}(k)$.

2.2   Distributed Particle Filter

In DPF [6], all sensor nodes are connected to form a network. Each sensor node uses its measurements to update the weight of the particle on the local filter and calculates the expectation and covariance. Transmit $p^{-1} \cdot u$ and covariance p^{-1} to the nodes who direct connected to it. After multiple iterations, the value on each node is consensus. The average consensus algorithm is a popular distributed algorithm to calculate the arithmetic average of $(a_l)_{l=1}^N$ values [5]. In the average consensus algorithm, each node initializes its consensus state $a_l(0) = a_l$, and runs the following equation iteratively.

$$a_l(t) = a_l(t-1) + \epsilon \sum_{j \in \aleph_l} (a_j(t-1) - a_l(t-1)) \tag{3}$$

where $\aleph_l$ represents the set of neighbor nodes for node l, t is the number of iterations, and the parameter ϵ is a number between 0 and $\frac{1}{\triangle_{max}}$. $\triangle_{max}$ is the maximum degree in the sensor networks topology graph. At the beginning of the iteration, node l sends its previous state $a_l(t-1)$ to its neighbor nodes, and also receives the previous state $a_j(t-1)$ of other nodes. It use Eq. (3) to update its state. And the state values of all nodes will converge. It use the consensus values to update the particles weights on the fusion filter. After the weights are updated, the estimated value of the target can also be obtained by calculating the expectations. Distributed particle filtering mainly has the following three advantages: (1) There is no centralized unit; (2) The sensor node does not need global knowledge of the networks topology; (3) Each node only needs to exchange information with its neighboring nodes.

2.3   Color Characteristics

Feature-based color particle filtering is often used to track any type of target [17]. Color, gradient, and other features are often used to track targets. Specifically, color may be the most popular feature of video surveillance and tracking targets. Because it is more effective than motion or geometry under conditions of partial occlusion, rotation, range, and resolution changes. At the end of each filter iteration, its estimates will be used to update the model, and a parameter a is used to control the speed of model update.

$$hist(k+1) = (1-\alpha)hist(k) + \alpha hist_{E[x_k]}(k) \tag{4}$$

where $hist$ represents the color histogram.

3   Distributed Color Particle Filter

DCPF runs two localized particle filters at each node, one is a local filter, and the other is a fusion filter. The implementation of the local filter is very similar to the standard particle filter. The local filter only processes the local observation value $z^{(l)}(1:k)$, as described in Sect. 3.1. The fusion filter uses the local distribution $P(x(k) \mid z^{(l)}(1:k)$ and $P(x(k) \mid z^{(l)}(1:k-1))$ to estimate the global posterior distribution $P(x(0:k) \mid z(1:k))$, as described in Sect. 3.2.

3.1   Local Filter

The model of the local filter is shown as follows

$$\text{State model :} x(k) = f(x(k-1)) + \xi(k) \tag{5}$$

$$\text{Observation model :} z^{(l)}(k) = hist^{(l)}(k) \tag{6}$$

$$\begin{pmatrix} r \\ g \\ b \end{pmatrix} = frame \begin{pmatrix} yMin:yMax & xMin:xMax & 1 \\ yMin:yMax & xMin:xMax & 2 \\ yMin:yMax & xMin:xMax & 3 \end{pmatrix} \tag{7}$$

$$hist = [imhist(r, hist_{bin}); imhist(g, hist_{bin}); imhist(b, hist_{bin})] \tag{8}$$

where $xMin, xMax, yMin, yMax$ are the minimum and maximum values on the x and y axes of a frame. r, g, b represent the color channel of the frame respectively. And $hist_{bin}$ is the size of the bin on each channel. $Hist$ is the color histogram. Each node has a local filter. The details of the local filter are as follows:

(i) First, the particles $\chi_j^{(l,lf)}$, weights $\omega_j^{(l,lf)}$ and color histograms $hist_{target}$ on the local filter of each node are initialized. The particles are initialized in world coordinates instead of the coordinates in the video frame, and the state of the target in the video frame needs to be transferred to the world state according to the conversion matrix. And the conversion matrix H can be calculated by the position of the camera.

(ii) According to the proposal distribution $q(\chi_j(k) \mid \chi_j(0:k-1), z^{(l)}(1:k))$ propagate particles $\chi_j^{(l,lf)}(k-1) \to \chi_j^{(l,lf)}(k)$

(iii) Estimate the position of the target in this node according to the propagated particle $\chi_j^{(l,lf)}(k)$ and the not updated weight $\omega_j^{(l,lf)}(k-1)$

$$\text{Expection} : v^{(l)}(k) = \sum_{j=1}^{n} \chi_j^{(l,lf)}(k)\omega_j^{(l,lf)}(k-1) \tag{9}$$

$$\text{Covariance} : r^{(l)}(k) = \sum_{j=1}^{n} \omega_j^{(l,lf)}(k-1)(\chi_j^{(l,lf)}(k)-v^{(l)}(k))(\chi_j^{(l,lf)}(k)-v^{(l)}(k))^T \tag{10}$$

(iv) Use the measurements $z^{(l)}(k)$ to update the weight of node l. Because of its importance, many literatures have been devoted to optimizing these weights. The method proposed by Zhang et al. [8] use correlation filter to optimize the weight of particles. The method proposed by Rincon et al. [15] uses linear Kalman to optimize the target color coefficient to get the desired weight. The method proposed by Zhang et al. [18] using the particle swarm genetic optimization method to optimize particles and remove useless particles. In this paper, the color histogram is used to optimize the weight of particles.

1) First, use the conversion matrix H to find the particle $\chi_j^{(l,lf)}$ in the video frame.

2) Using the converted particle state Eq. (5) and measurement model Eq. (6) to obtain the color histogram of the particle and normalize it.

$$hist = hist/sum(hist) \tag{11}$$

3) Calculate the particle color histogram and target color histogram according to the proposed distribution. Then combine the following equation to calculate the particles' weight.

$$\omega_j^{(l,lf)}(k) \propto \omega_j^{(l,lf)}(k-1) \frac{P\left(z^{(l)}(k) \mid \chi_j^{(l,lf)}(k)\right) P\left(\chi_j^{(l,lf)}(k) \mid \chi_j^{(l,lf)}(k-1)\right)}{q\left(\chi_j^{(l,lf)}(k) \mid \chi_j^{(l,lf)}(0:k-1), z^{(l)}(1:k)\right)} \tag{12}$$

(v) Estimate the position of the target in node l according to the propagated particle $\chi_j^{(l,lf)}(k)$ and the updated weight $\omega_j^{(l,lf)}(k)$.

$$\text{Expection} : u^{(l)}(k) = \sum_{j=1}^{n} \chi_j^{(l,lf)}(k)\omega_j^{(l,lf)}(k) \tag{13}$$

$$\text{Covariance} : p^{(l)}(k) = \sum_{j=1}^{n} \omega_j^{(l,lf)}(k)(\chi_j^{(l,lf)}(k) - u^{(l)}(k))(\chi_j^{(l,lf)}(k) - u^{(l)}(k))^T$$

$$(14)$$

3.2   Fusion Filter

The process of obtaining the global posterior distribution $P(x(0:k) \mid z(1:k))$ by the fusion filter is similar to the local filter.

(i) First, the local filter on each node is initialized. The particle $\chi_j^{(l,ff)}$ and the weight $\omega_j^{(l,ff)}$ are similar to the local filter.

(ii) According to the proposal distribution $q(\chi_j(k) \mid \chi_j(0:k-1), z^{(l)}(1:k))$ propagate particles $\chi_j^{(l,ff)}(k-1) \rightarrow \chi_j^{(l,ff)}(k)$

(iii) When all nodes get $u^{(l)}(k), p^{(l)}(k), v^{(l)}(k), r^{(l)}(k)$ at time k, the consensus algorithm is executed.

Initialize the consensus state for node l $X_{c1}^{(l)}(0) = (p^{(l)}(k))^{-1}, x_{c2}^{(l)}(0) = (p^{(l)}(k))^{-1}u^{(l)}(k), X_{c3}^{(l)}(0) = (r^{(l)}(k))^{-1}, x_{c4}^{(l)}(0) = (r^{(l)}(k))^{-1}v^{(l)}(k)$. Then use Eq. (3) to achieve consensus. Once the consensus is reached, the average value of $u^{(l)}(k), p^{(l)}(k), v^{(l)}(k), r^{(l)}(k)$ can be obtained by the following equation

$$p(k) = \frac{1}{N} \lim_{t \to \infty} \left\{ X_{c1}^{(l)}(t)^{-1} \right\} \tag{15}$$

$$u(k) = \lim_{t \to \infty} \left\{ X_{c1}^{(l)}(t)^{-1} \times x_{c2}^{(l)}(t) \right\} \tag{16}$$

$v(k)$ and $r(k)$ are calculated in the same way as $u(k)$ and $p(k)$.

(iv) The calculation of the weight is shown as follow

$$\omega_j^{(l,ff)}(k) = \omega_j^{(l,ff)}(k-1) \frac{\mathcal{N}\left(\chi_j^{(l,ff)}(k); u(k), p(k)\right) P\left(\chi_j^{(l,ff)}(k) \mid \chi_j^{(l,ff)}(k-1)\right)}{\mathcal{N}\left(\chi_j^{(l,ff)}(k); v(k), r(k)\right) q\left(\chi_j^{(l,ff)}(k) \mid \chi_j^{(l,ff)}(k-1), z(k)\right)}$$

$$(17)$$

It can be seen from Eq. (17) that the calculation of the weight must know the proposal distribution. Use $P(\chi_j^{(l,ff)}(k)\chi_j^{(l,ff)}(k-1))$ as the proposal distribution, which is called SIR filter. It is also the most common and most convenient method. This article uses this proposal distribution.

(v) If the particle degradation is severe, then resample.

Algorithm 1: Fusion filter

Input: given particle set $(\chi_j^{(l,ff)}(k-1), \omega_j^{(l,ff)}(k-1))_{j=1}^n$ –fusion filter's
     particles and their weights

Output: $(\chi_j^{(l,ff)}(k), \omega_j^{(l,ff)}(k))_{j=1}^n$ –the updated particles and
     weights of the fusion filter.

1.for $l = 1 : N$,

2.do $(u^{(l)}(k), v^{(l)}(k), p^{(l)}(k), r^{(l)}(k))$
     =local filter$((\chi_j^{(l,ff)}(k-1), \omega_j^{(l,ff)}(k-1))_{j=1}^n, z^{(l)}(k))$

3.end for

4.fusion $u^{(l)}(k), P^{(l)}(k)_{l=1}^N$ to calculate $u^{(l,ff)}(k), P^{(l,ff)}(k)$

5.for $j = 1 : n, do,$

  • Generate particles by proposal distribution sampling $\chi_j^{(l,ff)}(k)$

  • Use weight update Eq (12) to calculate $\omega_j^{(l,ff)}(k)$

6.end for

7.observe $(\chi_j^{(l,ff)}(k), \omega_j^{(l,ff)}(k))_{j=1}^n$ –if degraded then
     = resample $(\chi_j^{(l,ff)}(k), \omega_j^{(l,ff)}(k))_{j=1}^n$

8. $(\chi_j^{(l,lf)}(k), \omega_j^{(l,lf)}(k))_{j=1}^n = (\chi_j^{(l,ff)}(k), \omega_j^{(l,ff)}(k))_{j=1}^n$

4 Experimental Results

We track a single target in a distributed camera network. The number of camera nodes is $N = 4$, non-full connection mode. In some existing distributed particle filter target tracking methods, the processed data is not video data. There are other methods using video frame data for target tracking [9,14,19], the effect is excellent. However, this methods are not distributed, and only processes a single video. The method proposed by Kamal et al. [7] is distributed, the effect is excellent, but the method is based on Kalman, not particle filtering. We use a centralized CPF to evaluate the performance of the DCPF and track pedestrians in multiple environment. Set the target state $x(k) = [X(k), Y(k), \dot{X}(k), \dot{Y}(k)]^T$, the initial velocity is $[1, 15]$, $\epsilon = 0.325$, Q_ξ is covariance of process noise. In order to simplify the calculation, a linear equation is used.

$$f = \begin{bmatrix} 1 & 0 & 1 & 0 \\ 0 & 1 & 0 & 1 \\ 0 & 0 & 1 & 0 \\ 0 & 0 & 0 & 1 \end{bmatrix} \quad Q_\xi = \begin{bmatrix} 10 & 0 & 0 & 0 \\ 0 & 10 & 0 & 0 \\ 0 & 0 & 1 & 0 \\ 0 & 0 & 0 & 1 \end{bmatrix} \tag{18}$$

$z^{(l)}(k) = hist^{(l)}(k)$, $hist^{(l)}(k)$ is the color histogram of node l at time k.

    The video dataset in Scene 1 is provided by [9], where the size of the video frame is 640*480, a total of 58 frames of data are used, and track a person whose upper body is blue. After the 30th frame, the target is occluded by other objects in the C4 camera, and the target does not appear in the field of view of C3 after the 5th frame. The other observation is good. The video dataset in Scene 2 is provided by [20], the size of the video frame is 360*288, a total of 58 frames

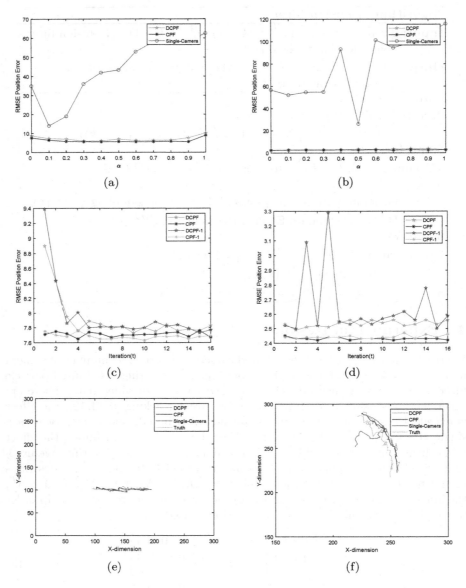

Fig. 2. Scene 1 are the Figs. 2(a), (c), (e) on the left, and Scene 2 are the Fig. 2(b), (d), (f) on the right. Figure 2(a) $n = 500, hist_{bin} = 8, t = 12$, different color coefficient α, error of different methods. Figure 2(c) $hist_{bin} = 8$, $\alpha = 0$, DCPF, CPF represent the error of $n = 500$ repeated running 10 times with different iterations. DCPF-1, CPF-1, represent the error of $n = 300$ repeated running 10 times with different iterations. Figure 2(e) $n = 500$, $hist_{bin} = 8, t = 12$, trajectories of different methods, among which single-camera selects the trajectory with the smallest error. Figure 2(b), (d), (f) is the same as Fig. 2(a), (c), (e).

of data are used. And the person is tracked whose upper body is white. The four cameras have good measurements, and there is no target lost. The tracking result is shown in Fig. 2.

As shown in Fig. 2(a,d), under the condition of different color coefficient α, DCPF has always maintained a low error, while the single camera is greatly affected. The number of DCPF iterations is 12, the number of network nodes is 4, and the result has converged, so the error of DCPF is close to CPF. As shown in Fig. 2(b,e), under the same conditions, the errors of DCPF are similar to CPF. The error among different particle numbers is small. DCPF is stable when the number of particles is 500. When the number of particles is 300, the DCPF algorithm is relatively unstable and error-prone. It can be seen from Fig. 2(c,f) that the single camera is easily affected by the environment, while the DCPF still maintain good performance. The curves of different colors represent the trajectories of the targets in real world coordinates which based on different algorithms. DCPF can keep consistent with the real trajectory in most α and environment, while the single-camera algorithm needs the optimal α and specific environment to keep consistent with the real trajectory. It can be seen from Fig. 2 that DCPF can keep stable and have small errors in various camera networks environment.

5   Summary

We propose a new distributed color particle filter, which combines color features in the measurement model to obtain more accurate performance. We use local filters and fusion filters to propagate particles, update the weights, and estimate the position of the target. Our measurements is the video frame data. Using the color histogram, we can accurately estimate the position of the target in the picture. In this way, the position of the target in the video frame can be obtained at every moment. Besides, distributed color particle filtering can effectively solve the problems of occlusion, and the target is missing in some cameras. Although the performance of the distributed color particle filter proposed in this paper is excellent, it cannot effectively solve the scale problem. If the size of the target changes rapidly in all cameras, DCPF may fail. In the future, we will work to expand to the problem of target size changes.

Acknowledgment. This work is supported in part by the National Natural Science Foundation of China under Grand (No. 6180200).

References

1. Li, S., Da Li, X., Zhao, S.: 5G Internet of Things: a survey. J. Ind. Inf. Integr. **10**, 1–9 (2018)
2. Wolf, M., Schlessman, J.: Distributed smart cameras and distributed computer vision. In: Bhattacharyya, S.S., Deprettere, E.F., Leupers, R., Takala, J. (eds.) Handbook of Signal Processing Systems, pp. 361–377. Springer, Cham (2019). https://doi.org/10.1007/978-3-319-91734-4_10

3. Cetin, M., et al.: Distributed fusion in sensor networks. IEEE Signal Process. Mag. **23**(4), 42–55 (2006)
4. Gu, D., Sun, J., Hu, Z., Li, H.: Consensus based distributed particle filter in sensor networks. In: 2008 International Conference on Information and Automation, pp. 302–307. IEEE (2008)
5. Kamal, A.T., Farrell, J.A., Roy-Chowdhury, A.K.: Information weighted consensus filters and their application in distributed camera networks. IEEE Trans. Autom. Control **58**(12), 3112–3125 (2013)
6. Mohammadi, A., Asif, A.: Distributed particle filter implementation with intermittent/irregular consensus convergence. IEEE Trans. Signal Process. **61**(10), 2572–2587 (2013)
7. Kamal, A.T., Bappy, J.H., Farrell, J.A., Roy-Chowdhury, A.K.: Distributed multi-target tracking and data association in vision networks. IEEE Trans. Pattern Anal. Mach. Intell. **38**(7), 1397–1410 (2015)
8. Zhang, T., Liu, S., Changsheng, X., Liu, B., Yang, M.-H.: Correlation particle filter for visual tracking. IEEE Trans. Image Process. **27**(6), 2676–2687 (2017)
9. Bolme, D.S., Beveridge, J.R., Draper, B.A., Lui, Y.M.: Visual object tracking using adaptive correlation filters. In: 2010 IEEE Computer Society Conference on Computer Vision and Pattern Recognition, pp. 2544–2550. IEEE (2010)
10. Collins, R.T., Lipton, A.J., Fujiyoshi, H., Kanade, T.: Algorithms for cooperative multisensor surveillance. Proc. IEEE **89**(10), 1456–1477 (2001)
11. Presti, L.L., Sclaroff, S., La Cascia, M.: Path modeling and retrieval in distributed video surveillance databases. IEEE Trans. Multimed. **14**(2), 346–360 (2011)
12. Onel, T., Ersoy, C., Deliç, H.: On collaboration in a distributed multi-target tracking framework. In: 2007 IEEE International Conference on Communications, pp. 3265–3270. IEEE (2007)
13. Yu, J.Y., Coates, M.J., Rabbat, M.G., Blouin, S.: A distributed particle filter for bearings-only tracking on spherical surfaces. IEEE Signal Process. Lett. **23**(3), 326–330 (2016)
14. Han, B., Comaniciu, D., Zhu, Y., Davis, L., et al.: Incremental density approximation and kernel-based Bayesian filtering for object tracking. CVPR **1**, 638–644 (2004)
15. Martínez-del Rincón, J., Orrite, C., Medrano, C.: Rao-blackwellised particle filter for colour-based tracking. Pattern Recognit. Lett. **32**(2), 210–220 (2011)
16. Arulampalam, M.S., Maskell, S., Gordon, N., Clapp, T.: A tutorial on particle filters for online nonlinear/non-gaussian Bayesian tracking. IEEE Trans. Signal Process. **50**(2), 174–188 (2002)
17. Khan, Z., Balch, T., Dellaert, F.: A rao-blackwellized particle filter for eigentracking. In: Proceedings of the 2004 IEEE Computer Society Conference on Computer Vision and Pattern Recognition, 2004. CVPR 2004, vol. 2, pp. II. IEEE (2004)
18. Zhang, Y., Wang, S., Li, J.: Improved particle filtering techniques based on generalized interactive genetic algorithm. J. Syst. Eng. Electron. **27**(1), 242–250 (2016)
19. Han, B., Zhu, Y., Comaniciu, D., Davis, L.: Kernel-based Bayesian filtering for object tracking. In 2005 IEEE Computer Society Conference on Computer Vision and Pattern Recognition (CVPR 2005), vol. 1, pp. 227–234. IEEE (2005)
20. Fleuret, F., Berclaz, J., Lengagne, R., Fua, P.: Multicamera people tracking with a probabilistic occupancy map. IEEE Trans. Pattern Anal. Mach. Intell. **30**(2), 267–282 (2007)

Towards a Trusted Collaborative Medical Decision-Making Platform

Hamza Sellak[✉], Mohan Baruwal Chhetri, and Marthie Grobler

CSIRO's Data61, Melbourne, Australia
{Hamza.Sellak,Mohan.BaruwalChhetri,Marthie.Grobler}@data61.csiro.au

Abstract. Traditionally, shared decision-making has been considered as a one-off dyadic encounter between a patient and physician within the confines of the consultation room. In practice, several stakeholders are involved, and the decision-making process involves multiple rounds of interaction and can be influenced by different biopsychosocial, cultural, spiritual, financial, and legal determinants. Thus, there is an opportunity for developing an innovative digital platform for distributed collaborative medical decision-making. However, given the sensitive nature of data and decisions, there are several challenges associated with safeguarding the consent, privacy, and security of the contributors to the decision-making process. In this paper, we propose a conceptual framework and reference architecture for a trusted collaborative medical decision-making (TCMDM) platform that addresses some of these challenges by using a consensus building mechanism built on top of blockchain technology. We illustrate how the TCMDM platform functions by using a real-life use case scenario of an early stage breast cancer patient collaborating with other stakeholders to reach consensus on the best treatment plan.

Keywords: Shared decision-making · Clinical encounter · Consensus building · Privacy · Accountability · Trust · Blockchain

1 Introduction

As the number of diagnostic and therapeutic options increase, so does the complexity of medical decision-making. *Informed consent* and *shared decision-making* (SDM) are two important processes that promote patient participation and autonomy in such situations. Informed consent is a legal process that refers to an individual's explicit authorisation of a medical intervention, especially in treatments involving high risks. SDM is an ethical approach that seeks to enhance patient choice by incorporating their preferences into the healthcare decisions, particularly when multiple comparable medical choices exist [1,17]. Traditionally, SDM has been considered as a one-off dyadic encounter between patient and physician within the confines of a consultation room. However, in reality, there are several other stakeholders involved, and the decisions are influenced by different biopsychosocial, cultural, financial and legal determinants, and shaped over time as opposed to single, discrete encounters.

© ICST Institute for Computer Sciences, Social Informatics and Telecommunications Engineering 2021
Published by Springer Nature Switzerland AG 2021. All Rights Reserved
H. Gao et al. (Eds.): CollaborateCom 2020, LNICST 350, pp. 407–418, 2021.
https://doi.org/10.1007/978-3-030-67540-0_25

As an illustrative scenario, consider an *early stage breast cancer* patient having to choose between mastectomy and lumpectomy with radiation [17]. Based on clinical evidence, both treatment options produce indistinguishable cure rates for early stage breast cancer. The patient can collaborate with a number of stakeholders including her family physician, surgeon, radiation oncologist, medical oncologist, family member/s, well-wisher/s and legal representative to select the option that best meets her emotional and situational needs. Enabling consensual and collectively accepted decisions is paramount to ensure that the patient gets the best possible medical option that encompasses her personal preferences while also considering the healthcare providers' recommendations and family members' wishes. This is not possible in a single clinical encounter and requires an **iterative consensus building process**. Moreover, the patient and other stakeholders may not want to reveal their preferences over the treatment options for a number of different reasons. Hence, the collaborative medical decision-making (CMDM) process has to be **secure, privacy-preserving and confidentiality-preserving**. The recent case of a new-born's death after *"lotus birth"*[1] further highlights the challenges associated with CMDM in such cases. It is believed that babies born through lotus birth have improved blood circulation and a stronger immune system however medical evidence does not support this [2]. Following the death, there were conflicting claims from the hospital and the parents. The hospital claimed that the mother rejected specifically suggested medical treatment options, including vaccinations of and transfer to a special care nursery for the baby, while performing alternative treatments on the baby without the knowledge of the hospital staff. This incident raises additional questions related to the way informed consent and CMDM are being implemented in practice including: (a) who takes responsibility if the selected treatment option does not yield the desired outcome, or worse still, results in an adverse outcome (**accountability**), and (b) which decision in the clinical encounter is the cause of the adverse outcome (**traceability and auditability**). Finding answers to such questions requires having documented evidence about the sequence of decisions taken over the entire clinical encounter (**provenance**).

We posit that there is a need for a practical, digital CMDM platform that (a) *supports multiple stakeholders'* participation, (b) *maintains privacy, confidentiality and security* of sensitive data and decisions, (c) *enhances trust* among stakeholders through transparent communication while permitting decisional authority to lie with one or more of them, (d) *maintains provenance, accountability and traceability*, and (e) *supports second or third opinions* to provide participants additional reassurance about the final decision. To the best of our knowledge, none of the existing CMDM models address all of the above. Therefore, we propose a novel conceptual framework and reference architecture for a trusted collaborative medical decision-making (TCMDM) platform that addresses some of these challenges. In particular, we propose a *consensus building mechanism* to support multiple stakeholders in reaching a consensual and collectively accepted decision that encompasses the patient's preferences, the healthcare provider's

---

[1] https://bit.ly/35yffK4.

recommendation and the family members' wishes. Integration of the consensus building mechanism with blockchain technology ensures trusted, secure, privacy-preserving and confidentiality-preserving interactions leading to the final decision. The digital platform shifts the CMDM process from the confines of a consultation room to a distributed digital environment in which stakeholders are able to freely express their preferences over the available treatment options.

The rest of the paper is organised as follows. Section 2 discusses related work on medical decision-making models and tools, consensus-building in group decision-making and blockchain technology in healthcare. Section 3 outlines the key requirements associated with CMDM. Section 4 discusses the proposed blockchain-based architecture of the TCMDM platform that addresses some of the challenges identified in Sect. 3. Section 5 concludes the paper with a discussion of future work.

2 Background and Related Work

In this section, we summarize the background and related work that provide the groundwork for this research study.

2.1 Medical Decision-Making Models

Medical decision-making models can be divided into four types based on the physician-patient relationship. In the *paternalistic* or *beneficience* model [12], the physician takes on a *directive, expert* role, while the patient has a *passive, dependant* role. The physician has full control of the clinical encounter and uses his skill and expertise to diagnose the patient, and recommend the tests and treatments without taking the patient preferences into consideration. The patients' participation is restricted to *providing consent* for the treatment.

In the *autonomous* or *informed* model, the patient takes on the role of autonomous decision maker while the physician's role is limited to that of a technical expert who provides relevant information to the patient but not their personal opinions. Having complete information about the treatment options, the patient determines the most appropriate medical intervention. The 'flip side" of this model is the *professional-as-agent* model in which the physicians either elicit the patients' preferences or assume that they are familiar with them. Having complete information about the patient's preferences, the physician becomes the sole decision maker without sharing any information with the patient. However, the physician's treatment preferences do not count. Both models were proposed to deal with the information asymmetry between patients and physicians.

In the *interpretive* model, the physicians not only share relevant information about the treatment options and the associated risks and benefits but also help the patients elucidate and articulate their preferences and make suggestions about what treatment options would help to realize those preferences. However, the physicians do not express their own treatment preferences and the final decision lies in the hands of the patients [8].

In contrast to the above models, *collaborative decision-making* involves information sharing, preference sharing and decision sharing between the physician, the patient, and any other participants, including family members and other physicians. The physician shares information about the patient's condition and the treatment alternatives (no treatment, non-surgical treatment, and surgical treatment) and may recommend the best course of action. The patient also shares preferences over the different treatment alternatives. All participants take steps to build consensus about the preferred treatment and a mutually acceptable agreement is reached between all parties on the treatment [4].

Of the four models presented, the collaborative decision-making model has the greatest emphasis on active participation by both parties and joint satisfaction and investment in the decision outcomes.

2.2   Medical Decision-Making Tools

Most medical decision-making tools that assist physicians and patients when using any of the previously mentioned decision-making models are either electronic or paper-based tools. The electronic tools can either be targeted at the physicians or at both the physicians and patients. The ones designed for physicians are aimed at providing decision-support based on state-of-the-art medical knowledge, evidence based medicine (EBM) or clinical practices guidelines (standard of care), or decision-automation and/or optimisation by generating patient-specific assessments and/or recommendations. They leverage Big Data, artificial intelligence and machine learning technologies to facilitate decision-support and decision-optimisation. Examples include sBest Practice (BMJ) (https://bestpractice.bmj.com/), Dynamed (https://www.dynamed.com/), and Up-to-date (https://www.uptodate.com/). Tools meant to be jointly used by the physicians and patients include decision-aids (https://decisionaid.ohri.ca/AZinvent.php), decision-boxes (https://dartmed.dartmouth.edu/spring08/html/disc_drugs_we.php), evidence summaries (https://www.uptodate.com/), question prompt lists (https://www.cancer.nsw.gov.au/patient-support/what-i-need-to-ask) and communication frameworks (https://askshareknow.com.au/). Some of these tools are aimed at providing information to patients (e.g., decision aids) while others provide ways of initiating or structuring conversations (e.g., question prompt lists and communication frameworks).

The main shortcomings of these tools are (a) they are predominantly designed for one-off clinical encounters taking place in a consultation room, (b) they make the assumption that CMDM is a dyadic encounter and only support the patient and the consulting physician. However, it should be noted that these tools can be used by some of the stakeholders (mainly the patient and physician) in different stages of the iterative CMDM process for making some of the interim decisions.

2.3   Consensus Building

Group decision-making (GDM) is a frequent, yet often challenging, human activity in which multiple stakeholders attempt to make a decision collectively.

Normally, GDM can be seen as the task of fusing the individual preferences of a group of decision makers to obtain the group's decision/s. In practice, the individual opinions may differ substantially, and therefore, the collective decision/s made in such conditions might not be positively accepted by all participants. Therefore, a *consensus building mechanism* is required to increase the agreement level within the group before proceeding to the selection phase, i.e., the last step in the decision-making process in which a selection criterion is applied to determine the final collective decision(s) [9]. Within this process, the individual preferences are first aggregated using an appropriate *aggregation operator* and the current level of agreement in the group is computed using a predefined *consensus measure*. The consensus degree is then compared with a preset minimum *consensus threshold*. If the consensus degree exceeds the threshold, the group moves on to the selection phase, otherwise, a procedure is applied to increase the level of agreement in the following consensus building round (e.g., an interactive feedback process to suggest modifying farthest opinions from consensus or through automatic adjustment of decision makers' preferences).

Different consensus reaching mechanisms have been developed in various decision contexts and many application areas are being actively explored such as education, manufacturing, resource management and energy planning [15]. Yet, to the best of our knowledge, there is little or no work on the use of consensus building for CMDM. A number of studies have suggested applying a GDM model to help patients, physicians, and their families collaboratively select a medical treatment among a set of available options [5]. However, such models limit CMDM to a single round of interaction, hence not guaranteeing that everyone's preferences are considered. Iterative CMDM is necessary to ensure that the final decision reflect the patient's personal preferences, their doctors' recommendations and their families' wishes.

2.4 Blockchain Technology in Healthcare

Blockchain technology is essentially a decentralized, distributed ledger that records the provenance of digital assets [10]. It has a number of inherent properties that address some of the challenges associated with CMDM. Transactions recorded on the blockchain are secured through the use of advanced cryptographic techniques (*security*). Once inserted in the blockchain, transactions become permanent and cannot be modified retroactively, even by the authors (*immutability*). Participants with authority have the ability to verify and track transactions recorded in the blockchain (*transparency*). The use of permissioned blockchains enables selective visibility of transactions by restricting access to the blockchain network to permitted users only (*privacy*). Smart contracts enable fine-grained access control to transaction data based on access rights, thereby further ensuring privacy of participants (*privacy*). Blockchain allows the tracking, and documenting in real-time, of complex interactions between numerous diverse stakeholders through a time-stamped workflow (*traceability & auditability*). It also supports the use of pseudo-identities to retain the anonymity of participants (*privacy*).

In recent years, several applications leveraging blockchain technology have been proposed in the healthcare domain. *Health record management* is one of the most popular blockchain-enabled healthcare applications, and includes the management of electronic health records, electronic medical records and personal health records. Specific solutions developed in this area include health data sharing between healthcare professionals [11], and providing patients a unified view of their health records across multiple healthcare providers [14]. Blockchain technology has also been proposed for *clinical trials* to ensure peer verifiable integrity of clinical trial data, improve the accuracy of data analytics and encourage data sharing and collaboration [16]. Blockchain-enabled trusted, privacy-preserving *wearable data marketplace* is another innovative healthcare application that enables the exchange of real-time, user-generated wearable data between wearable owners and health data consumers [6]. Recently, blockchain technology has also been proposed for *CMDM* using the notion of *Proof-of-Familiarity* [18]. However, similar to most traditional CMDM approaches, it also makes the very strong assumption that a clinical encounter is a one-off interactions. This limits its practicability in real-world clinical encounters.

3   Ecosystem

In this section, we analyse the major contributor roles to CMDM, and then discuss the main requirements that a **trusted** CMDM platform should fulfil.

3.1   Contributor Roles

- **Patient.** In the patient-centred approach, the patient is a key contributing role in the CMDM process. A competent adult *patient* always maintains final decisional authority in CMDM, i.e., the right to accept or reject any clinical intervention. In our scenario, this is the early stage breast cancer patient.
- **Core Medical Team.** As the complexity of the medical treatment increases, so too does the need for coordination of care and sharing of knowledge between multidisciplinary health professionals [7]. In our scenario, the core medical team comprises of one or more general practitioners, specialist nurses, pathologists, radiologists, surgeons, medical and radiation oncologists.
- **Non-Core Medical Team.** In addition to the core medical team, additional non-core contributors are also involved in CMDM, including genetic/hereditary counsellors, physiotherapists, psychiatrists, psychologists, plastic surgeons, palliative carers and social workers.
- **Power of Attorney.** In cases where the patient is not capable of making the decision (due to incapacitation or old age, or when the patient is a minor), the authority is delegated to a family member or surrogate [17].
- **Family Members/Carers.** While autonomy focuses on the notion of self-governance and the right of a patient to have final decisional authority (also referred to as *atomistic/individualistic autonomy*), the patient does not exist in isolation and is part of a larger relational network leading to the concept of

relational autonomy [3]. Therefore, any decisions concerning the patient have to incorporate the preferences of those who care for and support the patient, including family members and care givers.

- **Hospital Representative.** To ensure that there are no financial, logistical or legal obstacles to implement the CMDM process, the hospital (or clinic) should have an appropriate representative.
- **Insurance Agent.** The insurance company (represented by an agent) is a key stakeholder in CMDM since its policies can determine the medical decision that is acceptable to the patient, the patient's family and the medical team.
- **Fellow Patients.** The patients may communicate and gain experiential knowledge from *fellow patients* who are (a) either in or have been through similar medical conditions, and (b) have had successful or failed medical interventions.
- **Clinical Evidence Base.** In today's age of big data, insurmountable amounts of medical knowledge are being published and disseminated around the world. The aggregate state of this knowledge is referred to as *medical* or *clinical evidence*, and is used by doctors to deliver diagnosis, prognosis and treatment.
- **Unregulated Information Sources.** Patients have easy access to vast amounts of medical information, often unregulated and potentially inaccurate and clinically unsound, which can contribute to their decision preferences.

3.2   Requirements

- **Iterative CMDM.** Depending upon the complexity of the medical condition, there are multiple stakeholders involved in CMDM, and different biopsychosocial, spiritual and cultural determinants can influence the CMDM outcome, and several professional, clinical and legal guidelines that have to be adhered to. It is unlikely that a consensual, collectively accepted decision encompassing the patient's preferences, the healthcare providers' recommendations, and the family members' wishes, will be reached in one round of interaction. Therefore, CMDM should incorporate an *iterative consensus building mechanism*.
- **Privacy preserving CMDM.** Participants want assurance that their preferences are kept private during the CMDM process. For example, patients may not want to share their spiritual and cultural values with their doctor but may be willing to reveal it to their counsellor. Similarly, they may be willing to share their health data with the medical team but not with their family members. Family members may also not want to reveal their preferences over the different treatment options to the patient, or to other family members. Thus, the CMDM mechanism has to be *privacy-preserving* and participants should have assurances that their data and decisions are only shared with authorised parties in conformance to their sharing policies.
- **Confidentiality preserving CMDM.** Preference data as well as health data exchanged during the CMDM process are highly confidential. Hence, CMDM should be *confidentiality-preserving*, i.e., it should strictly enforce

data sharing on a *need-to-know* basis and only with the owner's consent. For example, if a physician attending to a patient wants a second opinion from another expert, he/she should share the health data only with the consent of the patient, and without revealing any personally identifiable information (PII). Similarly, if a patient wants a second opinion, they may only share the consultation outcome without revealing the PII about the consulting physician. This requirement is related to the above-mentioned privacy-preserving requirement.

– **Secure CMDM.** All contributors need assurance that the interactions will take place securely and there will be no tampering with the data/decisions exchanged during the CMDM process. Therefore, CMDM should eliminate any risks of confidential data leakage by implementing advanced mechanisms to enforce data protection and security. This requirement is also closely related to that of privacy/confidentiality-preservation.

– **Traceable and auditable CMDM.** It should be possible to audit the interactions between participants if and when required. Evidence-based medicine essentially relies on trial and error, and poor decision-making is unfortunately common in medicine [13]. Hence, in order to advance the state of knowledge, it is important to support audit of the CMDM process. Additionally, to deal with situations where the medical intervention does not yield the desired outcome, or worse still, results in an adverse outcome, it is important to have a *traceable and auditable CMDM*.

4   Proposed High-Level Architecture

This section outlines a high-level reference architecture for the proposed TCMDM platform and explains the main interactions involved in *consensus building*.

4.1   Platform Components

Figure 1 shows the high-level platform architecture. The platform is built on top of a *private* or *permissioned* blockchain owned by a trusted *consortium* of hospitals, insurance companies, pharmaceutical companies, medical research institutions and regulatory bodies. The key platform components are summarized below:

– **Identity Manager (IM).** IM provides interfaces for registration of the participants in the CMDM process including patients, core medical teams, non-core medical teams, family members, legal representatives, and insurance companies. It enables pseudo-identities to be used, where appropriate, to retain the anonymity of the participants. It also provides authorisation to these participants to access various functions on the blockchain such as which party has access to which transactions.

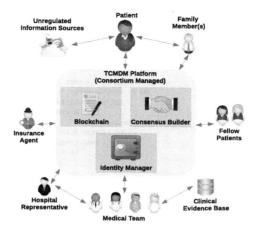

Fig. 1. High-level platform architecture.

– **Consensus Builder (CB).** CB is responsible for managing the entire
  CMDM life cycle including consensus building, selection and contracting. It
  uses an appropriate consensus building mechanism to generate a decision that
  is a consensual and collectively accepted decision.
– **Transaction Ledger (TL).** TL is responsible for recording the provenance
  of the CMDM process. Every interaction between the CB, the patient and all
  other contributors are recorded in the blockchain using a time-stamped workflow. This includes the interim group decisions as well as the final group decision that is acceptable to the patient. This ensures auditability and enables
  regulatory oversight or evidence for legal remedies in the case of negligence
  or an undesirable outcome resulting from the medical intervention.

4.2   Component Interactions

We refer to the first scenario presented in Sect. 1 to illustrate how the proposed
TCMDM platform could be used in practice. We consider a case in which the
patient goes through the CMDM process to decide the best possible treatment
option. The key contributors and components are described below:

– **Patient** (P). P is an early stage breast cancer patient who has to select a
  treatment option in collaboration with PC, MCT, FP, and IC.
– **Principal Clinician** (PC). PC is the lead clinician who works closely with P
  and collaborates with MCT, FP, and IC to help P reach a medical decision.
– **Multidisciplinary Cancer Team** (MCT). MCT is a group of $m(\geq 1)$
  health professionals including oncologists, radiologists, surgeons, and specialist nurses, that collaborates with PC, FP, and IC to help P with an expert
  medical decision. For simplicity, we consider MCT as a single entity.

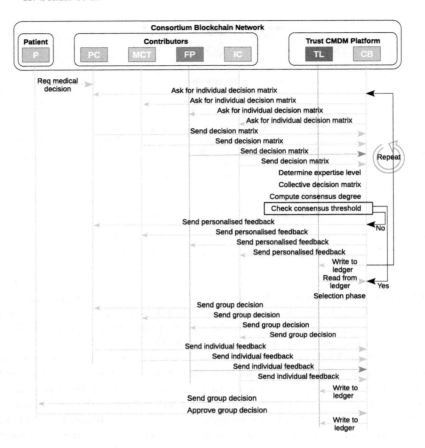

Fig. 2. Interaction diagram of the CMDM process using the TCMDM platform.

- **Fellow Patients** (*FP*). *FP* is a group of $n(\geq 1)$ (ex)breast cancer patients who have previously experienced at least one of the treatment options. *FP* helps *P* with a non-medical decision based on their own personal experience. For simplicity, we represent *FP* as a single entity.
- **Insurance Company** (*IC*). *IC* is responsible for analysing and finalising the health-related financial plans, and collaborates with *PC*, *MCT*, and *FP* to help *P* with the financial aspects of the medical decision.
- **Consensus Builder** (*CB*). *CB* is the main component that enables collaboration between *PC*, *MCT*, *FP*, and *IC* and supports them while deciding the best treatment option for *P*.
- **Transaction Ledger** (*TL*). *TL* is the permissioned blockchain integrated to enable trust between *P*, *PC*, *MCT*, *FP*, and *IC*, and give them comfort that the collaboration process is secure, their privacy is preserved, and only authorized entities have access to their personal interactions.

Figure 2 illustrates the main interactions using the expertise-based consensus reaching mechanism proposed in [15]. At the beginning of each interaction round, CB asks PC, MCT, FP, and IC to send their opinions and preferences in the form of *individual decision matrices*; rows represent treatment options, columns represent evaluation criteria, and each element represents the quantitative/qualitative value assigned to a treatment option against a criterion. A fixed deadline may apply for each round to prevent delays and ensure progress. CB analyses these decision matrices, determines the expertise levels, and uses this information to reflect the influence each contributor should have on the *collective decision matrix*, i.e., an aggregation of the individual decision matrices and represents the group's opinion. The collective decision matrix is used to compute the current *consensus degree*. If the *minimum consensus threshold* is met, the group moves to the selection phase; else, another discussion round is required and CB sends feedback to each individual contributor asking them to consider updating some of their preferences to help reach consensus. This process is repeated until either the group meets the minimum consensus threshold requirement or the number of consensus rounds exceeds the preset limit. At the end of each consensus building round, CB records all elicited and computed data on TL including contributors' identities, individual and collective decision matrices, expertise levels, consensus degree and feedback. If the group reaches agreement and moves to the selection phase, a ranked list of treatment options is presented to P to help her decide the best possible medical option that encompasses her personal preferences as well as the recommendations of PC, MCT, FP, and IC. The TL ensures that all data are kept secured and only authorised entities have access to them. This way, *the platform ensures that CMDM is iterative, trusted, secure, privacy-preserving, confidentiality-preserving and auditable.*

5   Conclusion

In this paper, we presented our vision for a trusted collaborative medical-decision making (TCMDM) platform that shifts medical decision-making from the confines of a consultation room and into a distributed digital environment. We enumerated the main requirements for realising practical CMDM including trusted, secure, privacy-preserving, confidentiality-preserving and distributed data and decision-sharing, as well as data and decision provenance. Based on these requirements, we proposed a conceptual framework and reference architecture that combines consensus building with blockchain technology to address some of the highlighted challenges. As future work, we will explore some of the human factors that might encourage or deter the uptake of digital health services such as the TCMDM platform.

Acknowledgments. This research was financially supported by CSIRO's Precision Health Future Science Platform.

References

1. Barr, P., Elwyn, G., Scholl, I.: Achieving patient engagement through shared decision-making. In: The Wiley Handbook of Healthcare Treatment Engagement: Theory, Research, and Clinical Practice, pp. 531–550 (2020)
2. Bonsignore, A., Buffelli, F., Ciliberti, R., Ventura, F., Molinelli, A., Fulcheri, E.: Medico-legal considerations on "lotus birth" in the Italian legislative framework. Italian J. pediatr. **45**(1), 39 (2019)
3. Chan, B., Chin, J., Radha Krishna, L.: The welfare model: a paradigm shift in medical decision-making. Clin. Case Rep. Rev. **1**(9) (2015)
4. Charles, C., Gafni, A., Whelan, T.: Shared decision-making in the medical encounter: what does it mean? (or it takes at least two to tango). Soc. Sci. Med. **44**(5), 681–692 (1997)
5. Chen, T.Y.: An interactive method for multiple criteria group decision analysis based on interval type-2 fuzzy sets and its application to medical decision making. Fuzzy Optim. Decis. Making **12**(3), 323–356 (2013)
6. Colman, A., Chowdhury, M.J.M., Baruwal Chhetri, M.: Towards a trusted marketplace for wearable data. In: 2019 IEEE 5th International Conference on Collaboration and Internet Computing (CIC), pp. 314–321. IEEE (2019)
7. Devitt, B., Philip, J., McLachlan, S.A.: Team dynamics, decision making, and attitudes toward multidisciplinary cancer meetings: health professionals' perspectives. J. Oncol. Pract. **6**(6), e17–e20 (2010)
8. Emanuel, E.J., Emanuel, L.L.: Four models of the physician-patient relationship. Jama **267**(16), 2221–2226 (1992)
9. Herrera, F., Herrera-Viedma, E., verdegay, J.: A model of consensus in group decision making under linguistic assessments. Fuzzy Sets Syst. **78**(1), 73–87 (1996)
10. Lo, S.K., et al.: Analysis of blockchain solutions for IoT: a systematic literature review. IEEE Access **7**, 58822–58835 (2019)
11. Mettler, M.: Blockchain technology in healthcare: the revolution starts here. In: 2016 IEEE 18th International Conference on E-health Networking, Applications and Services (Healthcom), pp. 1–3. IEEE (2016)
12. O'Grady, L., Jadad, A.: Shifting from shared to collaborative decision making: a change in thinking and doing. J. Participatory Med. **2**(13), 1–6 (2010)
13. Pow, R.E., Mello-Thoms, C., Brennan, P.: Evaluation of the effect of double reporting on test accuracy in screening and diagnostic imaging studies: a review of the evidence. J. Med. Imaging Radiat. Oncol. **60**(3), 306–314 (2016)
14. Roehrs, A., da Costa, C.A., da Rosa Righi, R.: Omniphr: a distributed architecture model to integrate personal health records. J. Biomed. Inform. **71**, 70–81 (2017)
15. Sellak, H., Ouhbi, B., Frikh, B., Ikken, B.: Expertise-based consensus building for MCGDM with hesitant fuzzy linguistic information. Inf. Fusion **50**, 54–70 (2019)
16. Shae, Z., Tsai, J.J.: On the design of a blockchain platform for clinical trial and precision medicine. In: 2017 IEEE 37th International Conference on Distributed Computing Systems (ICDCS), pp. 1972–1980. IEEE (2017)
17. Whitney, S.N., McGuire, A.L., McCullough, L.B.: A typology of shared decision making, informed consent, and simple consent. Ann. Internal Med. **140**(1), 54–59 (2004)
18. Yang, J., Onik, M.M.H., Lee, N.Y., Ahmed, M., Kim, C.S.: Proof-of-familiarity: a privacy-preserved blockchain scheme for collaborative medical decision-making. Appl. Sci. **9**(7), 1370 (2019)

Code Prediction Based on Graph Embedding Model

Kang Yang[1], Huiqun Yu[1,2(✉)], Guisheng Fan[1,3(✉)], Xingguang Yang[1], and Liqiong Chen[4]

[1] Department of Computer Science and Engineering,
East China University of Science and Technology, Shanghai, China
{yhq,gsfan}@ecust.edu.cn
[2] Shanghai Key Laboratory of Computer Software Evaluating and Testing, China,
Shanghai, China
[3] Shanghai Engineering Research Center of Smart Energy, Shanghai, China
[4] Department of Computer Science and Information Engineering,
Shanghai Institute of Technology, Shanghai 200235, China

Abstract. Code prediction aims to accelerate the efficiency of programmer development. However, its prediction accuracy is still a great challenge. To facilitate the interpretability of the code prediction model and improve the accuracy of prediction. In this paper, the source code's Abstract Syntax Tree (AST) is used to extract relevant structural paths between nodes and convert them into training graphs. The embedded model can convert the feature of the node sequence in the training graph into a vector that is convenient for quantization. We calculate the similarity between the candidate value and the parent node vector of the predicted path to obtain the predicted value. Experiments show that by using prediction data to increase the weight of related nodes in the graph, the model can extract more useful structural features, especially in Value prediction tasks. Adjusting the parameters embedded in the graph can improve the accuracy of the model.

Keywords: Big code · Graph embedding · Code prediction

1 Introduction

Recent years, with the development of technology, code prediction has attracted more and more attention in the field of Big Code. The emergence of code prediction technology can effectively improve the efficiency of programmers. Because the code has obvious repetitive characteristics [1], the analysis of the source code file can successfully extract the characteristic information of the code. Then, we bring them into the probability model or deep neural model to predict the missing code.

Traditionally, for probability models such as n-gram, the model uses the node's $n-1$ tokens to predict the probability of the nth token. The probability

H. Gao et al. (Eds.): CollaborateCom 2020, LNICST 350, pp. 419–430, 2021.
https://doi.org/10.1007/978-3-030-67540-0_26

prediction model is simple and effective, but the accuracy is poor. These technologies [2] appeared in early research in this field. Subsequently, in order to improve the accuracy of prediction, Raychev et al. [3] improved the code prediction model by combining technologies in the NLP field. Since the AST of the programming language is also a natural language with a rich structure, NLP technology can be used to extract more characteristic information. The RNN, Bi-RNN, Attention mechanism [4] and other NLP technologies are fused into the code prediction model by researchers. These models can effectively complete the prediction task, but the interpretability of the model is poor, and the training process takes a long time.

Node2vec is a simple neural network algorithm that can extract graph structural features. This model can embed the structural path between graph nodes as a fixed-length vector which contains feature information between graph structure nodes, and is beneficial to the quantitative calculation of the model. Compared with the DeepWalk [5] algorithm, the random walk method of Node2vec is biased. Adjust the model parameters (p, q), the structure of node extraction can be inclined to Depth-First-Search(DFS) or Breadth First Search(BFS). Due to the rich structural characteristics of the code, we can convert the AST node path of the source code into training graph. The Node2vec model can convert the training graph into node sequences, and convert and embed them into vectors.

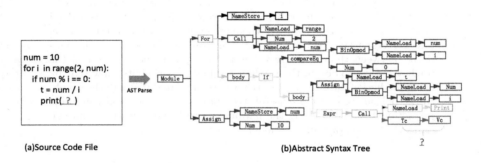

(a)Source Code File   (b)Abstract Syntax Tree

Fig. 1. Examples of Python programs and their corresponding AST

As can be seen in Fig. 1, a source code for finding prime numbers is transformed into an AST, where each node may have one attribute: type or value. Only terminal nodes contain type and value. The code lacks the output of the last line, and our task is to predict the node's type and value. We extract a large amount of AST's paths and convert them into a node graph to facilitate the extraction of related structural information. Finally, we embed the node path to quantify the prediction task and find the most suitable prediction value t.

In this paper, to facilitate effective code prediction. The main contribtions of this paper are as follows:

- Extract effective training graph structure. We extract the path of the AST terminal node from the source code and filter out irrelevant paths. Each set of paths will be converted into a related training graph. This process effectively reduces the node range of the training graph and can speed up the prediction task.

– The task prediction accuracy has been improved. Compared with the state-of-the-art experiment results, the experiment shows that increasing the weight of the edges associated with the prediction data and adjusting the parameters can effectively improve the prediction accuracy.

This paper continues as follows. The Sect. 2 introduces some related works, the basic problem definition and the proposed method are explained in Sect. 3. Section 4 describes the experiment and the experimental results. Section 5 summarizes this paper and outlines future work prospects.

2   Related Works

Probabilistic models [6] are widely used in the field of code prediction because of the characteristics of simple and effective learning. The n-gram probabilistic model [7] derived from statistical natural language processing, it has been proven to successfully capture source code repeatability and predictable regularity. Besides, the model can perform code profiling, transplantation and design auxiliary coding equipment. Vincent J [8] compared the effects of n-gram model and deep learning model in code prediction tasks, and the results show that deep learning is not optimal. DeFreez et al. [9] uses the method of control-flow-graph. Traditional machine learning algorithms, such as decision trees, conditional random fields [10]are used in code prediction.

In addition, the neural network model [11]is also an effective prediction method in code prediction model. Veselin et al. [12] uses neural networks to automatically learn code from large dynamic JavaScript codes. The paper shows that the neural network not only uses token-level information, but also uses code structure information. Adnan Ul et al. [4] used the Bi-LSTM model training to split the source code identifiers, reducing the number of identifier core libraries, thereby improving the prediction accuracy of the code. Pointer networks [4] are widely used in code prediction tasks. The LSTM-based Attention Hybrid Pointer Network, which proposes a soft attention [13] or memory mechanism to alleviate the gradient disappearance problem in standard RNN [14]. Jian Li [15] proposed parent pointer hybrid network to better predict the OoV words in code completion.

In the past ten years, many graphics embedding methods have been proposed, including DeepWalk [5], Node2vec [16], and SDNE [17]. These methods have no obvious difference in the conversion method of embedded word vectors. However, the way of extracting paths is different. The Node2vec algorithm proposes an bias random walk algorithm based on the weights of graph node edges, which can fuse the structural information of node's DFS and BFS. The powerful function of GNN [18] in modeling the dependency relationship between graph nodes, which makes graph analysis related research fields achieve better results.

3   The Proposed Approach

3.1   Problem Definition

Definition 1 (Train-Graph G). Train-Graph $G = (F, A, Path)$ is a graph converted from the AST's node path of the source file. F represents a set of n source code files, $F = \{f_1, f_2, f_3, ...f_n\}$. A represents a set of Abstract Syntax Tree (AST) which is transformed by context-free grammer $A = \{a_1, a_2, a_3, ...a_n\}$. These AST files contain a majority of node structural feature information. The *Path* contains AST's terminal node Up path and Down path. Finally, all node's paths are converted into a training graph G.

Definition 2 (Node Combination (T, V)). (T, V) is the predicted combination of missing nodes. It is calculated by the similarity between the parent node of the prediction node and the candidate value in Train-Graph G. T represents s candidate values in TYPE task, $T\_cad = \{T_1, T_2, T_3, ...T_s\}$. V represents k candidate values in VALUE task, $V\_cad = \{V_1, V_2, V_3, ...V_k\}$. T is the maximum value calculated by predicting the similarity between the parent node P_node and T, $T = S(G, T_i, P\_node)$. V is the maximum value calculated by predicting the similarity between the parent node and V, $V = S(G, V_i, P\_node)$.

For example, as shown in Fig. 1, we predict the missing node $= (T, V)$ in the last line of code. In the corresponding AST, it can be obtained that the parent node is *Call* in Fig. 1(b).

After embedding graph nodes, the model calculates the similarity between the candidate value and the *Call* vector of the parent node of the prediction node. Because the structural characteristics of the node are more similar to the parent node *Call*, the more likely the two are connected in AST. So the TYPE prediction task is to calculate $T = S(G, T_i, Call)$. Determine the type of the prediction node, then we use the type as the new parent node and calculate $V = S(G, V_j, T)$ to predict the VALUE task. This prediction process is transformed into the prediction of the node, as shown in the following Eq. (1), (2).

$$\exists i \in \{1, 2, 3, ...s\} : \arg\max_{T_i} S(G, T_i, P\_node) \tag{1}$$

$$\exists j \in \{1, 2, 3, ...k\} : \arg\max_{V_j} S(G, V_j, P\_node) \tag{2}$$

Therefore, the solution is to calculate the vector similarity of each candidate value with parent node of the prediction node. Finally the prediction token (T, V) is obtained.

3.2   Model Framework

In this Section, we will introduce the framework of our model in detail. Figure 2 shows the main architecture of our proposed model.

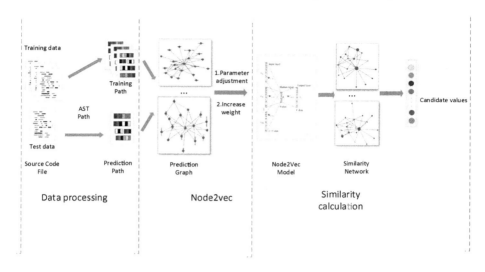

Fig. 2. Model framework

3.3 Data Processing

We parse source file to AST node's path and use the parent node of the prediction data to filter all extracted paths. In other words, each prediction data gets a corresponding set. Data processing can effectively use the information of existing prediction nodes to filter nodes that are not related to the prediction data.

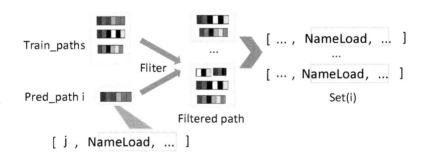

Fig. 3. Data processing

As shown in Fig. 3, the parent node *Nameload* of the prediction node is used to filter the training data. Finally, we convert these node paths into training graphs, and each prediction path *Set* has a corresponding training graph.

3.4   Graph Embedding Model

Node2vec is a graph embedding method that comprehensively considers the DFS and BFS node of the graph. It is regarded as an extended algorithm of DeepWalk.

Random Walk: Node2vec obtains the neighbor sequence of vertices in the graph by biased random walk, which is different from DeepWalk. Given the current vertex v, the probability of visiting the next vertex $c_i = x$ is:

$$P(c_i = x|c_{i-1} = v) = \begin{cases} \frac{\pi_{vx}}{Z} & if (v, x) \in E \\ 0 & otherwise \end{cases} \tag{3}$$

where π_{vx} is the unnormalized transition probability between nodes v and x, and Z is the normalizing constant.

Search Bias α: The simplest method of biased random walk is to sample the next node according to the weight of the edge. However, this does not allow us to guide our search procedure to explore different types of network neighborhoods. Therefore, the biased random walk should be the fusion of DFS and BFS, rather than the mutual exclusion of the two method, combining the structural characteristics and content characteristics between nodes.

The two parameters p and q which guide the random walk. As shown in Fig. 4, we suppose that the current random walk through the edge (t, v) reaches the vertex v, edge labels indicate search biases α. The walk path now needs to decide on the next step. The method will evaluates the transition probabilities π_{vx} on edges (v, x) leading from v. Node2vec set the transition probability to $\pi_{vx} = \alpha_{pq}(t, x) * w_{vx}$, where

$$\alpha_{pq}(t, x) = \begin{cases} \frac{1}{p} & if \quad d_{tx} = 0 \\ 1 & if \quad d_{tx} = 1 \\ \frac{1}{q} & if \quad d_{tx} = 2 \end{cases} \tag{4}$$

and w_{vx} is the edge weight between nodes.

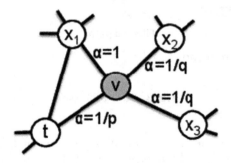

Fig. 4. The next step out of node v

The range of parameters p and q can control the path structure of node extraction. When $p = 1$ and $q = 1$, the walk mode is equivalent to the random walk in DeepWalk.

Similarity Calculation: We use Word2vec's SkipGram algorithm to convert the node path into a vector of fixed dimensions. We use the cosine function to calculate the structural coincidence between nodes. The candidate node with the largest cosine value also contains the most of the same node structure feature.

The calculation function of the similarity between nodes is shown in Eq. 5:

$$cos(\theta) = \frac{V_1 \cdot V_2}{\|V_1\| \|V_2\|} = \frac{\sum_{i=1}^{n} V_{1i} \times V_{2i}}{\sqrt{\sum_{i=1}^{n} (V_{1i})^2} \times \sqrt{\sum_{i=1}^{n} (V_{2i})^2}} \quad (5)$$

The pseudo code for our entire model: **Algorithm 1**

Algorithm 1: Graph Embedding Code Prediction

Input: Train Data $TD=(DownPath, UpPath)$,number n;
  Prediction Data $TeD = (DownPath, UpPath)$;
  Canditite value $Cand_Value$ and number s ;
Output: Predicted Value $pred\_value$;
1 $Parent\_node \leftarrow$ Each $TeD's$ parent node
2 **for** $i = 1 : n$ **do**
3 $\quad |$ $Set \leftarrow$ parent_node in TD_i
4 **end**
5 $G(i) \leftarrow$ Produce graph by Set's node
6 $G'(i) \leftarrow$ Increase the $G(i)$ weight of edges related to Ted
7 $Embed\_model \leftarrow$ Node2vec$(G'(i), p, q)$, adjusting parameters p, q
8 **for** $j = 1 : s$ **do**
9 $\quad |$ $V1 = Embed\_model(parent\_node)$
10 $\quad |$ $V2 = Embed\_model(Cand\_Value_j)$
11 $\quad |$ $SimScore \leftarrow Cos(V1, V2)$
12 **end**
13 $pred\_value \leftarrow Max(SimScore)$

---

4 Experiment SetUp

The experiment is mainly divided into four parts. First, we introduce the data set. Then, discuss the prediction tasks of TYPE and VALUE. Finally, we discuss the experimental results. The experimental hardware environment is Intel(R) Xeon(R) CPU E5-2660 v4 @ 2.00 GHz; RAM 32.00 GB.

4.1 DataSet

In the experiment, we collect Python data sets from the Github repository. The data set contains 10,000 training data files and 500 prediction data files. These Python source files have high star mark in Github and are public available.

4.2   TYPE Prediction

In Table 1, we extract the node types in all source files, a total of 132 types. However, there is a big difference between the maximum and minimum numbers. For example, *CompareLtELtELtE* appears only once in the source data, which has little effect on other candidate values that appear tens of thousands of times.

Table 1. TYPE nodes type

	Types	Size
1	NameLoad	$1.2*10^6$
2	attr	$1.1*10^6$
3	AttributeLoad	$8.4*10^5$
4	Str	$5.1*10^5$
...	...	...
132	CompareLtELtELtE	1

Table 2. VALUE nodes type

	Types	Size
1	Self	$2.8*10^5$
2	None	$4.0*10^4$
3	0	$3.8*10^4$
4	1	$3.4*10^4$
...	...	...
$5.1*10^5$	Sysbench-read cleanup on %s	1

For the prediction task of TYPE, we mainly learn the structural characteristics of the nodes in the training graph. Besides, finding the relative path of the parent node in the prediction data. We increase the weight of these path edges in the training graph, which can effectively affect the structural feature extraction. The adjustment of parameters is mainly related to the parent node of the prediction node.

4.3   Value Prediction

As shown in Table 2, the number of node values in the training code source file is $5.1*10^5$, and most of them are random Strings defined by programmers. The number of unique node values in the data set is too large and there is a large gap in the number, which cannot be fully applied to the neural language model. Therefore, we choose the most frequent value of $K = 1000$ in each training set to build a type of global vocabulary. The training graph filtered by each parent node has related K candidate values. For example, *self* appears the most frequently in source file, it will not be a candidate for *num*.

The prediction task of VALUE is much more difficult than the prediction task of TYPE. First, the prediction candidate value of the VALUE task reaches 1,000 more. Secondly, for the artificially defined word names of programmers, it is difficult to train effective structural feature information in the training data. Especially the node whose type is *Str*. However, during the experiment, we found that it is not necessary to bring every candidate value into the model for calculation. Because the value of the parent node is the type of prediction node will help us filter candidate values.

4.4   Experimental Results

First of all, we introduce prediction accuracy to evaluate the performance of our proposed model, which can be described as Eq. (6):

$$Accuracy = \frac{number\ of\ correct\ prediction\ node}{total\ number\ of\ prediction\ node} \tag{6}$$

The experimental results compared to the-state-of-art [16] in the same data set are shown in the table below: Nw,Np: The model does not increase the weight of related nodes, nor adjust the parameters. Nw,Wp: The model does not increase the weight of related nodes, but has adjustment parameters. Ww,Wp: The model increases the weight of related nodes and adjusts the parameters.

Table 3. Comparison of final results

	TYPE	VALUE
Attentional LSTM	71.1%	–
Pointer mixture network	–	62.2%
Nw, Np	47.8%	38.6%
Nw, Wp	70.2%	47.8%
Ww, Wp	**75.4%**	**63.6%**

Compared with the state-of-art model experimental results, our model prediction results are better. As can be seen from the Table 3, the adjustment of parameters (p, q) can effectively integrate the DFS and BFS structural information of the prediction nodes of the training graph. Secondly, artificially increasing the weight of the relevant edges of the predicted node path can affect the node selection during the random walk of the path. This process can effectively distinguish codes with similar structures.

Table 4. The effect of path length on accuracy

Length path	3	4	5	6	7
Type Acc	67.1%	71.8%	**75.4%**	74.3%	72.8%
Value Acc	52.5%	58.4%	61.6%	**63.6%**	62.3%

Extracting different path lengths has an impact on the prediction accuracy of the model. In Table 4, it can be seen that the longer the path length, the accuracy will not always increase. Because the longer the extracted path, the more overlap between the paths. The prediction accuracy of the model will decrease.

It can be seen from the Table 3 that for the TYPE prediction task, by adjusting the parameters (p, q), the structural feature method of fusing DFS and BFS is more effective than adjusting the weight of the relevant edges of the node path. Therefore, it can be seen that for TYPE prediction tasks, the structural information between nodes is more important than the characteristics of some edge nodes. However, in the prediction task of VALUE, because the relevant edge nodes can help the model effectively reduce the prediction candidate value, increasing the weight of the relevant path is greater than the accuracy of adjusting the parameters.

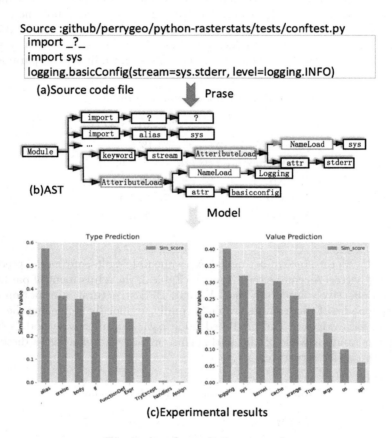

Fig. 5. A code prediction example

As shown in Fig. 5, an actual prediction example, this is the most common configuration test code in Python, which lacks the name of an imported package. In the training graph, the structure of *alias* and *import* are very related and they are appear in pairs. So the type can be accurately predicted to be *alias*. After completing the prediction of the node type, we can narrow the prediction range of the value by type *alias*. At the same time, we will increase the weight of the relevant node edges to facilitate the algorithm to extract the relevant path. As shown in the above example, the missing source code is the name of the calling package. Obviously the name will be called later in the next code line, so we should pay attention on the structure path of the package call of $AttributeLoad-nameLoad-CadWord$. And the set $CadWord = \{logging, sys\}$ in the source code of this example. This method can accelerate the calculation of the entire model, and the accuracy is high.

5 Conclusion and Future Work

In this paper, we use the node information of the source code AST to construct a training graph that contains a large amount of node structure information.

The code prediction task is divided into two parts: one is node TYPE prediction, and the other is node VALUE prediction. Due to the small number of candidate values, the prediction task of TYPE is relatively simple, and the node structure can be effectively extracted in the training graph. For the prediction task of VALUE prediction, although the candidate value table is very large, we can filter the candidate values when the type of the node is known. The appropriate parameters (p, q) can effectively integrate the DFS and BFS information between nodes. On the other hand, increasing the edge weights of related nodes can also improve the accuracy of prediction. However, for VALUE prediction of type Str, the accuracy of model prediction is very low.

In future work, we will consider the OoV problem of the value prediction task. Besides, we will test the model on other data sets and compare the results with other algorithms.

Acknowledgments. This work is partially supported by the NSF of China under grants No. 61702334 and No. 61772200, the Project Supported by Shanghai Natural Science Foundation No. 17ZR 1406900, 17ZR1429700 and Planning project of Shanghai Institute of Higher Education No. GJEL18135.

References

1. Allamanis, M., Barr, E.T., Devanbu, P.T., Sutton, C.A.: A survey of machine learning for big code and naturalness. ACM Comput. Surv. **51**(4), 81:1–81:37 (2018)
2. Nguyen, T.T., Nguyen, A.T., Nguyen, H.A., Nguyen, T.N.: A statistical semantic language model for source code. In: Joint Meeting of the European Software Engineering Conference and the ACM SIGSOFT Symposium on the Foundations of Software Engineering, pp. 532–542 (2013)
3. Raychev, V., Vechev, M.T., Yahav, E.: Code completion with statistical language models. In: ACM SIGPLAN Conference on Programming Language Design and Implementation, PLDI 2014, pp. 419–428 (2014)
4. Ul-Hasan, A., Ahmed, S.B., Rashid, S.F., Shafait, F., Breuel, T.M.: Offline printed Urdu Nastaleeq script recognition with bidirectional LSTM networks. In: 12th International Conference on Document Analysis and Recognition, pp. 1061–1065 (2013)

5. Perozzi, B., Al-Rfou, R., Skiena, S.: Deepwalk: online learning of social representations. In: The 20th ACM SIGKDD International Conference on Knowledge Discovery and Data Mining, pp. 701–710 (2014)
6. Allamanis, M., Barr, E.T., Bird, C., Sutton, C.A.: Suggesting accurate method and class names. In: Proceedings of the 2015 10th Joint Meeting on Foundations of Software Engineering, pp. 38–49 (2015)
7. Maddison, C.J., Tarlow, D.: Structured generative models of natural source code. In: Proceedings of the 31th International Conference on Machine Learning, pp. 649–657 (2014)
8. Vincent J, Devanbu, P.T.: Are deep neural networks the best choice for modeling source code? In: Proceedings of the 2017 11th Joint Meeting on Foundations of Software Engineering, pp. 763–773 (2017)
9. DeFreez, D., Thakur, A.V., Rubio-Gonzalez, C.: Path-based function embedding and its application to error-handling specification mining. In: Proceedings of the 2018 ACM Joint Meeting on European Software Engineering Conference and Symposium on the Foundations of Software Engineering, pp. 423–433 (2018)
10. Raychev, V., Vechev, M.T., Krause, A.: Predicting program properties from "big code". In: Proceedings of the 42nd Annual ACM SIGPLAN-SIGACT Symposium on Principles of Programming Languages, pp. 111–124 (2015)
11. White, M., Vendome, C., Vasquez, M.L., Poshyvanyk, D.: Toward deep learning software repositories. In: 12th IEEE/ACM Working Conference on Mining Software Repositories, pp. 334–345 (2015)
12. Veselin, R., Martin, T.V., Andreas, K.: Predicting program properties from "big code". Commun. ACM **62**(3), 99–107 (2019)
13. Vinyals, O., Fortunato, M., Jaitly, N.: Pointer networks. In: Advances in Neural Information Processing Systems 28: Annual Conference on Neural Information Processing Systems 2015, pp. 2692–2700 (2015)
14. Xu, K., et al.: Show, attend and tell: Neural image caption generation with visual attention. In: Proceedings of the 32nd International Conference on Machine Learning, ICML 2015, Lille, 6–11 France July 2015, pp. 2048–2057 (2015)
15. Graves, A.: Supervised Sequence Labelling with Recurrent Neural Networks, Studies in Computational Intelligence, vol. 385. Springer (2012)
16. Li, J., Wang, Y., Lyu, M.R., King, I.: Code completion with neural attention and pointer networks. In: Proceedings of the Twenty-Seventh International Joint Conference on Artificial Intelligence, pp. 4159–4165 (2018)
17. Grover, A., Leskovec, J.: node2vec: scalable feature learning for networks. In: Proceedings of the 22nd ACM SIGKDD International Conference on Knowledge Discovery and Data Mining, pp. 855–864 (2016)
18. Wang, D., Cui, P., Zhu, W.: Structural deep network embedding. In: Proceedings of the 22nd ACM SIGKDD International Conference on Knowledge Discovery and Data Mining, pp. 1225–1234 (2016)

Technical Implementation Framework of AI Governance Policies for Cross-Modal Privacy Protection

Yuxiao Lei[1], Yucong Duan[1(✉)], and Mengmeng Song[2]

[1] School of Computer Science and Cyberspace Security, Hainan University,
Haikou, China
duanyucong@hotmail.com
[2] School of Tourism, Hainan University, Haikou, China

Abstract. AI governance has been increasingly preeminent in digital world which presents a practical challenge of combining technical digital implementation with legal governance in real world. This work proposes a strategical bridging framework between AI governance policies and technologies on the fundamental content expression modals of data, information and knowledge, and crossing modals processing of DIKW content/resources through formalization abstract or intangible policies and their interactions as concrete technical tangible executions of modal transformations, essential computation and reasoning. The essential content oriented formalization, computation and reasoning of features, policies and associations bridging legal pursue and technical details is unified based our proposed existence computation and relationship defined everything of semantics mechanisms, which we further proposed as essence computation and reasoning for multiple modal and cross modal content. We demonstrate the feasibility and effectiveness of our essential content oriented multiple modals analysis and modeling through construction of a systemic privacy protection framework in the background of the whole process of perception, storage, transition and processing of user generated content in virtual communities. This solution framework seamlessly integrates originally discrete policies and demands on fairness, personal security, financial security, peace and quiet, autonomy, integrity against commodification and reputation at the technical execution level in terms of specific activities and concrete actions inside and cross data modal, information modal and knowledge modal.

Keywords: DIKW graph · Virtual community · Privacy protection · Right to privacy · Value of privacy

Supported by Natural Science Foundation of China Project (No. 61662021 and No. 72062015), Hainan Provincial Natural Science Foundation Project No. 620RC561, Hainan Education Department Project No. Hnky2019-13 and Hainan University Educational Reform Research Project No. HDJWJG03.

H. Gao et al. (Eds.): CollaborateCom 2020, LNICST 350, pp. 431–443, 2021.
https://doi.org/10.1007/978-3-030-67540-0_27

1   Introduction

The automated decision-making of artificial intelligence system is fast and convenient than, When the decision is related to people, it may lead to prejudice and discrimination against human individuals [1] and privacy violations [2].

The definition of privacy is related to human pensonal values. Nowadays, more and more users choose to share life in virtual community, or communicate with people with the same interest in data, information, knowledge and other forms of content. As a result, Virtual trace ($T_{virtual}$) left by users' browsing and User-generated content (UGC) posted by users themselves on virtual community are also included in the category of privacy. $T_{virtual}$ can reflect the user's own character and behavior [3], UGC can reflect the private resources of user which are not influenced by the outside world [4].

$T_{virtual}$ includes user browsing history, purchase history, and interaction history, and UGC includes text, video, voice and any other types of content that users post in virtual community. There exists a connection between Tvirtual and UGC that $T_{virtual}$ includes the behavior of users posting UGC.

A study on the privacy status of personal online social networks in the 10 most visited online social networking sites (OSNs) in the world indicated that many users ignore the privacy risks on OSN, and traditional privacy protection methods have limited effects on the protection of personal privacy [5]. In the four links of the circulation of private resources: Sensing, Storage, Transfer and Processing, The development of privacy protection technology lags behind the development of privacy acquisition technology. It's necessary for both virtual community and users to strengthen their awareness of privacy protection.

There are three parties involved in the circulation of privacy resources: Generator (User), Acquirer (Visitor) and Communicator (Virtual Community). The different party have different rights to privacy in the different links of circulation. The interaction and balance between the rights to privacy of the three parties combined with DIKW graph technology to maximize privacy protection. DIKW graph is a multi-layered architecture for processing typed resources, which can be divided into four parts: Data graph, Information graph, Knowledge graph and Wisdom graph. The DIKW graph technology can be used to optimize the processing efficiency of the integration of storage, transmission and calculation [6] and privacy data protection [7].

2   Privacy Resources of Users in Virtual Community

The privacy resource ($Privacy_{DIK}$, P_{DIK}) can be extracted from the $T_{virtual}$ and UGC in virtual community. Through the semantic formalization of key elements of P_{DIK} [8], P_{DIK} is classified into three types: $Data_{DIK}$, $Information_{DIK}$ and $Knowledge_{DIK}$, according to the differences of P_{DIK}'s own attributes, such as Eq. (1). All of the P_{DIK} represents a feature of users that can be used to analyze and define users.

$$Privacy_{DIK} = \{Data_{DIK}, Information_{DIK}, Knowledge_{DIK}\} \tag{1}$$

2.1   Privacy Resources in the DIKW Graph

Data, Information and Knowledge

$Data_{DIK}$. $Data_{DIK}$ is a directly observed discrete element that has no meaning without context and is not associated with a specific human purpose. In this paper, $Data_{DIK}$ refers to the attributes of the UGC that the user posts in virtual community, such as a photo, a paragraph of text, and the user's profile, name, age, degree, and so on.

$Information_{DIK}$. $Information_{DIK}$ or I_{DIK} is used to explore, analyze, and express the interaction between two entities, which can be either a person or other objects. In virtual community, I_{DIK} records the relationship $R(User, E_{associated})$ between users and entity directly connected to users, as well as the relationship $R(E, E_{other})$ between a entity and other entities. Relationships can define everything at the semantic level [9].

$Knowledge_{DIK}$. $Knowledge_{DIK}$ or K_{DIK} is derived from D_{DIK} and I_{DIK} through structural and formal derivation, and is further improved on the basis of I_{DIK}. I_{DIK} represents the relationship between individual entities. K_{DIK} summarized the relationship between entities of the same type on the basis of I_{DIK}.

K_{DIK} is the reasult of a induction and prediction of individual behaviors. K_{DIK} has two basic: the validity probability of $K_{DIK}(K_{DIK}(Val))$ and the precision probability of $K_{DIK}(K_{DIK}(Pre))$. $K_{DIK}(Val)$ refers to the probability of K_{DIK} predicting user behavior and psychology successfully. $K_{DIK}(Pre)$ indicates the abundance of related content contained in K_{DIK} for the same event. For example, $K_{DIK2}(Pre)$ is greater than $K_{DIK1}(Pre)$ as followed:

K_{DIK1} = "The user need Commodity$_A$"

K_{DIK2} = "The user need Commodity$_A$ at Time$_B$"

Properties of Privacy Resource. In the case of two or more private resource contents with logical conflicts exists in DIKW graph. For example, virtual communities provide users with an environment separated from reality, in which users may create a virtual image that does not conform to their own real image. The $T_{virtual}$ and UGC is partly in line with the user's actual self image, and partly in line with the virtual image that users imagine in their minds.

As shown in Eq. (2), P_{DIK} are classified into $P_{consistent}$ and $P_{inconsistent}$ after traversal comparison with other P_{DIK}. If the result is ture, the $P_{consistent}$ belongs to $P_{inconsistent}$, which has no logical conflict with other P_{DIK}. If the result is false, the combination of P_{DIK1} and the conflict P_{DIK2} $P_{consistent}$ belongs to $P_{inconsistent}$, And there is logical conflict between the P_{DIK} in $P_{inconsistent}$.

$$TraverseCompare(P_{DIK1}, P_{DIK})$$

$$= \begin{cases} TURE \rightarrow P_{DIK1} \in P_{consistent} \\ FALSE \rightarrow (P_{DIK1}, P_{DIK2}) \in P_{inconsistent} \end{cases} \tag{2}$$

Group Privacy and Intimacy Group. P_{DIK} is not only belong to individuals, but also can belong to group. It is possible to dig out the P_{DIK} of individual's family members, friends, neighbors, and other groups based on the P_{DIK} of individuals.

Group privacy [10] exists in two or more entities E_1, E_2, ,E_n, which can be classified into Group relationship privacy ($GP_{relation}$) and Group content privacy ($GP_{content}$) according to their attributes. Entities in Group privacy constitute an intimate group ($G_{Intimacy}$) of mutual protection of privacy. $G_{Intimacy}$ is not limited to a collection of multiple individuals who are related, but can also be people of the same race, gender and age.

$GP_{relation}$ is the retention of R, which is the relationship between multiple entities. Users do not want to be known about their relationships with other users in $G_{Intimacy}$ by others; or for some purpose, not to be known by others. $GP_{content}$ refers to the P_{DIK} shared by $G_{Intimacy}$, in which the importance of different individuals on P_{DIK} may vary due to individual differences but remain within a certain range.

2.2   Handle of Privacy Resources

There are a lot of duplicate and invalid P_{DIK} in virtual community. It is necessary to organize $P_{DIK}K$ before DIKW graph modeled, which includes extract, transform and load (ETL), transferring P_{DIK} from virtual community as a source to the destination of the DIKW graph.

Extraction of Privacy Resources. The extraction of P_{DIK} includes extracting the content of P_{DIK} from homogeneous or heterogeneous source. Based on the self-subjectivity of privacy, different users have different reservation degree of P_{DIK}, and the standards of P_{DIK} extraction are different.

Users have different degrees of reservation to different $P_{DIK}(D_{Res})$. The higher the value of $P_{DIK}(D_{Res})$, the higher the retention of P_{DIK}. When the value of $P_{DIK}(D_{Res})$ is higher than the threshold D_W, the P_{DIK} is classified $Secret_{DIK}$ as shown in Eq. (5), which and will be abandoned in the process of extracting P_{DIK}.

$$Secret_{DIK} = \{P_{DIK}|P_{DIK}(D_{Res}) > D_W\} \tag{3}$$

As shown in Eq. (4), the function Reserve is constructed to calculated $P_{DIK}(D_{Res})$ based on the source of P_{DIK} ($P_{DIK}(Source)$) and the behavior record group associated with the P_{DIK} ($Inter(P_{DIK})$) which storaged in $InformationGraph_{DIK}$.

$$P_{DIK}(D_{Res}) = Reserve(P_{DIK}(Source), Inter(P_{DIK})) \tag{4}$$

Among them, $P_{DIK}(Source)$ includes $T_{virtual}$ and UGC, and $T_{virtual}$ belongs to passive resources, UGC belongs to active resources. In general, the $P_{DIK}(D_{Res})$ of P_{DIK} from $T_{virtual}$ is higher than P_{DIK} from UGC.

$Inter(P_{DIK})$ includes positive behavior $Inter^{pos}$ and negative behavior $Inter^{neg}$ in the protect of P_{DIK}, which are respectively positively and negatively correlated with the value of $P_{DIK}(D_{Res})$.

Transition of Privacy Resources. P_{DIK} can be converted by the Transition module to a new privacy resource $(P_{DIK}{}^{new})$. There are three kinds of transition, First-order Transition, Multi-order Transition and Technical Transition in Transition module.

$P_{DIK}(In)$ denote the degree of entry of each P_{DIK}, which is the degree generated by the other P_{DIK}. $P_{DIK}(Out)$ denote the degree of exit of each P_{DIK}, which is the way generated by the other P_{DIK}.

First-order Transition. First-order Transition to generate a new $P_{DIK}{}^{new}$ from a single P_{DIK}, Which includes same-type transition and cross-type transition among D_{DIK}, I_{DIK} and K_{DIK}.

Multi-order Transition. Multi-order Transition is also known as Associative Transition, refer to generating $P_{DIK}{}^{new}$ by combining several P_{DIK}. There is no limit to the type and number of P_{DIK} and $P_{DIK}{}^{new}$ in a Multi-order transition.

$P_{DIK}{}^{(1)}$ represents an initial P_{DIK}; $P_{DIK}{}^{(2)}$ represents a P_{DIK} connected to $P_{DIK}{}^{(1)}$; $P_{DIK}{}^{3}$ represents a new $P_{DIK}{}^{new}$ generated by combining $P_{DIK}{}^{(1)}$ with one or more $P_{DIK}{}^{(2)}$.

Technical Transition. First-order and Multi-order Transition are simple P_{DIK} transition based on common sense reasoning, while other P_{DIK} transition require the assistance of technology and other resource contents, which is Technical Transition. Technical transition of P_{DIK} has different difficulty, and is not necessarily able to complete. The difficulty of Technical transition is related to the entity E involved in the transition. As shown in Eq. (5), the function Difficulty is constructed to compute the difficulty of transitioning P_{DIK} to a $P_{DIK}{}^{new}$ $(T_{Difficulty})$. The content of E includes the technology(E_{tech}) and other resources ($E_{resource}$). When the $T_{Difficulty}$ is infinite, it means that P_{DIK} can not be transitioned to $P_{DIK}{}^{new}$ with only entity E.

$$
\begin{aligned}
T_{Difficulty} &= Difficulty(P_{DIK}, P_{DIK}{}^{new}, E) \\
&= Difficulty(P_{DIK}, P_{DIK}{}^{new}, E_{tech}, E_{resource})
\end{aligned}
\tag{5}
$$

Load of Privacy Resources. Load of P_{DIK} means insert P_{DIK} into final target storage medium. After extraction and transition, the DIKW graph will be modeled based on all P_{DIK} by classifying P_{DIK} into D_{DIK}, I_{DIK}, K_{DIK} and storaged separately on Data graph, Information graph, Knowledge graph, which constitute DIKW graph of user [11], such as Eq. (6).

$$DIKWGraph = \{DataGraph, InformationGraph, KnowledgeGraph\} \tag{6}$$

In addition, load of P_{DIK} also includes the modeling of the group DIKW graph shown in Eq. (7), which taking $G_{Intimacy}$ as a entity E and $GP_{content}$ of $G_{Intimacy}$ as P_{DIK}. There exists relation between $DIKWGraph^G$ and the DIK-WGraph of the user in $G_{Intimacy}$ but is reserved as $GP_{relation}$. And the realtion between individuals of $G_{Intimacy}$ is also reserved as $GP_{relation}$. $GP_{relation}$ is stored in $InformationGraph^G$.

$$DIKWGraph^G = \left\{ DataGraph^G, InformationGraph^G, KnowledgeGraph^G \right\} \quad (7)$$

2.3   The Value of Privacy

Privacy is a big category, in a narrow sense, it includes the individual's control of self-resources. In a broad sense, it represents many different interests and values, including fairness, personal security, financial security, peace and quiet, autonomy, integrity against commodification and reputation.

Fairness. In the automated decision-making of AI system, different individuals should be treated fairly. Privacy protection is an important part of ensuring fairness of decision-making of AI system.

Such as Eq. (8), the function Fairness is constructed to compute fair index $V_{Fairness}$ of AI system. $V_{Fairness}$ is true means the decision-making of AI system meet the need for fairness and is legal.

$$V_{Fairness} = Fairness(P_{DIK(G)}, U_{price}) \quad (8)$$

Where $P_{DIK(G)}$ represents the group of P_{DIK} participating in decision-making of AI system that should not include any P_{DIK} will affect the decision-making. U_{price} represents the price that different individuals need to pay. For example, the U_{price} of normal people and people with disabilities in the same event is different. Therefore the AI system should try to balance the costs of the two through additional conditions when making relevant decisions to ensure the fairness for everyone.

In addition, the AI system should also be equipped with an additional "application-verification-approval" mechanism to correct decision errors caused by the untimely update of DIKW graph.

Personal Security. P_{DIK} related to personal security of user includes travel trajectory, home address, commuting time and so on. The leakage of P_{DIK} will increase the possibility of users being attacked by potential attackers.

As shown in Eq. (9), the function PS as followed is used to calculate the personal security index V_{PS}. When V_{PS} is higher than the threshold $V_{PS}{}^W$, it is proved that the personal safety of the user can be guaranteed and the decision behavior of AI system is leagal.

$$V_{PS} = PS(E, P_{DIK(G)}) \quad (9)$$

Where the visiting entity E includes the purpose of E ($E_{purpose}$) and the indentity of E (E_{ID}). It is part of the decision-making work of AI system to verify the identity of the visitor and determine the $P_{DIK(G)}$ transmitted to the visitor based on ($E_{purpose}$).

Financial Security. In the process of financial security protection, different from personal security, the AI system not only need to verify the identity of the visitor, but also need to consider the group privacy attributes of financial security. The target of harcker that can threatens the security of property is not a specific user, but the user with the most property in a $G_{Intimacy}$.

When two $G_{Intimacy}$ contain the same user, the privacy disclosure of $G_{Intimacy1}$ will affect the privacy protection of $G_{Intimacy2}$. For example, the attacker can infers the rich gathering area based on the home address of the user with the highest assets in $G_{Intimacy1}$, which will affect the financial security of $G_{Intimacy2}$ composed of the user and his neighbors.

As shown in Eq. (10), the function FS is constructed to calculates the financial security index (V_{FS}) of user. When V_{FS} is higher than the threshold $V_{FS}{}^W$, the financial security of user in the decision-making process of AI system can be guaranteed, and the decision-making behavior of AI system is legal.

$$V_{FS} = FS(E, P_{DIK(G)}, GP_{content}) \qquad (10)$$

Peace and Quiet. In the virtual community, many users maintain a virtual image which is different from own real image in the real world, and trying to keep it that way. In the virtual community, many users maintain a virtual image that is different from the real image in the real world, and want to maintain this state without being disturbed. Users do not want others in the virtual community to know their identities in the real world, nor do they want contacts in real life to know their ID numbers in the virtual world. The existence of the virtual community provides a "window" for many users to escape from the real world. When privacy is protected, Virtual communities can give users a state of peace and quiet that they want but can not have in the real world.

The mutually exclusive privacy resource group $P_{inconsistent}$ is a conflict created by a user's desire for two different identities. As shown in Eq. (11), The function PQ is constructed to calculate V_{PQ} based on $P_{inconsistent}$ When V_{PQ} is higher than the threshold $V_{PQ}{}^W$, The dual identity of user will not be disturbed and the decision-making of AI system is legal.

$$V_{PQ} = PQ(P_{inconsistent}, P_{DIK(G)}) \qquad (11)$$

Autonomy. Autonomy means that individuals are free and able to act and choose and do what they want. Privacy and autonomy are important to individual growth. Nowadays, with the development of big data technology, the problem of "technology crossing the boundary" has arisen in the collection and use of individual P_{DIK}, which infringes the autonomy of users.

The recommend system is an important part of AI system. The recommend system provides users with appropriate customized services based on *Knowledge$_{DIK}$* of users, but the customized services should not be limited to the most suitable for the user calculated by the AI system. While using big data technology, the user's right to choose autonomously should be guaranteed.

The success rate of recommend($R_{recommend}$) to user of AI system can reflect the user's acceptance of the recommend system. As in Eq. (12) , the function Autonomy is constructed to calculate the autonomy index $V_{Autonomy}$. The AI system will recommend different lines to different users according to $V_{Autonomy}$.

$$V_{Autonomy} = Autonomy(R_{recommend}) \qquad (12)$$

Integrity Against Commodification. Commodification refers to the behavior that treat individual, life or human nature as a pure commodity which puts money above personal life. Privacy protection is also a protection of the individual. The individual should not be treated differently because of age, race, education level or economic class in particular systems such as health care and law.

As shown in Eq. (13), the function IAC is constructed to calculate V_{IAC} in the decision-making process of AI system. If and only if V_{IAC} is true, the decision-making process conforms to the requirements of integrity against commodification, and the decision-making behavior is legal. Where $E_{purpose}$ represents the different decision-making system such as law system. V_{IAC} is calculated by comparing $E_{purpose}$ and $P_{DIK(G)}$.

$$V_{IAC} = IAC(E_{purpose}, P_{DIK(G)}) \qquad (13)$$

Reputation. Reputation is closely related to privacy, and the defamation of others is an invasion of privacy. Defamation refers to make an incorrect characterization or association of user based on true or false P_{DIK}, so that the reputation or psychological, emotional health of user are affected.

Equation (14) is used to calculate the reputation index $V_{Reputation}$, If $V_{Reputation}$ is higher than $V_{Reputation}{}^{W}$, it means that the decision-making process of AI system will not affect the reputation of user, and the process is legal.

$$V_{Reputation}{}^{W} = Reputation(E_{purpose}, E_{ID}, P_{DIK(G)}) \qquad (14)$$

3   The Rights in Circulation of Privacy Resources

The circulation of privacy resources in decision-making process has four circulation links: Sensing, Storage, Transfer, and Processing. The rights to privacy in circulation includes the right to know, the right to participate, the right to forget and the right to supervise.

3.1 Rights to Privacy

Right to Know. The right to know refers to the individual's right to know and obtain P_{DIK}. The right to know is not unlimited but differentiated according to different participants.

The right to know includes *Know(course)* and *Know(content)*. *Know(course)* is the right to know about the circulation of P_{DIK}, including $Sensing^K$, $Storage^K$, $Transfer^K$, $Processing^K$. *Know(content)* is the $P_{DIK(G)}$ calculated in Eq. (15), which represents what the participant have the right to know($P_{DIK(G)}{}^{Know}$).

$$P_{DIK(G)}{}^{Know} = Know(E_{ID}, E_{purpose}, process) \tag{15}$$

As shown in Eq. (15), $P_{DIK(G)}{}^{Know}$ is calculated based on the identity and purpose of participant. Besides, *Know(content)* of the same participant in different processes is also different.

Right to Participate. The right to participate refers to the right of the method participant participating in the management and decision-making of P_{DIK}. As shown in Eq. (16), the function Participate is constructed to calculate the content of right to participate(*Participation*), Such as the form of participating, the number of participating and the deadline of participating.

$$Participation = Participate(E_{ID}, E_{purpose}, process) \tag{16}$$

Right to Forget. The right to forget refers to remove old P_{DIK} ($P_{DIK}{}^{old}$) and unvalue P_{DIK} ($P_{DIK}{}^{unvalue}$) from the DIKW graph. $P_{DIK}{}^{old}$ is the P_{DIK} that is replaced by a new P_{DIK} over time. $P_{DIK}{}^{unvalue}$ is the P_{DIK} whose value is less than the storage cost.

A forgetting period T_{forget} is set to prevent the influence of $P_{DIK}{}^{old}$ on decision-making of AI system and the drag of $P_{DIK}{}^{unvalue}$ on virtual community. Every period of T_{forget} is passed, the virtual community will take place a systematic forgetting of P_{DIK}.

Right to Supervise. The right to supervise in the process circulation of P_{DIK} can be divided into logic supervision(S_{logic}), value supervision(S_{value}) and right supervision S_{right}. The right of supervision is the threshold of the decision-making process of AI system, only the supervision result of each participant in each process is true, the decision-making behavior is legal. The subject of supervision can be any interested participant.

Logic Supervision. Logicl supervision is mainly to supervise the common basic logic errors. For example, the number of votes is greater than the number of voters, which a logic error occurred in the decision-making process. The result of S_{logic} is false and the decision-making result is legal and not recognized.

Value Supervision. There are seven values of privacy: fairness, personal security, financial security, peace and quiet, autonomy, integrity against commodificatione and reputation, which is the content needs supercised of value supervision. When $V_{Fairness}$ and V_{IAC} are true, and the others greater than their respective thresholds, S_{value} is true and the decision-making process of AI system is legal.

Right Supervision. Right supervision is to supervise whether the use of privacy rights by participants exceeds the limit in each process. S_{right} includes S_{know}, $S_{paticipate}$ and S_{forget} in all four links of circulation.

3.2   The Circulation of Private Resources

Sensing. The sensing process occurs between Generator and Communicator. Virtual communities extract P_{DIK} from $T_{virtual}$ and UGC, and model a DIKW graph of user based on P_{DIK}. The rights to privacy involved in the Sensing process are: $Know_G$, $Know_C$, $Participate_C$, $Supervise_{(G-C)}$, $Supervise_C$.

Storage. The storage process is that Communicator storage the DIKW graph of different types P_{DIK} in a medium that can be accessed and restored. Some of the privacy rights involved include: $Participate_C$, $Forget_C$, $Forget_G$, $Supervise_C$.

Transfer. Transfer is the process that Communicator transfers P_{DIK} on the DIKW graph to Acquirer. Some of the privacy rights involved include: $Know_A$, $Know_G$, $Participate_A$, $Supervise_{(C-A)}$, $Supervise_{(A-C)}$, $Supervise_{(G)}$. Among them, $Know_G$ and $Supervise_{(G)}$ are the inherent rights of user. In practice, it possible for user not to exercise their rights, but both of these rights still exist.

Processing. Processing is the process that Acquirer exploits and develops a P_{DIK} obtained from virtual community. The privacy rights involved are as follows: $Participate_A$, $Supervise_{(G-A)}$, $Supervise_A$.

4   Privacy Protection

The significance of privacy protection is to provide guidance that can reduce the privacy risk and enable the AI system to make effective decisions in resource allocation and system control.

The privacy protection has a three-layer decision mechanism. The first layer is the supervision mechanism, each participant has the right to supervise in the process of P_{DIK} circulation. The second layer of privacy protection is anonymous mechanism, which protects P_{DIK} generated by sigle P_{DIK}. The third layer is partition mechanism, which protects P_{DIK} generated by multiple P_{DIK}.

4.1   Anonymous Protection Mechanism

Data Anonymous Protection. Data anonymous protection can be used to protect particular $Data_{DIK}$. For example, a negative or positive HIV test result(D_{DIK1} = "HIV=negative/positive") is represented by the value of parameter A ($D_{DIK1}{}^A$ = "A = 0/1") in the trasfer process of P_{DIK}.

The professional visitor with eligible for access ($Visitor_{profession}, Visitor_{pro}$) has the ability to restore $D_{DIK1}{}^A$ back to D_{DIK1}. Whereas a Hacker who is not qualified to access D_{DIK1} does not have the ability even if he obtained $D_{DIK1}{}^A$ through improper means. Data anonymous protection can reduce the risk of P_{DIK} leakage at the data level.

Information Anonymous Protection. Information anonymous protection is to hide the relationship between user and other entities through anonymity. For example, as shown followed, $I_{DIK1}{}^A$ generated by I_{DIK1} after anonymization:

I_{DIK1} = "User A tested positive for HIV"
$I_{DIK1}{}^A$ = "User XX tested positive for HIV"

The anonymization of I_{DIK1} does not affect $Visitor_{pro}$ with medical research purposes to use of private resources of patient. But it's difficult for hacker to connect $I_{DIK1}{}^A$ with a specific user. Information anonymous protection contributes to reduce the risk of patient privacy leakage.

Knowledge Anonymous Protection. Knowledge Anonymous Protection is that the attribute $K_{DIK}(Val)$ of K_{DIK} is concealed. The K_{DIK} trasferred by by Communicator to Acquirer is a collection of all possible K_{DIK} rather than the most possible single K_{DIK}. The solution of target K_{DIK} based on the visitor's purpose will diverge to varying degrees, and the solution space of K_{DIK} will expand with multiple $K_{DIK1}{}^A$. The validity of all K_{DIK} is the same for visitor, and visitor will provide difference services based on different $K_{DIK}{}^A$, which can ensure choice autonomy of user.

4.2   Partition Protection Mechanism

It is known that the P_{DIK} on DIKW graph can be transformed to $P_{DIK}{}^{new}$. If the calculated $T_{Difficulty}$ of P_{DIK} according to Eq. (5) is infinite for visitor, the P_{DIK} will be allowed to transfer to Acquirer by Communicator. Beacause the visitor has no ability to transform P_{DIK}, and will not cause the threat of irrelevant P_{DIK} disclosure.

The decision to protect P_{DIK} based on $T_{Difficulty}$ does not apply to all situations, because in some cases the calculation of $T_{Difficulty}$ is too troublesome. If both P_{DIK} can meet the needs of visitor in decision-making of AI system, the P_{DIK} with a small $P_{DIK}(Out)$ will be choosen as output. The smaller the $P_{DIK}(Out)$, the smaller the possibility of P_{DIK} being transformed to $P_{DIK}{}^{new}$, and the smaller the risk of leakage of irrelevant P_{DIK}.

5   Conlusion

$T_{virtual}$ and UGC left by user in virtual community is kinds of privacy resources and can be classified into data resources, information resources and knowledge resources which constitute the DIKW graph. Then the protection of $T_{virtual}$ and UGC in the law is seriously lagging behind. The application of big data technology in and $T_{virtual}$ and UGC has created a "technology crossing the boundary" problem but neither the operators of the virtual community nor the users themselves are aware of this problem.

There are four links of circulation of privacy resources: Sensing, Storage, trasfer and processing. The four links are completed by one or two of the three participants: Generator (User), Communicator (Virtual community) and Acquirer (Visitor). A legal framework of AI Governance for privacy resources protection in virtual community is established by clarifying the content of rights to privacy of each participant in each link. The content includS the privacy allowed to be known, the form of participation, the time when private to be forgotten, and so on. And the supervision mechanism is used to ensure that participant does not exceed the scope of their privacy rights. In addition, anonymous and partition mechanisms have also been applied to the protection of private.

References

1. Bozdag, E.: Bias in algorithmic filtering and personalization. Ethics Inf. Technol. **15**(3), 209–227 (2013)
2. Mittelstadt, B., Allo, P., Taddeo, M., Wachter, S., Floridi, L.: Bias in algorithmic filtering and personalization. Ethics Inf. Technol. **15**(3), 209–227 (2016). https://doi.org/10.1007/s10676-013-9321-6
3. Girardin, F., Calabrese, F., Fiore, D., Ratti, C., Blat, J.: Digital footprinting: uncovering tourists with user-generated content. IEEE Pervasive Comput. **7**(4), 36–43 (2008)
4. Krumm, J., Davies, N., Narayanaswami, C.: User-generated content. IEEE Pervasive Comput. **7**(4), 10–11 (2008)
5. Ulrike, H.: Reviewing person's value of privacy of online social networking. Internet Res. **21**(4), 384–407 (2011)
6. Song, Z., et al.: Processing optimization of typed resources with synchronized storage and computation adaptation in fog computing. Wirel. Commun. Mob. Comput. **2018**, 3794175:1–3794175:13 (2018)
7. Duan, Y., Lu, Z., Zhou, Z., Sun, X., Wu, J.: Data privacy protection for edge computing of smart city in a DIKW architecture. Eng. Appl. Artif. Intell. **81**(MAY), 323–335 (2019)
8. Daries, J., et al.: Privacy, anonymity, and big data in the social sciences. IEEE Pervasive Comput. **57**(9), 56–63 (2014)
9. Duan, Y.: Towards a Periodic Table of conceptualization and Formalization on Concepts of State, Style, Structure, Pattern, Framework, Architecture, Service, etc. Based on Existence Computation and Relationship Defined Everything of Semantic. SNPD (2019)

10. Mittelstadt, B.: From individual to group privacy in big data analytics. Philos. Technol. **30**(4), 475–494 (2017). https://doi.org/10.1007/s13347-017-0253-7
11. Duan, Y., Sun, X., Che, H., Cao, C., Yang, X.: Modeling data, information and knowledge for security protection of hybrid IoT and edge resources. IEEE Access **7**, 99161–99176 (2019)

Toward Sliding Time Window of Low Watermark to Detect Delayed Stream Arrival

Xiaoqian Zhang[1] and Kun Ma[2(✉)] (iD)

[1] School of Information Science and Engineering, University of Jinan,
Jinan 250022, China
965326116@qq.com

[2] Shandong Provincial Key Laboratory of Network Based Intelligent Computing,
University of Jinan, Jinan 250022, China
ise_mak@ujn.edu.cn

Abstract. Some emergency events such as time interval between input streams, operator's misoperation, and network delay might cause stream processing system produce unbounded out-of-order data streams. Recent work on this issue focuses on explicit punctuation or heartbeats to handle faults and stragglers (outlier data). Most parallel and distributed models on stream processing, such as Google MillWheel and Apache Flink, require hot replication, logging, and upstream backup in an expensive manner. But these frameworks ignore straggler processing. Some latest frameworks such as Google MillWheel and Apache Flink only process disorder on an operator level, but only point-in-time and fixed window of low watermarks are discussed. Therefore, we propose a new sliding time window of low watermarks to detect delayed stream arrival. Contributions of our methods conclude as adaptive low watermarks, distinguishing stragglers from late data, and dynamic rectification of low watermark. The experiments show that our method is better in tolerating more late data to detect stragglers accurately.

Keywords: Stream processing · Watermark · Out-of-order data · Stragglers · Late data

1 Introduction

Fault tolerance, low latency, balancing correctness, and checkpoints are features of streaming processing system [9]. Data arrival time does not strictly reflect its event time. It is important to distinguish whether the late data is simply delayed (due to the network delay and unbalanced allocation of the system resources) on the wire or actually lost somewhere [1]. If the stragglers (also called outlier data or slow streams) cannot be processed timely, the correctness of the processing result will be affected. Besides, stragglers decrease the running speed and increase error probability. However, current streaming processing systems has

H. Gao et al. (Eds.): CollaborateCom 2020, LNICST 350, pp. 444–454, 2021.
https://doi.org/10.1007/978-3-030-67540-0_28

little measures to address this fault tolerance issue [14,18]. In our paper, we try to detect delayed stream arrival in stream processing.

Logging, hot replication [13] (where every node has two copies), and upstream backup [8,12] (where messages are replied when tasks failed) are common method in stream fault recovery. Discretized stream (D-Stream) is abstraction layer of Spark Streaming system, which allows checkpointing and parallel recovery [6] and recover from fault to tolerate stragglers [18]. The method relies on micro batches in the technical stack of Spark system. An out-of-order processing framework is proposed to prevent ordering constraints [11] by AT&T Labs. x Google MillWheel proposed a low watermark strategy to solve this problem for the incoming stream and checkpoints its progress at fine granularity as part of this fault tolerance [1]. Low watermark is a limit of timestamp of all processing data. Apache Flink proposed the window mechanism of low watermark [5] to advance low watermarks of Mill-Wheel. They use maximum allowed delay to allow data delay over a period of time. But these methods cannot distinguish straggler and late data. Apache Storm uses Trident (the high-level transaction abstraction) to guarantee exactly-once semantics for record delivery to avoid out-of-order data [15]. It requires strict transaction ordering to operate, and is based on a continuous operator model that performs recovery by replication (several copies of each node) or upstream backup (nodes buffer sent messages and replay them to a new copy of a failed node). This will cause extra hardware cost and take a long time to recover due to the transaction framework. More importantly, neither replication nor upstream backup handles stragglers. A straggler must be treated as a failure or fault in upstream backup, while it will slow down both replicas with synchronization protocols [14].

We propose a new method, adaptive sliding time window of low watermark, to make late data, stragglers and normal data from streams different. The method can reduce the number of late data and improve the accuracy of result. Unlike Low Watermark point-in-time of MillWheel, adaptive sliding time window of low watermark is a window. Adaptive low watermark, distinguishing stragglers from late data, and automatic correction of low watermark are unique compared to other methods.

The chapters of this paper are distributed as follows. Section 2 describes the latest work on out-of-order stream, low watermark and improved adaptive watermark. Section 3 introduces how to distinguish late data, stragglers and normal data. Firstly, we expand the point-in-time of watermark to sliding time window of low watermark. Secondly, using the window to define late data, straggler and normal data. Finally, we propose the delayed stream arrival detection algorithm. The experimental result, displayed in Sect. 4, shows the advantage of our method. Section 5 outlines brief conclusions and future research.

2 Related Work

2.1 Low Watermark

It is different between the order of arrival of tuples and the order of source [16]. Network delay and delivery errors cause disorder. Stragglers (also known as slow data) are inevitable in large clusters. Therefore, it is necessary to detect delayed stream arrival for fault recovery. There is a mechanism which allows windows to stay open in additional time to process out-of-order stream data.

Low watermark is a timestamp that limit the timestamp of future data [1]. The low watermark of MillWheel tracks all events waiting to be solved in the distributed system to distinguish two situations. First, there is no late data arrival if the low watermark advances past time t without the data arriving. Second, there is late data with out-of-order stream arrival, which can solve timing issues in real-time computing. Given a computation A, the oldest work in A is a timestamp relevant to the oldest uncompleted record in A. The recursive definition of low watermark of A based on data flow is as follows.

Low watermark of A = min (oldest work of A, low watermark of C: C outputs to A). A and C are computations.

If there are no input streams, the low watermark is equipped to oldest work values.

2.2 Adaptive Watermark

Adaptive Watermark is a new generation strategy that can adaptively decide when to generate a watermark and which timestamp without any prior adjustments [2]. ADWIN (Adaptive Window) is used to detect concept drifts [7]. When the ADWIN detects concept drift, the algorithm will calculate dropped rate (computed by lateElements/totalElements). If this ratio is smaller than the threshold l (late arrival threshold), a new watermark will generate. When a new watermark is generated, the dropped rate is resetted. Adaptive Watermark can adapt to changes in data arrival frequency and delayed arrival rate, which realizes the balance between latency and accuracy. However, some fixed parameters like l are setted empirically, which may cause high-latency when processing data. Apache Flink provides abstractions that allow the developers to assign their own timestamps (timestamp assigner) and emit their own watermarks (watermark generator) [5]. That can be individualized by developers. In Apache Flink, the two watermark generation methods are periodic and punctuated [3]. Programmer can realize their thought by overwriting the *extractTimestamp* function. Allowed lateness is also a particularly improved feature of Apache Flink in dealing with disorderly events [4]. By default, when a watermark passes through the window, late data will not be discarded. In an improved way, allowed lateness allows for a period of time within an acceptable range (also measured by event time) to wait for the arrival of data before a watermark exceeds end-of-window in order to process the data again.

3 Delayed Stream Arrival Detection

3.1 Sliding Time Window of Low Watermarks

We propose the sliding time window of low watermark to distinguish different data types and receive data. Figure 1 shows the sliding time window of

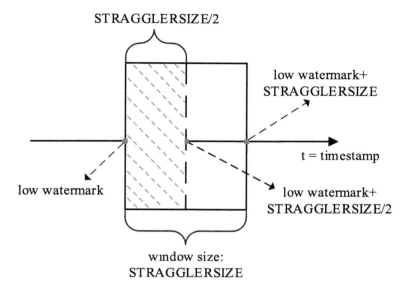

Fig. 1. Sliding time window of low watermark

low watermark in detail. As we can see from Fig. 1, the left of the window is *lowwatermark*, which represent a bound on the timestamps of future data. The size of window is $STRAGGLERSIZE$. The range of sliding time window is $[lowwatermark, lowwatermark + STRAGGLERSIZE)$. $lowwatermark + STRAGGLERSIZE/2$, the middle of the window, is an important boundary used to differentiate data types. Specific data types refer to Sect. 3.2.

3.2   Definition of Late Data and Straggler

We divide the data into four types, normal pending data, normal data, late data, and stragglers, by sliding time window of low watermark. We can get them by comparing the timestamp of processing data and the range of the sliding time window.

Normal pending data: data with timestamp belonging to the interval $[lowwatermark + STRAGGLERSIZE, +\infty)$; Normal data: data with timestamp belonging to the interval $[lowwatermark + STRAGGLERSIZE/2, lowwatermark + STRAGGLERSIZE)$; Stragglers: data with timestamp belonging to the interval $[lowwatermark, lowwatermark + STRAGGLERSIZE/2)$; Late data: data with timestamp belonging to the interval $(-\infty, lowwatermark)$.

Every time, late data and normal pending data are outside the time window, stragglers and normal data are in the time window. Every new data appears as normal pending data and the timestamp of new data is after $lowwatermark + STRAGGLERSIZE$. All the types of data are shown as Fig. 2, where A is the late data, B is the stragglers, C is the normal data and D is the normal pending data.

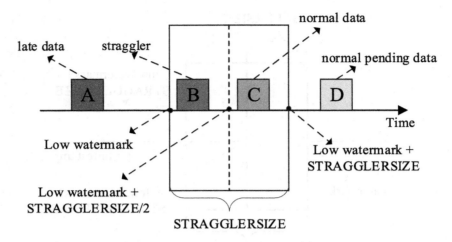

Fig. 2. Data types

3.3 Delayed Stream Arrival Detection

In order to generate a new watermark and keep it updated when new elements processed, we use m, the maximum allowed delay, to represent. And dynamic m (Table 1) is proposed to adapt the real-time stream account for high accuracy and low latency. When the number of stream per second (with *dataRate* for short) bigger than a certain threshold (with $DATARATETHRESHOLD$ for short), dynamic m will change automatically. All the parameters we used to dynamic rectification low watermark are shown in Table 1.

Table 1. Parameters used for watermark generators

Parameter	Description
m	Maximum allowable delay. $m \in [1, SLIDEWINDOWSIZE]$
CHANGERATE	Maximum allowable delay change ratio, CHANGERATE $\in$ $(0, 1]$. Default is 0.01
DATARATETHRESHOLD	Threshold of data per second. Default is 5
dataRate	Number of data per second in real-time stream

And Algorithm 1 describes how m change with $CHANGERATE$ in detail. For a continuous stream, there is a list of collecting new data when it is processed. We get *dataRate* by counting the number of data in the list. Then, we will compare *dataRate* and $DATARATETHRESHOLD$ to decide how to change m. When *dataRate* is higher than $DATARATETHRESHOLD$, m will increase by $CHANGERATE$, otherwise m will reduce. The new m will be used to generate next watermark. The Algorithm 1 shows the detail of the change of m.

Algorithm 1. Adaptive watermark generation

Require:
A data stream S

Ensure:
1: $m \leftarrow$ 350L; watermark $\leftarrow$ 0;
2: $dataRate \leftarrow 5$; $DATARATETHRESHOLD \leftarrow 0.01$;
3: extractedTimestamp = $-\infty$; lastExtractedTimestamp = $-\infty$;
4: $dataCount = 0$; N = 1000; SLIDEWINDOWSIZE = 500L;
5: List: processedELments $\leftarrow$ NULL;
6: **for** each $e \in S$ **do**
7:     watermark = extractedTimestamp - m;
8:     emit(watermark)
9:     processedELments $\leftarrow$ e;
10:     **if** processedELments.size () > 0 **then**
11:         lastExtractedTimestamp = processedElments.get(processedELments.size ()-1)
12:         **for** each $element \in processedELments$ **do**
13:             **if** lastExtractedTimestamp $>=$ lastExtractedTimestamp - N && lastExtractedTimestamp $<=$ lastExtractedTimestamp **then**
14:                 $dataCount$ ++
15:             **end if**
16:         **end for**
17:         $dataRate = dataCount/$(N/1000)
18:         $dataCount \leftarrow 0$
19:     **end if**
20:     **if** $dataRate > DATARATETHRESHOLD$ **then**
21:         $m = m + m * CHANGERATE$
22:         if m $>$ SLIDEWINDOWSIZE **then**
23:             m = SLIDEWINDOWSIZE
24:         **else**
25:             $m = m - m * CHANGERATE$
26:             if m $<= 1$ **then**
27:                 m = 1
28:             **end if**
29:         **end if**
30:     **end if**
31: **end for**

The initial values of m, $CHANGERATE$ and $DATARATETHRESHOLD$ must be specified when generating watermark with dynamic m. In our experiment, the default values for these parameters are 350, 0.01 and 5. Figure 3 is an example to explain how the watermark works. When the first data which et (event time) is 19213 comes, m is 350, $dataRate$ is 0, $dataRate$ is smaller than $DATARATETHRESHOLD$ and the watermark is $19213 - 350 = 18863$. When second data with et 19356 arrives, $dataRate$ is 1, which less than 5, and m is reduce to $350 * (1 - 0.01) = 346$, therefore, watermark is $19356 - 346 = 19010$. The third, fourth and fifth data are similar to above. When the sixth data arrive, $dataRate$

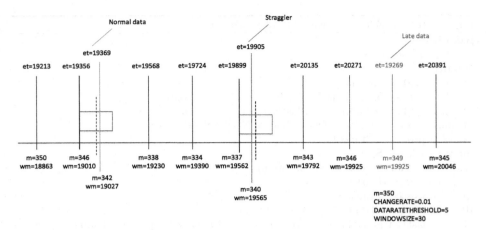

Fig. 3. Example for watermark generation and different data types (Color figure online)

is equal to $DATARATETHRESHOLD$ and m will increase, $334 * (1 + 0.01) = 337$, hence, watermark is $19899 - 337 = 19562$. We can see from our method that m can change dynamically. It can better adapt to the constantly changing in real-time stream. The different data types can be clearly seen from Fig. 3. According to the definition of data types, the blue is normal data. The green is straggler. And the red is late data.

4   Experiments

4.1   Setup and Dataset

Setup: We have implemented low watermark sliding time window with dynamic m on top of Apache Flink v1.9.0. The experiment runs on the computer with 8 GB memory and i5-8500 CPU at 3 GHz.

Dataset: We use three different scale of data sets to compare watermark with dynamic m and static m, the DBES 2012 [10], DBES 2015 [2] Grand Challenge and Out-of-Order dataset from Internet of Things [17]. **1)** The *DBES* 2012 data set, which has 9564 tuples with 3225 out-of-order data, comes from the high-tech manufacturing equipment. **2)** The *DBES* 2015 data set is composed of reports of taxi trips including starting location, drop-off location, the time and amount of payment. After deduplicating the value, there are 10427 tuples with 8384 out-of-order data. **3)** The Out-of-Order data set is generated from standard commercial equipment, networks and protocols commonly used in IoT applications. S-8 to S-10 records the WLAN information, whose unordered data is higher than D-1 to D-5, recording the UMTS information. We choose S-10 for our experiment, which has the most data in S-8 to S-10. This data set, including 18386 out-of-order data, has 29999 tuples.

We use the timestamp of every tuple from the three data sets to distinguish different data type, especially stragglers and late data. For the *DBES* 2012 data set, we choose timestamp which is generated by the embedded PC upon the creation of the given event. For the *DBES* 2015, we choose the timestamp of starting point. And for the Out-of-Order data set, we use S-Client-Detection-Time. Figure 4 shows the top 100 tuples of every data set. (a) is *DBES* 2012, (b) is *DBES* 2015 and (c) is S-10. We can see the trend, timestamp increases overall, but there is some disorder, of every data set clearly from the Fig. 4.

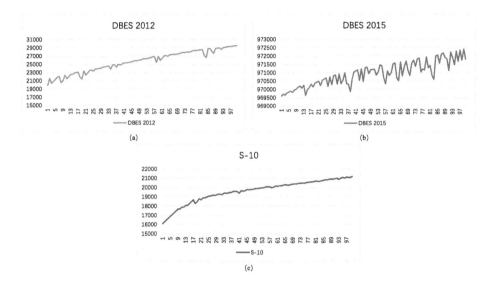

Fig. 4. Trend of data sets

4.2   Comparisons

The ratio of late data (with *Dropped* for short) and the ratio of stragglers are two metrics in our experiment. We compare the two method of static m and sliding time window of low watermark with dynamic m. And we choose two parameters of m and $CHANGERATE$ to change. The results of two different methods are shown in Table 2.

We can see the ratio of dropped and the ratio of straggler from the Table 2. For the static m method, we use 0 and 350 to represent how long allow delay. For the dynamic m, we use an initial value of 350, which can compare with 350 ms delay of static m. From the Table 2, the ratio of *Stragglers* in static m is null, which means the static method cannot find stragglers. On the other hand, our approach not only can find stragglers, but also can decrease the ratio of dropped. That illustrates dynamic m could adapt the change of real-time stream better and the window could find stragglers. For the different $CHANGERATE$, it has

Table 2. Comparison results of different methods

Dataset	Method	m	CHANGERATE	Dropped (%)	Stragglers (%)
DBES 2012	Original	–	–	33.72	–
	Static m	0	–	9.45	–
		350	–	0.31	–
	Sliding time window of low watermark	Dynamic	0.01	0.25	61.23
		Dynamic	0.1	0.25	61.47
DBES 2015	Original	–	–	80.40	–
	Static m	0	–	35.26	–
		350	–	17.16	–
	Sliding time window of low watermark	Dynamic	0.01	12.15	21.31
		Dynamic	0.1	12.11	21.28
S-10	Original	–	–	61.29	–
	Static m	0	–	14.47	–
		350	–	0.14	–
	Sliding time window of low watermark	Dynamic	0.01	0.05	49.15
		Dynamic	0.1	0.05	49.15

almost the same dropped rate and stragglers rate. That is because m is limited by $WINDOWSIZE$ and trend of data sets. The specific analysis of result is as follows.

For the method of static m, different m has different dropped ratio. Every dropped ratio has lower latency than original dropped ratio, but it doesnot distinguish stragglers. And we can see the result of 350 ms delay is better than 0ms, which certifies allowing delay can reduce dropped ratio. As we can see from Table 2, for the $DBES$ 2012, original dropped ratio is 33.72%. When $m = 0$, dropped ratio is 14.47% and when $m = 350$ is 0.31%. $DBES$ 2015 and S-10 data sets have similar result. The larger m the lower dropped ratio. As a result of allowed delay, it can process more data.

For the dynamic m, we can see from the Table 2 that dropped ratio of dynamic m is lower than static m. Moreover, our method can distinguish stragglers. $CHANGERATE$ determines the magnitude of each change, but it isnot the only factor. Figure 4 describes that the trend of data set also has influence on the result. In Fig. 4 (a), large changes occur in an intervals in $DBES$ 2012. Uncertain changes will have some impact on the results, but it has small influence. From the Fig. 4 (a), we can see the trend of $DBES$ 2012 data set is that there is a big change in every other period of time, which has small impact. The larger the value of $CHANGERATE$, the higher the tolerance of the system.

Hence, when $CHANGERATE$ is 0.1, straggler ratio is higher than it is 0.01. From the Fig. 4 (b), we can see the trend of $DBES$ 2012 data set is that there are more fluctuations. Too much out-of-order data and irregular changes will have some impact on the results. A little change of watermark will influence the result. That results in the greater $CHANGERATE$ the lower straggler rate. In the Fig. 4 (c), the timestamp grows steadily. Stable increasing timestamp has almost no influence in different $CHANGERATE$. Moreover, occasional changes between nearby data provide a guarantee for stream processing. The above reasons make the results the same in different $CHANGERATE$.

5 Conclusion

In this paper, we propose a new method called sliding time window of low watermark to generate a watermark dynamically and distinguish stragglers and late data. The number of data in unit time, the threshold and the ratio of change is three parameters to decide what the value of new watermark will be. Then we use the new watermark and the sliding time window to distinguish normal data, late data and stragglers. Our method adapts the perplexing change of stream better than static m watermark generation, as the experiment result show. Moreover, detailed data classification will be beneficial to stream processing.

Acknowledgments. This work was supported by the National Natural Science Foundation of China (61772231), the Shandong Provincial Natural Science Foundation (ZR2017MF025), the Project of Shandong Provincial Social Science Program (18CHLJ39), the Science and Technology Program of University of Jinan (XKY1734 & XKY1828), and the Project of Independent Cultivated Innovation Team of Jinan City (2018GXRC002).

References

1. Akidau, T., et al.: MillWheel: fault-tolerant stream processing at internet scale. Proc. VLDB Endow. **6**(11), 1033–1044 (2013)
2. Awad, A., Traub, J., Sakr, S.: Adaptive watermarks: a concept drift-based approach for predicting event-time progress in data streams. In: EDBT, pp. 622–625 (2019)
3. Carbone, P., Ewen, S., Fóra, G., Haridi, S., Richter, S., Tzoumas, K.: State management in Apache Flink®: consistent stateful distributed stream processing. Proc. VLDB Endow. **10**(12), 1718–1729 (2017)
4. Carbone, P., et al.: Large-scale data stream processing systems. In: Zomaya, A.Y., Sakr, S. (eds.) Handbook of Big Data Technologies, pp. 219–260. Springer, Cham (2017). https://doi.org/10.1007/978-3-319-49340-4_7
5. Carbone, P., Katsifodimos, A., Ewen, S., Markl, V., Haridi, S., Tzoumas, K.: Apache Flink: stream and batch processing in a single engine. Bull. IEEE Comput. Soc. Tech. Comm. Data Eng. **36**(4), 28 (2015)
6. Chen, Q., Liu, C., Xiao, Z.: Improving MapReduce performance using smart speculative execution strategy. IEEE Trans. Comput. **63**(4), 954–967 (2013)

7. Grulich, P.M., Saitenmacher, R., Traub, J., Breß, S., Rabl, T., Markl, V.: Scalable detection of concept drifts on data streams with parallel adaptive windowing. In: EDBT, pp. 477–480 (2018)
8. Hwang, J.H., Balazinska, M., Rasin, A., Cetintemel, U., Stonebraker, M., Zdonik, S.: High-availability algorithms for distributed stream processing. In: 21st International Conference on Data Engineering (ICDE 2005), pp. 779–790. IEEE (2005)
9. Iqbal, M.H., Soomro, T.R.: Big data analysis: Apache storm perspective. Int. J. Comput. Trends Technol. 19(1), 9–14 (2015)
10. Jerzak, Z., Heinze, T., Fehr, M., Gröber, D., Hartung, R., Stojanovic, N.: The debs 2012 grand challenge. In: Proceedings of the 6th ACM International Conference on Distributed Event-Based Systems, pp. 393–398. ACM (2012)
11. Li, J., Tufte, K., Shkapenyuk, V., Papadimos, V., Johnson, T., Maier, D.: Out-of-order processing: a new architecture for high-performance stream systems. Proc. VLDB Endow. 1(1), 274–288 (2008)
12. Nagano, K., Itokawa, T., Kitasuka, T., Aritsugi, M.: Exploitation of backup nodes for reducing recovery cost in high availability stream processing systems. In: Proceedings of the Fourteenth International Database Engineering & Applications Symposium, pp. 61–63. ACM (2010)
13. Shah, M.A., Hellerstein, J.M., Brewer, E.: Highly available, fault-tolerant, parallel dataflows. In: Proceedings of the 2004 ACM SIGMOD International Conference on Management of Data, pp. 827–838. ACM (2004)
14. Shoro, A.G., Soomro, T.R.: Big data analysis: Apache spark perspective. Glob. J. Comput. Sci. Technol. 15, 7 (2015)
15. Toshniwal, A., et al.: Storm@ twitter. In: Proceedings of the 2014 ACM SIGMOD International Conference on Management of Data, pp. 147–156. ACM (2014)
16. Traub, J., et al.: Scotty: efficient window aggregation for out-of-order stream processing. In: 2018 IEEE 34th International Conference on Data Engineering (ICDE), pp. 1300–1303. IEEE (2018)
17. Weiss, W., Jiménez, V.J.E., Zeiner, H.: A dataset and a comparison of out-of-order event compensation algorithms. In: IoTBDS, pp. 36–46 (2017)
18. Zaharia, M., Das, T., Li, H., Hunter, T., Shenker, S., Stoica, I.: Discretized streams: fault-tolerant streaming computation at scale. In: Proceedings of the Twenty-Fourth ACM Symposium on Operating Systems Principles, pp. 423–438. ACM (2013)

A Novel Approach for Seizure Classification Using Patient Specific Triggers: Pilot Study

Jamie Pordoy$^{(\boxtimes)}$ ⓘ, Ying Zhang ⓘ, and Nasser Matoorian

School of Computing and Engineering, University of West London, London W5 5RF, UK
jamiepordoy@hotmail.com, {ying.zhang,nasser.matoorian}@uwl.ac.uk

Abstract. With advancements in personalised medicine, healthcare delivery systems have moved away from the one-size-fits-all approach towards tailored treatments that meet the needs of individuals and specific subgroups. As nearly one-third of those diagnosed with epilepsy are classed as refractory and are resistant to antiepileptic medication, there is a need for a personalised method of detecting epileptic seizures. Epidemiological studies show that up to 91% of those diagnosed identify one or more triggers as the causation of their seizure onset. These triggers are patient-specific and can affect those diagnosed in different ways dependent on each person's idiosyncratic tolerance and threshold levels. Whilst these triggers are known to induce seizure onset, only a few studies have even considered their use as a preventive component. Therefore, this pilot study investigates the use of patient-specific triggers (PST) in diagnosed epileptics, and whether they can be used as an additional modality when detecting seizures. This study used a precision medicine approach with artificial intelligence (AI), to train and test several patient-specific algorithms that classified epileptic seizures based on the PST of each participant. Experimental results show accuracy, sensitivity, and specificity scores of 94.73%, 96.90% and 93.33% for participant 1 and 96.87%, 96.96% and 96.77% for participant 2, respectively.

Keywords: Multi-modal · Machine learning · Epilepsy · Seizure · Patient-specific

1 Introduction

Epilepsy is a prevalent neurological condition that effects an estimated 70 million people worldwide [1]. An overload of electrical activity between communicating neurons causes a temporal imbalance of neurological activity, culminating in the occurrence of an unprovoked seizure, often leaving an individual with a loss of anatomical motor functions and clarity of memory [2]. An estimated 30% of those diagnosed are classed as refractory and are resistant to anti-epileptic drugs (AEDs) [3]. Those who are resistant have no form of defence and are at a higher risk of triggering a convulsive seizure which can lead to an acute cardiac and respiratory dysfunction [4].

A sudden unexpected death in epilepsy (SUDEP) is the most frequent direct cause of epilepsy-related deaths, predominately affecting those who are resistant or have poorly

H. Gao et al. (Eds.): CollaborateCom 2020, LNICST 350, pp. 455–468, 2021.
https://doi.org/10.1007/978-3-030-67540-0_29

controlled chronic epilepsy. A study by Lambert *et al.* [5] identified 58% of SUDEP cases are nocturnal and occur once an individual has been asleep and experienced a generalised tonic-clonic (GTC) seizure. As the underlying cause of SUDEP remains unknown and without treatment at a therapeutic level, recent case studies have suggested that onset, and in turn SUDEP, could be triggered by several predisposed risk and trigger factors [6].

As observed by Hesdorffer *et al.* [7], the most significant risk factor is an increase in the frequency of GTC seizures, as this can lead to a cardiac and respiratory dysfunction. Patients with epilepsy (PWE) who experience ≥3 GTC seizures per year are 15 times more likely to have a fatal epilepsy-related event such as SUDEP. Other frequent risk factors include partial seizures, missing doses of AEDs and an intelligence quotient (IQ) < 70 [8].

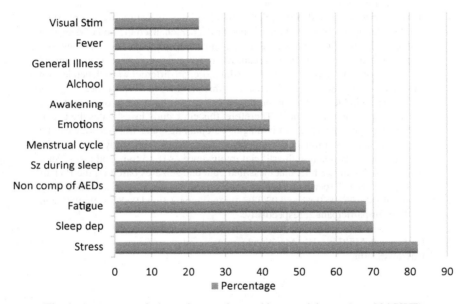

Fig. 1. Percentage of trigger factors observed in *n*-participants ($n = 104$ PWE)

In addition, it is important to ask whether there are specific triggers (precipitants) that increase the probability of onset and whether these could be used in conjunction with the aforementioned risk factors to improve on current methods of seizure detection [9]. Although seizures are sporadic and seemingly random, studies show there are patient-specific triggers (PST) that increase the likelihood of onset [10]. As defined by Aird and Gordon [11], PST can be categorised as seizure inducing or seizure triggering events. Those classed as seizure inducing (lights, noises, and patterns), are caused by environmental or endogenous events and cause a transient lowering of the seizure threshold level [12]. Seizure triggering events (sleep deprivation, stress, and fatigue) are risk factors that vary based on each person's specific threshold and tolerance levels. A study by Ferlisi and Shorvon [13] interviewed 104 PWE to identify the frequency of seizure precipitants (triggering factors).

Results show that seizure triggering events are more frequent, with an estimated 91% of participants experiencing one or more triggers prior to a seizure [13]. The distribution of these triggers is illustrated in Fig. 1, with stress, sleep deprivation, fatigue, and a non-compliance of AEDs as the four most common causes of seizure onset with percentage scores of 82%, 70%, 68% and 54% respectively.

These results reflect the findings of Nakken *et al.* [14], who also identified stress, sleep deprivation and fatigue as the most frequent triggers, with 592 (51%) PWE listing at least one trigger as the causation of their onset [14]. Furthermore, a study by Balamurugan *et al.* [15], analysed 405 PWE, observing that 86.9% experience at least one trigger prior to onset. These results show a noncompliance of AEDs as the most frequent trigger (40.98%), followed by stress (31.35%), sleep deprivation (19.75%) and fatigue (15.30%).

1.1 Patient-Specific Triggers

Precision medicine is a newly adapted paradigm of healthcare that allows medical treatments to be idiosyncratically tailored towards the needs of individuals or specific subgroups [16]. Whilst precision medicine in epilepsy is still relatively unexplored, a recent study by Porumb *et al.* [17], combined precision medicine and machine learning for hypoglycemic event detection from ECG wavelets. By applying a patient-specific approach, a classification model was trained using data from a single participant, which was then tested on unseen data from the same participant. The results demonstrate the potential application of patient-specific classification, with models attaining an accuracy measure of 84.8%, 88.5%, 89.9% and 78.3% for participants 1–4 respectively.

Similarly, a study by Ince *et al.* [18], has explored patient-specific classification of cardiac cycles to detect ventricular ectopic beats (VEBs) and supra-VEBs (SVEBs). Using fully connected feed-forward neural networks that were optimally designed for each persons' idiosyncrasies, this studies patient-specific classification models have surpassed many state of the art algorithms with accuracy and sensitivity scores of 98.3% – 84.6% and 97.4% – 63.5% for VEBs and SVEBS respectively.

Given the successful detection of hypoglycemic anomalies and the use of patient-specific classification in other medical fields, it is our feeling that this method of detection can be used to classify the PST observed in diagnosed epileptics prior to seizure onset.

2 Methodology

This paper presents a preliminary pilot study that investigates the practical application of PST when classifying epileptic seizures. We believe this investigation is supported by the notion that PST are preceding events that are responsible for initiating or precipitating a seizure [14]. Due to the impact of the current global crisis (Covid-19), we cannot conduct a full-scale clinical trial, which in turn has reduced the size of our dataset. However, as this is a pilot study that focuses on patient-specific classification, we decided to proceed using the participants we had available.

Classification models were developed using the Python programming language and coded in Jupyter notebook. The Python libraries used for the development include TensorFlow, Keras and Scikit-learn.

This pilot study will be used as a preliminary component to test the validity of PST and deem whether further research with a full-scale clinical trial is feasible. As far as we know, this is a novel concept and there is no existing research that has used these specific triggers for patient-specific classification of epileptic seizures.

2.1   Data Acquisition

For this study we collected data from two participants with epilepsy, one female (participant 1) in her late twenties and one male (participant 2) in his mid-thirties. Of the 300 days participant 1 was observed, we documented 17 positive instances (seizures), whilst participant 2 was observed for a shorter duration of 248 days, of which 22 positive instances were recorded.

Based on Ferlisi and Shorvon's findings [13], we decided to record the trigger factors of sleep deprivation, stress, and fatigue.

Whilst it has not been observed as a commonly occurring trigger, we have decided to record exercise as an additional PST, as we can see a clear correlation with stress reduction, which in turn could help to prevent or reduce the frequency of seizure onset [19].

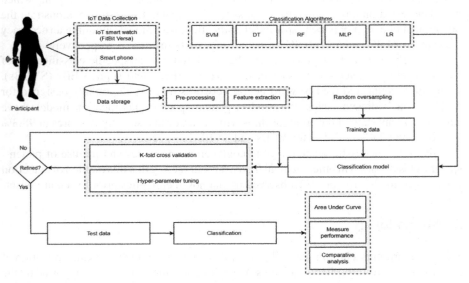

Fig. 2. Pilot study workflow; SVM = support vector machine; DT = decision tree; RF = random forest; MLP = multi-layer perceptron; LR = logistical regression; IoT = Internet of Things.

2.2 Pilot Study Workflow

Figure 2 shows a high-level diagrammatic representation of this study's methodology, which consists of four phases, data collection; data preparation; machine learning and finally a comparative analysis. Both participants were issued with an IoT-enabled smart watch (Fitbit Versa) and smart phone (Samsung SM-N950P) for the duration of the study. We deployed an android based CRUD application with pre-installed quantitative questionnaires on each smart phone, which was then used to record daily measures of stress and fatigue. We used a Fitbit Versa to calculate the metabolic equivalents (METs) of participants, as it provided an accurate measure for the number of minutes that active movement occurred [20]. We also used the embedded accelerometer and heart rate sensor to measure fluctuations in heart rate variability (HRV) and body movement to estimate the quality and duration of a participants sleep cycle [21].

The following sections briefly describe the components that were used to measure the PST and the machine learning models used for patient-specific classification.

2.3 Resampling the Dataset

Accurately classifying rare events can be problematic as the frequency of occurrence can leave a dataset highly imbalanced. For this study, 548 instances were recorded, 509 negative instances where no seizure was documented and 39 positive instances. To account for the varying disparity between positive and negative instances, we used random over-sampling. This sampling technique balanced our datasets distribution as duplicate instances from the minority class were added to our training dataset [22].

2.4 Perceived Stress Scale

Perceived stress scale (PSS) is a widely used stress assessment tool that measures a subject's stress levels [23]. Although the PSS is generally calculated at the end of each month, for this study, we used a modified variant that measures stress daily. The PSS is comprised of 10 questions that require a numerical response from 1 to 5, where $1 :=$ no stress and $5 :=$ maximum stress. To calculate the PSS score, the value responses for questions 4, 5, 7 and 8 are reversed so that question $1 \leftrightarrow 5, 2 \leftrightarrow 4, 3 \leftrightarrow 3, 4 \leftrightarrow 2, 5 \leftrightarrow 1$ [24].

The values for each question are than combined and divided by 10. The PSS scale also accounts for perception, as two participants who encounter the same set of events can accumulate different scores.

2.5 Rating of Fatigue

The Rating of Fatigue (RoF) is a measurement tool that tracks the intensity of fatigue. RoF uses a linear scale of 11 numerical intervals (0–10) with a response of 0 indicating no fatigue whilst a 10 indicates total fatigue and exhaustion. At the start of each day, the pre-installed RoF app shows a set of diagrammatic and descriptive components that guide participants when measuring fatigue [25].

2.6 Classification Algorithms and Techniques

This section provides a summary of the supervised learning classification models and techniques used for this study's comparative analysis. Classification is an instance of supervised learning, where a classification model (classifier) observes a set of input features and makes a prediction on unseen data that shares the same features. Classification models predict binomial outcomes such as yes/no, true/false and positive/negative to categorise new observations.

K-Fold Cross Validation
• K-fold cross validation segments a dataset into k-subsets of approximately equal size, and in turn each classification model is then trained using k-1 subsets where the remaining subset is used to validate the classifiers performance on unseen data [26]. This process in then repeated for k-iterations, and each iteration uses a different subset for validation, whilst the previous validation subset becomes a training subset of k-1. For this study, we used a k-fold cross validation technique where $k = 10$, to evaluate each classifiers performance on unseen data.

Naive Bayes
• Stemming from Bayes theorem of probability, naive Bayes (NB) classifiers are a family of probabilistic classification algorithms which assume that each predictor is an independent entity that equally contributes to the outcome of the target class [27]. NB classifiers can calculate the posterior probability of event A, given the occurrence of event B [28]. This can be expressed mathematically as

$$P(c|x) = \frac{P(x|c)P(c)}{P(x)} \tag{1}$$

Where $P(c|x)$ represents the posterior probability, $P(c)$ the prior probability of class c, $P(x)$ the probability of each predictor and $P(x|c)$ the probability of a predictor given the occurrence of class c [29]. As NB assumes that features are independent, only the variances of each training label need to be calculated, instead of the entirety of the covariance matrix. This enables a NB classifier to use a small quantity of training data when predicting the mean and variance for each predictor, which is ideal for this studies patient-specific approach.

Support Vector Machines
• Support vector machines (SVM) are supervised learning models used to increase the predictive accuracy for classification and regression analysis. SVM classifies data by finding the optimum position for a hyperplane in n-dimensional space, where n represents the number of features, and each feature has a specific co-ordinate value. The features closest to the hyperplane act as support vectors and are used to determine the orientation of the hyperplane. This enables the hyperplane to separate n-classes of training data by the maximum distance, leading to maximal generalization and improved performance [30].

Logistical Regression
• Logistic Regression (LR) is a statistical, supervised learning model used to calculate

binary and binomial response data. LR computes the probabilistic relationship between a dependent, dichotomous variable (0, 1) and one or more independent predictors (variables).

$$z = b_0 + b_1 \cdot x_1 + \cdots b_n \cdot x_n \tag{2}$$

A sigmoid function transforms a linear equation into a logistic equation that converts any input values between negative and positive infinity to a value between 0 and 1 [31]. The equation for LR is derived from a generalised linear equation of independent predictors where e represents the natural logarithm base, b and x are the classifiers parameters, and P is the probability of 1 [32].

$$p = \frac{1}{1 + e^{-(b_0 + b_1 \cdot x_1 + \cdots b_n \cdot x_n)}} \tag{3}$$

Decision Tree with Gini Index
• Decision Trees (DT) are hierarchical decision analysis structures that use a series of interconnected nodes to classify a decision and its consequences. A generalised tree structure consists of single root node, a series of decision nodes and leaf nodes. Decision nodes represent the consequences of an action and have two or more branches that connect to the leaf nodes, whereas the leaf nodes represent the final classification decision of that action. For this study we used the Gini Index to measure impurity, as it enables the most relevant decision nodes to be closer to the root node. As the tree structure traverses downwards, the level of uncertainty surrounding each decision decreases, ensuring a more accurate method of classification [33]. The Gini index is calculated using the following formula where the sum of the squared probabilities for each class is deducted from 1 [34].

$$Gini\ Index = 1 - \sum_{i=1}^{c} (p_i)^2 \tag{4}$$

Random Forest
• Random Forest (RF) is an ensemble classifier that trains multiple decision trees in parallel for increased measures of performance [35]. RF classifiers use a combination of Breiman's "bagging" and random feature selection to form a process called majority voting where each tree classifies input data to identify the most frequently occurring class (prediction). This method of classification exhibits good generalisation, and often outperforms other classification models when measuring accuracy [36]. RF classifiers can be expressed mathematically as

$$\hat{f} = \frac{1}{B} = \sum_{b=1}^{B} \hat{f}_b\left(x^{'}\right) \tag{5}$$

Where $\hat{f}$ represents the final prediction, B the number of trees used, b the current tree and x the training sample used to teach the classifier.

Multi-layer Perceptron
Artificial neural networks (ANN) are information processing paradigms that share performance characteristics with the human biological nervous system [37].

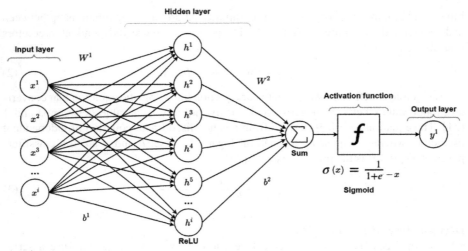

Fig. 3. Multi-layer perceptron architecture; x_i (Input layer) = 6 nodes; h_i (Hidden layer) = 12 nodes; y_i (Output layer) = 1 node; b_i = bias; w_i = weights; ReLU = rectified linear unit;

Multi-layer perceptron's (MLP) are a type of feed-forward neural network that analyses the relationship between a series of independent input variables and a set of dependent output variables. MLP are a modified variation of the standard two layered perceptron that uses three or more layers of neurons with nonlinear activation functions to process complex computations. The MLP used in this study is shown in Fig. 3 and can be expressed mathematically as

$$y_i = f\left(\sum\nolimits_{i=1}^{n} w_i \cdot w_{ih} + \theta_h\right) \tag{6}$$

Where y_i stands for the hidden layer, x_i represents node i of the input layer, w_i is the connection weight between node i of the input layer and node h of the hidden layer. The number of input layer nodes is expressed as n, the bias values of node h is θ_h and the network's Sigmoid function is expressed as $f(u)$ [38].

3 Experiments

As defined by literature [39], the following four metrics are the optimum base measures used to assess the performance of binomial classification models. These base measures are calculated from a series of experiments using a set of positive and negative instances, where TP = true positive, FP = false positive, FN = false negative, TN = true negative. Accuracy (Acc) is used to measure the number of correctly classified instances by the total number of instances; sensitivity (Sen) measures the number of true positive instances by all positive instances; specificity (Spe) measures the number of true negative instances by the number of negative instances and the positive predicted value (Ppv) is the rate of positive instances that are positive. The mathematical formulas for each base measure are expressed below in algorithms 7–10.

$$Acc = \frac{TP + TN}{TP + TN + FP + FN} \tag{7}$$

$$Sen = \frac{TP}{TP + FN} \tag{8}$$

$$Spe = \frac{TN}{TN + FP} \tag{9}$$

$$Ppv = \frac{TP}{TP + FP} \tag{10}$$

To account for the imbalances in our dataset and the use of over-sampling, we constructed a multi-point receiver operating characteristic (ROC) curve of probability. A ROC curve plots a classifiers TP rate against its FP rate at multiple decision thresholds. We then measured the area under curve (AUC) to assess how efficient each classification model was at distinguishing between positive and negative predictions. To calculate AUC, we used the following formula [40]:

$$\hat{A} = \frac{S_0 - n_0(n + 1)/2}{n_0 \cdot n_1} \tag{11}$$

Where n_0 represents the number of positive instances, n_1 the number of negative instances, and $S_0 = \sum r_i$ where r_i is the rank of i th instance [41]. The following classification experiments were conducted on three separate datasets, D1 (participant 1), D2 (participant 2) and D3 (D1 + D2). For D3, we combined the data from both participants to see if it would improve the performance of our patient-specific classifiers (Fig. 4).

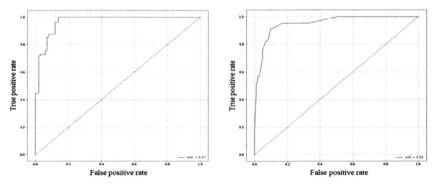

Fig. 4. ROC curve and AUC score for D1 (left) & D3 (Right) using a multi-layer perceptron

4 Experimental Results

The following section illustrates the experimental results recorded for this pilot study. Table 1 summarises the comparative analysis of our classification models for participant 1, with the MLP outperforming the other classifiers with an accuracy measure of 94.73%, sensitivity of 96.29% and an AUC measure of 0.970. Table 2 illustrates the classification results for participant 2, with the MLP once again outperforming the other classification models.

Table 1. Results of each classification model for participant 1

Model	Acc (%)	Sen (%)	Spe (%)	AUC	Ppv (%)
MLP	94.73	96.29	93.33	0.970	96.55
LR	94.70	95.00	94.44	0.947	95.55
NB	94.11	88.76	92.22	0.944	89.01
SVM	93.85	88.75	97.77	0.933	90.72
DT	92.94	91.25	94.44	0.931	92.39
RF	89.44	88.50	90.36	0.942	88.23

Table 2. Results of each classification model for participant 2

Model	Acc (%)	Sen (%)	Spe (%)	AUC	Ppv (%)
MLP	96.87	96.96	96.77	0.987	96.70
SVM	91.20	95.45	87.23	0.913	95.34
DT	91.17	96.87	86.11	0.915	96.87
LR	90.44	90.90	90.00	0.905	91.30
RF	86.75	82.14	91.25	0.961	87.95
NB	85.84	82.45	89.28	0.859	83.33

The following results show how each classifier performed when the datasets for both participants were combined. Once again, the MLP outperformed the remaining classifiers, with accuracy scores of 94.11%, sensitivity of 92.15% and an AUC measure of 0.952 (Table 3).

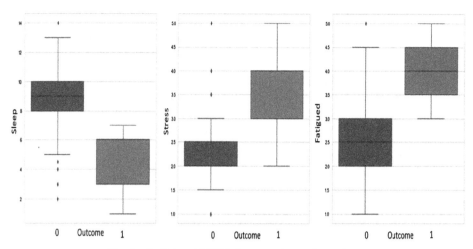

Fig. 5. Participant 1 PST; 0 = no seizure; 1 = seizure.

Table 3. Classification results for dataset D3 (participant 1 + participant 2)

Model	Acc (%)	Sen (%)	Spe (%)	AUC	Ppv (%)
MLP	94.11	92.15	96.07	0.952	92.45
LR	92.48	92.10	92.85	0.925	92.25
NB	92.15	90.12	94.44	0.923	89.47
SVM	91.66	91.17	92.15	0.917	91.26
DT	91.17	88.88	93.75	0.913	88.23
RF	89.54	90.80	88.23	0.958	90.60

5   Discussion

This section summarises the findings and contributions made from this pilot study. Experimental results indicate that PST can successfully train a classification algorithm using a single person's data, and then successfully classify the occurrence of seizure onset using the same persons unseen data. The machine learning approach used for this study successfully classified the fluctuations seen in PST prior to their onset.

For datasets D1, D2 and D3 the MLP outperformed the other classification models regarding Accuracy and AUC. Our findings support the notion that onset is influenced by idiosyncratic triggers as shown in Fig. 5 and Fig. 6 Although there have only been a few studies that assess the correlation between sleep and those diagnosed with epilepsy, our findings indicate seizure onset was more likely to occur when participant 1 had ≤ 6 h of sleep and participant 2 ≤ 8 h. Stress and fatigue also show a correlation with the frequency of onset, with participant 1 having a stress score of 3.5 or above in 88% of recorded seizures. A lower correlation was observed for participant 2, with 64% of seizures having a stress score of 3.5 or higher. Although participant 2 was less affected by stress, a higher fatigue score was observed throughout, with 77% of the 22 seizures recorded having a fatigue score of 4 or above. It was also observed that participant 2 had a stress score of 3.5 or above in 64% of total observations compared to participant 1's 48%.

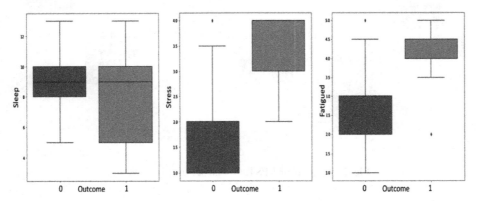

Fig. 6. Participant 2 PST; 0 = no seizure; 1 = seizure.

5.1 Limitations

One concern about the findings of this study was the sample size. Due to Covid-19, our sample size and participant availability was greatly affected, leaving this pilot study with 2 available participants. Whilst this is a preliminary pilot study that used patient-specific models for early detection of seizure onset, a larger sample size would further validate our initial hypothesis and the practical application of PST.

5.2 Future Research

The use of PST should be considered for future research as an additional sensing modality regarding non-EEG detection methods. Current multi-sensor modalities tend to focus on the use of biometric sensors such as electrocardiogram (ECG) to formulate predictions, as they allow for biometric fluctuations to be measured in real time. Using PST in conjunction with standard sensing modalities could account for the varying diversities seen in different types of epilepsy and reduce the frequency of false alarms.

6   Conclusion

This pilot study has undertaken a preliminary investigation into whether PST from the same participant can be used to train and test a classification model. Results indicate that both participants were more likely to experience a seizure if they had < 6–8 h of sleep and/or a stress and fatigue factor ≥ 3.5. To our knowledge, this is the first pilot study that has proposed the use of PST for epilepsy detection, and the results presented here warrant further investigation in the form of a full-scale clinical trial.

References

1. Thijs, R.D., et al.: Epilepsy in adults. Lancet **393**(10172), 689–701 (2019)
2. Iasemidis, L.D., et al.: Adaptive epileptic seizure prediction system. IEEE Trans. Biomed. Eng. **50**(5), 616–627 (2003)
3. Laxer, K.D., et al.: The consequences of refractory epilepsy and its treatment. Epilepsy Behav. **37**, 59–70 (2014)
4. Barot, N., Nei, M.: Autonomic aspects of sudden unexpected death in epilepsy (SUDEP). Clin. Auton. Res. **29**(2), 151–160 (2018). https://doi.org/10.1007/s10286-018-0576-1
5. Lamberts, R.J., et al.: Sudden unexpected death in epilepsy: people with nocturnal seizures may be at highest risk. Epilepsia **53**(2), 253–257 (2012)
6. DeGiorgio, C.M., et al.: Ranking the leading risk factors for sudden unexpected death in epilepsy. Front. Neurol. **8**, 473 (2017)
7. Hesdorffer, D.C., et al.: Combined analysis of risk factors for SUDEP. Epilepsia **52**(6), 1150–1159 (2011)
8. Van de Vel, A., et al.: Non-EEG seizure detection systems and potential SUDEP prevention: state of the art: review and update. Seizure **41**, 141–153 (2016)
9. Pack, A.M.: SUDEP: what are the risk factors? Do seizures or antiepileptic drugs contribute to an increased risk? Epilepsy Curr. **12**(4), 131–132 (2012)
10. Manford, M., et al.: An analysis of clinical seizure patterns and their localizing value in frontal and temporal lobe epilepsies. Brain **119**(1), 17–40 (1996)
11. Aird, R.B., Gordon, N.S.: Some excitatory and inhibitory factors involved in the epileptic state. Brain Dev. **15**(4), 299–304 (1993)
12. Smith, S.J.M.: EEG in the diagnosis, classification, and management of patients with epilepsy. J. Neurol. Neurosurg. Psychiatry **76**(2), ii2–ii7 (2005)
13. Ferlisi, M., Shorvon, S.: Seizure precipitants (triggering factors) in patients with epilepsy. Epilepsy Behav. **33**, 101–105 (2014)
14. Nakken, K.O., et al.: Which seizure-precipitating factors do patients with epilepsy most frequently report? Epilepsy Behav. **6**(1), 85–90 (2005)
15. Balamurugan, E., et al.: Perceived trigger factors of seizures in persons with epilepsy. Seizure **22**(9), 743–747 (2013)
16. Ginsburg, G.S., Phillips, K.A.: Precision medicine: from science to value. Health Aff. **37**(5), 694–701 (2018)
17. Porumb, M., et al.: Precision medicine and artificial intelligence: a pilot study on deep learning for hypoglycemic events detection based on ECG. Sci. Rep. **10**(1), 170 (2020)
18. Ince, T., et al.: A generic and robust system for automated patient-specific classification of ECG signals. IEEE Trans. Biomed. Eng. **56**(5), 1415–1426 (2009)
19. Pimentel, J., et al.: Epilepsy and physical exercise. Seizure **25**, 87–94 (2015)
20. Fitbit.: What are active minutes? https://help.fitbit.com/articles/en_US/Help_article/1379. Accessed 2 May 2020
21. Fitbit.: What should I know about sleep stages? https://help.fitbit.com/articles/en_US/Helparticle/2163. Accessed 21 May 2020
22. Xu, X., et al.: Over-sampling algorithm for imbalanced data classification. J. Syst. Eng. Electron. **30**(6), 1182–1191 (2019)
23. Cohen, S., Janicki-Deverts, D.: Who's stressed? Distributions of psychological stress in the United States in probability samples from 1983, 2006, and 2009. J. Appl. Soc. Psychol. **42**(6), 1320–1334 (2012)
24. Cohen, S., et al.: A global measure of perceived stress. J. Health Soc. Behav. **24**(4), 385–396 (1983)

25. Micklewright, D., et al.: Development and validity of the rating-of-fatigue scale. Sports Med. **47**(11), 2375–2393 (2017). https://doi.org/10.1007/s40279-017-0711-5
26. Refaeilzadeh, P., et al.: Cross-validation. In: Liu, L., Özsu, M.T. (eds.) Encyclopedia of Database Systems. Springer, Boston
27. Yusa, M., Utami, E.: Classifiers evaluation: comparison of performance classifiers based on tuples amount. In: 4th International Conference on Electrical Engineering, Computer Science and Informatics (EECSI), pp. 1–7. IEEE, Yogyakarta, Indonesia (2017)
28. Zhang, H., Li, D.: Naïve Bayes text classifier. In: IEEE International Conference on Granular Computing, pp. 708–708. IEEE, Fremont (2007)
29. Jing, N., et al.: Information credibility evaluation in online professional social network using tree augmented naïve Bayes classifier. Electron. Commer. Res. 1–25 (2019). https://doi.org/10.1007/s10660-019-09387-y
30. Mavroforakis, M.E.: Theodoridis, S.: Support vector machine (SVM) classification through geometry. In: 13th European Signal Processing Conference, pp. 1–4. IEEE, Antalya (2005)
31. Lammertyn, J., et al.: Logistic regression analysis of factors influencing core breakdown in 'Conference' pears. Postharvest Biol. Technol. **20**(1), 25–37 (2000)
32. Sun, Y., et al.: Application of logistic regression with fixed memory step gradient descent method in multi-class classification problem. In: 6th International Conference on Systems and Informatics (ICSAI), pp. 516–521. IEEE, Shanghai (2019)
33. Tahsildar, S.: Gini Index for Decision Trees. https://blog.quantinsti.com/gini-index/. Accessed 09 July 2020
34. Decision Tree Flavors: Gini Index and Information Gain. https://www.learnbymarketing.com/481/decision-tree-flavors-gini-info-gain/. Accessed 05 May 2020
35. Misra, S., Li, H., He, J.: Machine Learning for Subsurface Characterization, 1st edn. Elsevier, Cambridge (2020)
36. Sinnott, R.O., Duan, H., Sun, Y.: Chapter 15 - a case study in big data analytics: exploring twitter sentiment analysis and the weather. In: Buyya, R., Calheiros, R.N., Dastjerdi, A.V. (eds.) Big Data: Principles and Paradigms, 1st edn. Elsevier, Burlington (2016)
37. Balaji, S.A., Baskaran, K.: Design and development of artificial neural networking (ANN) system using sigmoid activation function to predict annual rice production in Tamilnadu. Int. J. Comput. Sci. Eng. Inf. Technol. **3**(1), 13–31 (2013)
38. Matarat, K., et al.: Comparison of classification algorithms for movie reviews. In: 16th International Conference on Electrical Engineering/Electronics, Computer, Telecommunications and Information Technology (ECTI-CON), pp. 826–829. IEEE, Pattaya (2019)
39. Canbek, G., et al.: Binary classification performance measures/metrics: a comprehensive visualized roadmap to gain new insights. In: 2nd International Conference on Computer Science and Engineering (UBMK), pp. 821–826. IEEE, Antalya (2017)
40. Hand, D.J., Till, R.J.: A simple generalisation of the area under the ROC curve for multiple class classification problems. Mach. Learn. **45**, 171–186 (2001)
41. Huang, J., Ling, C.X.: Using AUC and accuracy in evaluating learning algorithms. IEEE Trans. Knowl. Data Eng. **17**(3), 299–310 (2005)

Cooperative Pollution Source Exploration and Cleanup with a Bio-inspired Swarm Robot Aggregation

Arash Sadeghi Amjadi[1]($\boxtimes$), Mohsen Raoufi[1], Ali Emre Turgut[1],
George Broughton[2], Tomáš Krajník[2], and Farshad Arvin[3]

[1] Mechanical Engineering Department, Middle East Technical University,
Ankara, Turkey
arash.amjadi@metu.edu.tr
[2] Artificial Intelligence Centre, Faculty of Electrical Engineering,
Czech Technical University, Prague, Czechia
tomas.krajnik@fel.cvut.cz
[3] Swarm and Computational Intelligence Lab (SwaCIL), Department of Electrical
and Electronic Engineering, The University of Manchester, Manchester, UK

Abstract. Using robots for exploration of extreme and hazardous environments has the potential to significantly improve human safety. For example, robotic solutions can be deployed to find the source of a chemical leakage and clean the contaminated area. This paper demonstrates a proof-of-concept bio-inspired exploration method using a swarm robotic system based on a combination of two bio-inspired behaviors: aggregation, and pheromone tracking. The main idea of the work presented is to follow pheromone trails to find the source of a chemical leakage and then carry out a decontamination task by aggregating at the critical zone. Using experiments conducted by a simulated model of a Mona robot, the effects of population size and robot speed on the ability of the swarm was evaluated in a decontamination task. The results indicate the feasibility of deploying robotic swarms in an exploration and cleaning task in an extreme environment.

Keywords: Swarm robotics · Aggregation · Bio-inspired · Exploration

1 Introduction

Exploration of extreme environments came to the focus of the robotic research community [15] since, as shown by the relatively recent Fukushima disaster, most standard robots fail to operate reliably in extreme environments with chemical and radiation contamination [23]. For instance, in multi-robot systems, wireless communication is widely used to transfer data between robots [13]. Radiation sources can hamper wireless connections, rendering standard multi-robot teams ineffective. To overcome this problem, one can use short-range communication

© ICST Institute for Computer Sciences, Social Informatics and Telecommunications Engineering 2021
Published by Springer Nature Switzerland AG 2021. All Rights Reserved
H. Gao et al. (Eds.): CollaborateCom 2020, LNICST 350, pp. 469–481, 2021.
https://doi.org/10.1007/978-3-030-67540-0_30

techniques, common in robotic swarm research, such as an effective inter-robot connection protocol [17], which is more resilient to extreme conditions.

In the multi-robot research community, researchers mainly tend to distribute the robot tasks among large numbers of simple robots rather than assigning a complex duty to a single robot because of the numerous advantages of multi-robot systems [6]. For instance, a group of cooperating robots can carry out a task at a higher speed when compared to a single robot [4]. A collective decision made by a group of robots will be more reliable than the decision of a single robot in most applications. In addition, multi-robot systems are more fault-tolerant, and improvements in efficiency can be achieved simply by scaling up the number of robots in the swarm. For example, sensor noise and uncertainties in extreme environments can heavily influence a single robot's decisions. On the other hand, a multi-robot platform can compensate for this uncertainty by processing and combining the information of multiple agents [10].

Moreover, emergent swarm behaviors [28] are also robust to noise and uncertainty of an individual robot's sensors. By having this point in mind, robot interaction rules and methods are required in order to establish a protocol by which a number of robots can cooperate with each other and complete a task. In many cases, such protocols contain high-level data transfer among robots and decision centers [8]. Bio-inspired behaviors shown by social animals like ants, bees, and termites can be utilized to establish social behavior among multiple robots [12]. The observation of effective behaviors in social animals encourages us to choose a deeper and detailed approach to apply a similar scenario for robots.

Aggregation is a behavior that is observed in many social animals varying from insects [7] to amoeba [26]. This behavior can be defined as the gathering of individuals around an area with optimal ecologic properties. By forming this aggregation, they gain new abilities like building a habitat, deterring much stronger predators, and defeating larger prey [24]. Aggregation is observed in two types: cue-based aggregation [20], in which nest members aggregate by following an external cue, and self-organized aggregation, where nest members aggregate without any external guidance and regardless of their environment. For exploration in extreme environments, physical cues (e.g., chemicals or radiation) are crucial; therefore, cue-based aggregation is more relevant.

The BEECLUST method, introduced in [27] is one of the most popular bio-inspired aggregation principles. By performing a detailed analysis of honeybees, which follow cue-based aggregation around a zone with a more optimal temperature [14], the BEECLUST aggregation method [27] was chosen as a basis to implement our exploration system. This bio-inspired aggregation was studied intensively in many research works [3,5].

In terms of communication challenges in swarm robotics, one of the most distinct phenomena of social insects' interactions is their communication method. Social insects often use the environment itself as a communication medium by spreading organic substances (pheromones) that can indicate a multitude of environmental features from the presence of food to dangerous animals. The interaction of pheromones and individual agents leads to efficient swarm behaviors capable of solving chaotic and complex situations. The principles of pheromone communication were utilized by swarm robots in various research studies [2,18,21], and its potential for robotic applications has been demonstrated. For instance, artificial pheromone trails were employed for aiding swarm aggregation in [22]. They proposed a state-of-the-art artificial pheromone system that could emulate environmental effects on pheromone distribution more than previous studies on simulating artificial pheromones. It was concluded that the emulation of realistic spatio-temporal development of pheromone trails brings new insights for the interaction of swarm members, environment, and released pheromones.

In this paper, a method for exploring environments with extreme conditions, e.g., chemical or radiation leakage, is developed using robot swarms. The proposed method combines the BEECLUST algorithm and pheromone following behavior. The main goal of this paper is to make robots detect and clean the source of chemical leakage in order to decontaminate an entire environment. In particular, we investigate the impact of population size, speed of motion, the intensity of environmental cues (contamination magnitude), and coherency of the group, on the efficiency of the cleaning operation.

The rest of the paper is organized as follows: in the second section, the proposed exploration method is described. In the third section, the robotic platform and experiments are demonstrated. In the fourth section, the results of the experiments are analyzed and discussed. Finally, in the last section, the paper is concluded, and future work on this exploration method is explained.

2   Localization and Cleanup Method

In the proposed localization method, a combination of cue-based aggregation [27] and pheromone following behavior [21] was developed (shown in Fig. 1). Robots tend to reach the source of leakage by following the intensity gradient of the cue (contamination magnitude). This chemotaxis behavior has been shown in nature e.g., ants foraging. The main goal is to reach the source and start the cleaning task with the presence of other robots. They stop their motion and start the cleaning task when they detect another robot. Therefore, robots are always in one of three stages: 1) follow a chemical gradient, 2) avoid walls, and 3) stop and clean. It should be mentioned that the states of the original BEECLUST method are: 1) move forward, 2) avoid walls, 3) stop and wait. Hence, as it is apparent from the comparison of the BEECLUST method and the proposed method, the first state of the BEECLUST method is replaced by going towards the center of cue. Besides, the cleaning task is added to the last state of the BEECLUST method.

In the proposed method, after a robot-robot collision occurs, robots wait for a while, depending on the cue intensity that they sense. During this waiting period, they clean the contamination beneath them. Details about cleaning are provided in the next section. When the waiting time is over, they continue their initial task i.e., following a chemical gradient. The duration of this cleaning time, w_s, relies on the cue's intensity, where robots find other robots. Hence, this cleaning time is calculated using the following equation:

$$w_s = w_{max} \frac{\bar{S}_c^{\,2}}{\bar{S}_c^{\,2} + 25000},$$

(1)

where w_{max} is the maximum waiting time and $\bar{S}_c$ is the average reading from left and right sensors, $\bar{S}_c = \frac{s_r + s_l}{2}$, $0 \le \bar{S}_c \le 255$.

When the waiting time is over, robots make a turn of θ degrees where θ is a random variable with uniform distribution in the range of $[90°\ 180°]$ in both the clockwise and counter-clockwise directions. After the random turn, robots continue to follow the chemical gradient in order to reach its center.

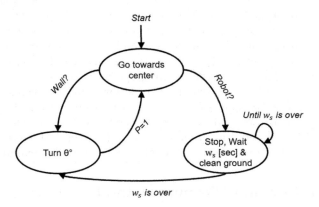

Fig. 1. Proposed exploration method. The ovals indicate stages and arrows indicate transitions.

3   Realization of the Method

3.1   Robotic Platform

The proposed exploration scenario was implemented using Mona robots [1]. It is a miniature wheeled robot with a diameter of 0.08 m. Mona was modeled [25] in Webots software to mimic the real Mona robot, and the proposed method was applied via a simulation model of Mona. Figure 2 shows the model used in Webots for Mona and the real Mona robot. IR proximity sensors are utilized in modeled Mona to detect collision and distinguish a wall from a robot. For robots to discriminate wall from other robots, when a collision occurs, each robot during collision sends an IR pulse to its neighbor. If the robot receives back an IR pulse from its neighbor, then it has collided with a robot. Otherwise, the collision has been made with a wall. This method of differentiating robots from wall was inspired by [27].

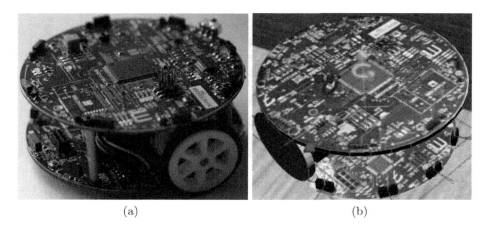

(a)   (b)

Fig. 2. (a) Mona, an open-source low-cost robot developed for swarm robotics and (b) simulated model of Mona in Webots.

In order to sense the cue, as was mentioned previously, two sensors beneath the right and left wheels of robots are added. In the simulation, two sensors facing towards the ground were used, and after calibration, the value changes in the domain of [0 255] where 0 is a region with no chemical leakage, and 255 is the region with the maximum possible intensity of leakage.

To describe the algorithm by which robots follow the cue, two sensors are utilized under the left and right wheels of Mona in order to measure the cue intensity under these wheels. The value of these sensors are used in order to determine the desired rotation of Mona in a direction that cue increases so robots will move towards the zone with highest chemical leakage. To achieve this objective, required value of right wheel speed, N_r, and left wheel speed, N_l, were computed using the equations below:

$$N_l = \frac{s_l - s_r}{\alpha} + \beta \quad \text{and} \quad N_r = \frac{s_r - s_l}{\alpha} + \beta, \tag{2}$$

where s_l is the value extracted from the left cue sensor, and s_r is the value of the right sensor. The coefficient α adjusts the sensitivity of motors to the sensor values' difference in a way that with lower values of α robots will react dramatically to small differences between two sensors. In this paper, α is set to 2. The last parameter β is the biasing coefficient of motors with a possible range of [0 10]. For the modeled Mona, $\beta = 6$ is equal to speed of 0.08 ms^{-1} (speed of one Mona length per second) and consequently, $\beta = 3$ is equal to speed of 0.04 ms^{-1} (speed of half Mona length per second).

Interaction with Environment: To simulate the environment, a square shaped arena with the length of 2.85 m is considered which encompasses a circular shaped white region (contaminated zone) with a diameter of 2.227 m. The intensity of chemical leakage decreases linearly as it gets far from the center of

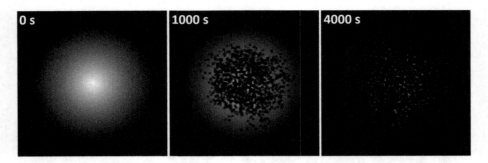

Fig. 3. Evolution of cue during 4000 s with $N = 30$ robots.

cue. Figure 3 shows the described cue at $t = 0$ s. To demonstrate the cleaning task in a greater details, the equation below is utilized while robots are in their waiting state:

$$\bar{S}_n = \bar{S}_c - (8 - \sqrt{p^2 + q^2}) \quad , \tag{3}$$

where $\bar{S}_c$ is the average value of cue measured by sensors, and $\bar{S}_n$ is the value of cue after cleaning. p and q are variables in a range of $[-4\ 4]$, and they were utilized to give gradient decrease shape to the cleaned area. As it is apparent, the cleaning level will have its maximum value of 8 per second when $p = 0$ and $q = 0$ so the center of the squared shaped cleaned region, will be darker and the cleaning level will have its minimum value of 2.34 per second when $|p| = |q| = 4$. Therefore, the outer layers will be cleaned less in comparison with the center. This equation will be executed once in a second while a robot is waiting. Figure 3 illustrates how a cleaned cue will look like in $t = 1000$ s and $t = 4000$ s.

3.2   Experimental Setup

Webots software is used to provide results for the proposed method, which is a simulation software for robotic platforms [19]. In Webots software, a supervisor is utilized to set initial conditions of robots and track simulation to record results. At the beginning of each simulation, robots are located randomly with uniform distribution by the supervisor. Also, they get random rotation with uniform distribution. The reason that the initial state of robots is randomized is to make results independent from robots' initial positions and rotations. Besides, random positioning of the initial state of robots is done with a different seed each time that simulation is repeated.

The effect of two parameters - speed of robot and population of the swarm - on the performance of the exploration are investigated. Experiments have conducted in five different population sizes of $N = \{10, 20, 30, 40, 50\}$ robots each one with two different speeds of a robot length per second ($v_r = 0.08$ ms^{-1}) and a half robot length per second ($v_r = 0.04$ ms^{-1}). Each set of experiments was repeated six times for the duration of 4000 s.

Table 1. Parameters and variables used in paper

Parameter	Description	Value/range
N	Number of robots	$\{10, 20, 30, 40, 50\}$ robots
v_r	Robot's linear velocity	$\{0.04, 0.08\}$ ms^{-1}
r_c	Radius of the cue center	0.7 m
t	Time of experiment	$[0 \quad 4000]$ s
w_s	Waiting time of robots	$[0 \quad 21.7]$ s
w_{max}	Maximum waiting time	30 s
s_r	Right sensor value	$[0 \ 255]$
s_l	Left sensor value	$[0 \ 255]$
N_r	Speed of right wheel	$[0 \ 10]$
N_l	Speed of left wheel	$[0 \ 10]$

Table 1 shows the list of parameters and variables that were used in this work. Three metrics were selected to evaluate the performance of robots: the change of average cue during the time, number of robots inside a circular zone with a radius of r_c, and finally, the coherency of robots. Additionally, to analyze the effect of the system parameters on each of the above-mentioned metrics, the ANOVA (Analysis of Variance) test was utilized, whose result will be presented subsequently. For each test, we have considered time, population, and the speed of robots as factors. In the following subsections, greater details about these three metrics is provided.

Average Cue vs Time: Average cue intensity is measured once in a second, hence by considering the overall experiment duration, 4000 samples of cue level are collected in each test case, and the median of measured cue intensities are used for six experiments in order to evaluate the swarm performance. The primary purpose of this measurement was to determine the impact of change in population size on the performance of robots. Besides the population size, the effect of robot speed on the overall performance is also studied.

Ratio of Robots Close to Cue Center: The ratio of robots that are within a circle with same center as cue and with radius of $r_c = 0.7$ m is computed each second for $N = \{10, 30, 50\}$ robots and $v_r = 0.08$ ms^{-1} in order to evaluate the performance of robots. Also, the cleaning pattern of robots can be recognized by this metric.

Coherency: The average distance of robots from each other is calculated to evaluate coherency for three populations with equal speeds. This observation is to figure out how much robots' behavior depends on each other during the experiments. Coherency also measures the level of cooperation among robots.

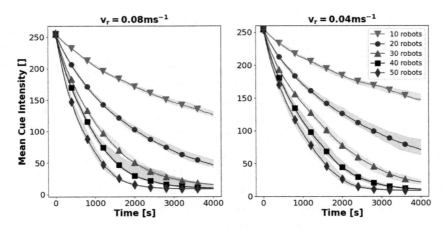

Fig. 4. Mean intensity of the cue vs time for various swarm population and robot speeds. The shaded areas around the plots indicate the min and max values.

4   Results and Discussion

To provide an example of how a chemical leakage disappears, Fig. 3 shows the chemical leakage state at $t = \{0, 1000, 4000\}$ s for $N = 30$ robots with $v_r = 0.08$ ms^{-1}. At $t = 4000$ s, it could be said that cue has been disappeared. In the following paragraphs, analyses of the results of conducted experiments are provided.

Average Cue vs Time: Figure 4 demonstrates results of two experiment cases. In each case, experiments were conducted with five different populations where each population was repeated six times with different random seeds, and the median of six experiments was plotted. At first glance, it could be deduced that according to the results, robots with $v_r = 0.08$ ms^{-1} could decrease the intensity of chemical leakage faster in comparison with robots that move by the velocity of $v_r = 0.04$ ms^{-1}. It is also apparent that as the number of robots increases, so does the vanishing speed of chemical leakage. Therefore, similar to the results of aggregation experiments in [3], the growth of the population significantly improved swarm performance. It can be seen from Fig. 4 that the best performance among test cases belongs to $N = 50$ robots with $v_r = 0.08$ ms^{-1} where the weakest performance belongs to $N = 10$ robots with $v_r = 0.04$ ms^{-1}.

The results were also statistically analyzed to find the most effective factor. Table 2 shows the results of the ANOVA test, that reveal all parameters (time, population size, and speed of robot) significantly affect the system ($p \leq 0.05$), however, speed of robot was the most significant factor ($F = 67.144$). On the other hand, the least effective factor was the time ($F = 5.032$) that shows the cleaning process is less time-dependent.

Table 2. Results of ANOVA test for the cue intensity

Factors	p-value	F-value
Time	0.000	5.032
Population	0.000	39.838
Speed	0.000	67.144

Ratio of Robots at the Source of Leakage: The ratio of robots that are close to source of leakage with a distance of less than 0.7 m for $N = \{10, 30, 40\}$ robots and $v_r = 0.08$ ms^{-1} to all robots of that experiment case were plotted in Fig. 5(a). This ratio decreases during time and remains constant at about $t = 1500$ s for $N = \{30, 50\}$ robots. For $N = 10$ robots, it can be observed that the ratio of robots close to the center of cue does not change considerably. If $t = 1500$ s (the time that ratio remains unchanged) is observed in Fig. 4 for $N = \{30, 50\}$ robots and $v_r = 0.08$ ms^{-1}, it can be understood that for both population the intensity of cue has been reduced more than 50% by that time. So as proposed, there is an obvious connection between this metric and the previous one. When cue almost disappears, robots do not aggregate close to the center anymore and move to places with higher intensity of cue.

As proposed in metrics section, the cleaning pattern of robots can be determined from Fig. 5(a) since it can be seen that the ratio of robots which are close to the center of the cue (with a distance less than $r_c = 0.7$ m) increases significantly while the cue has not been decreased by 50%. After the noticeable increase in the ratio of robots close to the cue center, it decreases to reach its steady-state. From this manner of the ratio of robots close to the cue center, it can be concluded that at first thousand seconds, most of the robots are within the radius of $r_c = 0.7$ m, and they distribute from the center as the leakage vanishes. Therefore, robots clean the cue from inside, starting with the highest leakage points and continuing to spots with less leakage. When the cue almost disappears, they do their random walk, and at that moment, measurements in all presented plots remain steady.

In terms of studying the impact of the population size on the ratio of robots close to the cue center, it can be observed that as the population increases, the steady-state ratio decreases since the center of cue vanished more and robots tend to spread all over the environment randomly. However, the steady-state ratio will be higher for smaller populations since there are still some visible parts of the cue. Besides comparing the steady-state ratio, the slope of plots before remaining unchanged also heavily depends on the population. That way, the higher population will have sharper decreases before their steady-state since they will clean the cue quicker and more.

Coherency: Figure 5(b) demonstrates the coherency of three population sizes of $N = \{10, 30, 50\}$ robots with $v_r = 0.08$ ms^{-1}. The similar point of test cases is that they verge to a steady-state after a while. On the other hand, they differ in the slope of verging to this steady value as for higher population sizes, coherency

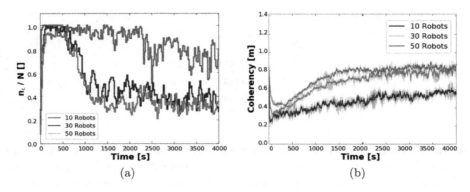

Fig. 5. (a) Ratio of robots within a distance of $r_c = 0.7$ m from the source of leakage for $N = \{10, 30, 50\}$ robots with speed of $v_r = 8.0$ cm s^{-1}, and (b) Median coherency measured in meters vs time for $N = \{10, 30, 50\}$ robots with $v_r = 8.0$ cm s^{-1}. Shades area indicates the maximum and minimum coherency.

Table 3. Results of ANOVA test for the coherency

Factors	p-value	F-value
Time	0.000	5.214
Population	0.000	19.088
Speed	0.000	5.305

reaches its steady-state faster. To define a steady-state for coherency, we shall consider $N = 50$ robots and compare its results with previous metrics. In Fig. 4, it can seen that for $N = 50$ robots with $v_r = 0.08$ ms^{-1} speed, after $t = 2500$ s the average cue intensity does not change considerably. If the same moment is considered in Fig. 5(b), it can be seen that at that moment, coherency also does not change considerably. Therefore it can be deduced that whenever chemical leakage vanishes, coherency and average cue reach their steady-state.

There are similar features between Fig. 5(a) and Fig. 5(b). Primarily, in both figures for $N = 10$ robots, plots do not change considerably, and as the population grows, plots in both figures fluctuate more. For $N = \{30, 50\}$ robots, plots in both figures remain unchanged after $t = 1500$ s. To analyze this behavior, when coherency remains unchanged, there is no longer cooperation among robots. At that moment, the ratio of robots close to the center of the cue also remains constant, so the robots are not following cue anymore. Hence, they mainly tend to perform random walk. Consequently, from these observations, it can be concluded that the more the cue disappears, the tendency of robots for random walk increases, and their cooperation decreases.

In terms of comparing the effect of population size on coherency, it can be seen that as the population grows, the robots' coherency changes more dramatically. It can be observed that for $N = 10$ robots, coherency does not alter noticeably.

In contrast, for $N = 50$ robots, coherency encounters a notable decrease meaning that the robots get very close to each other while the cue is not cleaned.

The results in this section were also statistically analyzed (Table 3). The results showed that all the factors significantly affected the coherency of the swarm system ($p \leq 0.05$). However, the population was the most significant factor ($F = 19.088$) in the coherency of the robot's motion.

Although the primary goal of this work was to demonstrate an application of swarm robotics in an extreme environment, such a system will also be useful in other applications e.g., agri-robotics [11], industrial machinery [9] and outdoor localization [16].

5 Conclusion and Future Work

In this paper, a chemical leakage localization and cleanup scenario was implemented using the simulated swarm of robots. The proposed method is based on a bio-inspired aggregation scenario for swarm robots. Impacts of population and speed on swarm performance were studied. The evolution of cooperation between robots and the ratio of robots present near the source of the leakage is also studied to track the robots' interaction during the process. As the results revealed, robot cooperation decreases as leakage vanishes. Therefore, these results can be a significant guide for developing the real-world application of swarm robotics in an extreme environment.

Acknowledgements. This work was partially supported by OP VVV project Research Center for Informatics, code CZ.02.1.01/0.0/0.0/16_019/0000765, and the UK EPSRC projects RAIN (EP/R026084/1) and RNE (EP/P01366X/1).

References

1. Arvin, F., Espinosa, J., Bird, B., West, A., Watson, S., Lennox, B.: Mona: an affordable open-source mobile robot for education and research. J. Intell. Robot. Syst. **94**, 1–15 (2018)
2. Arvin, F., et al.: ΦClust: pheromone-based aggregation for robotic swarms. In: 2018 IEEE/RSJ International Conference on Intelligent Robots and Systems (IROS), pp. 4288–4294. IEEE (2018)
3. Arvin, F., Turgut, A.E., Krajník, T., Yue, S.: Investigation of cue-based aggregation in static and dynamic environments with a mobile robot swarm. Adapt. Behav. **24**(2), 102–118 (2016)
4. Bayındır, L.: A review of swarm robotics tasks. Neurocomputing **172**, 292–321 (2016)
5. Bodi, M., Thenius, R., Szopek, M., Schmickl, T., Crailsheim, K.: Interaction of robot swarms using the honeybee-inspired control algorithm BEECLUST. Math. Comput. Model. Dyn. Syst. **18**(1), 87–100 (2012)
6. Brambilla, M., Ferrante, E., Birattari, M., Dorigo, M.: Swarm robotics: a review from the swarm engineering perspective. Swarm Intell. **7**(1), 1–41 (2013)

7. Camazine, S., Deneubourg, J.L., Franks, N.R., Sneyd, J., Bonabeau, E., Theraula, G.: Self-Organization in Biological Systems, vol. 7. Princeton University Press, Princeton (2003)
8. Cao, Y.U., Fukunaga, A.S., Kahng, A.: Cooperative mobile robotics: antecedents and directions. Auton. Robots 4(1), 7–27 (1997)
9. Correll, N., Martinoli, A.: Multirobot inspection of industrial machinery. IEEE Robot. Autom. Mag. 16(1), 103–112 (2009)
10. Fox, D., Burgard, W., Kruppa, H., Thrun, S.: Collaborative multi-robot localization. In: Förstner, W., Buhmann, J.M., Faber, A., Faber, P. (eds.) Mustererkennung 1999. Informatik aktuell. Springer, Heidelberg (1999). https://doi.org/10.1007/978-3-642-60243-6_2
11. Grieve, B.D., et al.: The challenges posed by global broadacre crops in delivering smart agri-robotic solutions: a fundamental rethink is required. Glob. Food Secur. 23, 116–124 (2019)
12. Halloy, J., et al.: Social integration of robots into groups of cockroaches to control self-organized choices. Science 318(5853), 1155–1158 (2007)
13. Hassan, M.A.A.: A review of wireless technology usage for mobile robot controller. In: Proceeding of the International Conference on System Engineering and Modeling (ICSEM 2012), pp. 7–12 (2012)
14. Heran, H.: Untersuchungen über den temperatursinn der honigbiene (apis mellifica) unter besonderer berücksichtigung der wahrnehmung strahlender wärme. Zeitschrift für vergleichende Physiologie 34(2), 179–206 (1952)
15. Huang, X., Arvin, F., West, C., Watson, S., Lennox, B.: Exploration in extreme environments with swarm robotic system. In: 2019 IEEE International Conference on Mechatronics (ICM), vol. 1, pp. 193–198. IEEE (2019)
16. Kowadlo, G., Russell, R.A.: Robot odor localization: a taxonomy and survey. Int. J. Robot. Res. 27(8), 869–894 (2008)
17. Liu, Z., West, C., Lennox, B., Arvin, F.: Local bearing estimation for a swarm of low-cost miniature robots. Sensors 20(11), 3308 (2020)
18. Mayet, R., Roberz, J., Schmickl, T., Crailsheim, K.: Antbots: a feasible visual emulation of pheromone trails for swarm robots. In: Dorigo, M., et al. (eds.) ANTS 2010. LNCS, vol. 6234, pp. 84–94. Springer, Heidelberg (2010). https://doi.org/10.1007/978-3-642-15461-4_8
19. Michel, O., Cyberbotics Ltd.: Webots^{TM}: professional mobile robot simulation. Int. J. Adv. Robot. Syst. 1(1), 39–42 (2004)
20. Morrell, L.J., James, R.: Mechanisms for aggregation in animals: rule success depends on ecological variables. Behav. Ecol. 19(1), 193–201 (2007)
21. Na, S., et al.: Bio-inspired artificial pheromone system for swarm robotics applications. Adapt. Behav. 1–21 (2020)
22. Na, S., Raoufi, M., Turgut, A.E., Krajník, T., Arvin, F.: Extended artificial pheromone system for swarm robotic applications. In: Conference on Artificial Life (ALIFE), pp. 608–615. MIT Press (2019)
23. Nagatani, K., et al.: Emergency response to the nuclear accident at the fukushima daiichi nuclear power plants using mobile rescue robots. J. Field Robot. 30(1), 44–63 (2013)
24. Parrish, J.K., Edelstein-Keshet, L.: Complexity, pattern, and evolutionary trade-offs in animal aggregation. Science 284(5411), 99–101 (1999)
25. Raoufi, M., Turgut, A.E., Arvin, F.: Self-organized collective motion with a simulated real robot swarm. In: Althoefer, K., Konstantinova, J., Zhang, K. (eds.) TAROS 2019. LNCS (LNAI), vol. 11649, pp. 263–274. Springer, Cham (2019). https://doi.org/10.1007/978-3-030-23807-0_22

26. Rappel, W.J., Nicol, A., Sarkissian, A., Levine, H., Loomis, W.F.: Self-organized vortex state in two-dimensional dictyostelium dynamics. Phys. Rev. Lett. **83**(6), 1247 (1999)
27. Schmickl, T., et al.: Get in touch: cooperative decision making based on robot-to-robot collisions. Auton. Agents Multi-Agent Syst. **18**(1), 133–155 (2009)
28. Schranz, M., Caro, G.A.D., Schmickl, T., Elmenreich, W., Arvin, F., Şekercioğlu, A.: Swarm intelligence and cyber-physical systems: concepts, challenges and future trends. Swarm Evol. Comput. **60**, 100762 (2020)

A Tamper-Resistant and Decentralized Service for Cloud Storage Based on Layered Blockchain

Fuxiao Zhou[ID], Haopeng Chen[✉][ID], and Zhijian Jiang[ID]

Shanghai Jiao Tong University, Shanghai, China
{zhoufuxiao,chen-hp,jiangzhijian}@sjtu.edu.cn

Abstract. With the ever-increasing scale of data, IP traffic of data centers will gradually suffer from a serious shortage. Centralized clouds may no longer deliver satisfactory storage services. To alleviate network pressure, transmitting source data to the edge of network instead of the remote cloud is becoming more and more important. Based on the blockchain technology, we propose a decentralized cloud storage service with tamper-resistant function. Our service provides users with metadata storage and management capabilities. In order to overcome the inherent defects of blockchain, in our design, the blockchain network has a layered and collaborative structure. After the storage capacity problem is solved, our service allows users to upload digests of source data to make the query function more user-friendly. For privacy protection, we subjoin the access management function as well. Finally, the experimental results show that the performance and availability of the blockchain network are able to support our service to provide efficient functions to users.

Keywords: Cloud storage · Collaboration · Decentralization · Tamper · Blockchain · Metadata

1 Introduction

The emergence of cloud computing has tremendously changed the way people deal with data [1]. A growing number of users are willing to process their data on the cloud. The use of cloud-based storage services for storing data has also become a popular alternative to traditional local storage systems [2]. As a result, storage as a service (STaaS) is raising widespread concern. However, with the advent of big data era, cloud storage has also been facing new challenges. As estimated by Cisco Global Cloud Index [3], nearly 850 zettabytes data will be generated by all people, machines and things by 2021. Simultaneously, the global data center IP traffic will only reach 20.6 zettabytes by that time. Though most of the ephemeral data will be consumed instantly, the remaining data which is worth saving still occupies a large part. Given the scarcity of data center IP traffic, it is arduous to upload all the relatively important data to cloud centers.

© ICST Institute for Computer Sciences, Social Informatics and Telecommunications Engineering 2021
Published by Springer Nature Switzerland AG 2021. All Rights Reserved
H. Gao et al. (Eds.): CollaborateCom 2020, LNICST 350, pp. 482–493, 2021.
https://doi.org/10.1007/978-3-030-67540-0_31

Under the restriction of scarce network resources, centralized clouds may no longer own the ability to deliver satisfactory storage services to users. Fortunately, edge or fog computing might help bridge this gap [3]. Given a finite network bandwidth, the large volume of data which needs to be stored and shared can be transmitted to the edge of the network instead of the remote cloud, so as to solve the data transfer bottleneck. While the metadata [4] with which users can locate and inspect the data actually needed should be uploaded and administrated. For the purpose of storing metadata on the cloud, several requirements ought to be satisfied.

Ensuring data integrity and validity is essential. To protect data from tampering, some archival storage systems such as Amazon Glacier [5] follow a Write Once, Read Many (WORM) paradigm, which treats all archive data as read-only. Even if supported by high fault tolerance strategies, there is a certain risk that data can be maliciously modified without notification [6, 7].

When users propose to apply a traditional centralized cloud storage service, it means that the supplier of the service is considered to be a trusted third-party. Nevertheless, this is a questionable proposition. In addition to the trust crisis, a centralized service may inevitably lead users to suffer from other intrinsic defects such as performance bottleneck caused by central server and the risk of data unavailability caused by data center failure.

Briefly, a decentralized cloud storage service secured from tampering is quite in need. As it happens, the prevalence of Bitcoin [8] has raised great attention, we found the underlying technology blockchain just in line with our needs. If only its inherent drawbacks overcome, with key characteristics like decentralization, persistency, anonymity and auditability [9], blockchain can play a pivotal role in this scenario. For instance, in Bitcoin blockchain, the information once entered can never be erased. As each node in the Peer-to-Peer (P2P) network is required to store the entire history information, which restricts the capacity substantially.

Under this condition, we develop a decentralized service based on blockchain for cloud storage in order to supply users with a tamper-resistant function. Supported by security mechanisms, our service will take charge of management and preservation of the meta information. When actually used, data can be positioned, checked and verified with the information extracted from our service. To cross the barrier of capacity limit and scalability, we come up with a layered and collaborative structure. In our design, the mainchain stores indexes of metadata while subchains allow users to upload metadata and digest of source data.

In this paper, we have made the following contributions:

- A simple storage strategy to relieve the contradiction between huge data volume and limited network resources of cloud.
- A design of decentralized cloud storage service based on blockchain which provides powerful tamper resistance and access management function.
- A layered structure design of blockchain to expand the storage capacity without sacrificing scalability.

The rest of this paper is organized as follows: Sect. 2 presents the detailed design of our tamper-resistant and decentralized service. Section 3 gives an introduction of workflows, together with the deployment strategy. In Sect. 4, we evaluate our service based on the simulation by Hyperledger Fabric. Section 5 discusses the related works. Finally, we end with a conclusion in Sect. 6.

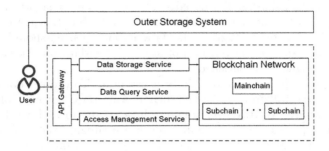

Fig. 1. Service architecture

2   Design

2.1   Overview

Due to the application scenario, we implement our ideas as a form of cloud service. Collaborating with outer storage systems, our service is capable of providing reliable functions to users. Figure 1 demonstrates the service architecture. The layered design of blockchain network increases the storage capacity and enhances

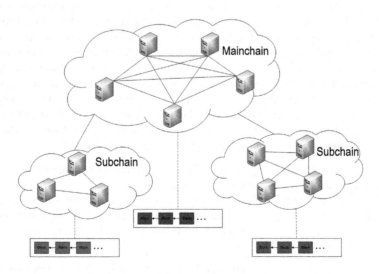

Fig. 2. Layered and collaborative blockchain architecture

scalability. Subchains can be added in any position according to actual demand. All subservices keep contact with the blockchain network for tamper-resistent, persistency, fault-tolerance and other features.

2.2   Blockchain Network

Blockchain network is a core part of our design. In order to ensure the tamper-resistent ability of our service, any data that needs to be persisted should under the protection of a reliable mechanism. Although it provides a strong guarantee for security, the traditional blockchain cannot fully meet our requirements for two reasons.

The first reason is capicity limit. Because of the decentralization feature, blockchain is a peer-to-peer network which requires every single node to keep a copy of all blocks and stay in synchronization with each other at any time. This constraint limits the amount of data across the network extremely. For this reason, uploading large data to the traditional blockchain network is not allowed. However, with compact data, the query experience will be very unpleasant.

The second reason is the scalability defect of blockchain. It is well known that high scalability is hard to achieve in a blockchain network because of consensus issue. So putting all operations on a single blockchain network is easy to cause insufficient throughput. With a traditional blockchain network, our service will lose the ability to make a guarantee of concurrency.

Because of the considerations above, we design a two-layer structure for our blockchain network. Of course, with more layers, the blockchain network may achieve higher capacity and scalability. We found it unworthy owing to it brings much more complexity at the same time. The architecture is shown in Fig. 2. There is a mainchain and serval subchains in our network.

The unique mainchain is responsible for direct interactions with the data query service. The mainchain maintains a record of mapping from subchains to node IP address list as well. When users query data, the data query service can get the metadata from any IP address in the list of corresponding subchain. Which node will be chosen is determined by the network delay and the remaining bandwidth of each node.

Subchains take charge of direct interactions with the data storage service and access management service. A special subchain is designed for the access management service to store authentication information of users. Each of the subchains has a unique id and any change of its node IP address will be synchronized to the mainchain. In order to expand capacity, a new subchain can be added in any position.

2.3   Data Storage Service

When users upload new data or update old data, the data storage service will be called. A complete data insert operation involves uploading the metadata and the digest of source data to a subchain and transmitting the index to the mainchain.

We design a unified data structure to store metadata and digest of source data in subchains. As shown in (1), we use six parameters to represent specific information.

$$UUID : \{validation, metadata, digest, reader\_set, writer\_set\} \qquad (1)$$

- *UUID*: UUID is a hash value calculated from the metadata uploaded by the user.
- *validation*: validation is a boolean value whose default value is true.
- *metadata*: metadata is uploaded by users.
- *digest*: digest is also uploaded by users. It is a relatively detailed description of the source data and is represented by a string.
- *reader_set*: readerset is a set of user_id.
- *writer_set*: similar to reader_set, writer_set is used for access management as well.

As shown in (2), we design another unified data structure to store indexes of metadata in the mainchain. In general key-value databases like RocksDB [10], the insert operation of a record will cover the old record with the same key. This can also be considered a form of tampering. In order to solve this problem, in our design, key is allowed to appear repeatedly. For each key, the mainchain maintains an array of pairs composed of UUID and subchain_ID. subchain_ID is globally unique. Obviously, the data structure is particularly compact to relieve the capacity burden of the mainchain.

$$KEY : [\{UUID, subchain\_ID\}, ...] \qquad (2)$$

The data storage service's interfaces were shown in (3), (4) and (5).

$$UUID \; PUT\_S(subchain\_ID, metadata, digest) \qquad (3)$$

PUT_S is used to insert records into subchains. It accepts three parameters include subchain_ID, metadata and digest. The UUID calculated from metadata will be returned. Another usage of UUID will be explained in the next subsection.

$$void \; PUT\_M(KEY, UUID, subchain\_ID) \qquad (4)$$

PUT_M is usually called after PUT_S, to upload the UUID and subchain_ID to the mainchain. It accepts three parameters include KEY, UUID and subchain_ID. As a suggestion, KEY should be a string related to the contents of source data for the convenience of inquiries. After PUT_M executes successfully, the record will be visible globally.

$$void \; REMOVE(subchain\_ID, UUID) \qquad (5)$$

REMOVE is designed for deleting records in subchains. It accepts two parameters include subchain_ID and UUID. When it is called, if there exists a record with the same UUID in the corresponding subchain, its validation value will be turned to false.

2.4   Data Query Service

When users need to perform a query operation, the data query service will be called. Different from the data storage service, the data query service does not interact directly with subchains. This mechanism will bring greater security to our service and will eliminate the impact of subchain failure on the overall situation. Since the mainchain usually contains more nodes, the throughput of query is easier to meet requirements.

$$result\ GET(KEY, UUID) \tag{6}$$

The interface exposed by the data query service is shown in (6). Between the two parameters accepted by GET, UUID is optional. With both KEY and UUID, the result returned will be a pair of metadata and digest. Users have already obtained the UUID while inserting records into subchains.

$$[\{UUID, \{metadata, digest\}\}, ...] \tag{7}$$

In the case of UUID default, the return value will be an array, as shown in (7). Due to the UUID is unknown, the data query service is unable to provide accurate queries and chooses to return all records of the KEY instead.

2.5   Access Management Service

Until now, records are public. All users are free to access any record they want, which will definitely cause privacy problems. As a result, we subjoin the access management service for privacy protection.

The implementation of this service is based on the two parameters in records which are stored in subchains. They are reader_set and writer_set. Records can only be accessed by users in their reader_set. Before each return of data query service, the reader_set will be checked. Compared with reader_set, writer_set is much more important. Users in the writer_set own the permission to modify parameters except UUID and metadata in a record, including reader_set and writer_set. Consequently, it should be cautious to add any user to the writer_set.

$$group\_id\ CREATE\_GROUP([user\_id, ...]) \tag{8}$$

To save storage space, the record size should be compressed as much as possible. Group_id is designed to represent a batch of users. The group_id can be used as a special user_id. As mentioned before, a special subchain is built to store authentication information of users. With the help of this subchain, building user groups is not a difficult task.

$$void\ UPDATE\_GROUP(group\_id, [user\_id, ...]) \tag{9}$$

The access management service exposes four interfaces as shown in (8), (9), (10) and (11). CREATE_GROUP is used to create a new group. It accepts a set of user_id and returns a unique group_id. The first user in the set will have the administrator privilege.

$$void \ UPDATE\_RSET(KEY, UUID, [user\_id, ...]) \tag{10}$$

UPDATE_GROUP is used to modify an existing group. Only the first user in the existing group will be allowed to update the user_id group.

$$void \ UPDATE\_WSET(KEY, UUID, [user\_id, ...]) \tag{11}$$

UPDATE_RSET and UPDATE_WSET are used to update the reader_set and the writer_set of a record stored in subchains. The writer_set will be checked before each operation executed.

3   Workflow and Deployment

In this section, we discuss the usage of our service, including general workflows and the deployment strategy.

3.1   Read Workflow

We will give an explanation of the read workflow as follows.

a) **Search indexes with KEY** Assigned a KEY, indexes stored in the main-chain can be find. If no indexes were found, return directly.
b) **Filter indexes by UUID** If UUID was assigned, use it to filter the indexes. Otherwise, skip this step. If no records left, return.
c) **Query records in subchains** With the help of indexes, query records in the corresponding subchains.
d) **Check reader_set** For each record, check its reader_set. If the user had the read access, append metadata and digest to the result.
e) **Return with result** If the result was not empty, return with it. Otherwise, just return.

In the read workflow, users do not interact directly with subchains. The reason is that subchains are usually composed of relatively few nodes, consequently, more vulnerable. This design can effectively isolate errors. Even if some subchains failed, the biggest loss is just missing some data.

3.2   Write Workflow

The write workflow is shown as follows.

a) **Select subchain**: Select the target subchain. In order to economize on network resources, the nearest subchain should be chosen.
b) **calculate UUID** Calculate the hash value of the metadata as UUID.

c) **Query index with KEY and UUID** With KEY and UUID, query the index from the mainchain. If not found, skip to step *g*.

d) **Query record** With the index, query the record in the corresponding subchain. If not found, skip to step *g*.

e) **Check writer_set** Check the writer_set. If the user did not own the write access, return.

f) **Check repeat** With UUID, check if there is already a valid record with the same UUID and the user does not have write access. If so, change a subchain until fulfilling the requirement and update the subchain_ID. If no such subchain found, return.

g) **Insert record** Insert a new record to the selected subchain.

h) **Upload index** Upload the index to the mainchain.

i) **Return with UUID** Return with the UUID.

In the write workflow, choosing subchain is a key point. For taking up less public bandwidth and lower latency, the nearest one should been selected. Another benefit is that when a subchain works normally, it means that there is no large area collapse occurring nearby. The metadata acquired by users is more valuable, for the source data has a high probability of being available.

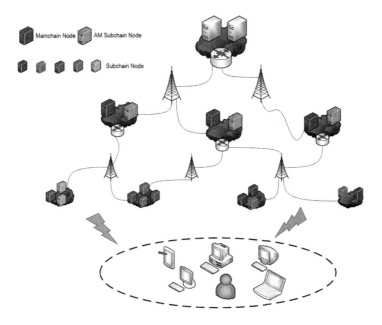

Fig. 3. Deployment strategy

3.3   Deployment Strategy

Our service is designed specifically for the cloud environment. Due to the complex network structure of cloud, the deployment of our service ought to meet certain principles as well. Figure 3 illustrates the deployment strategy of our service.

The mainchain nodes should be deployed closer to the cloud center, to cover a larger range. Nodes of subchains should be adjacent to the outer storage systems which are deployed in edge nodes. However, nodes of the special subchain designed for the access management service called AM Subchain Node ought to be neighbors of the mainchain nodes, for lower latency.

4   Evaluation

We built a little private blockchain that contains four peer-to-peer nodes based on Hyperledger Fabric. Through simulation, we evaluated the performance of our blockchain network and the availability of our service.

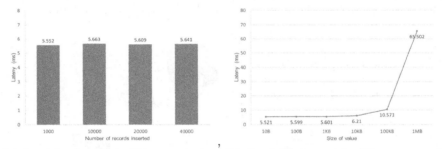

(a) Read latency with different number of records inserted

(b) Read latency with different record size

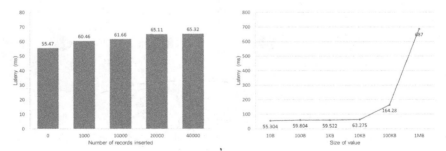

(c) Write latency with different number of records inserted

(d) Write latency with different record size

Fig. 4. Read and write latency of records with different number or size

4.1 Read Latency

To measure the latency of read operations, we randomly selected 1000 records to read in each node. In order to eliminate random errorterm, the result is the average value of all 4000 measured latency values. As shown in Fig. 4(a), with different number of records inserted, from 1000 to 40000, the read latency keeps around 5.6 ms. This performance is enough to guarantee the user experience of our service.

The size of records may also affect read latency. We inserted records with different value size before starting the reading test. The result is shown in Fig. 4(b). When the value size increases from 10 B to 10 KB, the change of read latency is not obvious. When the value size reaches 100 KB and 1 MB, it causes a large growth in read latency.

4.2 Write Latency

Similarly, we measured the write latency. As shown in Fig. 4(c) and Fig. 4(d), the result is average value as well. Despite the write latency is nearly 10 times higher than the read latency, 65 ms is still in an acceptable range. It is worth noting that as the value size increases, the latency of write operations will become a fairly large value. Therefore, both for performance and capacity considerations, the size of digest uploaded by users should be limited to under 10 KB.

4.3 Availability

A significant advantage of decentralized systems is high availability. Suppose we built a system with a mainchain and two subchains, the mainchain has four nodes while each subchain has three nodes. Table 1 lists the probability of each chain being inaccessible when a certain number of arbitrary nodes have crashed.

Due to all chains own at least three nodes, in the case where the number of crashed nodes is less than three, the availability of any chain will not be

Table 1. Failure probability of each chain when random nodes crash

Number	Mainchain	Subchain_1	Subchain_2
<3	0	0	0
3	0	0.83%	0.83%
4	0.48%	3.33%	3.33%
5	2.38%	8.33%	8.33%
6	7.14%	16.67%	16.67%
7	16.67%	29.16%	29.16%
8	33.33%	46.67%	46.67%
9	60%	70%	70%
10	100%	100%	100%

influenced. From the results we can discover that although the mainchain merely has one more node than subchains, it is more robust evidently. If half of the nodes crashed, the inaccessible probability of each chain is under 10%. Even if only one tenth of the nodes is working properly, the mainchain still has a 40% inventory possibility. Briefly, the blockchain network has extremely high availability and adding more nodes will greatly enhance it.

5   Related Works

The InterPlanetary File System (IPFS) [11] is a peer-to-peer distributed file system that seeks to connect all computing devices with the same system of files. With the help of a distributed hashtable [12], an incentivized block exchange, and a self-certifying namespace, IPFS has overcome shortcomings of centralized storage systems like network congestion and waste of storage space. However, IPFS has no tamper resistance which is quite important in our scenario.

BigchainDB [13] is a decentralized database with large scale. BigchainDB inherits features of modern distributed databases and adds blockchain characteristics. Of course, getting this property comes at a price. BigchainDB is an implementation of private blockchain. Compared to our service, adding new nodes in BigchainDB is troublesome. Although with quite high throughput and capacity, it is not realistic to deploy BigchainDB in the edge environment.

S. Ali and G. Wang [14] applied blockchain to PingER [15], a worldwide end-to-end Internet performance measurement project. With the help of permissioned blockchain and Distributed Hash Tables (DHT), the data storage and access framework stores metadata of the file on the blockchain, whereas the actual files are stored at multiple locations. Compared to our service, it specifies the storage location that will not help reduce network overhead. Besides, it does not support uploading digest, thence it is more suitable to store specific data like daily PingER data that does not need complicated query function.

6   Conclusion

With the explosive growth in data volume, cloud storage service is facing a serious situation that the network traffic of data centers is gradually unable to meet user needs. In order to solve this problem, storing source data in edge nodes as well as uploading metadata to clouds is a good choice.

However, storing metadata in traditional cloud storage systems has serval shortages. Blockchain, an underlying technology of Bitcoin can help to solve these problems. Based on blockchain, we propose a decentralized cloud storage service with high tamper-resistant.

For the purpose of expanding storage capacity, a layered and collaborative structure of blockchain has been designed. The blockchain network of our service allows users to upload metadata and digest of source data to subchains. To deal with query operations, the mainchain preserves indexes of records. Another function provided by our service is access management. We demonstrate the read and write workflow of our service. A deployment strategy is also recommended in this paper. To evaluate the performance and availability, we simulate the blockchain network by Hyperledger Fabric. In conclusion, the blockchain network is capable of providing the features we need.

Acknowledgment. This paper is supported by Project 213.

References

1. Joseph, A.D., Katz, R., Konwinski, A., Gunho, L., Patterson, D., Rabkin, A.: A view of cloud computing. Commun. ACM **53**(4), 50–58 (2010)
2. Waibel, P., Matt, J., Hochreiner, C., Skarlat, O., Hans, R., Schulte, S.: Cost-optimized redundant data storage in the cloud. Serv. Oriented Comput. Appl. **11**(4), 411–426 (2017). https://doi.org/10.1007/s11761-017-0218-9
3. Cisco Visual Networking Index: Forecast and methodology, 2016–2021. http://www.cisco.com
4. Duval, E., Hodgins, W., Sutton, S., Weibel, S.L.: Metadata principles and practicalities. D-lib Mag. **8**(4), 1082–9873 (2002)
5. Amazon Glacier. https://aws.amazon.com/glacier/. Accessed Nov 2017
6. Renner, T., Müller, J., Kao, O.: Endolith: a blockchain-based framework to enhance data retention in cloud storages. In: 2018 26th Euromicro International Conference on Parallel, Distributed and Network-Based Processing (PDP), pp. 627–634. IEEE (2018)
7. Lee, B., Awad, A., Awad, M.: Towards secure provenance in the cloud: a survey. In: Proceedings of the 8th International Conference on Utility and Cloud Computing, pp. 577–582. IEEE Press (2015)
8. Nakamoto, S., et al.: Bitcoin: a peer-to-peer electronic cash system (2008)
9. Zheng, Z., Xie, S., Dai, H.-N., Chen, X., Wang, H.: Blockchain challenges and opportunities: a survey. Int. J. Web Grid Serv. **14**(4), 352–375 (2018)
10. Borthakur, D.: Under the hood: building and open-sourcing RocksDB. Facebook Engineering Notes (2013)
11. Benet, J.: IPFS-content addressed, versioned, P2P file system. arXiv preprint arXiv:1407.3561 (2014)
12. Naor, M., Wieder, U.: A simple fault tolerant distributed hash table. In: Kaashoek, M.F., Stoica, I. (eds.) IPTPS 2003. LNCS, vol. 2735, pp. 88–97. Springer, Heidelberg (2003). https://doi.org/10.1007/978-3-540-45172-3_8
13. McConaghy, T., et al.: BigchainDB: a scalable blockchain database. White Paper, BigChainDB (2016)
14. Ali, S., Wang, G., White, B., Cottrell, R.L.: A blockchain-based decentralized data storage and access framework for PingER. In: 2018 17th IEEE International Conference on Trust, Security and Privacy in Computing and Communications/12th IEEE International Conference on Big Data Science and Engineering (TrustCom/BigDataSE), pp. 1303–1308. IEEE (2018)
15. Matthews, W., Cottrell, L.: The PingER project: active Internet performance monitoring for the HENP community. IEEE Commun. Mag. **38**(5), 130–136 (2000)

Multi-UAV Adaptive Path Planning in Complex Environment Based on Behavior Tree

Wendi Wu[1], Jinghua Li[2], Yunlong Wu[2,3(✉)], Xiaoguang Ren[2,3(✉)], and Yuhua Tang[1]

[1] State Key Laboratory of High Performance Computing (HPCL), College of Computer, National University of Defense Technology, Changsha 410073, Hunan, China
[2] Tianjin Artificial Intelligence Innovation Center (TAIIC), Tianjin 300457, China
[3] Artificial Intelligence Research Center (AIRC), National Innovation Institute of Defense Technology (NIIDT), Beijing 100071, China
ylwu1988@nudt.edu.cn, rxg_nudt@126.com

Abstract. In this paper, we consider a scenario where multiple tracking unmanned aerial vehicles (UAVs) pursue a target UAV in a complex environment. Consider the fast airspeed of the UAV, the path planning needs to be finished in a limited time. Moreover, the complex environment may involve diverse geographical areas, which raises the challenges for the path planning algorithms. For the first challenge, we will adopt the real-time algorithms to keep the efficiency of path planning. For the challenge of environment diversity, we involve the behavior tree (BT) model and propose a BT-organized path planning (BT-OPP) method aiming at achieving adaptive scheduling of different path planning algorithms in different geographical areas. Furthermore, in order to take the advantages of multiple tracking UAVs, we propose a virtual-target-based tracking (VTB-T) method which can make the tracking UAVs pursue the target UAV collaboratively. The effectiveness of the proposed BT-OPP method and the VTB-T method are verified by analysis and numerical results for different system configurations, showing that a substantial target tracking efficiency improvement may be achieved in comparison with the benchmark.

Keywords: Multi-UAV target tracking · Behavior tree · Real-time path planning

This work was supported in part by the National Key Research and Development Program of China under Grant No. 2017YFB1301104, and in part by the National Natural Science Foundation of China under Grant No. 61906212 and Grant No. 61802426.

H. Gao et al. (Eds.): CollaborateCom 2020, LNICST 350, pp. 494–505, 2021.
https://doi.org/10.1007/978-3-030-67540-0_32

1   Introduction

In recent years, the rapid development of unmanned aerial vehicles (UAVs) related technology has made it widely used in various fields such as industry, security, military, and scientific research [1]. Particularly, in the military field, UAVs have replaced manned aircraft in large numbers of important tasks such as intelligence, surveillance, and reconnaissance (ISR) [2]. As the decreasing manufacturing cost of the UAVs, it has led to their wide use in civil and commercial applications such as area exploration, surveillance, package delivery, etc. [3]. In UAV applications, target tracking as a typical task, is often applied to the national security and other aspects. The task requires the UAVs to face a complex environment for path planning in order to achieve a more efficient target tracking, and the path planning algorithm may face two main challenges. Firstly, the target moves fast and the UAV needs to respond in a limited time and plan a feasible path in real-time. Secondly, the environment is diverse, and the UAV may pass through many different geographical areas during the process of target tracking, which requires different area-oriented path planning algorithms can switch adaptively.

Traditional path planning algorithms are difficult to meet the real-time requirements, for example, the Dijkstra algorithm is a classical algorithm for single source path planning which adopts the breadth-first search to find the path [4]. The Dijkstra algorithm will eventually get a shortest path, but it is not a real-time algorithm. There are also more algorithms currently being researched for the real-time challenge. Based on the theory of Dijkstra algorithm, the A* algorithm introduces a heuristic search which can largely improve the efficiency of path planning [5,6]. The D* algorithm proposed in 1994 is an effective path planning in the unknown and dynamic environment, and only checks changes in the shortest path to the next or adjacent node when moving towards the target point in a reverse search mode. Based on the D* algorithm, Koenig proposed the D*Lite algorithm which can achieve fewer re-planning times and faster response to sudden obstacles. In 2004, Likhachev et al. implemented the ARA* algorithm which added the idea of anytime based on the A* algorithm [7]. When the time constraint is relaxed, the global optimal path can be obtained. On the contrary, if the time constraint is limited, a time-related sub-optimal path will be obtained. The anytime algorithm can allow us to obtain a planned path under any time constraint. In 2005, combined the ARA* algorithm with the D*Lite algorithm, the AD* algorithm is proposed to realize anytime path planning for unknown areas [8]. In 2012, Sun et al. designed the I-ARA* algorithm which runs on the basis of the ARA* algorithm [9]. In the I-ARA* algorithm, the planned results of the last iteration is efficiently reused, and the path planning for dynamic targets is finally realized through the idea of re-planning in motion.

To address the second challenge of environment diversity, a single algorithm cannot be adapted to multiple situations, and require a mechanism for combining with different algorithms to achieve good results. Therefore, an appropriate robot control architecture needs to be considered to integrate multiple path planning algorithms. In traditional robot control architecture, finite state machine (FSM), hierarchical finite state machine (HFSM), subsumption model, and decision tree can realize the robot controlling, but the above control architectures have certain drawbacks in the four aspects: implementability, maintainability, scalability, and reusability [10]. Behavior tree (BT), as a tree model consisting of hierarchical nodes, is used to construct robot behaviors for adapting to the switching among tasks [11]. BT is an effective method for creating modular and reactive complex systems, and can solve the challenge of environment diversity by constructing control nodes and execution nodes to decompose complex tasks into sub-tasks [12,13].

Based on the above discussion, in order to address the problem of target tracking in the complex environment, the two above challenges need to be simultaneously considered. Therefore, in this paper, we propose to implement the adaptive scheduling of different real-time path planning algorithms based on the BT model. The main contributions of this paper are as follows:

- A multi-UAV target tracking scenario is considered which involves a complex environment containing no obstacle area, known area and unknown area. When the tracking UAVs catch up with the target UAV, or the target UAV leaves a specified area, the tracking task is finished.
- In order to overcome the challenges of real-time and diversity brought by the above scenario, this paper proposes a BT-organized path planning (BT-OPP) method which can achieve the adaptive scheduling of different path planning algorithms for different areas.
- In order to fully take the advantages of multiple tracking UAVs, this paper proposes a virtual-target-based tracking (VTB-T) method which can significantly increase the success rate of target tracking.
- The effectiveness of the proposed method is verified through simulation experiments. From the experimental results, it shows that the BT-OPP method and the VTB-T method can largely improve the target tracking efficiency.

The rest of the paper is organized as follows. Firstly, the task scenario considered by this paper is presented in Sect. 2. Then, Sect. 3 and Sect. 4 propose the details of the BT-OPP method and the VPB-T method. Simulation results are provided and discussed in Sect. 5. Finally, we conclude this paper in Sect. 6.

2   Scenario Description

In this paper, we consider a multi-UAV tracking task in the complex environment, as shown in Fig. 1. In this scenario, multiple tracking UAVs catch up with a target UAV which flies through multiple diverse areas. We divide the environment into three areas which are no obstacle area, known area, and unknown area, respectively. In the no obstacle area, the path planning algorithm does not need to consider the influence of obstacles, and the tracking UAV can fly straight to the target UAV. In the known area, the positions of obstacles are known and can be considered into path planning. For the unknown area, the obstacles are not completely known or temporarily changes, the path planning algorithm needs to fully considers the effects of unknown obstacles. We model the environment as a two-dimensional grid map and assume that the movement of the UAV is omnidirectional. For example, when the UAV has an unit airspeed, the moving distance of the UAV in each time step is fixed as a grid. In addition, this paper further assumes that UAVs are flying at a fixed altitude.

Fig. 1. A scenario where multiple UAVs track a target UAV which flies through a complex environment.

At the initial moment, the tracking UAVs are located at the starting point of the leftmost area, and the target UAV is located at a certain position in no obstacle area. When the mission is received, the tracking UAVs begin to plan the path with the current position of the target UAV, and starts to move after the calculation is completed. The target UAV flies from left to right from the current position, passing through the three areas in turn. The tracking UAVs update the position of the target UAV in each time step, and then plan the new path. In the no obstacle area, the tracking UAVs may fly straight to the position of the target UAV. When the target UAV flies into the known area, the tracking UAVs will switch to use a more adaptive path planning algorithm, such as the I-ARA* algorithm. Similarly, if the target UAV files from the known area to the unknown area, the tracking UAVs may use an algorithm which is suitable for the unknown-area path planning, such as the D*Lite algorithm. If the target UAV successfully passes the whole area, the tracking task is considered to fail.

3   BT-Organized Path Planning (BT-OPP) Method

3.1   BT Design

In order to realize the adaptive path planning in complex environment, this paper adopts the BT model to organize the path planning algorithms which meets different requirements of the environment. The design of the BT structure is shown in Fig. 2. The ellipse denotes the condition node and the blue rectangle represents the action node. Starting from the root node, the BT firstly ticks the condition node that judge whether the target UAV has been successfully caught. If the target UAV fails to be caught (return failure), it will traverse to the child node to determine whether the target UAV is in the area under consideration. If it is not in the effective area, it returns failure. If the target UAV is in the effective area, its position will be obtained and transmitted to the tracking UAVs. When in the no obstacle area, tracking UAVs can fly straight to the target UAV. Otherwise, the I-ARA* algorithm or D*Lite algorithm will be adopted adaptively. The BT iterates through all nodes every 1 s until the end of the target tracking task.

Fig. 2. BT design that organizes three different path planning algorithms. (Color figure online)

3.2 Action Node Design

As shown in Fig. 2, the action nodes carry out path planning algorithms according to specific geographical areas. The *Straight* action node controls the tracking UAV to fly straight to the target UAV. The calculation cost of path planning is almost negligible, and the planned path is highly efficient. For the *I-ARA** action node, it will load the I-ARA* algorithm and provide a feasible path under anytime limit. The longer the time, the better the path, so that the real-time performance of the algorithm can be guaranteed. It repeatedly uses the planned results and reduces the computational cost of path planning as much as possible. Each time the BT is scheduled to the I-ARA* action node, a new starting position and target position will be assigned to it. The I-ARA* algorithm may use the information after the last calculation to re-plan a new path, realizing target tracking in the known and dynamic environment.

In the unknown area, the environment is not completely known by the tracking UAVs, which may leads to the appearance of unexpected obstacles. At this time, the *D*Lite* action node is selected. The D*Lite algorithm is used to deal with the path planning in unknown environment. The core of the D*Lite algorithm is assuming that the unknown area is all free space. On this basis, the path planning is implemented in an incremental way, and the shortest distance to the target position is found in a limited time. The traditional D*Lite algorithm is a path planning algorithm for static targets, while the improved D*Lite algorithm can re-plan a new path to a dynamic target every time BT is traversed.

4 Virtual-Target-Based Tracking (VTB-T) Method

In order to take the advantages of multiple UAVs, we propose a virtual-target-based tracking (VTB-T) method. This method is based on the idea of multiple hunters "rounding up" a prey. One hunter is selected as the main hunter, and the remaining hunters plan their paths according to the possible positions of the prey at the next moment. The possible positions are called virtual targets by our method and its distribution diagram is shown in Fig. 3.

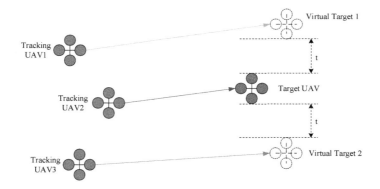

Fig. 3. Virtual targets selection in multi-UAV target tracking task.

As shown in Fig. 3, UAV1, UAV2 and UAV3 move towards the target UAV. At the beginning, UAV2 is closest to the target UAV and acts as the main tracking UAV. UAV1 and UAV3 will plan the paths for two virtual targets, respectively. In each time step, the distance between the tracking UAVs and the target UAV is calculated. If the shortest distance between the tracking UAVs and the target UAV changes, the corresponding main tracking UAV also needs to change, and the remaining tracking UAVs carries out path planning for the virtual targets. Otherwise, the tracking state remains unchanged.

In our considered scenario, each tracking UAV will be allocated a serial number i, $i \in [1, N]$, where N represents the number of tracking UAVs. The multi-UAV target tracking mainly focuses on selecting the best virtual target allocation for each tracking UAV. The main steps are shown in Algorithm 1. The inputs of the algorithm are the current positions of the tracking UAVs $p = \{p_1, p_2, \ldots, p_N\}$, where $p_i = (x_i, y_i)$, $x_i, y_i \in \mathbb{R}$, $i \in [1, N]$ denotes the position of the tracking UAV i, and x_i, y_i represent its positions on the x-axis and y-axis. $p_{target} = (x_{target}, y_{target})$ denotes the position of the target UAV, and w,

h are the width and height of the grid map G, respectively. The outputs of the algorithm are $q = \{q_1, q_2, \ldots, q_n\}$, the planned positions of the virtual targets, where $q_i = (x_i, y_i), x_i, y_i \in \mathbb{R}, i \in [1, N]$ represents the virtual target of the tracking UAV i.

Algorithm 1. Virtual-Target-Based Tracking (VTB-T) Method

Input: $p = \{p_1, p_2, ..., p_N\}$: position vector of the tracking UAVs, $p_i = (x_i, y_i)$; N: number of tracking UAVs; $p_{target} = (x_{target}, y_{target})$: the target UAV position; G: grid map; h: height of map G; w: width of map G; t: interval.
Output: $q = \{q_1, q_2, ..., q_N\}$: the positions of the virtual targets.
1: $k_{min} = \text{CALCULATENEARESTUAV}(p, p_{target})$
2: $q_{k_{min}} = p_{target}$
3: $N_{above} = \text{CALCULATETARGETNUMBER}(N, p_{target}, h, w)$
4: **for** $i \in [1, N]$ **and** $i \neq k_{min}$ **do**
5:     $x = x_{target} + |N_{above} - i| - 1$
6:     $y = y_{target} + t(i - N_{above})$
7:     **while** $G[x][y]$ is obstacle and $x < w$ **do**
8:         $x = x + 1$
9:     **end while**
10:    $q_i = (x, y)$
11: **end for**

In the beginning of Algorithm 1, we need to find the nearest tracking UAV to the target UAV which is calculated by function *CalculateNearestUAV()*, and k_{min} represents the index of the nearest tracking UAV. The details of *Calculate-NearestUAV()* is shown in Algorithm 2. On line 2 of Algorithm 1, $q_{k_{min}}$ represents the position of the target UAV. In addition to the tracking UAV k_{min}, there are still $N - 1$ remaining UAVs need to be assigned virtual targets, which will be processed by function *CalculateTargetNumber()* implemented by Algorithm 3.

Algorithm 2. Calculate the nearest tracking UAV

Input: $p = \{p_1, p_2, ..., p_N\}$: position vector of the tracking UAVs, $p_i = (x_i, y_i)$; N: number of tracking UAVs; $p_{target} = (x_{target}, y_{target})$: the target UAV position.
Output: k_{min}: the index of the nearest tracking UAV.
1: **function** $\text{CALCULATENEARESTUAV}(p, p_{target})$
2:     $d_{min} = \infty$, $k_{min} = -1$
3:     **for** each $p_i \in p$ **do**
4:         Calculate the distance d_i between the p_i and p_{target}
5:         **if** $d_i < k_{min}$ **then**
6:             $k_{min} = i$
7:         **end if**
8:     **end for**
9:     **return** k_{min}
10: **end function**

In order to achieve the rounding effect of multiple tracking UAVs, the virtual targets need to be placed above and below the target UAV. On lines 2 and 3 of Algorithm 3, half of the tracking UAVs are allocated above the target UAV, and the other half are allocated below. On line 3, the tracking UAV that is nearest to the target UAV is excluded. On lines 4 to 10 of Algorithm 3, if the upper or lower boundaries of the map are encountered, the upper and lower UAVs between target UAV increase or decrease by the same amount.

Algorithm 3. Calculate the number of virtual targets above and below the target UAV

Input: N: number of tracking UAVs; $p_{target} = (x_{target}, y_{target})$: the target UAV position; $height$: height of map; $width$: width of map.

Output: N_{above}: number of virtual targets above.

1: **function** CALCULATETARGETNUMBER($N, p_{target}, height, width$)

2:     $N_{above} = \lfloor \frac{N}{2} \rfloor$

3:     $N_{below} = N - N_{above} - 1$

4:     **if** $N_{above} > y_{target}$ **then**

5:         $N_{below} = N_{below} + N_{above} - y_{target}$

6:         $N_{above} = y_{target}$

7:     **end if**

8:     **if** $N_{below} > height - y_{target} - 1$ **then**

9:         $N_{above} = N_{above} + N_{below} - height + y_{target} + 1$

10:         $N_{below} = height - y_{target} - 1$

11:     **end if**

12:     **return** N_{above}

13: **end function**

Back to the step of Algorithm 1, after arranging the number of virtual targets above and below the target UAV, we need to allocate the positions of the virtual targets. On lines 4–9 of Algorithm 1, the tracking UAV $k_{\min}$ that has been allocated is excluded, leaving $N - 1$ tracking UAVs to be allocated. Lines 5 and 6 calculate the virtual target i between top and bottom of the target UAV. Lines 7 and 8 indicate that when the allocated virtual target is an obstacle, the position of the current virtual target is shifted to the right by one unit distance until it is free. If the allocated virtual target reaches the right boundary of the map, it will be fixed at the right boundary. Finally, line 10 sets the corresponding virtual targets to each tracking UAV.

5   Simulation Results

In this section, we may verify the effectiveness of our proposed methods. Firstly, in Subsect. 5.1, we may give the comparison between the BT-organized path planning (BT-OPP) method and the original path planning (OPP) method.

In order to take advantages of the multi-UAV system, Subsect. 5.2 will test the validation of our virtual-target-based tracking (VTB-T) method.

5.1 The Performance Comparison Between the BT-OPP and OPP Methods

In order to verify the effectiveness of the BT-OPP method, we consider a complex scenario where the target UAV flies through a complex environment, as shown in Fig. 1. The first area is a no obstacle area which can be seen as an open field. The second area is a known and obstacle area. In the third area, it has obstacles but unknown. In this scenario, the tracking UAV will pursue the target UAV until it gets to the end position of the map.

In order to verify the tracking performance of the BT-OPP method, we may select the D*Lite algorithm as the OPP method and the benchmark. For the BT-OPP method, straight, I-ARA*, and D*Lite algorithms are used to plan the paths for different areas, respectively. A trajectory comparison between the BT-OPP and OPP methods are shown in Fig. 4(a) and Fig. 4(b), respectively. In this testing, the airspeed of the tracking UAV is the same as that of the target UAV. From the result, we may see that the length of the planned path by the BT-OPP method is apparently shorter than the OPP method. Furthermore, the tracking UAV with the BT-OPP method finally catches up with the target UAV successfully, but the OPP method may not.

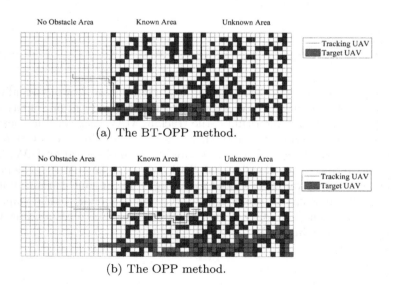

(a) The BT-OPP method.

(b) The OPP method.

Fig. 4. Trajectories planned by the BT-OPP method and the OPP method.

We may further analyze the efficiency of the BT-OPP and OPP methods. Let d and a denote the initial distance and airspeed ratio between the tracking

UAV and the target UAV, respectively. We may change d and a, then compare the final distance between the tracking UAV and the target UAV, as shown in Fig. 5. If the final distance is greater than 0, it means the target UAV can not be caught. From the figure, we may see that the final distance increases with the increase of the initial distance, but the tracking efficiency of the BT-OPP method is obviously higher than that of the OPP method. When $a \geq 1.5$, the final distance of the BT-OPP method was always 0, which means the target UAV can be always caught.

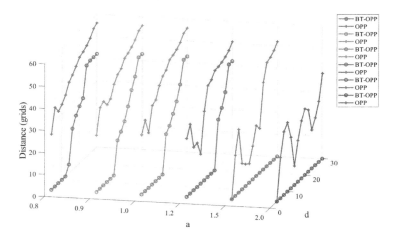

Fig. 5. Final distance comparison between the BT-OPP and the OPP methods in different settings.

5.2   Verify the Optimization Effects of the VTB-T Method

In this paper, a multi-UAV target tracking algorithm called VTB-T method is proposed, which aims at increasing the chance to catch up with the target UAV. We compare the flight trajectories under the BT-OPP methods with a single tracking UAV (Fig. 6(a)) and multiple tracking UAVs (Fig. 6(b)). In the multi-UAV case, the virtual target distribution is calculated by the VTB-T method.

In order to verify the optimization effect of the multiple tracking UAV case and the single tracking UAV case, this experiment designs the path planning efficiency comparison of the two cases. As shown in Fig. 7, the x-axis is the airspeed ratio of the tracking UAV to the target UAV, the y-axis is the time step index, and the z-axis is the time steps that the target UAV being caught. According to the results, it can be seen that under the same number of tracking UAVs, the tracking efficiency increases with the increase of a. Furthermore, with the same a, the more tracking UAVs there are, the faster the target will be caught.

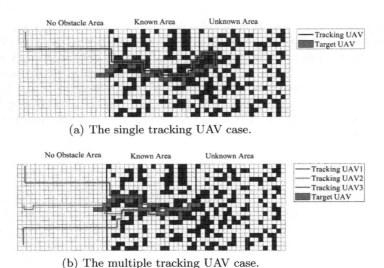

Fig. 6. Trajectory planned by the BT-OPP method with a single tracking UAV and multiple tracking UAVs.

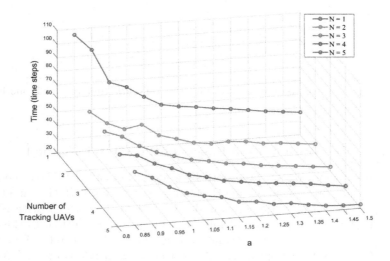

Fig. 7. Tracking efficiency comparison in different system settings.

6   Summary

This paper considered a multi-UAV target tracking scenario in a multi-geographical area, which involved the challenges of real-time path planning and environment diversity. We proposed to use BT to solve the two challenges, and introduced the BT-OPP method which could automatically select the appropriate path planning algorithms according to the environment conditions. Moreover, in order to increase the tracking performance, we proposed the VTB-T method that fully utilized the advantages of multiple tracking UAVs. The simulation results showed that the BT-OPP method could largely improve the target tracking performance more than the classical OPP method. When involved with the VTB-T method, the BT-OPP could further decrease the tracking time.

References

1. Wallace, L., Lucieer, A., Watson, C.S., Turner, D.: Development of a UAV-LiDAR system with application to forest inventory. Remote Sens. **4**(6), 1519–1543 (2012)
2. Spedicato, S., Notarstefano, G., Bülthoff, H.H., Franchi, A.: Aggressive maneuver regulation of a quadrotor UAV. In: Inaba, M., Corke, P. (eds.) Robotics Research. STAR, vol. 114, pp. 95–112. Springer, Cham (2016). https://doi.org/10.1007/978-3-319-28872-7_6
3. Wu, Y., Zhang, B., Yang, S., Yi, X., Yang, X.: Energy-efficient joint communication-motion planning for relay-assisted wireless robot surveillance. In: Proceedings of IEEE Conference on Computer Communications, pp. 1–9. IEEE (2017)
4. Dijkstra, E.W., Dijkstra, E.W., Dijkstra, E.W., Informaticien, E.U., Dijkstra, E.W.: A Discipline of Programming. Prentice-Hall, Englewood Cliffs (1976)
5. Zhang, H., Cheng, M.: Path finding using A* algorithm. Microcomput. Inf. **23**(6), 238–241 (2007)
6. Likhachev, M., Ferguson, D., Gordon, G., Stentz, A., Thrun, S.: Anytime dynamic A*: an anytime, replanning algorithm. In: Proceedings of International Conference on Automated Planning and Scheduling, pp. 262–271. AAAI Press (2005)
7. Likhachev, M., Gordon, G., Thrun, S.: ARA*: anytime A* with provable bounds on sub-optimality. In: Advances in Neural Information Processing Systems 16, pp. 767–774. MIT Press (2004)
8. Aine, S., Likhachev, M.: Anytime truncated D*: anytime replanning with truncation. In: Proceedings of Annual Symposium on Combinatorial Search, pp. 2–10. AAAI Press (2013)
9. Sun, X., Yeoh, W., Uras, T., Koenig, S.: Incremental ARA*: an incremental anytime search algorithm for moving-target search. In: Proceedings of International Conference on Automated Planning and Scheduling, pp. 243–251. AAAI Press (2012)
10. Hu, D., Gong, Y., Hannaford, B., Seibel, E.J.: Semi-autonomous simulated brain tumor ablation with RAVENII surgical robot using behavior tree. In: Proceedings of IEEE International Conference on Robotics and Automation, pp. 3868–3875. IEEE (2015)
11. Paxton, C., Hundt, A., Jonathan, F., Guerin, K., Hager, G.D.: CoSTAR: instructing collaborative robots with behavior trees and vision. In: Proceedings of IEEE International Conference on Robotics and Automation, pp. 564–571. IEEE (2017)
12. Gershenson, J., Prasad, G., Zhang, Y.: Product modularity: definitions and benefits. J. Eng. Des. **14**(3), 295–313 (2003)
13. Wu, Y., Ren, X., Zhou, H., Wang, Y., Yi, X.: A survey on multi-robot coordination in electromagnetic adversarial environment: challenges and techniques. IEEE Access **8**, 53484–53497 (2020)

Budget Constraint Task Allocation for Mobile Crowd Sensing with Hybrid Participant

Xin Wang[1,2], Peng Li[1,2(✉)], and Junlei Xiao[1,2]

[1] College of Computer Science and Technology,
Wuhan University of Science and Technology, Hubei, China
lipeng@wust.edu.cn
[2] Hubei Province Key Laboratory of Intelligent Information Processing
and Real-time Industrial System, Hubei, China

Abstract. In the Mobile crowdsensing (MCS) system, the task allocation problem is a crucial problem. In this paper, we focus on the task allocation problem for hybrid participants with a budget-constrained MCS system. There are two types of participants: mobile participants and static participants. Mobile participants have low cost, large numbers, and flexibility. However, most of the sensing data they submitted are of low quality. On the other hand, static participants, such as city cameras, roadside infrastructure, have high-quality sensing data. Despite the benefit of high quality, static participants have less coverage and high cost. Given a budget, the problem is how to assign the task to the two types of participants, such that the social welfare is maximized. To solve the problem, we propose a reverse auction-based task allocation method (ORA) to select winning bids round by round. Then, a Shapley value based online algorithm (OAA) is proposed to ensure the task is finished. Moreover, we consider the different types of participants to have a different probability to finish tasks. We exploit the semi-Markov model to calculate the probability that participants finished tasks. We prove that the proposed task allocation method has truthfulness and individual rationality. We conduct extensive experiments to evaluate the performance of our system, and the evaluation results demonstrate the remarkable effect of the proposed task allocation method.

Keywords: Mobile crowdsensing · Task allocation · Reverse auction · Shapely value · Semi-Markov model

1   Introduction

Mobile crowdsensing (MCS) [1] is a new sensing paradigm that uses the powerful sensing ability of mobile devices. Mobile devices such as smartphones are

---

This work is partially supported by the Hubei Provincial Natural Science Foundation of China (No. 2018CFB424).

H. Gao et al. (Eds.): CollaborateCom 2020, LNICST 350, pp. 506–517, 2021.
https://doi.org/10.1007/978-3-030-67540-0_33

carried by mobile users who have the potential to finish the sensing tasks. In the MCS system, the number of mobile users is huge and multiple mobile users could finish one task together. So mobile users always finish tasks at a low cost. The cost of users is critical because the platform publishes tasks always with budget constraints. The advantage of low cost means the platform could hire more users to guarantee the quality of sensing data. Because of this advantage of MCS, we can use MCS on some new systems to accomplish tasks, *e.g.*, monitoring of water distribution [2], traffic density estimation [3], and air pollution inference [4]. In this paper, we focus on the task allocation problem with the budget-constrained to maximize social welfare, which is called the maximize-social-welfare task allocation problem (MSW-PR). The MCS will be effective when the users submit high-quality sensing data. So how to allocate the sensing task to proper users is a crucial problem. Much former research studied this problem. In these studies, researchers' methods include the reverse auction [5] and semi-Markov-based prediction [6]. All of these studies assumed that there always have adequate users to participate. However, this is not valid in reality. Moreover, mobile users' performance is limited by the personal ability, weather, and stability of sensors on mobile devices, it is hard to guarantee the quality of sensing data. Accordingly, the task allocation problem is challenging when the number of mobile users is scarce and most of the mobile users have a low ability to perform tasks.

Generally, sensing tasks are location-sensitive [7]. The difficulty of the task varies with the location of the task, however, most previous studies only consider homogeneous participants. In fact, there are at least two types of participants in the MCS system, mobile participants, and static participants. Mobile participants have the advantage of low cost, flexible cover task locations. Static participants such as stationary cameras, roadside infrastructure, etc. Compared with the mobile participants, static participants have a high ability to perform the sensing task. However, the installation cost and maintenance cost is very high. Therefore, it is a non-trivial problem to reasonably allocate tasks to hybrid participants with the budget constraint.

To solve the above two challenges, we propose a reverse auction-based task allocation model with hybrid participants, including mobile participants, and static participants. The main contributions of our work are:

- We first propose a one-round auction (ORA) algorithm to select winning bids. Through this algorithm, we could determine which kinds of participants are proper for a certain sensing task. In the one round auction, we use the semi-Markov model to calculate the probability that the participant could submit high-quality sensing data.
- We propose the budget allocation strategy (BAS) based on Shapely value. We calculate the Shapely value of the different subregions. We will allocate more budget to this subregion which has a low Shapely value next sensing period to ensure the task completion rate. Based on the ORA and BAS algorithms, we propose an online auction algorithm to achieve the maximum social welfare goal.

- We prove that the proposed ORA and OAA algorithm has truthfulness and individual rationality.
- We conduct an extensive simulation. Through the real data simulation experiment, we demonstrate the remarkable effect of the proposed ORA and OAA algorithm.

2   Related Work

In recent years, many studies focus on user recruitment in MCS. Zhou *et al.* [5] proposed a highly credible incentive mechanism based on the reverse auction theory. This incentive minimizes the cost of the platform by controlling participants' bids through reverse auction and recruiting proper participants to finish tasks. In these works, researchers assume the number of users is always adequate. They ignore the situation of rare user participation. Moreover, in the real world, there is more than one kind of participant, but researchers only focus on a single kind of participant. So the existing studies cannot solve our MSW-PR problem. Many studies consider the budget constraints in MCS. Wang *et al.* [1] studied the coverage requirements and workload balancing requirements of mobile crowdsourcing and used dynamic programming to solve the problem in one-dimensional scenarios, and proposed greedy algorithms to solve the CBCR problem with submodule properties in two-dimensional scenarios. The above research concentrates on task allocation. Their studies all consider budget constraints. However, they allocate the budget equally, which can not guarantee the quality of the sensing task. In our work, we allocate the budget based on Shapely value, avoiding the equalitarianism which makes the budget allocation more reasonable.

3   System Description and Problem Definition

3.1   System Model

Our MCS system with hybrid participants includes a cloud platform, static participants, and mobile participants. In this paper, the platform divides the sensing region into subregions, denoted as $L = \{l_1, l_2, ..., l_m\}$. There are some tasks, denoted as $S = \{s_1, s_2, ..., s_k\}$. The platform can release multiple tasks in a certain sensing period t_i, where $t_i \in T$, $T = \{t_1, t_2, ..., t_n\}$ represents the set of sensing periods and t_i represents the i-th sensing period. Each sensing task is assigned with attributes, *e.g.*, the deadline D, and the subregion l which the task belongs. Participants could make multiple bids to different tasks, *e.g.*, participant i made j-th bid at sensing period t, denoted as b_{ij}^t. Mobile participants and static participants make a bid b_{ij}^t for the sensing task s_k^t they want to complete. For a certain user, the cloud platform can obtain the completion rate h_i of the participants i. Then, we use δ_{ijk}^t to represent participant i make j-th bid at sensing period t to participate in the sensing task k. Before the recruitment of participants, the completion rate of tasks can be estimated by the h_i. To improve the performance of participants, we have specified the minimum completion rate of each task $\{\mu_1, \mu_2, ..., \mu_k\}$. In the end, the cloud platform selects the proper participants $I = \{i_1, i_2, ..., i_n\}$ to complete the tasks. The final rewards p_i^t are paid by the platform to the participants.

3.2   Reverse Auction Based Participants Recruitment

The recruitment method of our reverse auction mainly includes the following five steps:

1. The cloud platform publishes k sensing tasks in a certain sensing period $t_i \in T$. Each sensing task $s_i \in S, S = \{s_1, s_2, ..., s_k\}$ needs to be completed within a time interval $[t_1, t_2]$, and the location of each task belongs to $L = \{l_1, l_2, ..., l_m\}$.
2. After participant i receives sensing task information, he will pack his bid price b_{ij}^t and the task he bid s_k^t into task-bid pair (b_{ijk}^t) and send to the platform. Then the platform receives the task-bid pairing, adding the participant's historical task-completion rate h_i into the task-bid pair, forming the task-bid-completion rate pair (b_{ijk}^t, h_i), which is also written as δ_{ijk}^t.
3. The cloud platform selects the winning bid pair through comparison of the task-bid-completion rate pair and adds the winning bid pair into the set of winners Φ.
4. After the winner gets the task, he/she starts to perform the task within the time interval of $[t_1, t_2]$, where t_1 is the beginning time. The winner must finish the task before the deadline, $t_2 = D$, and send the acquired sensing task data to the cloud platform.
5. The cloud platform receives the sensing task, pays p_i^t for the participants. The number of unsold tasks $|S^{remain}|$ on the sensing platform and the number of low-quality sensing tasks $|S^{low}|$ was calculated to calculate the task completion rate of different regions l_i.

We next define the utility u of the participant i:

$$u_i(b_{ij}^t) = \begin{cases} p_i^t - c_i^t & x_{ij} = 1 \\ 0 & x_{ij} = 0 \end{cases} \tag{1}$$

Among the participants' utility definition, the cost of a participant performing a task c_i^t is hard to reduce, so the utility mainly comes from the reward given by the platform. The reward is also the actual expenditure of the platform.

3.3   Problem Formulation

In this section, we solve the maximized social welfare task allocation problem (MSW-PR). For the MSW-PR problem, our goal is to maximize social welfare.

As thus, we could represent our system welfare to:

$$\sum_{t\in[T]}\sum_{i\in[I]}\sum_{j\in[J]} z_{ij}^t(v_i^t - p_i^t) + \sum_{t\in[T]}\sum_{i\in[I]}\sum_{j\in[J]} z_{ij}^t(p_i^t - c_i^t) \tag{2}$$

Where $z_{ij}^t = 1$ represents participant i provide high-quality sensing data to task j and $z_{ij}^t = 0$ represents participant i provide low-quality sensing data or provide nothing to task j.

Our MSW-PR problem can be formulated as:

$$\max e = \sum_{t\in[T]}\sum_{i\in[I]}\sum_{j\in[J]} z_{ij}^t(v_i^t - c_i^t) \tag{3}$$

subject to:

$$\sum_{i=1}^{l}\sum_{j=1}^{J} p_i^t x_{ij}^t \leq B^t \tag{4}$$

$$x_{ij}^t \in \{0,1\} \tag{5}$$

Where constraint (4) represents the payment p_i^t should be less than or equal to the budget B^t of a certain period t. Constraint (5) shows that x_{ij}^t is 0 or 1. $x_{ij}^t = 1$ represents the bid that is accepted by the cloud platform. Otherwise, it is rejected.

When selecting participants for the platform, we compare their weighted bids b_{ij}^t with their task completion rate h_i to determine the final winner of the task: $b_{ij}^{t}{}' = b_{ij}^t/h_i$.

The weighted bid price avoids the appearance of low-quality sensing data. In this way, we both consider the bid price and data quality in task allocation, which ensures the sensing data quality.

4   The MSW-PR Problem

In this section, we first propose a one-round auction algorithm (ORA). Next, we propose the semi-Markov-based participant task completion rate prediction method and the Shapely value-based budget allocation strategy (BAS). At last, we propose our online-auction algorithm (OAA) to solve the MSW-PR problem.

4.1   One-Round Auction Algorithm

First, calculate the task completion probability: $P_k^{\Phi} = 1 - \prod_{\delta_{ijk}^t \in \Phi \wedge k \in \delta_{ijk}^t} (1 - h_i)$.

Where Φ represents the win-bid set, h_i denotes the task completion probability of participant i, δ_{ijk}^t represents participant i make j-th bid at sensing period t to participate in the sensing task k.

Then, we calculate the utility u of the win-bid set Φ, where μ_k is the given minimum task completion probability of task s: $U(\Phi) \triangleq \sum_{i=1}^{k} \min\{P_k^{\Phi}, \mu_k\}$.

Through the reverse auction, we continuously add bid-pairs δ_{ijk}^t to the win-bid set Φ so that we can obtain the utility equation:

$$U_{\delta_{ijk}^t}(\Phi) = \sum_{s_i \in \delta_{ijk}^t} (\min\{P_k^{\Phi \cup \{\delta_{ijk}^t\}}, \mu_k\} - \min\{P_k^{\Phi}, \mu_k\}) \qquad (6)$$

The marginal contribution utility Eq. (6) ensures that these winning bids are the optimal choice of the current round t.

Algorithm 1. One-Round Auction Algorithm (ORA)

Input: Budget B, tasks s, bid-pair set Γ, winning bid-pair set Φ, task completion threshold θ

Output: winning bid-pair set Φ

1: Begin
2: $\Phi = \varnothing; \forall i \in [I], j \in [J], k \in [K]; x_{ij}^t = 0; z_{ij}^t = 0; p_i^t; B^t = B$
3: **while** $U(\Phi) < k\mu \&\& \sum_{i=1}^{n} p_i^t < B^t$ **do**
4: $\qquad \delta_{i*j*k}^t = \arg\min_{\delta_{ijk}^t \in \Gamma} \dfrac{U_{\delta_{ijk}^t}(\Phi)}{b_{ijk}^t}(h_i + \xi)$
5: $\qquad$ **if** $h_i > \theta$ **then**
6: $\qquad\qquad z_{i*j*}^t = 1$
7: $\qquad$ **else**
8: $\qquad\qquad z_{i*j*}^t = 0$
9: $\qquad \delta_{i'j'k}^t = \arg\min_{\delta_{i'j'k}^t \in \Gamma_{-\delta_{i*j*k}^t}} \dfrac{U_{\delta_{i'j'k}^t}(\Phi)}{b_{i'j'k}^t}(h_{i'} + \xi)$
10: $\qquad p_i^t = \dfrac{b_{i'j'k}^t U_{\delta_{i*j*k}^t}(\Phi')h_{i*}}{U_{\delta_{i'j'k}^t}(\Phi')h_i'}$
11: $\qquad$ update $U(\Phi) = U(\Phi) + U_{\delta_{i*j*k}^t}(\Phi)$
12: $\qquad$ update $\Phi = \Phi \cup \{\delta_{i*j*k}^t\}$
13: $\qquad$ update $\Gamma = \Gamma \backslash (\cup_{i \in [I]} \delta_{i*j*k}^t)$
$\qquad$ **return** Φ
14: End

The one-round auction algorithm is shown in Algorithm 1. During the iteration, we judge whether the utility of the winning bid-pair satisfies the needs of the platform. If the conditions are met, we will select the bid-pair δ_{i*j*k}^t with the highest marginal utility (line 3). Then, we mark the bid users as being recruited (line 4), where $\xi > 0$ avoids $h_i = 0$. When $h_i > \theta$, the threshold of task completion probability, we think the participant could provide high-quality sensing data and mark $z_{ij} = 1$. Otherwise, $z_{ij} = 0$ (line 5–7). We use the second price auction to pay the reward (line 9–10). After selecting the win-bid pair, we need to add the win-bid pair to the win-bid set Φ and update the utility of the win-bid set $U(\Phi)$ (line 11–12). Finally, since we only allow the participant i to complete one task in one sensing period, we need to exclude all bid-pairs for this participant (line 13).

4.2 Probability Calculation of High-Quality Sensing Data Methods for Hybrid Participants

For static participants, since the conditions (hardware conditions and operation mode) for completing tasks are almost the same each time, the completion probability of tasks can be calculated simply by the historical statistics: $h_i = num_i^h / num_i$. Where num_i^h is the historical high-quality sensing data count submitted by static participants, and num_i is the historical win-bid count. Both counts are from historical statistics.

However, for the mobile participants, it is impossible to use their historical task completion result to make statistics, since the condition of performing tasks may be different each time.

In this case, we choose to use the semi-Markov model to predict the participants' ability to provide high-quality sensing data in the current sensing period. First, we define the kernel part of the semi-Markov model: $Z_i^{bg}(t) = P(q_i^t = g, t_{ik} \leq T | q_i^{t-1} = b)$. $Z_i^{bg}(t)$ is the probability that participant i could provide high-quality sensing data at t-round sensing period, while he provides low-quality sensing data at $(t-1)$-round. $q_i^{t-1} = b$ and $q_i^t = g$ represent whether participant i provides good or bad sensing data, respectively. $t_{ik} < T$ represents that the participant i should finish the task k and send sensing data to the cloud platform within sensing time limit T.

To calculate $Z_i^{bg}(t)$, we define the probability $W_i^{bg}(t) = P(x | q_i^t = g, q_i^{t-1} = b)$, which represents that participant i sends sensing data within time limit T, where he provides low-quality sensing data at t-round sensing period and he provides high-quality sensing data at $t-1$-round.

Next, we define the probability $P_i^{bg} = P(q_i^t = g | q_i^{t-1} = b)$, which shows that the participant i provides high-quality sensing data at t-round sensing period while he provides low-quality sensing data at $(t-1)$-round.

Therefore, we could rewrite $Z_i^{bg}(t)$ by $W_i^{bg}(t)$ and q_i^{t-1}: $Z_i^{bg}(t) = P(q_i^t = g, t_{ik} \leq T | q_i^{t-1} = b) = W_i^{bg}(T)P_i^{bg}$ Similarly, we can get the probability of h_i that the participant can provide high-quality sensing data in the t-round under other circumstances. So we can get the probability that the participant can provide high-quality sensing data in the t-round:

$$h_i = P_i^g(t) = \frac{(Z_i^{bg}(t) + Z_i^{gg}(t))}{(Z_i^{bg}(t) + Z_i^{gg}(t) + Z_i^{bb}(t) + Z_i^{gb}(t))} \tag{7}$$

4.3 Budget Allocation Method Based on Shapley Value

In the MCS system, we expect participants to provide enough high-quality sensing data in a budget constraint. However, sensing tasks are very sensitive to the location. Some tasks may be difficult and not be completed well. This will seriously affect sensing data quality. For this problem, we use the Shapley value to allocate budgets to different regions.

We first define the concept of task completion rate for the regions.

Definition 1. *We define the task completion rate F_l, which is a measure of the task completion status of subregion l, and always have $F_l \leq 1$.*

After the sensing data is submitted by participants, the cloud platform calculates the number of all low-quality sensing data $\left|S_l^{low}\right|$ and the number of sensing tasks that no one bids for or fails to auction $\left|S_l^{remain}\right|$. Then we could calculate the task completion ratio with: $F_l = (|S_l| - \left|S_l^{low}\right| - \left|S_l^{remain}\right|)/\,|S_l|$.

Where $|S_l|$ is the count of all tasks s.

We employ the Shapley value method to allocate budget:

$$\zeta_l = \sum_{L' \subseteq L \setminus \{l\}} [U(L' \cup \{l\}) - U(L')] \frac{|L'|!(|L| - |L'| - 1)}{|L|} \tag{8}$$

$$B_l = B \frac{\zeta_l}{\sum_{l \in L} \zeta_l} \tag{9}$$

Where $L' \subseteq L \setminus \{l\}$, represent the subset of region set and $U(L')$ represent the utility of subset L'. After calculating the Shapley value, we can use Eq. (9) to allocate budget. Where B_l represent the budget B of subregion l.

Algorithm 2. Budget Allocation Algorithm (BAA)

Input: Total budget B, Location set L, task completion rate F_l
Output: subregions budget B_l

1: Begin
2: $B_l = \varnothing; \forall t \in [T], l \in [L]$
3: **for** $l = 1, 2, ..., L$ **do**
4:     use Eq. (17) to calculate the F_l for each subregion;
5:     use Eq. (18) to calculate the B_l
    return B_l, which is the budget of subregion l
6: End

4.4 Online Auction Algorithm

After proposed the one round auction algorithm ORA and the budget allocation strategy BAS, we now propose our online auction algorithm.

In our online auction algorithm, we calculated the Shapley value of different regions (line 3–7). It is noting that in the first period since there is no data of the previous sensing period, we cannot allocate the budget of each region according to the Shapley value, so we allocate the budget of each region equally (line 5). The winner in each region was calculated using the ORA algorithm (line 8) and the budget of subregion l is calculated by BAS (line 7).

Algorithm 3. Online Auction Algorithm (OAA)

Input: Budget of subregions B_l, Location set L, all bid-pair set Γ_t, win bid-pair Φ_t
Output: win bid-pair Φ_t
1: Begin
2:   $\Phi_t = \varnothing; \forall t \in [T], l \in [L], k \in [K], x_{ij}^t = 0, z_{ij}^t, p_i^t$
3: **for** $1 < t < T$ **do**
4:    **if** $t = 1$ **then**
5:       $B_l = \frac{B}{|L|}$
6:    **else**
7:       use algrithom BAS to caculate B_l
8:    $\Phi_t = ORA(B_l, \Gamma_t, \Phi_t)$
   return $\Phi_t, \forall t \in [T]$
9: End

---

5   Theoretical Analysis

In this section, we will prove that ORA and OAA have truthfulness and individually rational. In our subsequent experiment in Sect. 6, we analyze ORA by comparing the mixing ratio of different mobile participants and static participants. We select the winning bid-pair of current round t, from all the current auction pairs through reverse auction, and add it into the winner set Φ_t. For the OAA algorithm, in addition to the operation of the ORA algorithm, we also introduce the Shapley value to allocate budget B more reasonably.

5.1   Truthfulness

We give some definitions. Φ_t represents the win-bid pair set. Φ_t' the win bid pair set which remove the pair bid-pair δ_{ijk}^t from Γ. According to [8], we have $\Phi_t = \Phi_t'$. Then we have the following lemma.

Lemma 1. *Algorithm ORA is bid-monotonic.*

Proof. Since our reverse auction selects the win bid-pair for each round, the number of bid pairs we select for algorithm ORA will increase when the auction proceeds. For example, if participant i is the winner of auction t, his weighted bid must be the minimum value of this round of auction. Suppose that participant i now puts forward a lower bid than his former win bid, and we will put it into algorithm ORA:

$$\frac{U_{\delta_{ijk}^t}(\Phi)}{b_{ijk}^t}(h_i + \xi) < \frac{U_{\delta_{i'j'k}^t}(\Phi)}{b_{i'j'k}^t}(h_{i'} + \xi) \tag{10}$$

We find that participant i still win the t round auction. So our Lemma 1 is proved.

Lemma 2. *The reward payment of algorithm ORA is critical for all winning bids.*

Proof. We assume the current win-bid pair δ_{ijk}^t makes another bid $b_{i'j'k}^t$. If $b_{i'j'k}^t \geq p_i^t$, which means participant i's new bid is larger than the original bid's reward. Therefore, we have:

$$\frac{U_{\delta_{i'j'k}^t}(\varPhi_{t-1})}{b_{i'j'k}^t}(h_i + \xi) = \frac{U_{\delta_{i'j'k}^t}(\varPhi'_{t-1})}{b_{i'j'k}^t}(h_i + \xi)$$

$$\leq \frac{U_{\delta_{ijk}^t}(\varPhi'_{t-1})}{p_i^t}(h_i + \xi) \qquad (11)$$

$$< \frac{U_{\delta_{i^*j^*k}^t}(\varPhi_{t-1})}{b_{i^*j^*k}^t}(h_i + \xi)$$

We can see if participant i has a new bid $b_{i'j'k}^t \geq p_i^t$, he can not win in round t. We assume there is a participant i^* who has the same task completion probability with i' and he makes a bid lower than p_i^t. Then he can win in round t. So the reward for all winning bids is critical, and Lemma 2 holds.

Lemma 3. *Our Algorithm ORA and OAA is truthfulness.*

Proof. Because our algorithm ORA is bid-monotone (i.e., Lemma 1) and the reward for participant is critical value (i.e., Lemma 2) from algorithm OAA, so our algorithm ORA and OAA are both truthfulness.

5.2 Individual Rationality

Lemma 4. *Our reverse auction based participant allocation strategy is individually rational.*

Proof. We assume that the bid-pair δ_{ijk}^t won in round t, and the second lowest bid in round t is $\delta_{i^*j^*k}^t$. We have:

$$b_{ijk}^t < \frac{b_{ijk}^t U_{\delta_{i^*j^*k}^t}(\varPhi_{t-1})h_{i^*}}{U_{\delta_{ijk}^t}(\varPhi_{t-1})h_i} = \frac{b_{i^*j^*k}^t U_{\delta_{i^*j^*k}^{sj}}(\varPhi'_{t-1})h_{i^*}}{U_{\delta_{ijk}^t}(\varPhi'_{t-1})h_i} \leq p_i^t \qquad (12)$$

So if a participant i wants to win an auction, he have to make a real bid to reach $p_i^t \geq b_{ijk}^t$, which means the participant's income is non-negative.

6   Performance Evaluation

In this section, we use the simulation set and Chengdu/taxi trace set [9] to evaluate the performance of our algorithms OAA and ORA.

The bid price of mobile participants is generated between [10–15], and the bid price of static participants is defined as twice that of mobile participants [20–30]. The sensing period is always 30 mins. We use $\lambda = s/m$ to represent the ratio of mobile participants and static participants, where s and m represent the count of static participants and mobile participants, respectively.

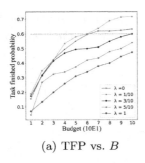
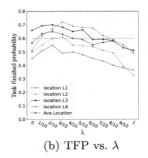

(a) TFP vs. B (b) TFP vs. λ

Fig. 1. Task finished probability under different λ and budget

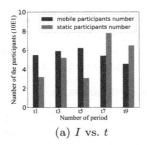
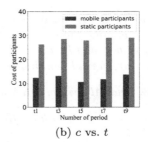

(a) I vs. t (b) c vs. t

Fig. 2. The number of the win bid participants and social cost

6.1 Evaluation of Offline Task Allocation Strategy and Online Task Allocation Strategy

To better evaluate our algorithm ORA, we introduce the task completion rate, which is the ratio of the complete task count and all task count. The following evaluation gave a minimum task completion rate ($\mu = 0.6$) and participant count ($i = 1000$). As shown in Fig. 1a, we can observe that with the increase of budget, the task completion rate has been increased when $\lambda < 3/10$, which is better than the situation of $\lambda = 0$. As shown in Fig. 1b, we can see that the task completion rate becomes bigger with λ increased. When λ is between $1/10$ and $3/10$, the task completion rate is better than there is no static participant. This is because the task completion probability of static participants is better than mobile participants.

Next, we evaluate the performance of the online task allocation strategy.

Win-Bid Participants: We can see the number of winners from Fig. 2a, in which the number of mobile participants is much higher than that of static participants because the bid price of mobile participants is much lower than static participants.

Average Social Cost of Participants: We can see the average social cost of participants from Fig. 2b. We can see that despite a large number of mobile participants, their social cost is much lower than that of static participants. This is also why we need to compare the social cost of mobile participants with static participants, that is, the social cost of mobile participants is much lower than that of static participants.

7   Conclusion

In this paper, we investigate the problem of task allocation strategy in mobile crowdsensing. First, we recruit participants and allocate tasks through a one-round auction algorithm (OAA). Then, according to the semi-Markov model, we propose a participant task completion probability prediction for mobile crowdsensing, where the platform could calculate the task completion probability for a certain round auction. Moreover, we propose an online auction algorithm using Shapely value. Through the use of Shapely value, we calculated the task finishing probability on different subregions and re-allocated the budget more reasonably. We conduct extensive simulations based on real traces: Chengdu/taxi. The results show that our online algorithm achieves a high task completion rate and one-round auction achieves the highest social welfare.

References

1. Wang, Z., et al.: Personalized privacy-preserving task allocation for mobile crowdsensing. IEEE Trans. Mob. Comput. **18**(6), 1330–1341 (2019)
2. Du, R., Fischione, C., Xiao, M.: Flowing with the water: on optimal monitoring of water distribution networks by mobile sensors. In: IEEE INFOCOM 2016 - The 35th Annual IEEE International Conference on Computer Communications, pp. 1–9 (2016)
3. Panichpapiboon, S., Leakkaw, P.: Traffic density estimation: a mobile sensing approach. IEEE Commun. Mag. **55**(12), 126–131 (2017)
4. Ma, R., et al.: Enhancing the data learning with physical knowledge in fine-grained air pollution inference. IEEE Access **8**, 88372–88384 (2020)
5. Zhou, R., Li, Z., Wu, C.: A truthful online mechanism for location-aware tasks in mobile crowd sensing. IEEE Trans. Mob. Comput. **17**(8), 1737–1749 (2018)
6. Wang, E., Yang, Y., Wu, J., Liu, W., Wang, X.: An efficient prediction-based user recruitment for mobile crowdsensing. IEEE Trans. Mob. Comput. **17**(1), 16–28 (2018)
7. Yang, Y., Liu, W., Wang, E., Wu, J.: A prediction-based user selection framework for heterogeneous mobile crowdsensing. IEEE Trans. Mob. Comput. **18**(11), 2460–2473 (2019)
8. Gao, G., Xiao, M., Wu, J., Huang, L., Hu, C.: Truthful incentive mechanism for nondeterministic crowdsensing with vehicles. IEEE Trans. Mob. Comput. **17**, 2982–2997 (2018)
9. Didi: Data Source: Didi Chuxing GAIA Initiative. https://outreach.didichuxing. com/app-vue/TTItrajectory?id=1001

A Unified Bayesian Model of Community Detection in Attribute Networks with Power-Law Degree Distribution

Shichong Zhang[1], Yinghui Wang[1], Wenjun Wang[1,3], Pengfei Jiao[2] (ORCID), and Lin Pan[4(✉)]

[1] College of Intelligence and Computing, Tianjin University, Tianjin, China
{zsc,wangyinghui}@tju.edu.cn
[2] Center for Biosafety Research and Strategy, Law School, Tianjin University, Tianjin, China
pjiao@tju.edu.cn
[3] College of Information Science and Technology, Shihezi University, Shihezi, China
wjwang@tju.edu.cn
[4] School of Marine Science and Technology, Tianjin University, Tianjin, China
linpan@tju.edu.cn

Abstract. Detecting community structure is an important research topic in complex network analysis. How to improve community detection results by using various features in the network is a very challenging problem. The scale-free and attributes of nodes are the two relatively independent aspects of the complex networks in the real world, the former is an inherent structural feature from the global perspective and the later can be used to significantly enhance community detection and community semantics. However, these two aspects are usually modeled and computed independently in previous methods. Based on that, we propose a novel unified Bayesian generative model which combines network topology and node attributes simultaneously to identify community structures via considering to model the scale-free feature. We propose the degree decay variable to preserve the power-law degree characteristic of the network. Specifically, this model composes of two closely correlated parts by a probabilistic transition matrix, one for network topology and the other for nodes attributes. Moreover, we develop a variational EM algorithm to optimize the objective function of the model. Experiments on synthetic and real networks show that our model has a better performance compared with some baselines on community detection in attribute networks.

Keywords: Community detection · Attribute network · Bayesian generative model · Variational EM algorithm · Power-law degree distribution

---

S. Zhang and L. Pan—Equal contribution.

© ICST Institute for Computer Sciences, Social Informatics and Telecommunications Engineering 2021
Published by Springer Nature Switzerland AG 2021. All Rights Reserved
H. Gao et al. (Eds.): CollaborateCom 2020, LNICST 350, pp. 518–529, 2021.
https://doi.org/10.1007/978-3-030-67540-0_34

1   Introduction

Complex networks are usually denoted and used to represent and analyze a variety of complex systems, such as social systems, biological systems, and ecology systems, etc. Community detection is key to understanding the structure of complex networks, and applications are diverse: from healthcare to regional geography, from human interactions and mobility to economics. Community detection is denoted to divide the nodes of the network into different groups where nodes within a group have more links or similar characteristics. Recent years, many approaches have been proposed to handle the community detection problem, which can be divided into different categories, such as hierarchical clustering algorithms [7], Modularity based optimized approaches, statistical inference [10], etc. A comprehensive and important review can refer to [5].

Based on current research, the real-world complex networks usually demonstrate a variety of features, such as the scale-free property and node attributes [13], which are the key factors that should be considered in the community detection. Many works have shown that making full use of the attribute information of nodes with the topological structure can significantly improve community detection results and it can be used to depict community profiles and semantics [4,16]. However, existing methods only focus on how to improve the community detection results and ignore the maintenance of network structure features. For instance, a generative model for joint identification of network communities and semantics [6] directly used the degree information of nodes, and it did not model the heterogeneity of nodes from the perspective of generation model. Besides node attributes, the power-law degree also an important feature that distributes naturally occurs in real-world networks. Thus how to preserve the heterogeneity of the degree of nodes to model network structures is conducive to community detection. The classical statistical model, such as the stochastic block model (SBM) [10], modeled the cluster structures at a block level and did not take the individuality of nodes into consideration. It treated nodes within a group equally and ignored the heterogeneity. Furthermore, the DC-SBM [1] and the power-law degree SBM [13] were proposed to generate the observed network from the perspective of a single node, i.e. from node-level. However, these methods only focused on the topological structure.

Most previous community detection methods considered only one aspect, and all the above methods ignore combining the scale-free feature and nodes attributes simultaneously for community detection. Actually, considering the heterogeneity of degree and node attributes simultaneously is worth studying in community detection, and faces the following challenges: 1) How to preserve the power-law degree characteristic during modeling the topology, and 2) How to effectively combine the node degree heterogeneity and node attributes to improve community detection results, i.e. jointly encode the varying degree distribution and node multiple attributes to make them complement each other. Only considering one aspect [1,13], modeling network structure or node attributes [16] separately or treating the nodes within a group equally [4] limit the performance for community detection in real-world complex networks.

Keeping the problem in our mind, in this paper, we propose a novel unified and principled Bayesian generated a model to handle the above problem for community detection. When modeling the network structure, we consider the naturally existing scale-free distribution characteristic to reduce the impact of ignoring the property of node degree on the results of community detection. Specifically, we take a global popular parameter for each node to generate the observed network via SBM. We also design a modified topic model for generating the attributes of nodes and semantics of communities and utilize a probabilistic transition matrix to relate network structure and attributes. Thus the model can preserve the power-law degree distributions and fully use node attributes at the same time. We finally propose a variational EM algorithm to optimize the objective function and apply our model on synthetic and real networks respectively. The experimental results show that our model achieves substantial gains compared with existing methods.

2   Related Work

In this section, we study two lines of related work: community detection methods based on attribute networks, and statistical network models especially the works for modeling the heterogeneity of nodes.

Recently, as node attributes have attracted extensive attention in the complex network analysis, some approaches have been proposed combining network topology and node attributes to improve the performance of community detection or community semantics. Ruan et al. [14] proposed a method using content information to determine the strength of the edges between nodes, which can be applied on graph clustering. Yang et al. [16] used a discriminative model combining node attributes and network topology to detect communities. Pool et al. [12] proposed a heuristic method to detect communities by optimizing the community score. Pei et al. [2] developed a model based on the nonnegative matrix trifactorization method to detect communities via modeling network structure and contents. Chen et al. [4] developed a Bayesian nonparametric attribute (BNPA) model to explore structural regularities in networks. This model combined links and node attributes for community detection via shared hidden variables and assumed network structures and node attributes shared the same community member-ships. These methods failed to give the relevant attributes of each community or semantic descriptions of communities. Jin et al. [9] by distinguishing words from either a background topic or some two-level topics (i.e. general and specialized topics) to help finding communities and provided a semantic community interpretation, but they focused more on words in networked contents. Though He et al. [6] introduced a generative model with two parts, one for communities and the other for semantics to explore network structure and explain the functional modules semantically, the method was only for the network of assortative structures and failed to detect generalized community.

Besides node attributes were considered as important factors in community detection, some research also noticed the characteristics of the network. Most of these methods based on the stochastic block model (SBM), which was a significant benchmark method to detect latent communities according to the observed edge in complex networks. SBM base on the assumption that the probability of links between different nodes only depends on the communities they belong to, so some similarity methods based on the cluster structures in a block-level did not consider the individuality of nodes. In other words, they treated nodes within a group equally. With the degree of heterogeneity becoming a problem worth studying in community detection, several models refined to generate the observed network from the perspective of a single node, i.e. from node-level. DC-SBM [1] introduced a Poisson-valued degree parameter for each node to handle the heterogeneity of node degree, thus networks could be split into heterogeneous. The power-law degree SBM [13] explicitly encoded the power-law feature of networks from the perspective of Bernoulli distribution. Besides, Newman et al. [11] developed a mixture model to explore the network structure, in which the nodes with the same link patterns were divided into the same groups. Thus a broad of structural signatures can be explored without any prior assumptions about the structure of the network. Shen et al. [8] focused on identifying the intrinsic structural rules in networks. In this model, the nodes within the same groups had a similar link preference to other groups.

So we combine the attributes and the power-law of the network to propose a Bayesian probability model for communities detection, let node degree heterogeneity and node attributes can be effectively utilized.

3   Unified Bayesian Model

In this section, we introduce the unified Bayesian model of community detection in attribute networks with power-law degree distribution (PLAC). We give a formal description of the proposed model and introduce the notations, preliminaries related to our model, and the generation process.

3.1   Notations and Preliminaries

Given an undirected attribute network G with N nodes and M attributes, the nodes and attributes are denoted as $V = (v_1, v_2...v_N)$ and $W = (\omega_1, \omega_2...\omega_M)$. The network can be represented by adjacency matrix $A^{N \times N}$ and attributes matrix $X^{N \times M}$, where $a_{ij} = 1$ if there is an edge from node v_i to node v_j and 0 otherwise, $X_{it} = 1$ if node i has the t-th attribute and 0 otherwise. The model contains three types of parameters: 1) Observed quantities: including the number of groups or communities K, the number of nodes N, the number of attributes M, the adjacency matrix A, and the attribute matrix X. 2) Latent quantities: including group labels z where z_i denotes the community membership of node v_i, the content memberships g where g_i denotes the topic labels of the v_i, and the degree decay variable δ_i of each node v_i. 3) Model parameters: including $\pi = (\pi_r)_{1 \times K}$, where $\pi_r = p(z_i = r)$ is the probability that node v_i belongs to community r. $B = (b_{rr'})_{K \times K}$, where $b_{rr'} = p(z_i = r, z_j = r')$

is the probability that node in community r links to node in community r'. $\eta = (\eta_{rs})_{K \times K}$, where $\eta_{rs} = p(g_i = s | z_i = r)$ denotes the probability that node v_i is in the s-th topic while it belongs to r-th community. $\theta = (\theta_{st})_{K \times M}$, where $\theta_{st} = p(x_{it} = 1 | g_i = s)$ denotes the s-th topic generates attribute ω_t. We try to infer the latent communities and its topics based on the observed networks.

3.2   The Generation Process

We define a unified Bayesian probabilistic generative model to handle network topology and node attributes simultaneously. To model the network structure, we need the link probability of node pairs. In classical stochastic block model, the probability of node v_i links to node v_j is b_{z_i, z_j}. Considering the fact of power-law feature in real-world networks, we define the degree decay variable δ_i to each v_i, and each δ_i is assigned an exponential prior $\text{Exp}(\lambda)$, i.e. $p(\delta_i | \lambda) = \lambda e^{-\lambda \delta_i}$. In order to capture the heterogeneous, we use $b_{z_i, z_j}^{1+\delta_i+\delta_j}$ instead.

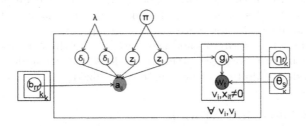

Fig. 1. The graphical representation of PLAC

To jointly model network structure and node attributes, we define the parameter $\eta = (\eta_{rs})_{K \times K}$. In our model, the nodes can be divided into K network communities or K attribute clusters, which called topics in LDA. Meanwhile, attribute topics and network communities may not align. So we use η_r denotes the general topics distribution in community r. When the community label of node v_i is obtained, i.e. $z_i = r$, η_{rs} is the probability that generating the topic of node v_i and $g_i = s$. Thus η provides the transition from communities to topics and correlates two parts closely. When we obtained the topic of node v_i, i.e $g_i = s$, we use θ_s to generate the attributes. Figure 1 is the graphical representation of the model, and the generation process is as follow.

(1) For each new node v_i, $i \in \{1, 2, ..., N\}$:
    (a) Sample a latent group assignment $z_i \sim Multinomial(\pi)$;
    (b) Sample the degree decay variable $\delta_i \sim Exp(\lambda)$;
    (c) Sample the topic assignment $g_i \sim Multinomial(\eta_{z_i})$
(2) For each node-pair $(v_i, v_j) \in \mathcal{N} \times \mathcal{N}$:
    (a) Sample edge $a_{ij} \sim Bernoulli(b_{z_i, z_j}^{1+\delta_i+\delta_j})$;
(3) For each of the t-th attribute ω_t with $x_{it} = 1$:

(a) Sample attribute $\omega_t \sim Multinomial(\theta_{g_i})$.

Then, the probability of the network G is generated by PLAC is:

$$
\begin{aligned}
p(A, & X, z, g, \delta \mid \pi, B, \lambda, \eta, \theta) \\
&= p(A \mid z, B, \delta)p(X \mid g, \theta)p(g \mid z, \eta)p(z \mid \pi)p(\delta \mid \lambda) \\
&= \prod_{a_{ij}=1} b_{z_i,z_j}^{1+\delta_i+\delta_j} \prod_{a_{ij}=0} (1 - b_{z_i,z_j}^{1+\delta_i+\delta_j}) \prod_i (\eta_{z_i g_i}) \prod_{x_{it}=1} (\theta_{g_i t}) \prod_i \lambda e^{-\lambda \delta_i} \pi_{z_i}
\end{aligned}
\tag{1}
$$

subject to $\sum_{r=1}^K \pi_r = 1$, $\sum_{s=1}^K \eta_{rs} = 1$, and $\sum_{t=1}^M \theta_{st} = 1$.

4   Optimization Algorithm

In this section, we use the variational EM algorithm and mean-field method [17] to optimize our model PLAC.

4.1   Variational E-Step

Calculating the observation likelihood $p(A, X \mid \pi, B, \lambda, \eta, \theta)$ is usually analytically intractable because of the integral over all latent variables. So there is no closed-form of joint posterior $p(z, g, \delta \mid A, X, \pi, B, \lambda, \eta, \theta)$. In the variational algorithm, we use a certain distribution family to approximate the real posterior. In our model and set:

$$
q(z, g, \delta) = \prod_i q(z_i)q(g_i)q(\delta_i)
\tag{2}
$$

We assume the variables are independent with each other. And we set

$$
q(z_i) = Multi(\phi_i), q(g_i) = Multi(\varphi_i), q(\delta_i) = 1(\bar{\delta}_i)
\tag{3}
$$

$q(z_i)$ and $q(g_i)$ are multinomial distributions respectively parameterized with ϕ_i and φ_i. $q(\delta_i)$ is a degenerated distribution with probability 1 at point $\bar{\delta}_i$.

The optimal result can be attained by maximizing the marginal likelihood of observation. It is obtained via minimizing the KL divergence between $q(z, g, \delta)$ and $p(z, g, \delta \mid A, X, \pi, B, \lambda, \eta, \theta)$. We equivalently transform minimization to the maximization of evidence lower bound (ELBO) of the marginal likelihood by using Jensen's inequality and Taylor approximation. The ELBO is:

$$
\begin{aligned}
\log p(A, & X, \mid \pi, B, \lambda, \eta, \theta) \geq \\
\ell(\phi, \varphi, \delta) &= \mathbb{E}_q \log p(A, X, z, g, \delta \mid \pi, B, \lambda, \eta, \theta) - \mathbb{E}_q \log q(z, g, \delta) \\
&\approx \sum_{a_{ij}=1} (1 + \bar{\delta}_i + \bar{\delta}_j) \sum_r \sum_{r'} \phi_{ir}\phi_{jr'} \log b_{rr'} - \sum_{a_{ij}=0} \sum_r \sum_{r'} \phi_{ir}\phi_{jr'} b_{rr'}^{1+\bar{\delta}_i+\bar{\delta}_j} \\
&+ \sum_i \sum_r \sum_s \phi_{ir}\varphi_{is} \log \eta_{rs} + \sum_{x_{it}=1} \sum_s \varphi_{is} \log \theta_{st} + \sum_i \sum_r \phi_{ir} \log \pi_r \\
&- \sum_i \sum_r \phi_{ir} \log \phi_{ir} - \sum_i \sum_s \varphi_{is} \log \varphi_{is} + N \log \lambda - \lambda \sum_i \bar{\delta}_i
\end{aligned}
\tag{4}
$$

Verifiably, Eq. (4) only concave to each variable ϕ, φ and $\bar{\delta}_i$, and not concave with $(\phi, \varphi, \bar{\delta}_i)$ simultaneously. So to optimize these variables, we use coordinate gradient ascent. The gradient of $\bar{\delta}_i$ is:

$$
\begin{aligned}
\frac{\partial \mathcal{O}(\bar{\delta}_i)}{\partial \bar{\delta}_i} = & \sum_r \phi_{ir} \sum_{r'} \log b_{rr'} \sum_{a_{ij}=1} \phi_{jr'} \\
& - \sum_r \phi_{ir} \sum_{a_{ij}=0} \sum_{r'} \phi_{jr'} b_{rr'}^{1+\bar{\delta}_i+\bar{\delta}_j} \ln(b_{rr'}) - \lambda
\end{aligned}
\tag{5}
$$

and the updating of ϕ_i and φ_i is given by:

$$
\begin{aligned}
\phi_{ir} \propto \pi_r \exp\Big(& \sum_{a_{ij}=1} \sum_{r'} (1 + \bar{\delta}_i + \bar{\delta}_j)\phi_{jr'} \log b_{rr'} \\
& - \sum_{a_{ij}=0} \sum_{r'} \phi_{jr'} b_{rr'}^{1+\bar{\delta}_i+\bar{\delta}_j} + \sum_s \varphi_{is} \log \eta_{rs} \Big)
\end{aligned}
\tag{6}
$$

$$
\varphi_{is} \propto \prod_{x_{it}=1} \theta_{st} \exp\Big(\sum_r \phi_{ir} \log \eta_{rs} \Big)
\tag{7}
$$

4.2   M-Step

In this part, we optimize the model parameters $(\pi, \lambda, B, \theta, \eta)$. We fix $\lambda = 0.01$ and update π, B, θ and η. The gradient of B is:

$$
\frac{\partial \mathcal{O}(b_{rr'})}{\partial b_{rr'}} = \frac{\sum\limits_{a_{ij}=1} (1 + \bar{\delta}_i + \bar{\delta}_j)\phi_{ir}\phi_{jr'}}{b_{rr'}} - \sum_{a_{ij}=0} (1 + \bar{\delta}_i + \bar{\delta}_j)\phi_{ir}\phi_{jr'} b_{rr'}^{\bar{\delta}_i+\bar{\delta}_j}
\tag{8}
$$

So the optimal π, η, θ are given by:

$$
\pi_k \propto \sum_i \phi_{ir}; \qquad \eta_{rs} \propto \sum_i \phi_{ir}\varphi_{is}; \qquad \theta_{st} \propto \sum_{x_{it}=1} \varphi_{is}
\tag{9}
$$

The predicted community label of each node v_i is obtained by maximizing the variational posterior distribution:

$$
z^* = \arg \min_r \phi_{ir}
\tag{10}
$$

where ϕ_{ir} is the optimal posterior parameter from Eq. (6). The implementation of the above algorithm is shown in Algorithm 1.

5   Experiments and Analysis

We evaluate the performance of PLAC on synthetic and real networks and compared with three other methods. And we use the Normalized Mutual Information (NMI) [3] as the evaluation index.

Algorithm 1. Inference for PLAC

Input: Initialize model parameters λ, B, π, η, θ and parameters ϕ, φ, $\bar{\delta}$; the community number K; stop criterion ε.
Output: Model parameters $\bar{\delta}^*$, π^*, B^*, η^*, θ^* and the community label z^*.

  1: **repeat**
  2:     Compute variational likelihood ℓ^{new} by Eq. (4).
  3:     $\ell^{old} = \ell^{new}$
  4:     **variational E-step**
  5:     update $\bar{\delta}$, ϕ, φ respectively by Eq. (5), Eq. (6), Eq. (7).
  6:     **M-step**
  7:     updte B, π, η, θ respectively by Eq. (8), Eq. (9).
  8: **until** $|\ell^{new} - \ell^{old}| \leq \varepsilon$
  9: predict the community index z_i of each node by Eq. (10).

5.1   Experiment on Synthetic Networks

The synthetic network is a random network in Newman's model [6]. The network contains 128 nodes divided into 4 disjoint communities with $z_{in} + z_{out} = 16$. As $\rho(= z_{in}/32) > \rho(= z_{out}/96)$, z_{in} (the internal edges) is greater than z_{out} (the external edges). For each v_i, a $4h$-dimensional binary attributes (i.e., x_i) are created to divide 4 attribute clusters in network. h_{in} denotes the number of attributes for every node v_i with $x_{it} = 1$ related to its community and h_{out} (noisy attributes) inversely denotes the number of attributes related to other communities, and $h_{in} + h_{out} = 16$.

We set $h = 50$, network topologies and node attributes share the same membership. Node attributes matrix and the community attributes matrix ($z_{out} = 3$) are shown in the Fig. 2. Firstly, We set $z_{out} = 8$, and h_{out} is changed from 0 to 9 with an increment 1. The larger h_{out} is, the vaguer the structure of node attributes will be.

We adapt PLD-SBM as the baseline method. Other comparison methods are NEMBP [6] and SCI [15]. As shown in Fig. 2(c), our model can fairly well exploit the node attributes information to improve the quality of community detection compare with other models. Then we set $h_{out} = 8$, and change z_{out} from 0 to 9 (the community structure is gradually blurred) with an increment 1 to study the ability to use network structures information. Our method also performs better than NEMBP and SCI. On the whole, Information of attributes and structures can be better used in our model. Figure 3 is the visualization of community detection results. And the results obviously are better compared with other methods.

5.2   Experiment on Real Networks

Five real networks we utilize are shown in Table 1. Texas, Cornell, Washington, and Wisconsin are four sub-networks of the WebKB network. Twitter is the largest sub-network (id629863) in Twitter data.

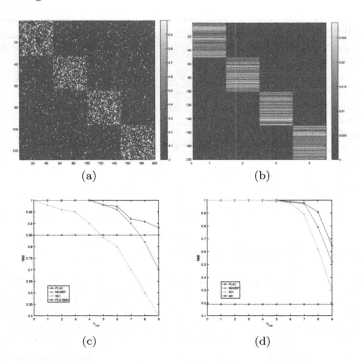

Fig. 2. (a) node attributes matrix. (b) community attributes matrix. The value of NMI of three methods on random networks with the change of (c) h_{out} from 0 to 9 and (d) z_{out} from 0 to 9.

We compare our model PLAC with the methods of three categories: 1) models based on network structures: DCSBM, PLD-SBM. 2) models based on nodes attributes: LDA. 3) models based on both structures and attributes: PCL-DC, NMMA, SCI, NEMBP. Results are shown in Table 2. Our model PLAC outperforms other models on Cornell, Washington and Wisconsin, achieves larger NMIs than most of the models on Texas and Twitter. It is mainly because that the attributes of each community are clear in Texas and the degree of nodes in Twitter does not obey the power-law distribution. Usually, the models based on

Table 1. Statistical characteristics of five real networks.

Datasets	N (nodes)	E (edges)	M (attributes)	K (communities)
Texas	187	328	1703	5
Cornell	195	304	1703	5
Washington	230	446	1703	5
Wisconsin	265	530	1703	5
Twitter	171	796	578	7

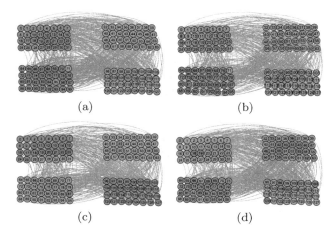

(a)   (b)

(c)   (d)

Fig. 3. The four blocks in each diagram indicate the true communities in network. Colors stand for predicted communities by PLAC and NEMBP. Nodes of same colors belong to same communities. The results of community detection: (a) PLAC, $z_{out} = 8$, $h_{out} = 9$. (b) NEMBP, $z_{out} = 8$, $h_{out} = 9$ (c) PLAC, $h_{out} = 8$, $z_{out} = 9$. (d) NEMBP, $h_{out} = 8$, $z_{out} = 9$. (Color figure online)

Table 2. NMI (%) of different models on five real networks

Models	NMI				
	Data				
	Texas	Cornell	Washington	Wisconsin	Twitter
DC-SBM	16.65	9.696	9.87	3.14	57.48
PLD-SBM	25.21	13.92	12.47	17.50	42.20
LDA	34.29	21.09	38.48	46.56	1.67
PCL-DC	10.37	7.23	5.66	5.01	52.64
NMMA	**41.57**	39.95	25.59	4.696	1.67
SCI	17.84	11.44	12.37	17.03	43.00
NEMBP	35.12	18.71	21.24	38.02	**59.73**
PLAC	34.29	**70.16**	**39.17**	**73.40**	44.02

structure and attributes outperform the models with only links or attributes. The results are further improved when considering the power-law degree distributions of each node in networks. Such as NEMBP uses the heterogeneity of node degree to model the network structures, which is associated with DC-SBM. But its optimized equation is dependent on cluster-level degrees rather than node-level rather than encodes degree distribution over each node like our model PLAC. So our method performs better than NEMBP on datasets that have the power-law feature.

5.3   A Case Study

Furthermore, we present a demonstration of community semantic analysis based on the widely used data set on community detection in attribute networks, LASTFM [6] from an online music system Last.fm, which includes $1,892$ users and $11,946$ attributes, and we set the number of communities is 38 as did in [6]. As represented in Fig. 4, we show the two examples of community interpretation. The first is a community of fans who may like heavy metal rock music, and the users in the second community prefer a more relaxed and relaxing style of pure music. We give that based on the parameters η and θ.

(a)   (b)

Fig. 4. Examples on community semantic in two different music domain based on the LASTFM data, the word sizes are proportional to the probability they belong to the corresponding topic: (a) topic 2 on 7th community. (b) topic 7 on 21th community

6   Conclusions

In this paper, we proposed a novel Bayesian probability model to detect communities in attribute networks with power-law degree distributions and token a variational EM algorithm to optimize the objective function of the model. We presented a demonstration of community interpretation by applying our model on a specific scene. It is similar to the existing methods that the number of communities needs to be given in advance, and we will focus on how to automatically determine the number of communities for this model in the future.

Acknowledgments. This work was supported by the National Natural Science Foundation of China (61902278), the National Key R&D Program of China (2018YFC0832101), the Tianjin Science and Technology Development Strategic Research Project under Grant (18ZXAQSF00110) and National Social Science Foundation of China (15BGL035).

References

1. Brian, K., Newman, M.E.J.: Stochastic blockmodels and community structure in networks. Phys. Rev. E Stat. Nonlinear Soft Matter Phys. **83**(2), 016107 (2011)

2. Chakraborty, Y.P.N., Sycara, K.: Nonnegative matrix tri-factorization with graph regularization for community detection in social networks. In: Twenty-Fourth International Joint Conference on Artificial Intelligence, pp. 2083–2089. AAAI Press (2015)
3. Chang, Z., Yin, X., Jia, C., Wang, X.: Mixture models with entropy regularization for community detection in networks. Phys. A: Stat. Mech. Appl. **496**, 339–350 (2018)
4. Chen, Y., Wang, X., Bu, J., Tang, B., Xiang, X.: Network structure exploration in networks with node attributes. Phys. A: Stat. Mech. Appl. **449**, 240–253 (2016)
5. Fortunato, S., Hric, D.: Community detection in networks: a user guide. Phys. Rep. **659**, 1–44 (2016)
6. He, D., Feng, Z., Jin, D., Wang, X., Zhang, W.: Joint identification of network communities and semantics via integrative modeling of network topologies and node contents. In: Thirty-First AAAI Conference on Artificial Intelligence (2017)
7. He, D., Yang, X., Feng, Z., Chen, S., Fogelman-Soulié, F.: A network embedding-enhanced approach for generalized community detection. In: Liu, W., Giunchiglia, F., Yang, B. (eds.) KSEM 2018. LNCS (LNAI), vol. 11062, pp. 383–395. Springer, Cham (2018). https://doi.org/10.1007/978-3-319-99247-1_34
8. Hua-Wei, S., Xue-Qi, C., Jia-Feng, G.: Exploring the structural regularities in networks. Phys. Rev. E Stat. Nonlinear Soft Matter Phys. **84**(2), 056111 (2011)
9. Jin, D., Wang, K., Zhang, G., Jiao, P., He, D., Fogelman-Soulie, F., Huang, X.: Detecting communities with multiplex semantics by distinguishing background, general and specialized topics. IEEE Trans. Knowl. Data Eng. **32**, 2144–2158 (2019)
10. Karrer, B., Newman, M.E.: Stochastic blockmodels and community structure in networks. Phys. Rev. E **83**(1), 016107 (2011)
11. Newman, M.E.J., Leicht, E.A.: Mixture models and exploratory analysis in networks. Proc. Natl. Acad. Sci. USA **104**(23), 9564–9569 (2007)
12. Pool, S., Bonchi, F., Leeuwen, M.V.: Description-driven community detection. ACM Trans. Intell. Syst. Technol. (TIST) **5**(2), 28 (2014)
13. Qiao, M., Yu, J., Bian, W., Li, Q., Tao, D.: Adapting stochastic block models to power-law degree distributions. IEEE Trans. Cybern. **49**(2), 626–637 (2019)
14. Ruan, Y., Fuhry, D., Parthasarathy, S.: Efficient community detection in large networks using content and links. In: International Conference on World Wide Web, pp. 1089–1098 (2013)
15. Wang, X., Jin, D., Cao, X., Yang, L., Zhang, W.: Semantic community identification in large attribute networks. In: Thirty-Second AAAI Conference on Artificial Intelligence, pp. 265–271 (2016)
16. Yang, T., Jin, R., Chi, Y., Zhu, S.: Combining link and content for community detection. In: ACM SIGKDD International Conference on Knowledge Discovery and Data Mining, pp. 927–936 (2009)
17. Yu, Z., Zhu, X., Wong, H.S., You, J., Zhang, J., Han, G.: Distribution-based cluster structure selection. IEEE Trans. Cybern. **47**(11), 3554–3567 (2017)

Workshop Track

A New Collaborative Scheduling Mechanism Based on Grading Mapping for Resource Balance in Distributed Object Cloud Storage System

Yu Lu[1], Ningjiang Chen[1,2(✉)], Wenjuan Pu[1], and Ruifeng Wang[1]

[1] School of Computer and Electronic Information,
Guangxi University, Nanning 530004, China
chnjgxu@edu.cn
[2] Guangxi Key Laboratory of Multimedia Communications and Network Technology,
Nanning 530004, China

Abstract. An algorithmic mapping of storage locations brings high storage efficiency to the storage system, but the loss of efficient scheduling makes systems prone to crashing at low usage. This paper uses the Ceph storage system as a research sample to analyze these issues and proposes a grading mapping adaptive storage resource collaborative optimization mechanism. This approach grading both the storage device and the storage content, and introduced random factors and influence factors as two-factors to quantify the grading mapping relationship between the two of them. This relation coordinates the storage systems' performance and reliability. In addition, a collaborative storage algorithm is proposed to realize balanced storage efficiency and control data migration. The experimental results show that in comparison with the inherent mechanism in the traditional Ceph system, the proposed cooperative storage adaptation mechanism for data balancing has increased the average system usage by 17% and reduces data migration by 50% compared to the traditional research approach.

Keywords: Cloud storage · Storage balance · Grading mapping · Ceph · Collaborative scheduling

1 Introduction

In the internet-based collaborative computing [1] mode, there are similarities between the operation and management of collaborative applications and the scheduling of nodes in distributed systems [2]. The performance of applications is affected by the construction and management of system infrastructure. In cloud storage, the storage mapping of data is related to the performance and the stability of the storage system. Distributed object storage systems with hash

© ICST Institute for Computer Sciences, Social Informatics and Telecommunications Engineering 2021
Published by Springer Nature Switzerland AG 2021. All Rights Reserved
H. Gao et al. (Eds.): CollaborateCom 2020, LNICST 350, pp. 533–549, 2021.
https://doi.org/10.1007/978-3-030-67540-0_35

algorithm mapping storage locations are widely used, have shown good performance. However, the lack of a central scheduling service makes the mapping of data to storage devices uncontrollable, so for balanced storage of data, collaborative scheduling between devices is required spontaneously.

Ceph [3] is a representative object-based storage system [4] (OBSS), which hasn't a dedicated metadata server [5] (MDS) to record the object-based storage device (OSD) locations of segmented objects. The OSD location uses a specific mapping algorithm CRUSH [6] to determine the locations of the objects and the replicas. When the data are searched or modified again, the mapping process can be done independently on each client. When the device is replaced or a new one is added, it is necessary for the storage system to adaptively calculate the storage location of the object to realize the recovery and balance, so easily scaling storage requirements for a huge scale of data.

In today's medium and large data centers, this feature reduces the data risks and lowers operating costs when replacing equipment. But the mapping algorithm [7] can't sense the state of the device where the data will be written, and if the device is overloaded, the data writes are denied, at which point the upper-level systems with Ceph as the base storage (such as Openstack [8], which provides block storage [15] primarily for applications such as its virtual machines) will crash.

In view of the above problems, existing researches mainly optimize CRUSH or adopt frequent migration to balance the system [9]. These are either too conservative and have a poor balance effect or are too aggressive and cause more performance problems. In view of these deficiencies, this paper analyzes the key factors of such problems using Ceph as the research platform, which differs from existing researches by using the collaborative work between storage nodes to control the process of mapping and migration of storage data. A graded mapping adaptation mechanism is proposed to make the data relevant to the device. This paper first devises influencing and random factors to capture the relationship between PGs(Placement Group) and OSDs(Object Storage Device), as well as a collaborative algorithm to enable a single OSD to use these factors for collaborative scheduling with other OSDs. The classification of OSD and PG also increases computational efficiency, manages migration direction, and reduces the operational difficulty. At the same time, the stability of the system can be greatly improved through effective migration.

The structure of this paper is as follows. In Sect. 2, we will introduce the Ceph storage process, analyze the problems and the factors that have been missed by existing researches, and then a new mechanism is proposed in Sect. 3. The experiments to verify the proposed mechanism are given in Sect. 4, and Sect. 5 conclude the paper.

2   Problem Analysis and Motivation

In the basic framework of Ceph, the Rados is responsible for the mapping process between the file to the storage device OSD [10]. This process has three basic steps: from file to object, from object to PG, and from PG to OSD. The relationship is shown in Fig. 1.

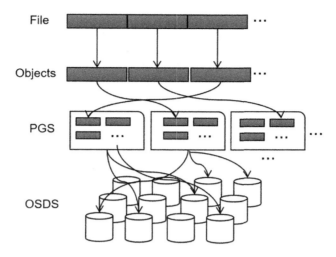

Fig. 1. Ceph storage mapping process

Step 1. $(ino, ono) \rightarrow oid$: This step splits the files. ino is the id of the file, ono is the unique identifier of the file slices, oid is the unique id of the object. Almost all object storage systems and slicing storage systems split files and store them in a similar mechanism.

Step 2. $hash(oid)\&mask \rightarrow pgid$: This step computes the PGs for the object. The object's unique identifier is used as the seed. In Ceph, this identifier is usually composed of the file name and object identifier. Then, a specific hash algorithm is used to generate the random number, and then the number of PGs is used to make the remainder operation. This allows objects to be randomly distributed to individual PGs. The number of PGs is confirmed when the system cluster is initialized. The conventional method is to use the number of redundancy as the reference when the number of PGs is initialized [11], as calculated by Eq. (1):

$$TotalPGs = (Number\_of\_OSD * 100)/replication \qquad (1)$$

Object mapping to PG is a completely random process in both research and production environments. Although the remainder operation is approximately evenly distributed, the difference between PGs is actually not as small as imagined as the randomness varies from the average value. We conducted an experimental analysis to explore this problem. This experiment uses the Ceph source code as the object of analysis to calculate the object distribution of 38TB of file data in 6,666 PGs(300 OSD according to Eq. 1 and the number of PGs in the 2 replica [12] mode). The file size is simulated as 1 GB. The OID is composed of a random file name and a numeric number, which is consistent with the system numbering. Figure 2 shows the PG number distribution of the theoretical deviation between the actual number of objects and the number of objects in the PG for the simulation environment, the horizontal coordinate is the difference between the size of a single PG and the average size of all PGs, and the vertical coordinate is the number of PGs. From the figure, the largest PG and the smallest PG differ by approximately 200 objects (4 MB for a single object). PG is the minimum storage unit considered in the Ceph system. Therefore, neither balancing efficiency nor data migration control can be improved if PG differences are not focused on in the research of balanced storage.

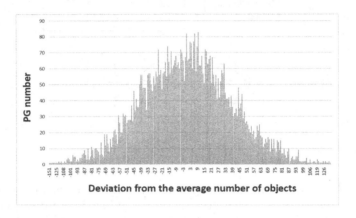

Fig. 2. Object deviation statistics in PG

Step 3. $crush(pgid) \rightarrow (osd1, osd2)$: This step is the OSD selection. It is implemented by using Ceph's default algorithm CRUSH which is a pseudorandom reproducible process. At present, most of researches work about storage balance focus on optimizing this step, including two typical methods, the first type is based on the size of the OSD capacity so that the average distribution of PG on each OSD, the other type is to dynamically adjust the number of PG on the OSD in the system runtime. Both approaches consider the PG to be uniform in size, the difference is that the former is based on the equalization of PG [13] the latter is based on the use of the classical MDS storage system [14] for equalization. However, as can be seen from the second step above, PGs do not have the same size, so these two methods can only perform relatively balanced work to some extent, without further considering the difference of PG. The size of the PG changes dynamically in real-time and is random.

Based on these conclusions, we know that distribution of average PG on OSD does not mean an average distribution of data and that the central scheduling of PG loses the advantage of mapping storage locations by algorithms. However, the mapping and parameter changes lead to data migration, which results in further performance problems and stability risks. Because the PG size difference is not considered, the existing researches are limited in their ability to solve the problem of storage balance; they incur performance costs. Therefore, this paper proposes a collaborative scheduling mechanism based on grading mapping, which gives the PG the ability to migrate autonomously. The balanced tasks can be coordinated among a number of OSDs, it can also provide accurate data migration solutions for PGs, ensuring that the amount of migrated data is controllable. Secondly, a model is established to evaluate the relationship between the PG and the OSD to avoid the scenarios in which a data migration triggers a cascade of further data migration. The main grading model guides the migration direction and conforms to the changes in the cluster's high performance and reliability for each period of the cluster's overall use. As a result, the proposed approach ensures high performance and high reliability of the system.

3   Collaborative Scheduling Mechanism Based on Grading Mapping

For the problems in Ceph analyzed in the previous section, the situation is that when an OSD node usage exceeds the average usage rate, the balancing mechanism can be triggered by the node itself, and the PG that needs to be migrated is evaluated according to its own size, usage rate, and the average cluster usage rate. Each PG that needs to be migrated can get the feedback from all the new OSD sets and find the best set to migrate, as shown in the figures below.

For the classic Ceph system, there are unresolved issues at each step:

(a) PG does not have the ability to migrate independently, even if a PG is selected, it cannot be migrated.
(b) Even with the ability to move, a random choice without direction may never find a choice that can be migrated.
(c) There is no evaluation standard that reflects whether the OSD combination can accept the PG.

To address the above problems, this paper proposes a novel solution that is introduced below.

3.1   Random Factor and Impact Factor

In the traditional Ceph storage system, the PG does not have the ability to redirect. The Reweight artificial modification mechanism is an external enhancement available in the BlueStore [16]. Therefore, in order to solve the balanced storage problem, the PG redirection capability is first given, so this capability is contained within the PG. This is accomplished by adding a random factor to the parameters. The addition of random factors interferes with the CRUSH selection results, and the changes in the incoming parameters allow the pseudo-random process to have more solutions. The principle of the random factor r^i is as follows:

$$R^i < OSDs >= CRUSH(pgid, r^i) \tag{2}$$

Where pgid is the unique identifier of the PG, r^i is the random factor, i is the number of selections, and R^i is the OSD combination selected by the r^i (the

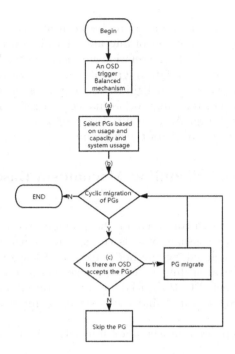

Fig. 3. The process of achieving storage balance in the Ceph system

master node and multiple copy storage nodes). For r^i, it can be randomly generated as a seed according to parameters such as time or memory space according
to the underlying implementation of each system. Because of the randomness of
CRUSH, the resulting set R^i also changes when r^i changes. The purpose of the
balanced storage that this paper focuses on is to make the usage rate of each
OSD of the whole system as close as possible to the average usage rate. If the
average usage rate of a system is assumed to be M, the index equalization rate
of a balanced storage condition of a system is evaluated. β can be calculated as
follows:

$$\beta = \frac{\sum_{j=1}^{n}(x_j - M)^2}{n} \tag{3}$$

where x_j is the usage rate of a single node in the cluster. The equalization rate
reflects the variance of cluster usage; it is a reflection of the difference in usage
of each node in the cluster. For the impact of a PG redirection on the whole
system, the calculation of the impact factor can be compared with the β before
the PG migration and the β after the PG migration. The system state after
the PG migration can also be judged, so the impact factor θ can be defined as
follows:

$$\theta = \begin{cases} -1 & x_r > 1 \quad or \quad x_r - M > \alpha \\ \dfrac{\sum_{j=1}^{n}(x_r - M)^2}{\sum_{j=1}^{n}(x_j - M)^2} & else \end{cases} \tag{4}$$

Here r is an OSD in a selection result set R, x_r is the usage rate of a single node in the migrated cluster, α is a set threshold. If the usage rate exceeds the threshold value compared with the average usage rate, the equalization is triggered. If the usage rate of an OSD in a group of OSDs exceeds 1 after the PG migration or the usage rate exceeds the average usage rate, the impact factor of this group is -1. This judgment prevents two undesirable outcomes. The first is the migration making other OSDs unavailable. The second is that the migration will not make the new node unavailable, but it will trigger another migration of the new OSD.

3.2   Grading Mapping Mechanism

The definition of the random factor gives the PG redirection ability. However, if there is no corresponding control and optimization method, when a PG needs to be migrated only the unknown random calculation can be performed and the most suitable combination can be found. This process is uncontrollable, and the calculation of the impact factor can also result in a substantial computational performance cost. Therefore, on the basis of random factors, a novel logical division OSD method is proposed, and the random number of random factors is used as the grading division method to ensure that the migration of PGs has direction and level. The advantage of OSD grading is that each pool can customize the rules for PG migration in and out. Because PG migration determines the amount of data migration, it is also possible to indirectly limit data migration through OSD grading. Further, we propose a grading approach for PGs, which combines PG grading with OSD grading. This approach converts the selection process of objects from PG→OSD to PG→grading→OSD. The grading of PG is matched with the grading of OSD. The OSDs grading is based on the number of selections of the random factors. Therefore, the PG level change must start from the lowest level of the OSD; it can be determined according to the random factor and the impact factor whether it can be migrated to a group of OSDs in this grading pool. If the low-level grading pool does not accept the PG according to the policy, the upgrade selects a higher-level OSD grading pool. The OSD grading pool is shown in Fig. 4.

In this way, when the average cluster usage is low, the PG can perform the equalization with less calculation. When the average cluster usage is high, more choices can be made. High performance is ensured at low system usage and high reliability at high usage rates. According to the above design concepts, the basic process of the OSD and PG dual grading strategy is as follows. The first step is the initialization phase, which consists of two parts: one is to pre-statically set the initial level of the OSD; the other is to initialize the PG level. Compared with the traditional random selection strategy of the CRUSH algorithm, the OSD

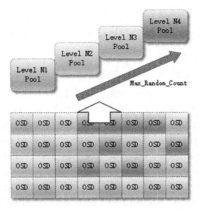

Fig. 4. Grading pool diagram

classification mechanism constrains the selectable range of PGs to one level. Considering the real-time changes of the PG size, it should be distributed in grading pools as evenly as possible in the PG level initialization. Therefore, this paper constructs a uniform hash ring according to the total size of each grading pool. The uniform hash is derived from the consistency hash. The maximum value of the hash ring is an integer, and the length of every hash segment is the same. The number of nodes on the hash ring is the same as the number of pools, and PG determines the grade by taking the hash value by pgid and then taking the surplus in relation to the number of pools. The principle is shown in Fig. 5.

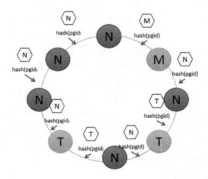

Fig. 5. PG initialization level with hashring

The example shown in the Figures is a grading method of three levels: M, N, and T. The size ratio of each logical pool is M:N:T = 1:5:2, so 8 nodes can be constructed according to the ratio. The entire hash ring is divided into 8 equal segments, 8 nodes are randomly scattered in 8 positions. Then, each PG performs a remainder operation on the ring maximum value according to the hash value calculated by its pgid. The remaining value falls into a point on the ring. Using a clockwise search, the first node level encountered is the PG level. The initialization algorithm of the PG is as follows Algorithm 1:

Algorithm 1. PG Initialization algorithm

Require: ;
    The set of all $OSDs\_Levels\_size$;
    The set of all PGs; Integer MAX_RING
  1: integer nodes_count;
  2: integer[level][ratio] level_array;
  3: level_array=Get_the_minimum_ratio($OSDs\_Levels$);
  4: **for** each $level \in level\_array$ **do**
  5:    nodes_count+=level[ratio];
  6: **end for**
  7: integer[nodes_count] nodes_array;
  8: **for** each $level \in level\_array$ **do**
  9:    **for** $i = 0$ to $level[ratio]$ **do**
10:      i=randomInt(nodes_count);
11:      **while** nodes_array[i] **do**
12:        i++;
13:      **end while**
14:      nodes_array[i]=level[level]
15:    **end for**
16: **end for**
17: initialize a hash ring by levels;
18: MAX_RING=MAX_RING÷nodes_count;
19: **for** each $pg \in all\,PGs$ **do**
20:    interval=(integer)hash(pgid)÷(MAX_RING)
21:    pg.level=nodes_array[interval] ;
22: **end for**

The second step is the collaborative balancing process, which is mainly composed of a PG selection, a PG redirection calculation, and a PG redirection to complete the migration. An example of the working process of this mechanism is shown in Fig. 6.

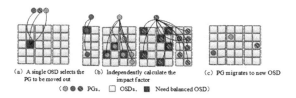

(a) A single OSD selects the PG to be moved out   (b) Independently calculate the impact factor   (c) PG migrates to new OSD

(◎ ● ● PGs.   □ OSDs.   ■ Need balanced OSD)

Fig. 6. PGs migration process

In the figure, the classification of the cluster is divided into three levels: red, blue, and green. The maximum random number is set to 2, 4, and 5. In (a), one node usage rate is higher than the average usage α, the mechanism is triggered in the OSD, and n PGs in the OSD are selected according to the mechanism.

(a) Select 3 PGs for migration according to the rules.
(b) Starts with the red PG, because the maximum random number of the first level is 2, so the red PG performs two random calculations to obtain 2 random factors. At the same time, two sets of OSD combinations in the first stage are obtained. The group OSD sends a migration request, assuming a reply of $[-1, a]$ (a!= −1). A returned value of −1 indicates that if the red PG migrates to the combination, a new node migration is triggered or the node is unavailable. A returned value of non-1 indicates that the combination accepted the migration request and the red PG completed the migration. Similarly, the green PG obtains $[-1, -1]$ in the combination of the two random factors of the first level, and the migration cannot be completed at the first level. As the result, the level rises to the second level. The maximum random number of the second level is 4, so 4 sets of OSD combinations are obtained with 4 random factors. Assuming that the four return values are $[-1, a1, a2, a3]$ ($-1<a1<a2<a3$), according to the calculation formula of the influence factor, we can know that the combined balance degree of a1 is greater than a3, and the migration of the smallest one of the three non-1s is taken. Finally, the blue PG also follows this approach.
(c) Set the new level and the new random factor to the new parameters of the three PGs, and the PG starts to migrate. After the migration, all of the node usage rates are less than the threshold α compared with the average usage rate, and no single point usage rate is prominent; therefore, the balance is achieved.

The algorithm is described as Algorithm 2:

Algorithm 1 is to initialize the parameters, the number of executions of this algorithmic process is only related to the number of PG, so the time complexity is $O(n)$. Algorithm 2 is a balanced storage algorithm, the number of executions of which is a pre-set parameter, and the time complexity of the algorithm is $O(n)$ too.

The mechanism uses two factors to plan and constrain the balancing, avoiding the uncertainty that occurs in collaborative balancing. Based on the grading mechanism, each migration of data is a joint decision of multiple nodes. The management model of a two-way migration strategy avoids the requirement to balance more nodes that are triggered by data migration. Since the balancing mechanism is decided first between nodes and then data is migrated, each migration is an efficient migration, reducing IO performance losses. The key performance of the mechanism lies in the calculation. When the cluster has a low usage rate, the balance can be completed in the low-level grading pool. In this case, the number of operations is small, and the performance is guaranteed.

Algorithm 2. Adaptive grading data balanced mechanism algorithm

---

Require: ;
   OSD;
   Average usage of this level M;
 1: PG[] pgs,i=0;
 2: List *Levels;
 3: balance=abs((M-OSD.ratio))*OSD.size;
 4: allPGs=sort_by_size_asc(OSD.PGs);
 5: **while** balance>0 **do**
 6:    pgs.push(allPGs[i]);
 7:    i++;
 8:    balance-=allPGs[i];
 9: **end while**
10: **for** each $pg \in pgs$ **do**
11:    map(r,θ) result;
12:    **while** Levels→hasNext() **do**
13:      **for** $i = 0$ to $Levels.max\_random\_count$ **do**
14:        r=Random();
15:        OSDs=CRUSH(pg,r);
16:        **for** each $osd \in OSDs$ **do**
17:          result.add(r,result.get(r)+θ)
18:        **end for**
19:      **end for**
20:      sort_by_θ_asc(result);
21:      **repeat**
22:        **if** *result.θ>0 **then**
23:          pg.level=Level.level;
24:          pg.r=*result.r;
25:        **end if**
26:      **until** !*result.hasNext()
27:    **end while**
28: **end for**

When the cluster has a high usage rate, it needs to consider higher levels. Finding the right combination decreases the performance, but ensures the validity of the calculation results, thus improving the reliability of the system.

4 Experiments and Evaluation

Experiments are conducted to compare the performance of the mechanism proposed in this paper, the Ceph system, and related research [9,12] examples in terms of data balancing and migration control by means of simulation. For the simulation system, this paper uses the CRUSH in the Ceph source code to re-implement a simulation system for writing and recording Ceph data based on the BS architecture. The architecture of the system is shown in Fig. 7.

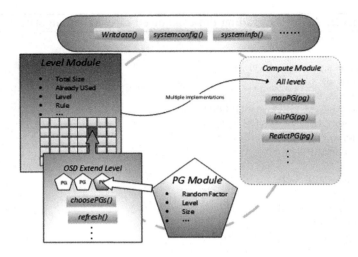

Fig. 7. Architecture of simulation system

The implementation of the simulation system is mainly divided into five modules. The uppermost layer is the interface module, which is used to configure the system, write data, obtain storage information, etc. Level Module is a grading pool class, which defines the capacity, used capacity, level, etc. OSD Module extends from Level Module, so OSD has the parent class's properties and independent PG operation methods. All calculations are performed in a single module that is completely decoupled to ensure the implementation of the algorithm for each control group.

4.1 Evaluation of the Effect of Storage Balance

The experimental comparison object of the balanced storage is mainly the research work of uniform PG and the traditional Ceph storage system. The experiments make the PG uniformly distributed across OSDs to simulate the native Ceph balancing method. Three control simulation environments are configured respectively: Normal, Partially small (single OSD capacity is half of the size), and Partially large (half the OSD size is 2 times the size). The average overall usage when testing the maximum writes R for the system is defined as:

$$R = max\_input/total\_size \tag{5}$$

The variance of the OSD usage is defined as:

$$V = \frac{\sum_{j=1}^{n}(u_j - R)^2}{n} \tag{6}$$

Where u_j is the single node usage rate.

Because it is difficult to use 100% of the algorithm map storage locations, the following approach is used as a reasonable alternative. When the usage rate of a single-point OSD reaches 100%, the system first returns the usage rate.

Experiments will be performed many times with different level distributions, different OSD counts, different OSD sizes, and different per-write volumes. Compared the mechanism proposed in this paper, the classic Ceph, and the simulation of the uniform PG distribution on OSDs(Simulated Ceph open source tool up-map). All experiments are in 2 replicate mode. Each time the system writes 3 times, the amount of testing data (until one of the nodes reaches 100%), the variance, and the usage rate at runtime are recorded. The experimental results are shown in Fig. 8. The vertical coordinates on the left are the variance values for each experimental control group, represented by a histogram. The right vertical coordinate is the system usage rate of each control group in each experiment, and the horizontal coordinate is the number of experiments.

From the figure, we can see that in all of the experiments, the comparison group using the mechanism proposed in this paper has an average utilization rate of more than 15% compared with the other two comparison objects. In some comparison groups, such as Partially small, it can reach 20%. In addition, the experiment using this mechanism maintains a very stable state in usage, and there is not much fluctuation. This is because each writes data is random, and the mechanism of this paper can be used to circumvent this randomness. In comparison, the other two approaches do not. Especially in the Partially small experimental environment, the other two approaches collapse when the overall usage rate is low. This is because the PGs in these approaches are not aware of the OSD's usage. The system using tool up-map is not completely better than the classic system. The reason is that the analyzed distribution of the PG is not equal to the data distribution. In terms of variance, the control group using the proposed strategy is lower than the other two approaches. The storage balance of the system is improved after using the new balancing mechanism.

4.2 Evaluation of Data Migration

In the previous section, we pointed out that although the method of using equalized write frequency has completed the function of equalized storage, it incurred substantial data migration. This causes the loss of system performance and made the systems unreliable. Compared with previous research, the algorithm of this paper focused on the transfer of a balanced object based on a single PG, rather than an entire OSD. In this experiment, this paper uses a quantitative writing method to write a fixed size of data each time [9]. The offline calculation to obtain the data migration will be used in this section. The evaluation compared the amount of data migration that occurred when writing the same amount of data due to Reweight in the classic Ceph system. The Reweight mechanism was not a well-balanced solution, so the amount of data migration was not compared. The results are shown in Fig. 9. The histogram shows the comparison of the migration of data between the 2 mechanisms in equalized storage at each

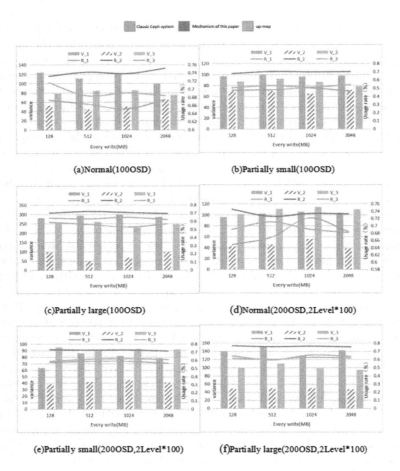

Fig. 8. Balanced storage capacity experiment

interval as usage increases, the table below shows the detailed migration data sizes.

As can be seen from the above figures, using the balanced storage strategy based on the write frequency, when the data are written into the system, no matter whether the system usage rate is high or low, a substantial amount of data migration occurs. The main reason is the differences among the PGs: this method seeks to migrate objects to achieve balanced storage, but the real migration is PGs. Consequently, data migration increases each time the data are written. In contrast, the algorithm strategy of this paper increases with the usage rate, because each node may trigger a new node to be balanced after each writes. However, at low usage rates, it does not intervene in the system to trigger data migration, ensuring equalization. While storing, it also ensures that the performance of the system is not affected. The migration of data increases at

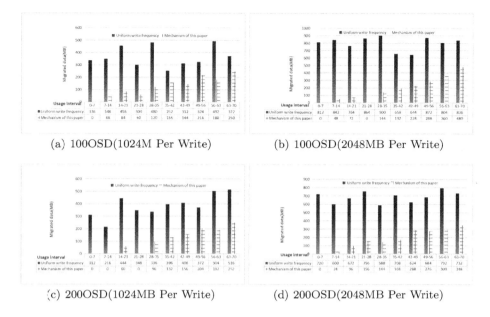

(a) 100OSD(1024M Per Write)   (b) 100OSD(2048MB Per Write)

(c) 200OSD(1024MB Per Write)   (d) 200OSD(2048MB Per Write)

Fig. 9. Comparisons of data migration

high usage rates; the migration amount across levels increases because the initial migration leaves many PGs in a low level. The strategy in this paper migrates 50% fewer data compared to the typical research approach and the reduced data migration ensures the reliability of the system.

4.3   Summary of the Experiments

It can be seen from the above experiments that the mechanism proposed in this paper has good performance in balanced storage, and at the same time, it has obvious advantages in data migration control compared to some typical previous researches. The mode of using collaborative work makes balancing spontaneous and efficient; it also makes the system more reliable. It is suitable for large-scale clusters. This is because the use of fewer OSD clusters means there are fewer OSDs per level after grading, so there are a limited number of different combinations for every PG.

5   Conclusions

In view of the storage imbalance problem in cloud storage systems, in this paper, a collaborative scheduling balanced storage mechanism based on a grading mapping model is proposed with Ceph as the research platform. The main contribution is to propose a novel collaboration mechanism to enable collaborative data balancing among storage nodes, which ensures balanced storage performance while reducing data migration costs. The proposed mechanism is independent and can be used with other algorithms mapping storage locations in the storage system to comprehensively solve the balanced storage and data migration problems. At the same time, many application scenarios such as data migration, cold data precipitation, etc. can be derived in the future. These are data balanced storage strategies with high application prospects. However, the work in this paper also has limitations, which is a goal for future work. First, the algorithm designed in this paper is very inefficient when the system is at high usage, because more attempts are needed and no good constraint can be found at present. Secondly, the structure of the storage system is simple and does not take into account structures such as Bucket in the algorithm, which is also the focus of future work.

Acknowledgment. This work is supported by the Natural Science Foundation of China(No. 61762008), the National Key Research and Development Project of China (No. 2018YFB1404404), the Major special project of science and technology of Guangxi(No.AA18118047-7), and the Guangxi Natural Science Foundation Project (No. 2017GXNSFAA198141).

References

1. Zhang, W., Flores, H., Pan, H.U.I.: Towards collaborative multi-device computing. In: 2018 IEEE International Conference on Pervasive Computing and Communications Workshops (PerCom Workshops). IEEE (2018)
2. Rump, F., Timm, B., Raphael, E.: Distributed and collaborative malware analysis with MASS. In: 2017 IEEE 42nd Conference on Local Computer Networks (LCN). IEEE (2017)
3. Aghayev, A., et al.: File systems unfit as distributed storage backends: lessons from 10 years of Ceph evolution. In: Proceedings of the 27th ACM Symposium on Operating Systems Principles. ACM (2019)
4. Zhou, J., et al.: Pattern-directed replication scheme for heterogeneous object-based storage. In: 17th IEEE/ACM International Symposium on Cluster 2017, Cloud and Grid Computing (CCGRID). IEEE (2017)
5. Kisley, R.V., Philip, D.K.: Distributed file serving architecture with metadata storage and data access at the data server connection speed. U.S. Patent No. 9,262,094. 16 February 2016
6. Huang, M., et al.: Research on data migration optimization of Ceph. In: 2017 14th International Computer Conference on Wavelet Active Media Technology and Information Processing (ICCWAMTIP). IEEE (2017)
7. Zhao, N., et al.: A reliable power management scheme for consistent hashing based distributed key value storage systems. Front. Inf. Technol. Electron. Eng. **17**(10), 994–1007 (2016)
8. Zhang, X., Gaddam, S., Chronopoulos, A.T.: Ceph distributed file system benchmarks on an openstack cloud. In: 2015 IEEE International Conference on Cloud Computing in Emerging Markets (CCEM). IEEE (2015)
9. Wang, L.: Optimizations on Ceph Cache Tiering. KylinCloud, Ceph day (2015)

10. Weil, S.A., et al.: RADOS: a scalable, reliable storage service for petabyte-scale storage clusters. In: Proceedings of the 2nd International Workshop on Petascale Data Storage: Held in Conjunction with Supercomputing 2007. ACM (2007)
11. Zhou, J., et al.: Pattern-directed replication scheme for heterogeneous object-based storage. In: 17th IEEE/ACM International Symposium on Cluster 2017, Cloud and Grid Computing (CCGRID). IEEE (2017)
12. Mseddi, A., Salahuddin, M.A., Zhani, M.F., et al.: Efficient replica migration scheme for distributed cloud storage systems. IEEE Trans. Cloud Comput. (2018)
13. D'atri, A., Bhembre, V., Singh, K.: Learning Ceph: unifed, scalable, and reliable open source storage solution. Packt Publishing Ltd. (2017)
14. Ou, J., et al.: EDM: an endurance-aware data migration scheme for load balancing in SSD storage clusters. In: 2014 IEEE 28th International Parallel and Distributed Processing Symposium. IEEE (2014)
15. Zhang, X., Wang, Y., Wang, Q., et al.: A new approach to double I/O performance for Ceph distributed file system in cloud computing. In: 2019 2nd International Conference on Data Intelligence and Security (ICDIS), pp. 68–75. IEEE (2019)
16. Aghayev, A., Weil, S., Kuchnik, M., et al.: File systems unfit as distributed storage backends: lessons from 10 years of Ceph evolution. In: Proceedings of the 27th ACM Symposium on Operating Systems Principles, pp. 353–369 (2019)

A Blockchain Based Cloud Integrated IoT Architecture Using a Hybrid Design

Ch Rupa[1(✉)], Gautam Srivastava[2], Thippa Reddy Gadekallu[3],
Praveen Kumar Reddy Maddikunta[3], and Sweta Bhattacharya[3]

[1] Department of Computer Science, VR Siddhartha Engineering College,
Vijayawada 520007, India
rupamtech@gmail.com
[2] Department of Mathematics and Computer Science, Brandon University,
Brandon, MB R7A 6A9, Canada
srivastavag@brandonu.ca
[3] School of Information Technology and Engineering, VIT - Vellore, Tamilnadu, India
{thippareddy,praveenkumarreddy,sweta.b}@vit.ac.in

Abstract. The Internet of Things (IoT) and its applications are gaining popularity in recent years due to features like ease of use and increased availability of the Internet. We can enhance the efficiency of IoT based systems by adding other advanced technologies like cloud infrastructure and blockchain technology. Through these enhancements IoT applications can be accessed at any time from any place. However, a database is required for storing the information of the application. Cloud infrastructure is an ideal solution for storing IoT based applications as it provides remote services such as storage, computation, and analysis. The major drawback of these applications are their inability to provide security and privacy for data. Blockchain technology helps in overcoming these drawbacks with features like immutability, transparency, and distributed structure. In this paper, a blockchain based cloud integrated IoT application is proposed that can assist to identify intruders through virtual monitoring. The main advantage of this application is that it can operate in areas where manual monitoring is challenging and data is stored in a blockchain-based tamper-free environment.

Keywords: Internet of Things · Cloud infrastructure · Centralized system · Blockchain technology · Distributed system

1 Introduction

Globally, attention to the Internet of Things (IoT) as well as Unmanned Aerial Vehicles (UAV) is growing tremendously. There are numerous applications based on IoT and UAV technologies such as in Healthcare, Vehicular, Surveillance, and in the Government sector [27–29]. According to the Information Handling Services (IHS) Markit report, in 2018, about 20 billion devices were connected

© ICST Institute for Computer Sciences, Social Informatics and Telecommunications Engineering 2021
Published by Springer Nature Switzerland AG 2021. All Rights Reserved
H. Gao et al. (Eds.): CollaborateCom 2020, LNICST 350, pp. 550–559, 2021.
https://doi.org/10.1007/978-3-030-67540-0_36

to the Internet, and about 1.2 billion IoT devices were installed. The numbers might be extended up to 125 billion by 2030 [30]. These analytics clearly show that IoT/UAV devices can be used to improve the services for society. The main features of the IoT/UAV technology are "CCCC" (Connections, Collection, Computation, Creation) which are interconnected to each other [22,33].

Global data transmission rates are expected to rise by about 20%-50% per year, on average [25]. Data protection from unauthorized access is an important task while using IoT/UAV devices. Previously the databases that are based on CRUD (Create, Read, Update, and Delete) operational model used to store the data [47]. Later cloud computing has gained popularity as it overcomes the disadvantages of databases. Cloud data can be accessed from any place and at any time [24,38,39]. Currently, one emerging technology, Blockchain, is impacting all database based applications because of its key features like immutability, transparency, distributed computing, and security [12,20,43,44]. Cyber attacks on the cloud is also on the rise [10,18].

Many sectors such as smart grid, Supply Chain Management, finance, cryptocurrencies, insurance, smart cities, and many others [15,32] are using blockchain technology to enhance credibility. The main reason behind the success of blockchain technology is consensus algorithms like Proof-of-Work (PoW), Proof-of-Stake (PoS), Practical Byzantine Fault Tolerance (PBFT), etc. Consensus algorithms play a major role in securing the sensitive details from the hackers; if any detail has to be modified in the blockchain, it has to be agreed upon by a majority of nodes through consensus. Consensus algorithms are the main reason behind the trustworthiness of nodes in blockchain networks. If any new node has to be added into the network, a majority of the nodes have to agree that the new node is trustworthy through consensus algorithms [26,49]. Another important concept in blockchain technology is smart contracts. Smart contracts are the codes which will be automatically executed in the blockchain network whenever a transaction occurs, to ensure that the transactions adhere to preconditions [48]. Blockchain technology [6,13] helps in reducing attacks on surveillance data collected from IoT/UAV devices. The information maintained by using blockchain technology along with cloud infrastructure can improve the security and privacy of data. Generally, any application communicates with the cloud to store collected data from the application gateway. Later, the data is processed and analyzed in real-time using batch analytics, visualization, and machine learning [36]. Various Cloud platforms like Open stalk, GCP (Google cloud platform), Microsoft Azure are available to design and develop such systems [14].

In the proposed system, we used the Open Stalk cloud platform and Ethereum in Ganache with a meta mask wallet for blockchain technology [21,35,37]. In this paper, we propose a methodology by considering an IoT based application such as a virtual circuit based vehicle monitoring system (VVMS). The collected data from the VVMS is stored, processed, and analyzed on the cloud. Instructions are received by the vehicle through an operator virtually, along with the VVMS responses for instructions are stored on the blockchain and on the cloud. Such an infrastructure can protect information from various cyber attacks like ransomware, known ciphertext attach, injection attacks [2,8], and other malware attacks [1,11].

The rest of the paper is organized as follows. Section 2 discusses about the literature survey related to the concepts and required architectures of the proposed system. Proposed system architecture and implementation results are discussed in Sect. 3. Section 4 gives an in-depth conclusion and future scope of this work.

2   Related Works

Anthi *et al.* [4] proposed a three-layer intrusion detection system [7,40] to defend IoT devices. The main functions of this system are the classification of device behavior and identification of suspected packets (malicious packets) and classify the attacks. The main drawback of this approach is that anyone can modify or update the data over the network like a man in the middle attack. This system cannot be operated at any time from any place as it is designed as a standalone system.

Ullah *et al.* [19] developed an IoT based system to detect lightweight attacks in terms of the Intrusion detection system (IDS). Here the authors used support vector machine-based supervised machine learning for identifying the injection attacks over the IoT network. Arshad *et al.* [5] discussed intrusion detection system (IDS) using IoT and mentioned the open challenges to achieve Intruder detections over IoT Infrastructure. In this paper, the authors proposed an IDS system by following the signature, specification, Anomaly and hybrid-based approach. The main limitation of this work is its lack of privacy.

Sabir *et al.* [42] proposed a driver-friendly interface system that can monitor the speed breakers and potholes by monitoring the roads. Crowdsourcing approach was used to design and develop the system. The main limitations of this system are the lack of privacy and preservation of the data which are captured by the proposed interface. Also there is no information about the data storage locations, access and computation specifications on the data.

Hu *et al.* [16] designed a smart vehicle system using IoT with embedded control. This system can automatically chase the light source and monitors the vehicle moving directions and its path. In the same way, Metlo *et al.* [31] proposed a system to track vehicles using GPS. In this work, authors have considered the GPS data as an input to the tracing algorithm to detect specific vehicles based on crowdsourced data. The main limitations of these works are lack of security to the information. Also there is no transparency and reliability in the proposed system.

Ali *et al.* [3] discussed the role of blockchain in the IoT. Here authors have explained clearly how blockchain-based systems can achieve the characteristics like security, immutability, and decentralization. This paper explained how the challenges faced by IoT systems that are based on centralized architecture can be solved by using blockchain (Table 1).

Table 1. Comparative analysis of existing approaches for IoT

Approaches	Characteristics					
	Is blockchain based?	Transparency	Immutable	Architecture	Visibility	Application
[34]	No	No	No	Centralized	Host	IDS
[23]	No	No	No	Centralized	Network	IDS
[17]	No	No	No	Centralized	Host	IDS
[9]	No	Yes	No	Distributed	Network	IDS
[41]	No	No	No	Centralized	Host	IDS
Proposed method	Yes	Yes	Yes	Distributed	Host	IDS through VMSS

3 Proposed Methodology

A proposed blockchain-based cloud-integrated IoT based application enhances the privacy of data while maintaining efficiency in both computation and energy use. The proposed system detects intruders by virtual monitoring and operating the vehicle. The design and development of the proposed IoT system uses Node MCU, Arduino, Motor drivers, camera, and mobile devices as connected devices (IoT). A gateway is required to transfer data from IoT connected devices to the cloud for maintaining the communication data. Therefore, Blynk App is used as a gateway and an Open stack cloud is used for the purpose of storing, computing, and analyzing the data. Moreover, Blynk APP uses to register the data into the system initially that helps in detecting the intruders. Blockchain technology helps to give privacy and preservation for the data with its unique features like transparency, immutability, and distributed nature[46]. Hence, in the proposed system, the instructions given to the IoT based vehicle and the vehicle responses are maintained in the blockchain as transactions.

Figure 1 shows the overview of the proposed system. This system consists of three modules that operate simultaneously such as the design and development of virtual circuit based IoT components as a first module. The second module is deploying the data on the cloud environment [45] and the third module is deploying the data on blockchain environment.

As part of the design and development of the first module, all IoT components (VMSS) such as Node MCU, Arduino, Motor Drivers, Wheels, and Camera are interconnected. These components are deployed using a python program. Initially, the user needs to give instructions to the vehicle through mobile devices regarding its movement like Left, Right, Straight, and Back. These instructions are processed over VMSS components. Later the surroundings are monitored with the help of the attached camera to the vehicle. Simultaneously, the captured data is compared with the registered data via BLYNK APP. If any intruder is detected, the data is stored in the cloud as well as in blockchain through the gateway i.e BLYNK APP and public chain blockchian architecture. Figure 2 shows the flow of transactions among the objects of the proposed system.

In the second module, the hash value of the received data from the vehicle will be computed using the SHA-1 algorithm. Later this hash value is compared with existing data hash value. If both the hash values are the same, then the

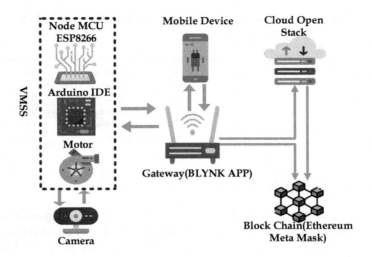

Fig. 1. Architecture of proposed methodology

Algorithm 1: Virtual circuit based IoT use case

---

1 Input: User's instructions as request (Left, Right, straight, Back)
2 Output: Vehicle movement & capture the data
3 User 'A' gives instructions to the vehicle through mobile devices
4 $x \leftarrow Req[A]$
5 Process the request based on deployed python program into Arduino connected
   VMSS
6 Motor Driver [vehicle] $\longleftarrow x$
7 Capture the data by the camera
8 **if** *intruder detects* **then**
9 $\quad|\quad$ y $\leftarrow Data[Intruder]$
10 **else**
11 $\quad|\quad$ continue
12 **end**

---

received data is considered as belonging to the intruder, otherwise to the authorized person or vice-versa. This depends on which data has to be taken to do registration at the initial stage. This response is forwarded to the user through the VMSS monitor and the hash value is sent to the mobile device via a gateway.

The data is deployed on the public blockchain for improving privacy, efficiency, and computational costs. The instructions from the user to the vehicle and the responses from the vehicle are maintained on a blockchain as block transactions. Here, an Ethereum based public blockchain is used that requires smart contracts & wallet balance to maintain the data on the block as transactions. Before making a transaction, the wallet balance (eth) is verified. Here, a web-based Metamask Wallet is used to manage (eth) balance. For each transaction,

Algorithm 2: Deployment of the data on Cloud

1 Input: Data from VMSS i.e 'y'
2 Output: Intruder/authorized data as a response
3 Establish the connection with the VMSS through the gateway (BLYNK APP)
4 Computes and maintains hash values of the suspicious data instructed by the User in cipher text using a public key
5 H[i] ← Hash (*intruderdata*)
6 Analyze the received data
7 **for** $i = 1; i <= length\ [data]; i + +$ **do**
8 | **if** $h\ [i] == Hash(y)$ **then**
9 | | Res ← Intruder otherwise Res ← authorized
10 | **else**
11 | **end**
12 **end**
13 Sends the response to VMSS monitor as well as hash value to mobile device

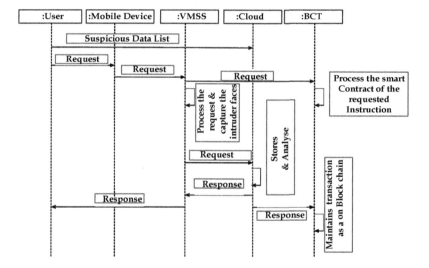

Fig. 2. Sequence of transactions in the proposed system

the wallet balance is reduced. After successful deployment, the transaction stores in a block. Later, this block joins into the blockchain. This block transaction is immutable, transparent, and distributed across all the connected nodes over this network.

Algorithm 3: Deployment of the data on Blockchain

---

1 Input : VMSS received Instructions from the users & vehicle Responses.
2 Output: Create immutable transactions over the blockchain and stores in a block
3 User 'A' gives instructions to the vehicle through mobile devices (VMSS)
4 $x \leftarrow Req[A]$
5 Deploy the smart contract over blockchain to maintains the instructions as transactions
6 **if** $balance[wallet] >= Minimum$ **then**
7 | Block [i]=Txn
8 **else**
9 | exit
10 **end**
11 Response received from the cloud then repeat step 5

---

4   Conclusion and Future Work

Blockchain-based cloud-integrated IoT applications like intruder detection systems improve performance, security, and efficiency. The hybrid design in this work improves certain factors of the system such as data privacy and portability. These can be achieved by using one-way hashing algorithms. It extends security and privacy due to the immutable property of the blockchain. Another factor of this application is data transparency among the connected nodes over the network due to its distributed nature included in the architecture by default. The proposed system can be used to monitor and record illegal activities. In the future, we would like to implement the proposed system using UAV. To identify the intruders, we would like to consider and verify the faces of the intruders. To design and develop the blockchain, an Ethereum based Ganache blockchain along with a metamask wallet is planned.

References

1. Alazab, M., et al.: A hybrid wrapper-filter approach for malware detection. J. Netw. **9**(11), 2878–2891 (2014)
2. Alazab, M., Layton, R., Broadhurst, R., Bouhours, B.: Malicious spam emails developments and authorship attribution. In: 2013 Fourth Cybercrime and Trustworthy Computing Workshop, pp. 58–68. IEEE (2013)
3. Ali, M.S., Vecchio, M., Pincheira, M., Dolui, K., Antonelli, F., Rehmani, M.H.: Applications of blockchains in the Internet of Things: a comprehensive survey. IEEE Commun. Surv. Tutor. **21**(2), 1676–1717 (2018)
4. Anthi, E., Williams, L., Słowińska, M., Theodorakopoulos, G., Burnap, P.: A supervised intrusion detection system for smart home IoT devices. IEEE Internet Things J. **6**(5), 9042–9053 (2019)
5. Arshad, J., Azad, M.A., Amad, R., Salah, K., Alazab, M., Iqbal, R.: A review of performance, energy and privacy of intrusion detection systems for IoT. Electronics **9**(4), 629 (2020)

6. Baza, M., Lasla, N., Mahmoud, M., Srivastava, G., Abdallah, M.: B-ride: ride sharing with privacy-preservation, trust and fair payment atop public blockchain. IEEE Trans. Netw. Sci. Eng. (2019, in press)
7. Bhattacharya, S., Kaluri, R., Singh, S., Alazab, M., Tariq, U., et al.: A novel PCA-firefly based XGBoost classification model for intrusion detection in networks using GPU. Electronics **9**(2), 219 (2020)
8. Ch, R., Gadekallu, T.R., Abidi, M.H., Al-Ahmari, A.: Computational system to classify cyber crime offenses using machine learning. Sustainability **12**(10), 4087 (2020)
9. Chaabouni, N., Mosbah, M., Zemmari, A., Sauvignac, C., Faruki, P.: Network intrusion detection for IoT security based on learning techniques. IEEE Commun. Surv. Tutor. **21**(3), 2671–2701 (2019)
10. Chiramdasu, R.: Extended statistical analysis on multimedia concealed data detections. J. 161–165 (2019). http://iieta.org/journals/isi24(2)
11. Djenouri, D., Badache, N.: Struggling against selfishness and black hole attacks in MANETs. Wirel. Commun. Mob. Comput. **8**(6), 689–704 (2008)
12. Dwivedi, A.D., Malina, L., Dzurenda, P., Srivastava, G.: Optimized blockchain model for Internet of Things based healthcare applications. In: 2019 42nd International Conference on Telecommunications and Signal Processing (TSP), pp. 135–139. IEEE (2019)
13. Dwivedi, A.D., Srivastava, G., Dhar, S., Singh, R.: A decentralized privacy-preserving healthcare blockchain for IoT. Sensors **19**(2), 326 (2019)
14. Ganapathy, S., et al.: A secured storage and privacy-preserving model using CRT for providing security on cloud and IoT-based applications. Comput. Netw. **151**, 181–190 (2019)
15. Hakak, S., Khan, W.Z., Gilkar, G.A., Imran, M., Guizani, N.: Securing smart cities through blockchain technology: architecture, requirements, and challenges. IEEE Netw. **34**(1), 8–14 (2020)
16. Hu, M.S., Chen, L.H.: The application of embedded control and IoT technology in the automatic light-chasing vehicles. In: 2019 IEEE Eurasia Conference on IOT, Communication and Engineering (ECICE), pp. 362–365. IEEE (2019)
17. Irfan, S., Rupa, C., Vinay, K., Veni, M.K., Rachana, R.: Smart virtual circuit based secure vehicle operating system. In: 2020 2nd International Conference on Innovative Mechanisms for Industry Applications (ICIMIA), pp. 386–390. IEEE (2020)
18. Iwendi, C., et al.: Keysplitwatermark: zero watermarking algorithm for software protection against cyber-attacks. IEEE Access **8**, 72650–72660 (2020)
19. Jan, S.U., Ahmed, S., Shakhov, V., Koo, I.: Toward a lightweight intrusion detection system for the Internet of Things. IEEE Access **7**, 42450–42471 (2019)
20. Jindal, A., Aujla, G.S., Kumar, N.: Survivor: a blockchain based edge-as-a-service framework for secure energy trading in SDN-enabled vehicle-to-grid environment. Comput. Netw. **153**, 36–48 (2019)
21. Khalid, U., Asim, M., Baker, T., Hung, P.C., Tariq, M.A., Rafferty, L.: A decentralized lightweight blockchain-based authentication mechanism for IoT systems. Cluster Comput. 1–21 (2020)
22. Khezr, S., Moniruzzaman, M., Yassine, A., Benlamri, R.: Blockchain technology in healthcare: a comprehensive review and directions for future research. Appl. Sci. **9**(9), 1736 (2019)
23. Li, D., Deng, L., Lee, M., Wang, H.: IoT data feature extraction and intrusion detection system for smart cities based on deep migration learning. Int. J. Inf. Manag. **49**, 533–545 (2019)

24. Li, X., Wang, Q., Lan, X., Chen, X., Zhang, N., Chen, D.: Enhancing cloud-based IoT security through trustworthy cloud service: an integration of security and reputation approach. IEEE Access **7**, 9368–9383 (2019)
25. Longstreet, P., Brooks, S.: Life satisfaction: a key to managing Internet & social media addiction. Technol. Soc. **50**, 73–77 (2017)
26. Ma, S., Deng, Y., He, D., Zhang, J., Xie, X.: An efficient NIZK scheme for privacy-preserving transactions over account-model blockchain. IEEE Trans. Dependable Secure Comput. (2020, in press)
27. Maddikunta, P.K.R., Gadekallu, T.R., Kaluri, R., Srivastava, G., Parizi, R.M., Khan, M.S.: Green communication in IoT networks using a hybrid optimization algorithm. Comput. Commun. **159**, 97–107 (2020)
28. Maddikunta, P.K.R., et al.: Unmanned aerial vehicles in smart agriculture: applications, requirements and challenges. arXiv preprint arXiv:2007.12874 (2020)
29. Maddikunta, P.K.R., Srivastava, G., Gadekallu, T.R., Deepa, N., Boopathy, P.: Predictive model for battery life in IoT networks. IET Intell. Transp. Syst. **14**(11), 1388–1395 (2020)
30. Markit, I.: The Internet of Things: a movement, not a market. Englewood, CO: IHS Markit (2017). Accessed 28 Dec 2018
31. Metlo, S., Memon, M.G., Shaikh, F.K., Teevno, M.A., Talpur, A.: Crowdsource based vehicle tracking system. Wirel. Pers. Commun. **106**(4), 2387–2405 (2019)
32. Mollah, M.B., et al.: Blockchain for future smart grid: a comprehensive survey. IEEE Internet Things J. (2020)
33. Niu, Y., Li, Y., Jin, D., Su, L., Vasilakos, A.V.: A survey of millimeter wave communications (mmWave) for 5g: opportunities and challenges. Wirel. Netw. **21**(8), 2657–2676 (2015)
34. Nobakht, M., Sivaraman, V., Boreli, R.: A host-based intrusion detection and mitigation framework for smart home IoT using openflow. In: 2016 11th International conference on availability, reliability and security (ARES), pp. 147–156. IEEE (2016)
35. Priya, K.L.S., Rupa, C.: Block chain technology based electoral franchise. In: 2020 2nd International Conference on Innovative Mechanisms for Industry Applications (ICIMIA), pp. 1–5. IEEE (2020)
36. Qiang, W.: Performance and security in cloud computing. J. Supercomput. **75**(1), 1–3 (2018). https://doi.org/10.1007/s11227-018-2671-4
37. Rasool, S., et al.: Blockchain-enabled reliable osmotic computing for cloud of things: applications and challenges. IEEE Internet Things Mag. (2020)
38. Reddy, G.T., Sudheer, K., Rajesh, K., Lakshmanna, K.: Employing data mining on highly secured private clouds for implementing a security-ASA-service framework. J. Theor. Appl. Inf. Technol. **59**(2), 317–326 (2014)
39. RM, S.P., et al.: Load balancing of energy cloud using wind driven and firefly algorithms in internet of everything. J. Parallel Distrib. Comput. **142**, 16–26 (2020)
40. RM, S.P., et al.: An effective feature engineering for DNN using hybrid PCA-GWO for intrusion detection in IOMT architecture. Comput. Commun. **160**, 139–149 (2020)
41. Rupa, C.: An integrated digital authentication mechanism for intrusion detection system. In: Big Data Analytics for Smart and Connected Cities, pp. 158–169. IGI Global (2019)
42. Sabir, N., Memon, A.A., Shaikh, F.K.: Threshold based efficient road monitoring system using crowdsourcing approach. Wirel. Pers. Commun. **106**(4), 2407–2425 (2019)
43. Salah, K., Rehman, M.H.U., Nizamuddin, N., Al-Fuqaha, A.: Blockchain for AI: review and open research challenges. IEEE Access **7**, 10127–10149 (2019)

44. Sharma, P.K., Kumar, N., Park, J.H.: Blockchain-based distributed framework for automotive industry in a smart city. IEEE Trans. Indu. Inform. **15**(7), 4197–4205 (2018)
45. Singh, S., Jeong, Y.S., Park, J.H.: A survey on cloud computing security: issues, threats, and solutions. J. Netw. Comput. Appl. **75**, 200–222 (2016)
46. Singh, S., Ra, I., Meng, W., Kaur, M., Cho, G.: SH-BlockCC: a secure and efficient Internet of Things smart home architecture based on cloud computing and blockchain technology. Int. J. Distrib. Sens. Netw. **15**(4), 1–18 (2019)
47. Truong, N.B., Sun, K., Lee, G.M., Guo, Y.: GDPR-compliant personal data management: a blockchain-based solution. arXiv preprint arXiv:1904.03038 (2019)
48. Wang, H., Qin, H., Zhao, M., Wei, X., Shen, H., Susilo, W.: Blockchain-based fair payment smart contract for public cloud storage auditing. Inf. Sci. **519**, 348–362 (2020)
49. Xiao, Y., Zhang, N., Lou, W., Hou, Y.T.: A survey of distributed consensus protocols for blockchain networks. IEEE Commun. Surv. Tutor. (2020, in press)

User Perspective Discovery Method Based on Deep Learning in Cloud Service Community

Lei Yu$^{(\boxtimes)}$, Yaoyao Wen, and Shanshan Liang

Department of Computer Science, Inner Mongolia University, Hohhot, China
yuleiimu@sohu.com

Abstract. The rapid development of cloud computing has promoted the coordinated integration of resources in various industries. In order to facilitate users' selection and invocation, more and more individuals and organizations have moved local application resources into the cloud service communities in the form of web services. In recent years, more and more people are interested in the emotional attitudes reflected in consumer reviews, but the sentiment analysis using the deep learning method to achieve evaluation of API (Application Programming Interface) services has received little attention. In order to explore the effective information of user's point of view data in the cloud service community, we propose an approach to analyze the user's opinion data using deep learning. We design three deep learning models of Long Short-Term Memory (LSTM), Bi-directional Long Short-Term Memory (Bi-LSTM) and Gated Recurrent Unit (GRU). The result shows that the accuracy rate and recall rate of Bi-LSTM model is higher than the LSTM and GRU. Finally, we evaluate the performance of the three deep learning models, and choose the optimal Bi-LSTM model as the model used by the cloud service community in the future. According to the parameter comparison experiment of Bi-LSTM model, we obtained the optimal tuning of the model, and the model achieved the accuracy of 89.68%.

Keywords: Sentiment classification · Deep learning · LSTM · Bi-LSTM · GRU

1 Introduction

With the rapid development of Chinese cloud computing industry, the number of web services and their users is increasing in the cloud service community [1]. On the one hand, the rich web services provide a wide range of options for target users to meet various service requirements. On the other hand, it also imposes a heavy burden on the service choices of target users, especially when many candidate services have the same or similar functions, service users not only need to know the ability of a particular web service meet functional requirements, but also need to know whether the service can provide a satisfactory user experience [1]. Suppliers use consumer reviews as a new marketing tool to build their credibility. The growing number of online user reviews in the cloud service community, it has been a huge challenge to extract useful information from

© ICST Institute for Computer Sciences, Social Informatics and Telecommunications Engineering 2021
Published by Springer Nature Switzerland AG 2021. All Rights Reserved
H. Gao et al. (Eds.): CollaborateCom 2020, LNICST 350, pp. 560–574, 2021.
https://doi.org/10.1007/978-3-030-67540-0_37

a large amount of comment data. Traditional dictionary-based methods and machine-based learning methods have not been able to deal with the emotional classification of massive reviews. Therefore, in order to recommend better services to users and increase the revenue of service providers, it's important to establish an efficient sentiment model.

In this paper, we use deep learning methods to classify user option data in the cloud service community. The more positive comments of user reviews, the higher the trust of the service, and then the service providers and developers can recommend the trusted services to users.

2   Related Work

As our society becomes more and more connected, the number of online comments in the cloud service community is growing exponentially each year. The online reviews are the direct expression of the user's point of view, and they greatly affect people's behavior. The method of opinion-based discovery in the cloud service community is mining based on user's review data. With the development and widespread application of deep learning technology, many researchers have begun to apply deep learning models to the field of sentiment analysis, it uses deep learning technology to build an emotion classification model and judges people's implicit emotional tendencies in comment sentences.

2.1   Research of Sentiment Analysis

The concept of deep learning first appeared in 2006, limited by large-scale data volume and high-performance hardware, it has not entered people's field of vision until the last 5 years. Deep learning simulates the mechanism of the human brain and builds neural networks to adapt to people's daily affairs. A simple deep learning framework is based on sample data, stacking modules with learning capabilities in multiple layers. Deep learning is a machine learning method with a deep neural network model. It is very good at processing complex high-dimensional data and has been widely used in many fields such as science, business, and natural language processing.

Sentiment analysis is a field of natural language processing. In the process of emotion classification, the high-level representation of deep learning can emphasize important category information while suppressing irrelevant background information. The field of sentiment analysis research has expanded from movie reviews to stock message boards to congressional debates, and the research results have been industrially deployed in systems that measure market reactions and summarize opinions from Web pages, discussion boards, and blogs. With these widely changing fields, researchers and engineers constructing sentiment classification systems need to collect and organize data in each new field they encounter, and annotate a corpus for each field. However, in different fields, the expression of emotions is different, and it is impractical to annotate the corpus for each area of possible interest, especially because the product characteristics will change over time. We study the domain adaptability of sentiment classification, pay attention to online reviews of different types of service products, and combine the most advanced neural language models with sentiment information. This is still an area that needs to be explored.

In sentiment analysis tasks, there are usually three methods based on lexicon, machine learning and deep learning. The method based on sentiment lexicon is a typical unsupervised learning. Kim et al. used existing sentiment lexicons to judge the polarity of texts by adding the sum of emotional vocabulary scores [2]. The establishment of an emotional dictionary requires a large amount of manual participation, which makes it rare to use sentiment lexicon for emotional classification. Machine learning is another important method of sentiment analysis. Pang et al. proposed using machine learning models for sentiment analysis [3], and tried using three classification algorithms: naïve Bayes, support vector machine and maximum entropy. Kale et al. proposed a semi-supervised model, they attempted to use untagged data for rating prediction [4]. Wan et al. proposed a text classification method based on KNN and support vector machine to improve the accuracy of classification [5]. Bing Liu et al. proposed a model for mining and summarizing product reviews [6], which can summarize comments based on their characteristics. However, supervise machine learning methods require a large number of manually labeled training corpora to improve learning ability of the model, which is hindered in practical applications.

The emergence of deep learning methods has better compensated for the shortcomings of the above methods, and has achieved many results in specific topic sentiment analysis tasks.

In 2013, Wang et al. described a neural network architecture that attempts to exploit the CNN and RNN architectures [7]. In 2014, Kim proposed using CNN for sentence modeling, and achieved good results on multiple data sets [8]. Zhang Xiang et al. proposed a CNN model based on character level and used for text sentiment classification [9]. In 2015, Zhang et al. proposed a two-way door-loop neural network model for calculating emotions in target in tweets [10]. Graves et al. completed the identification of the 2013 phone number using the LSTM system [11]. Yao et al. further analyzed the role of the Bi-LSTM model in Chinese word segmentation [12]. Chiu et al. combined Bi-LSTM with CNN and demonstrated the good performance of the model [13]. In 2018, Hao Zhe Lin et al. proposed a model of GRNN to capture the intrinsic link of literature-level sentiment classification evaluation [14].

In China, He Yan Xiang et al. established the MCNN model to classify Chinese microblogs and fuse the emoji matrix in the semantic expression of words, which has practical significance [15]. Liu Rui Mei et al. used a deep convolutional neural network to analyze the emotions of multimedia images [16]. Jing zhi Gang et al. proposed a new sentiment analysis model Bi-LSTMM-B, through the improvement of Bi-LSTM model and Bagging algorithm, which shows that the combination of deep learning model and integrated learning thought can improve the accuracy of sentiment analysis [17]. Xiang Zhang et al. used character- level convolutional networks (Conv Nets) for emotional classification of texts [9].

2.2 Research of Service Recommendations

As the distribution model of information systems shifts to the XaaS paradigm, microservices architecture is rapidly emerging, taking the RESTful principle as its API model of choice. Researchers have been studying service data mining on various platforms to promote research on service recommendations and service composition. However, researchers in the service computing field lack a cloud SaaS platform for hosting real-world services. The cloud service community we built is an independent SaaS platform that can connect API service providers and service users (including secondary developers and service users). The platform aggregates services into categories, and each category contains services with similar functions. The platform is committed to providing users with the most comprehensive and convenient services, as well as helping service providers open services and increase API calls. The platform has assembled more than 90 services required for application development, and unique services aggregate the platform's exclusive image processing, speech synthesis, ancient poetry synthesis, movie query, song query and other services. After registering an account, platform users can call all API services on the platform without jumping to a third-party website for operation. Users can also filter and accept services recommended by the platform, and can freely combine services. We use the cloud service community as a public data source, and users quickly generate and share data by using community information services. The cloud service community is a multi-information service domain consisting of multiple Mashups to meet comprehensive functional requirements and provide additional business value. Software service Web API usually provides black box functions to users through the Internet, which leads to a lack of reliability analysis of internal information.

With the development of popular computing concepts such as cloud computing, pervasive computing, and edge computing. Not only the underlying computing resources, but also enterprises, applications and data are open as Web services (usually Web API), making services on the internet unprecedented prosperity. Web services are black box software services, which provide software components as building blocks for enterprise application integration, effectively promoting the service deployment of cloud computing resource pool. The number of web services in the cloud resource pool are growing rapidly, so we need some new ways to make accurate service recommendations and choices. Providing accurate service recommendations to users is critical to improving the efficiency and success rate of application deployment in the cloud service community. In order to meet people's needs, Web service recommendation systems such as Web Service List, Web Service X, Programmable Web [18], Remote Methods [19] and Service-Repository [20] have appeared on the cloud service market. As the number of user comments increases, it is impossible for service provider to read all the comments. Therefore, it is important to mine user comments and extract valid information. In 2017, H. Wang et al. extracted the fine- grained value characteristics from customer reviews and identified the personalized distribution of each value feature [21], it shows the value preference of specific customers. They proposed a VFMine algorithm based on text mining, it can extract value characteristics from customer reviews.

According to the above analysis, we attempt to use deep learning method to extract fine-grained and value information from user comments and analyze the user's value preference.

2.3   LSTM and Bi-LSTM

According to the deliberate design, LSTM is able to avoid long-term dependence, it can remember the long-term information. It has strong timing signal processing capabilities. The difference between LSTM and RNN is that the LSTM has a gate structure. The gate is a structure that determines whether information can pass, it can control the flow of information, prevent information from fading away, and is more likely to capture long-term dependencies than RNN. The architecture of LSTM is shown on Fig. 1. It has three kinds of gate structures: input gate, output gate and forgetting gate. The distribution is represented by *it, ft* and *Ot,* Ct is the cell activation state, and the multiplication gate enables the LSTM memory cell to store and access information for a long time, thereby reducing the gradient disappearance problem. LSTM memory unit algorithm is as follows:

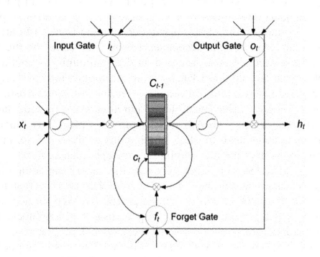

Fig. 1. LSTM single cell door structure

Forgetting gate decides when you need to forget the previous state.

$$f_t = \sigma\left(W_{xf}x_t + W_{hf}h_{t-1} + W_{cf}c_{t-1} + b_f\right) \qquad (1)$$

W represents the weight matrix, b is the offset term, σ is a nonlinear function applied to the hidden layer unit. The forgetting gate is used to output a value between 0 and 1 for each digit in the cell state c_{t-1} value. The next step is to decide what information to store in the cell state, which consists of two parts: in the first part, the input gate will decide which state to update.

$$i_t = \sigma(W_{xi}x_t + W_{hi}h_{t-1} + W_{ci}c_{t-1} + b_i) \qquad (2)$$

Next, the tanh layer adds a new cell state value vector c_t to the state, and we will combine them to create an update.

$$\tilde{c}_t = \tanh(W_{xc}x_t + W_{hc}h_{t-1} + b_c) \qquad (3)$$

Then we can update the old cell c_{t-1} to a new cell state c_t.

We multiply the old state by f_t to discard the information that needs to be discarded, and plus i_t multiply $\tilde{c}_t$ to get the cell state to get new information.

$$c_t = f_t \times c_{t-1} + i_t \times \tilde{c}_t \tag{4}$$

Finally, in the output gate, the sigmoid layer is used to determine which part of the information to output.

$$o_t = \sigma(W_{xo}x_t + W_{ho}h_{t-1} + W_{\infty}c_{t-1} + b_o) \tag{5}$$

We have $x_t^c = [x_t, h_{t-1}]$, so the formula $h_t = o_t \times tanh(c_t)$ can be written as follows:

$$f_t = \sigma\left(W_f x_t^c + b_f\right) \tag{6}$$

$$i_t = \sigma\left(W_i x_t^c + b_i\right) \tag{7}$$

$$\tilde{c}_t = \tanh\left(W_g x_t^c + b_o\right) \tag{8}$$

$$o_t = \sigma\left(W_o x_t^c + b_o\right) \tag{9}$$

In this paper, we superimposed two LSTM layers, each using 20% Dropout to prevent overfitting, and finally adding a fully connect layer to output the result. Different LSTM layers capture different fluctuations, so the model has the ability to perform higher-level timing expressions. The first layer LSTM needs to return the complete sequence data for the next layer. At the same time, in the second layer LSTM, we do not return the original sequence data, but return the calculated sequence data for the output layer. The model architecture is shown below (Fig. 2).

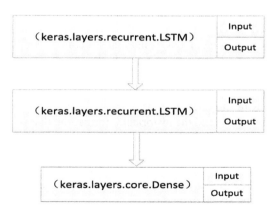

Fig. 2. Stacked neural network architecture

Because LSTM modeling has the problem of not being able to encode information from back to front, we use bidirectional LSTM to capture bidirectional semantic dependencies. Bi-LSTM is used to upgrade the input information block of the available network. It is composed of forward LSTM and backward LSTM. The structure is shown in Fig. 5. By using two times directions, Bi-LSTM can better understand the context. In this paper, after the vectorized representation of the text, the Bi-LSTM neural network model is used to learn the text vector to obtain the forward hidden state vector $\overrightarrow{h_t}$ and reverse hidden state sequence $\overleftarrow{h_t}$. By concatenating these two output sequence vectors, the final output vector is $h_t = \overrightarrow{h_t} \| \overleftarrow{h_t}$ (Fig. 3).

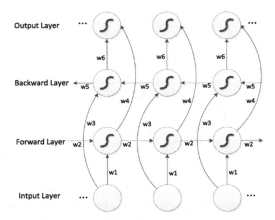

Fig. 3. Bidirectional LSTM

2.4 GRU

Because of the complex structure of both forgetting gates and input gates used by LSTM, Chung et al. proposed the GRU in 2014 [22], which preserves the resistance of LSTM to the vanishing gradient problem, but its internal structure is simpler than LSTM, so it is easier to train. Its internal structure is shown in the Fig. 4.

The GRU has an update gate z_t and reset gate r_t. The update gate is a combination of the input gate and the forget gate in the LSTM model. The output gate in the LSTM model functions is similar with the reset gate. The update gate z_t determines the integration of the new input information and the historical information, that is, how much previous information is to be retained, and the reset gate r_t determines the proportion of the state information in the model, that is, the new input information and the previous information can be well combined. GRU simultaneously combines the cell state and output into one state parameters. As the number of gates changes from 3 to 2, the training parameters decrease and the training speed increases.

Similar to LSTM, GRU uses the hidden state at time t-1 and the input time series value at time t to calculate the hidden state output at time t. The deduction formula of GRU neural network is:

$$z_t = \delta\left(W_z \times \left[h_{t-1}, x_t\right]\right) \tag{10}$$

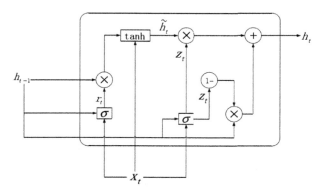

Fig. 4. A cell for GRU

$$r_t = \delta\big(W_r \times [h_{t-1}, x_t]\big) \tag{11}$$

$$\tilde{h}_t = tanh\big(W \times [r_t \times h_{t-1}, x_t]\big) \tag{12}$$

$$h_t = (1 - z_t) \times h_{t-1} + z_t \times \tilde{h}_t \tag{13}$$

h_{t-1} represents the output of the previous neuron, x_t represents the input of the current neuron, W_z represents the weight of the update gate, W_r represents the weight of the reset gate, $\tilde{h}_t$ represents the pending output value in this neuron, and h_t represents the output of the current neuron, δ represents the sigmoid function, and tanh represents

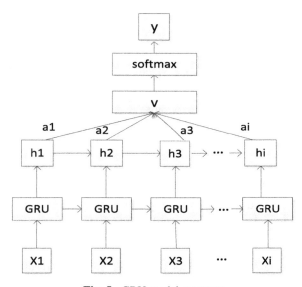

Fig. 5. GRU model structure

the hyperbolic tangent function. In this paper, we construct a conventional GRU neural network in the loop layer. The training flowchart of the model is shown in Fig. 5.

3   Experiment

We use the user comment data in the cloud service community as the experimental data set, it has 40411 comments. The community allows users to rate and comment on more than 90 different services such as gold inquiries, mobile phone number inquiries, and weather forecasts in the community. There are 10128 positive comments, 9573 neural comments, and 10710 negative comments. Since the data extracted from the cloud service community contains a lot of irrelevant information, we need to delete the irrelevant data and segment the data. Through data preprocessing, the total number of samples in the review data set is 24082, there are 8050 positive comments, 7902 neural comments and 8103 negative comments.

3.1   Dataset

Before constructing the user perspective propensity model, we divide the review text after data preprocessing and word vector representation into three parts: training set, verification set and test set. The distribution of various types is shown in Table 4. The training set is used to train the deep learning, the validation set is used to adjust the parameters of the model, and the test set is used to verify the results of the model (Table 1).

Table 1. Type assignment of a data set

Type of data	Number	Proportion (%)
Training set	12000	49.89
Validation set	7200	29.93
Test set	4855	20.18
Total	24055	100.00

3.2   Experiment Evaluation Criteria

The experiment designed two sets of comparative experiments, one compared to the traditional LSTM model and one compared to the GRU.

In order to select the best model among the three neural network models of Bi-LSTM, LSTM and GRU, we need to evaluate the performance of the three models. We use accuracy as an evaluation index, and use precision, recall and F1 values to compare the performance of the three deep learning models. The following describes each indicator (Table 2):

Table 2. Confusion matrix

		Predicted emotion		
		Positive	Neutral	Negative
Actual emotion	Positive	T1	F1	F2
	Neutral	F3	T2	F4
	Negative	F5	F6	T3

T_i (i = 1, 2, 3), it represents the number of samples whose sentiment classification is correctly predicted.

F_i (i = 1, 2, 3, 4, 5, 6), it represents the number of samples whose sentiment classification is incorrectly predicted.

1) Accuracy.

Accuracy represents the ratio of the number of samples correctly predicted and the total number of samples by the classification model.

$$Accuracy = \frac{\sum_{i-1}^{3} T_i}{\sum_{i=1}^{6} F_i + \sum_{i=1}^{3} T_i} \tag{14}$$

2) Precision.

Precision measures the ability of the classifier that not divide negative sample errors into positive samples.

$$Precision_{positive} = \frac{T1}{T1 + F1 + F2} \tag{15}$$

$$Precision_{neural} = \frac{T2}{T2 + F3 + F4} \tag{16}$$

$$Precision_{negative} = \frac{T3}{T3 + F5 + F6} \tag{17}$$

3) Recall.

The recall rate is the ability of the metric classifier to find all the correct classification samples.

$$Recall_{positive} = \frac{T1}{T3 + F3 + F5} \tag{18}$$

$$Recall_{neural} = \frac{T2}{T2 + F1 + F6} \tag{19}$$

$$Recall_{negative} = \frac{T3}{T3 + F2 + F4} \tag{20}$$

4) F1.

The F1 value takes into account the precision and recall of the prediction result, it is the weighted harmonic mean of precision and recall.

$$F - measure = 2\frac{precision \times recall}{precision + recall} \quad (21)$$

The value of the indicators obtained in the final training of the model are shown in Table 3.

Table 3. Classification results of the three models

Model	Accuracy	Recall	Precision	F1 value
LSTM	0.8741	0.8728	0.8751	0.8740
Bi-LSTM	0.8968	0.8766	0.8950	0.8854
GRU	0.8798	0.5746	0.8738	0.6905

It can be seen from the data in the table that the accuracy of the three neural network models of LSTM, Bi-LSTM and GRU exceeds 85%. The GRU model has fewer tensors and takes the shortest time, but the F1 value is the smallest. The Bi-LSTM model has exceeded the traditional LSTM model and GRU model, so this paper chooses Bi-LSTM model as the deep learning model used by the cloud service community in the future.

3.3 Model Parameter Comparison Experiment

In this paper, the Bi-LSTM model is used as an experimental model to study the effects of different parameters on the model. We designed a comparison experiment for the three parameters of Epoch value, Batchsize value and Dropout value.

Experiment 1. Examine the effect of Epoch parameters on the model result.

Epoch is the number of iterations of training. We keep the parameters unchanged other than Epoch. The accuracy and loss of the neural network model are show in Fig. 6.

Through the above training curve, we found that the training results of the model are good, the loss is decreasing, and the accuracy of the training set is close to 100%. We can see that as the number of iterations increases, the accuracy of the training set increases gradually, and the loss value gradually decreases. From this we can determine that the size of the Epoch value has an effect on the experiment results. In this paper, the Epoch value is 1000 and the accuracy of the training set remains basically unchanged.

Experiment 2. Investigate the impact of Batchsize parameters on the model result.

Batchsize is the amount of sample data that passed into the deep learning model each time. The size of the Batchsize value affects the optimization degree and speed of the experiment. If the value is too small, the network will not converge, and too large

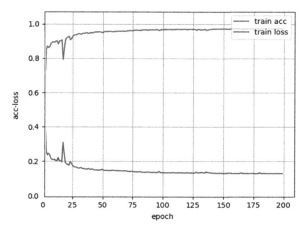

Fig. 6. Curve of accuracy and loss value variation

Table 4. Comparison of experiment results and usage time under different Batchsize

Batchsize	Average time (s/epoch)	Accuracy (%)
8	250	88.62
16	169	88.60
32	165	89.68
64	118	88.81
128	111	88.74
256	107	88.10

may cause a memory explosion. We set difference Batchsize value while keeping other parameters unchanged. The training results are shown in Table 4.

As we can see from the table, the larger the value of Batchsize is, the longer the average time of each iteration training is. In order to improve the memory utilization and the parallel efficiency of matrix multiplication, and considering the time factor, the Batchsize value selected in this paper is 64.

Experiment 3. Examine the effect of Dropout value on the model result.

The feedforward artificial neural network has multiple layers of nonlinear "hidden" units between its input and output. Dropout allows the model to randomly omit each hidden unit from the network during the training process, so that the hidden unit cannot depend on other existing ones. In this paper, we keep the other parameters unchanged other than dropout values. The training result is shown in Fig. 7.

As we can see from the figure, when dropout takes 0.8, the model obtains the minimum value, and the accuracy of the model is lowest. When taking 0.5, the accuracy is the highest. This shows that the deep learning model cannot implement abandonment

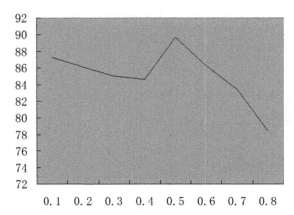

Fig. 7. Training results under different Dropout conditions

strategy for too many neural units during training, that is, the value of dropout cannot be set too large.

4   Conclusions

With the growing number and diversity of web services in the cloud service community, and the number of similarity-featured web services is innumerable, it's difficult to find the most suitable web service for users. In order to recommend more satisfactory services to users, we take users as the center and uses the deep learning method to analyze the online customer reviews of cloud service providers in the community. We designed the performance evaluation and comparison experiments of three deep learning models, and selected the best performance model as the model used by the developers of the cloud service community. The result shows that the performance of the Bi-LSTM model is higher than the LSTM and GRU models.

In future work, we plan to add the convolutional neural network, the deep learning methods used in this article are all cyclic neural networks, so we decide to add convolutional neural networks, they can recognize emotions in comments texts and increase the breadth of user opinions.

Acknowledgment. This work was supported by grants from National Natural Science Foundation of China under Grant (NSFC No. 61962040).

References

1. Qi, L., Xiang, H., Dou, W., Yang, C., Qin, Y., Zhang, X.: Privacy - preserving distributed service recommendation based on locality- sensitive hashing. In: 2017 IEEE International Conference on Web Services (ICWS), pp. 49–56, Honolulu, HI (2017)

2. Gamon, M., Aue, A.: Proceedings of the Workshop on Sentiment and Subjectivity in Text (2006)
3. Bo, P., Lee, L., Vaithyanathan, S.: Thumbs up?: sentiment classification using machine learning techniques. In: Proceedings of the ACL-02 Conference on Empirical Methods in Natural Language Processing. vol. 10. Association for Computational Linguistics (2002)
4. Liu, X.M., Kale, A., Wasani, J., Ding, C., Yu, Q.: Extracting ranking and evaluating quality features of web services through user review sentiment analysis. In: Proceedings of the 2015 IEEE International Conference on Web Services (ICWS 2015), pp. 153–160 (2015)
5. Wan, C.H., et al.: A hybrid text classification approach with low dependency on parameter by integrating K-nearest neighbor and support vector machine. Expert Syst. Appl. **39**(15), 11880–11888 (2012)
6. Wan, C.H., et al.: A hybrid text classification approach with low dependency on parameter by integrating K-nearest neighbor and support vector machine. Expert Syst. Appl. **39**(15) 11880–11888 (2012)
7. Wang, X., Jiang, W., Luo, Z.: Combination of convolutional and recurrent neural network for sentiment analysis of short texts. In: Proceedings of the 26th International Conference on Computational Linguistics, pp. 2428–2437 (2013)
8. Kim, Y.: Convolutional neural networks for sentences classification. arXiv preprint arXiv: 1408.5882 (2014)
9. Zhang, X., Zhao, J., Yann, L.: Character-level convolutional networks for text classification. Adv. Neural Inf. Process. Syst. **28**, 649–657 (2015)
10. Zhang, M., Zhang, Y., Vo, D.: Gated neural networks for targeted sentiment analysis (2015)
11. Graves, A., Mohamed, A., Hinton, G.: Speech recognition with deep recurrent neural networks. In: 2013 IEEE International Conference on Acoustics, Speech and Signal Processing. IEEE (2013)
12. Yao, Y., Huang, Z.: Bi-directional LSTM recurrent neural network for Chinese word segmentation. In: Hirose, A., Ozawa, S., Doya, K., Ikeda, K., Lee, M., Liu, D. (eds.) ICONIP 2016. LNCS, vol. 9950, pp. 345–353. Springer, Cham (2016). https://doi.org/10.1007/978-3-319-46681-1_42
13. Chiu, J.P.C., Eric, N.: Named entity recognition with bidirectional LSTM-CNNs. Trans. Assoc. Comput. Linguist. **4**, 357–370 (2016)
14. Lin, H., Fan, Y., Zhang, J., Bai, B.: PRNN: piecewise recurrent neural networks for predicting the tendency of services invocation. In: IEEE International Conference on Web Services (ICWS), pp. 56–64 (2018)
15. He, Y., Sun, S., Niu, F.: An emotional semantic enhanced deep learning model for microblog emotion analysis. Chin. J. Comput. (4) (2017)
16. Liu, R., Meng, X.: Emotional analysis of multimedia picture based on deep learning. Res. Audio-Vis. Educ. (4) (2018)
17. Jin, Z., Han, Y., Zhu, Q.: An emotional analysis model combining deep learning and integrated learning. J. Harbin Inst. Technol. **50**(11), 38–45 (2018)
18. Maximilien, E.M., Ranabahu, A.: The ProgrammableWeb: agile, social, and grassroot computing. In: International Conference on Semantic Computing (ICSC 2007), pp. 477–481, Irvine, CA (2007)
19. https://loopback.io/doc/en/lb3/Remote-methods.html#19
20. Weiping, L., Weijie, C., Li, L.: Fuliang, G.: A semantically enhanced service repository for service oriented application system development. In: 2009 World Conference on Services - II, pp. 41–48, Bangalore (2009)

21. Wang, H., Chi, X., Wang, Z., Xu, X., Chen, S.: Extracting fine-grained service value features and distributions for accurate service recommendation. In: 2017 IEEE International Conference on Web Services (ICWS), pp. 277–284, Honolulu, HI (2017)
22. Chung, J., Gulceher, C., Cho, K.H., et al.: Empirical evaluation of gated recurrent neural networks on sequence modeling. Eprint Arxiv (2014)

Author Index